# Lecture Notes in Computer Science    11395

Commenced Publication in 1973
Founding and Former Series Editors:
Gerhard Goos, Juris Hartmanis, and Jan van Leeuwen

More information about this series at http://www.springer.com/series/7412

Mihaela Pop · Maxime Sermesant · Jichao Zhao
Shuo Li · Kristin McLeod · Alistair Young
Kawal Rhode · Tommaso Mansi (Eds.)

# Statistical Atlases and Computational Models of the Heart

## Atrial Segmentation and LV Quantification Challenges

9th International Workshop, STACOM 2018
Held in Conjunction with MICCAI 2018
Granada, Spain, September 16, 2018
Revised Selected Papers

 Springer

*Editors*
Mihaela Pop
University of Toronto
Toronto, ON, Canada

Kristin McLeod
GE Healthcare
Oslo, Norway

Maxime Sermesant
Inria, Epione Group
Sophia-Antipolis, France

Alistair Young
King's College London
London, UK

Jichao Zhao
Auckland University
Auckland, New Zealand

Kawal Rhode
King's College London
London, UK

Shuo Li
University of Western Ontario
London, ON, Canada

Tommaso Mansi
Siemens Medical Solutions USA, Inc.
Princeton, NJ, USA

ISSN 0302-9743          ISSN 1611-3349  (electronic)
Lecture Notes in Computer Science
ISBN 978-3-030-12028-3          ISBN 978-3-030-12029-0  (eBook)
https://doi.org/10.1007/978-3-030-12029-0

Library of Congress Control Number: 2019931002

LNCS Sublibrary: SL6 – Image Processing, Computer Vision, Pattern Recognition, and Graphics

This Springer imprint is published by the registered company Springer Nature Switzerland AG
The registered company address is: Gewerbestrasse 11, 6330 Cham, Switzerland

# Preface

Integrative models of cardiac function are important for understanding disease, evaluating treatment, and planning intervention. In the recent years, there has been considerable progress in cardiac image analysis techniques, cardiac atlases, and computational models that can integrate data from large-scale databases of heart shape, function, and physiology. However, significant clinical translation of these tools is constrained by the lack of complete and rigorous technical and clinical validation as well as benchmarking of the developed tools. For doing so, common and available ground-truth data capturing generic knowledge on the healthy and pathological heart are required. Several efforts are now established to provide Web-accessible structural and functional atlases of the normal and pathological heart for clinical, research, and educational purposes. We believe that these approaches will only be effectively developed through collaboration across the full research scope of the cardiac imaging and modeling communities.

STACOM 2018 (http://stacom2018.cardiacatlas.org) was held in conjunction with the MICCAI 2018 international conference (Granada, Spain, Canada), following the past eight editions: STACOM 2010 (Beijing, China), STACOM 2011 (Toronto, Canada), STACOM 2012 (Nice, France), STACOM 2013 (Nagoya, Japan), STACOM 2014 (Boston, USA), STACOM 2015 (Munich, Germany), STACOM 2016 (Athens, Greece), and STACOM 2017 (Quebec City, Canada). STACOM 2018 provided a forum to discuss the latest developments in various areas of computational imaging and modeling of the heart, as well as statistical cardiac atlases. The topics of the workshop included: cardiac imaging and image processing, machine learning applied to cardiac imaging and image analysis, atlas construction, statistical modelling of cardiac function across different patient populations, cardiac computational physiology, model customization, atlas-based functional analysis, ontological schemata for data and results, integrated functional and structural analyses, as well as the pre-clinical and clinical applicability of these methods.

Besides regular contributing papers, additional efforts of the STACOM 2018 workshop were also focused on two challenges: the 3D Atrial Segmentation Challenge and the Left Ventricle Full Quantification Challenge, briefly described below. A total of 51 papers (i.e., regular papers and from the two challenges) were accepted to be presented at STACOM 2018, and are published by Springer in this LNCS proceedings volume.

### 3D Atrial Segmentation Challenge

Atrial fibrillation (AF) is the most common type of cardiac arrhythmia. The poor performance of current AF treatment is due to a lack of understanding of the atrial structure. Gadolinium-enhanced magnetic resonance imaging (i.e., late gadolinium enhancement, LGE) is widely used to study the extent of fibrosis (scars) across the atria. Direct segmentation of the atrial chambers from LGE images is very challenging. There is an urgent need for intelligent algorithms that can perform fully automatic atrial segmentation for the atrial cavity, to accurately reconstruct and visualize the atrial structure. This challenge has provided an open competition for wider communities to test and validate their methods for image segmentation on a large 3D clinical dataset. The exciting development is a very important step toward patient-specific diagnostics and treatment of AF.

The reader can find more information on the challenge website: http://atriaseg2018. cardiacatlas.org/.

### *Left Ventricle (LV) Full Quantification Challenge*

Accurate cardiac left ventricle (LV) quantification is among the most clinically important and most frequently demanded tasks for identification and diagnosis of cardiac diseases and is of great interest in the research community of medical image analysis. The LVQuan18 challenge:

- Focused on full quantification of the LV (i.e., all clinical significant LV indices regarding to the anatomical structure of LV were investigated in addition to the frequently studied LV volume)
- Provided a unified platform for participants around the world to develop effective solutions for LV full quantification
- Benchmarked all submitted solutions for the task of LV quantification and advance the state-of-art performance for computer-aided cardiac image quantification

The reader can find more information on the challenge website: https://lvquan18. github.io/.

We hope that the results obtained by these two challenges, along with the regular paper contributions, will act to accelerate progress in the important areas of heart function and structure analysis.

We would like to thank all organizers, reviewers, authors, and sponsors for their time, efforts, contributions, and financial support in making STACOM 2018 a successful event.

September 2018

Mihaela Pop
Maxime Sermesant
Jichao Zhao
Shuo Li
Kristin McLeod
Alistair Young
Kawal Rhode
Tommaso Mansi

# Organization

## Chairs and Organizers

### STACOM

| | |
|---|---|
| Mihaela Pop | Sunnybrook, University of Toronto, Canada |
| Maxime Sermesant | Inria, Epione Group, France |
| Alistair Young | University of Auckland, New Zealand |
| Kristin McLeod | GE Healthcare, Norway |
| Kawal Rhode | KCL, London, UK |
| Tommaso Mansi | Siemens Healthineers, USA |

### 3D Atrial Segmentation Challenge

| | |
|---|---|
| Jichao Zhao | University of Auckland, New Zealand |
| Zhaohan Xiong | University of Auckland, New Zealand |

### LV Full Quantification Challenge

| | |
|---|---|
| Shuo Li | Digital Imaging Group, Western University, London, Ontario, Canada |
| Wuefeng Xue | Western University, London, Ontario, Canada |

### Additional Reviewers

| | |
|---|---|
| Nicolas Cedilnik | Matthew Ng |
| Florin Ghesu | Tiziano Passerini |
| Akif Gulsun | Saikiran Rapaka |
| Fumin Guo | Avan Suinesiaputra |
| Florian Mihai Itu | Wuegeng Xue |
| Shuman Jia | Ingmar Voigt |
| Viorel Mihalef | |

## OCS - Springer Conference Submission/Publication System

| | |
|---|---|
| Mihaela Pop | Medical Biophysics, University of Toronto, Sunnybrook Research Institute, Toronto, Canada |

## Webmaster

| | |
|---|---|
| Avan Suinesiaputra | University of Auckland, New Zealand |

## Workshop Website

stacom2018.cardiacatlas.org

## Sponsors

We are extremely grateful for industrial funding support. The STACOM 2018 workshop and challenges received financial support from the following sponsors:

**SysAfib** (http://sysafib.org/)
**Nvidia** (http://nvidia.com)
**MedTech coRE** (https://www.cmdt.org.nz/medtechcore)
**Arterys** (http://arterys.com)

# Contents

**Regular Papers**

Estimating Sheets in the Heart Wall . . . . . . . . . . . . . . . . . . . . . . . . . . 3
  Tabish A. Syed, Babak Samari, and Kaleem Siddiqi

Automated Motion Correction and 3D Vessel Centerlines Reconstruction
from Non-simultaneous Angiographic Projections . . . . . . . . . . . . . . . . . . 12
  Abhirup Banerjee, Rajesh K. Kharbanda, Robin P. Choudhury,
  and Vicente Grau

Left Ventricle Segmentation and Quantification from Cardiac Cine MR
Images via Multi-task Learning. . . . . . . . . . . . . . . . . . . . . . . . . . . . . 21
  Shusil Dangi, Ziv Yaniv, and Cristian A. Linte

Statistical Shape Clustering of Left Atrial Appendages. . . . . . . . . . . . . . . . 32
  Jakob M. Slipsager, Kristine A. Juhl, Per E. Sigvardsen,
  Klaus F. Kofoed, Ole De Backer, Andy L. Olivares, Oscar Camara,
  and Rasmus R. Paulsen

Deep Learning Segmentation of the Left Ventricle in Structural CMR:
Towards a Fully Automatic Multi-scan Analysis . . . . . . . . . . . . . . . . . . . . 40
  Hakim Fadil, John J. Totman, and Stephanie Marchesseau

Cine and Multicontrast Late Enhanced MRI Registration
for 3D Heart Model Construction . . . . . . . . . . . . . . . . . . . . . . . . . . . . 49
  Fumin Guo, Mengyuan Li, Matthew Ng, Graham Wright,
  and Mihaela Pop

Joint Analysis of Personalized In-Silico Haemodynamics and Shape
Descriptors of the Left Atrial Appendage . . . . . . . . . . . . . . . . . . . . . . . . 58
  Jordi Mill, Andy L. Olivares, Etelvino Silva, Ibai Genua,
  Alvaro Fernandez, Ainhoa Aguado, Marta Nuñez-Garcia,
  Tom de Potter, Xavier Freixa, and Oscar Camara

How Accurately Does Transesophageal Echocardiography Identify
the Mitral Valve? . . . . . . . . . . . . . . . . . . . . . . . . . . . . . . . . . . . . . 67
  Claire Vannelli, Wenyao Xia, John Moore, and Terry Peters

Stochastic Model-Based Left Ventricle Segmentation in 3D
Echocardiography Using Fractional Brownian Motion . . . . . . . . . . . . . . . . 77
  Omar S. Al-Kadi, Allen Lu, Albert J. Sinusas, and James S. Duncan

Context Aware 3D Fully Convolutional Networks for Coronary
Artery Segmentation . . . . . . . . . . . . . . . . . . . . . . . . . . . . . . . . . . . . . .     85
    Yongjie Duan, Jianjiang Feng, Jiwen Lu, and Jie Zhou

Learning Associations Between Clinical Information and Motion-Based
Descriptors Using a Large Scale MR-derived Cardiac Motion Atlas . . . . . . .     94
    Esther Puyol-Antón, Bram Ruijsink, Hélène Langet, Mathieu De Craene,
    Paolo Piro, Julia A. Schnabel, and Andrew P. King

Computational Modelling of Electro-Mechanical Coupling in the Atria
and Its Changes During Atrial Fibrillation . . . . . . . . . . . . . . . . . . . . . . . .     103
    Sofia Monaci, David Nordsletten, and Oleg Aslanidi

High Throughput Computation of Reference Ranges of Biventricular
Cardiac Function on the UK Biobank Population Cohort . . . . . . . . . . . . . .     114
    Rahman Attar, Marco Pereañez, Ali Gooya, Xènia Albà, Le Zhang,
    Stefan K. Piechnik, Stefan Neubauer, Steffen E. Petersen,
    and Alejandro F. Frangi

Lumen Segmentation of Aortic Dissection with Cascaded
Convolutional Network . . . . . . . . . . . . . . . . . . . . . . . . . . . . . . . . . . . . .     122
    Ziyan Li, Jianjiang Feng, Zishun Feng, Yunqiang An, Yang Gao,
    Bin Lu, and Jie Zhou

A Vessel-Focused 3D Convolutional Network for Automatic Segmentation
and Classification of Coronary Artery Plaques in Cardiac CTA . . . . . . . . . .     131
    Jiang Liu, Cheng Jin, Jianjiang Feng, Yubo Du, Jiwen Lu, and Jie Zhou

Semi-automated Image Segmentation of the Midsystolic Left Ventricular
Mitral Valve Complex in Ischemic Mitral Regurgitation. . . . . . . . . . . . . . .     142
    Ahmed H. Aly, Abdullah H. Aly, Mahmoud Elrakhawy, Kirlos Haroun,
    Luis Prieto-Riascos, Robert C. Gorman Jr., Natalie Yushkevich,
    Yoshiaki Saito, Joseph H. Gorman III, Robert C. Gorman,
    Paul A. Yushkevich, and Alison M. Pouch

Atrial Scar Segmentation via Potential Learning
in the Graph-Cut Framework . . . . . . . . . . . . . . . . . . . . . . . . . . . . . . . . .     152
    Lei Li, Guang Yang, Fuping Wu, Tom Wong, Raad Mohiaddin,
    David Firmin, Jenny Keegan, Lingchao Xu, and Xiahai Zhuang

4D Cardiac Motion Modeling Using Pair-Wise Mesh Registration. . . . . . . .     161
    Siyeop Yoon, Stephen Baek, and Deukhee Lee

ISACHI: Integrated Segmentation and Alignment Correction
for Heart Images. . . . . . . . . . . . . . . . . . . . . . . . . . . . . . . . . . . . . . . . . .     171
    Benjamin Villard, Ernesto Zacur, and Vicente Grau

3D LV Probabilistic Segmentation in Cardiac MRI Using Generative
Adversarial Network . . . . . . . . . . . . . . . . . . . . . . . . . . . . . . . . . . . . . . .   181
   *Dong Yang, Bo Liu, Leon Axel, and Dimitris Metaxas*

A Two-Stage U-Net Model for 3D Multi-class Segmentation
on Full-Resolution Cardiac Data . . . . . . . . . . . . . . . . . . . . . . . . . . . . . .   191
   *Chengjia Wang, Tom MacGillivray, Gillian Macnaught, Guang Yang,
and David Newby*

Centreline-Based Shape Descriptors of the Left Atrial Appendage
in Relation with Thrombus Formation . . . . . . . . . . . . . . . . . . . . . . . . . .   200
   *Ibai Genua, Andy L. Olivares, Etelvino Silva, Jordi Mill,
Alvaro Fernandez, Ainhoa Aguado, Marta Nuñez-Garcia,
Tom de Potter, Xavier Freixa, and Oscar Camara*

**3D Atrial Segmentation Challenge**

Automatic 3D Atrial Segmentation from GE-MRIs Using Volumetric Fully
Convolutional Networks. . . . . . . . . . . . . . . . . . . . . . . . . . . . . . . . . . . . .   211
   *Qing Xia, Yuxin Yao, Zhiqiang Hu, and Aimin Hao*

Automatically Segmenting the Left Atrium from Cardiac Images Using
Successive 3D U-Nets and a Contour Loss. . . . . . . . . . . . . . . . . . . . . . .   221
   *Shuman Jia, Antoine Despinasse, Zihao Wang, Hervé Delingette,
Xavier Pennec, Pierre Jaïs, Hubert Cochet, and Maxime Sermesant*

Fully Automated Left Atrium Cavity Segmentation from 3D GE-MRI
by Multi-atlas Selection and Registration . . . . . . . . . . . . . . . . . . . . . . .   230
   *Mengyun Qiao, Yuanyuan Wang, Rob J. van der Geest, and Qian Tao*

Pyramid Network with Online Hard Example Mining for Accurate
Left Atrium Segmentation . . . . . . . . . . . . . . . . . . . . . . . . . . . . . . . . . . .   237
   *Cheng Bian, Xin Yang, Jianqiang Ma, Shen Zheng, Yu-An Liu,
Reza Nezafat, Pheng-Ann Heng, and Yefeng Zheng*

Combating Uncertainty with Novel Losses for Automatic Left
Atrium Segmentation. . . . . . . . . . . . . . . . . . . . . . . . . . . . . . . . . . . . . . .   246
   *Xin Yang, Na Wang, Yi Wang, Xu Wang, Reza Nezafat, Dong Ni,
and Pheng-Ann Heng*

Attention Based Hierarchical Aggregation Network for 3D Left
Atrial Segmentation. . . . . . . . . . . . . . . . . . . . . . . . . . . . . . . . . . . . . . . .   255
   *Caizi Li, Qianqian Tong, Xiangyun Liao, Weixin Si, Yinzi Sun,
Qiong Wang, and Pheng-Ann Heng*

Segmentation of the Left Atrium from 3D Gadolinium-Enhanced MR
Images with Convolutional Neural Networks . . . . . . . . . . . . . . . . . . . . . .    265
*Chandrakanth Jayachandran Preetha, Shyamalakshmi Haridasan,*
*Vahid Abdi, and Sandy Engelhardt*

V-FCNN: Volumetric Fully Convolution Neural Network
for Automatic Atrial Segmentation . . . . . . . . . . . . . . . . . . . . . . . . . . . .    273
*Nicoló Savioli, Giovanni Montana, and Pablo Lamata*

Ensemble of Convolutional Neural Networks for Heart Segmentation . . . . . .    282
*Wilson Fok, Kevin Jamart, Jichao Zhao, and Justin Fernandez*

Multi-task Learning for Left Atrial Segmentation on GE-MRI . . . . . . . . . . .    292
*Chen Chen, Wenjia Bai, and Daniel Rueckert*

Left Atrial Segmentation Combining Multi-atlas Whole Heart Labeling
and Shape-Based Atlas Selection. . . . . . . . . . . . . . . . . . . . . . . . . . . . . .    302
*Marta Nuñez-Garcia, Xiahai Zhuang, Gerard Sanroma, Lei Li,*
*Lingchao Xu, Constantine Butakoff, and Oscar Camara*

Deep Learning Based Method for Left Atrial Segmentation in GE-MRI. . . . .    311
*Yashu Liu, Yangyang Dai, Cong Yan, and Kuanquan Wang*

Dilated Convolutions in Neural Networks for Left Atrial Segmentation
in 3D Gadolinium Enhanced-MRI. . . . . . . . . . . . . . . . . . . . . . . . . . . . .    319
*Sulaiman Vesal, Nishant Ravikumar, and Andreas Maier*

A Semantic-Wise Convolutional Neural Network Approach for 3-D Left
Atrium Segmentation from Late Gadolinium Enhanced Magnetic
Resonance Imaging . . . . . . . . . . . . . . . . . . . . . . . . . . . . . . . . . . . . . . .    329
*Davide Borra, Alessandro Masci, Lorena Esposito, Alice Andalò,*
*Claudio Fabbri, and Cristiana Corsi*

Left Atrial Segmentation in a Few Seconds Using Fully Convolutional
Network and Transfer Learning. . . . . . . . . . . . . . . . . . . . . . . . . . . . . . .    339
*Élodie Puybareau, Zhou Zhao, Younes Khoudli, Edwin Carlinet,*
*Yongchao Xu, Jérôme Lacotte, and Thierry Géraud*

Convolutional Neural Networks for Segmentation of the Left Atrium
from Gadolinium-Enhancement MRI Images . . . . . . . . . . . . . . . . . . . . . .    348
*Coen de Vente, Mitko Veta, Orod Razeghi, Steven Niederer,*
*Josien Pluim, Kawal Rhode, and Rashed Karim*

Mixture Modeling of Global Shape Priors and Autoencoding Local
Intensity Priors for Left Atrium Segmentation. . . . . . . . . . . . . . . . . . . . . .    357
*Tim Sodergren, Riddhish Bhalodia, Ross Whitaker, Joshua Cates,*
*Nassir Marrouche, and Shireen Elhabian*

## Left Ventricle Full Quantification Challenge

Left-Ventricle Quantification Using Residual U-Net .................. 371
*Eric Kerfoot, James Clough, Ilkay Oksuz, Jack Lee, Andrew P. King,*
*and Julia A. Schnabel*

Left Ventricle Full Quantification Using Deep Layer Aggregation
Based Multitask Relationship Learning .......................... 381
*Jiahui Li and Zhiqiang Hu*

Convexity and Connectivity Principles Applied for Left Ventricle
Segmentation and Quantification................................ 389
*Elias Grinias and Georgios Tziritas*

Calculation of Anatomical and Functional Metrics Using Deep Learning
in Cardiac MRI: Comparison Between Direct
and Segmentation-Based Estimation ........................... 402
*Hao Xu, Jurgen E. Schneider, and Vicente Grau*

Automated Full Quantification of Left Ventricle with Deep
Neural Networks.............................................. 412
*Lihong Liu, Jin Ma, Jianzong Wang, and Jing Xiao*

ESU-P-Net: Cascading Network for Full Quantification of Left Ventricle
from Cine MRI................................................ 421
*Wenjun Yan, Yuanyuan Wang, Shaoxiang Chen, Rob J. van der Geest,*
*and Qian Tao*

Left Ventricle Full Quantification via Hierarchical Quantification Network... 429
*Guanyu Yang, Tiancong Hua, Chao Lu, Tan Pan, Xiao Yang, Liyu Hu,*
*Jiasong Wu, Xiaomei Zhu, and Huazhong Shu*

Automatic Left Ventricle Quantification in Cardiac MRI via Hierarchical
Refinement of High-Level Features by a Salient Perceptual
Grouping Model............................................... 439
*Angélica Atehortúa, Mireille Garreau, David Romo-Bucheli,*
*and Eduardo Romero*

Cardiac MRI Left Ventricle Segmentation and Quantification:
A Framework Combining U-Net and Continuous Max-Flow............. 450
*Fumin Guo, Matthew Ng, and Graham Wright*

Multi-estimator Full Left Ventricle Quantification Through
Ensemble Learning ........................................... 459
*Jiasha Liu, Xiang Li, Hui Ren, and Quanzheng Li*

Left Ventricle Quantification Through Spatio-Temporal CNNs ........... 466
*Alejandro Debus and Enzo Ferrante*

Full Quantification of Left Ventricle Using Deep Multitask Network
with Combination of 2D and 3D Convolution on 2D + t Cine MRI . . . . . . .   476
   *Yeonggul Jang, Sekeun Kim, Hackjoon Shim, and Hyuk-Jae Chang*

**Author Index** . . . . . . . . . . . . . . . . . . . . . . . . . . . . . . . . . . . . . . .   485

# Regular Papers

# Estimating Sheets in the Heart Wall

Tabish A. Syed$^{(\boxtimes)}$, Babak Samari, and Kaleem Siddiqi

School of Computer Science and Centre for Intelligent Machines,
McGill University, Montreal, Canada
{tabish,babak,siddiqi}@cim.mcgill.ca

**Abstract.** Models of sheets in the heart wall play an important role in
the visualization of myofiber geometry, in modelling mechanics and in cardiac electrophysiology. For example, the assumption of distinct speeds of
propagation in the directions of myofibers, the sheets in which they lie,
and the direction across them, can predict the arrival time of the conduction wave that triggers myocyte contraction. Almost all current analyses
based on DTI data use the third eigenvector of the diffusion tensor as an
estimate of the local sheet normal. This construction suffers from the limitation that the second and third eigenvector directions can be ambiguous
since they are associated with eigenvalues that are quite similar. Here we
present and evaluate an alternate method to estimate sheets, which uses
only the principal eigenvector. We find the best local direction perpendicular to the principal eigenvector to span a sheet, using the Lie bracket and
the minimization of an appropriately constructed energy function. We test
our method on a dataset of 8 ex vivo rat and 8 ex vivo canine cardiac diffusion tensor images. Qualitatively the recovered sheets are more consistent
with the geometry of myofibers than those obtained using all three eigenvectors, particularly when they curve or fan. Quantitatively the recovered
sheet normals also give a low value of holonomicity, a measure of the degree
to which they are orthogonal to a family of surfaces. Our novel fitting approach could thus impact cardiac mechanical and electrophysiological analyses which are based on DTI data.

**Keywords:** Cardiac DTI · Myofiber geometry · Sheets · Lie bracket

## 1 Introduction

Heart wall myofibers are thought to lie within laminar sheets, an organization
which exerts an important influence on the heart's mechanical and electrophysiological properties [3,8]. In work based on thick tissue sections from excised
canine hearts using scanning electron microscopy, as well as studies based on
contrast-enhanced MRI, it has been shown that these laminar sheets are geometrically prominent, and can be associated with large cleavage planes which
fall between them [4,6]. When working with intact excised hearts, sheet orientation is typically estimated using the three eigenvectors of a diffusion tensor

---

T. A. Syed and B. Samari—Equal contribution.

© Springer Nature Switzerland AG 2019
M. Pop et al. (Eds.): STACOM 2018, LNCS 11395, pp. 3–11, 2019.
https://doi.org/10.1007/978-3-030-12029-0_1

reconstruction. In the majority of such approaches the principal eigenvector is assumed to coincide with myofiber orientation, and is taken together with the second eigenvector to span the local sheet. The third eigenvector thus gives the normal to the sheet, with qualitative justification such as that given in [5]. A typical example is the tracking and reconstruction algorithm developed by Rohmer et al. for the visualization of laminar sheets in a cadaver human heart in [10].

Once estimated, sheet structure can be the basis for assessment of normal heart function. For example, Dou et al. investigate the hypothesis that ventricular thickening in humans occurs via laminar sheet shear and sheet extension, using diffusion and strain MRI [2]. In the context of electrophysiology Young and Panfilov put forth the hypothesis that an assumption of differing conduction speeds along fibers $(v_f)$, across them but in laminar sheets $(v_s)$ and between the laminar sheets $(v_n)$, can reliably predict the time of arrival of the conduction wave, at any particular location in the heart [13]. This conduction wave triggers the contraction of myocytes, which in turn determines the timing and dynamic shape of the heart beat.

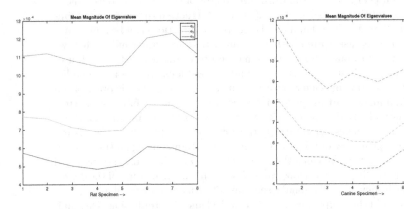

**Fig. 1.** The mean eigenvalues of the three eigenvectors of the diffusion tensor for 8 different rat (left) and canine (right) specimens.

Whereas the idea that the principal eigenvector gives a faithful estimate of local myofiber orientation has widespread support, the second and third eigenvectors can be associated with eigenvalues having similar values. Examples of this are shown for 8 rat and 8 canine heart datasets in Fig. 1. In one of the first reported studies of similar canine DTI datasets by Peyrat et al. in [9], a statistical analysis framework for atlas construction was proposed. The results showed consistency in fiber orientations associated with the principal eigenvector, but greater variability in sheet structure derived from the remaining two. Similar trends in terms of the relative magnitudes of the mean of the three eigenvalues were observed by Lombaert et al. in [7], where an atlas of 10 ex vivo human hearts was constructed, and detailed tensor statistics were reported. Here the second and third eigenvalues were found to have greater standard deviations as well as overlap in their distributions. Thus, in such situations, the directions of the second and third eigenvectors can be ambiguous and reliance on them for sheet estimation could be problematic.

In the present article, we propose and evaluate an alternate algorithm for sheet estimation from diffusion data which relies *only* on the principal eigenvector. The idea is to find the best local vector field orthogonal to the principal eigenvector in the sense of spanning a sheet. We use the Lie bracket along with a suitable energy function to estimate this vector field via energy minimization. Qualitative and quantitative results on 8 canine and 8 rat diffusion datasets demonstrate the potential of this approach and suggest its possible role in further mechanical and electrophysiological analyses of heart wall function.

We begin by reviewing mathematical tools used by us in Sect. 2. We then discuss the details of our method in Sect. 3 and present experimental results on canine and rat diffusion datasets in Sect. 4.

## 2    Background

### 2.1   Lie Bracket Background and Computation

Given two smooth vector fields $\mathbf{a}$ and $\mathbf{b}$ on a manifold $M$, the Lie bracket $\mathbf{L} = [\mathbf{a}, \mathbf{b}]$ is their commutator. With $f$ a smooth function defined on $M$ the commutator acts on $f$ as follows:

$$L(f) = \mathbf{a}(\mathbf{b}f) - \mathbf{b}(\mathbf{a}f).$$

In Cartesian coordinates we let $\mathbf{a} = a_i \dfrac{\partial}{\partial x^i}$ and $\mathbf{b} = b_i \dfrac{\partial}{\partial x^i}$ (with summation implied over repeated indices). The Lie bracket is then given by

$$L = \left( a_j \frac{\partial b^i}{\partial x^j} - b_j \frac{\partial a^i}{\partial x^j} \right) \frac{\partial}{\partial x^i}.$$

Therefore, given vector fields $\mathbf{a}$ and $\mathbf{b}$, we have

$$L = J_{\mathbf{b}}\mathbf{a} - J_{\mathbf{a}}\mathbf{b}, \tag{1}$$

where, $J_{\mathbf{a}}$ and $J_{\mathbf{b}}$ are the Jacobians of $\mathbf{a}$ and $\mathbf{b}$, respectively. In Cartesian coordinates the Jacobian of $\mathbf{a}$ is defined as

$$J_{\mathbf{a}} = \begin{pmatrix} \dfrac{\partial a^1}{\partial x^1} & \dfrac{\partial a^1}{\partial x^2} & \dfrac{\partial a^1}{\partial x^3} \\ \dfrac{\partial a^2}{\partial x^1} & \dfrac{\partial a^2}{\partial x^2} & \dfrac{\partial a^2}{\partial x^3} \\ \dfrac{\partial a^3}{\partial x^1} & \dfrac{\partial a^3}{\partial x^2} & \dfrac{\partial a^3}{\partial x^3} \end{pmatrix}. \tag{2}$$

Given the two vector fields $\mathbf{a}$ and $\mathbf{b}$, we estimate the value of $\mathbf{L}$, which is itself a vector field, using Eq. (1). This equation gives the Lie bracket an interpretation as the vector displacement while moving along a parallelogram spanned by $\mathbf{a}$

and **b** at a point. In other words, we start at the point $p$ and then move along streamlines of **a**, **b**, $-$**a** and then $-$**b**. The vector displacement to $p$ is the Lie bracket. The magnitude of **L** at a point $p$ can therefore be used as a measure of deviation from a sheet spanned locally by fields **a** and **b**, around $p$. For vector fields spanning sheets we expect the parallelograms to close, and therefore the magnitude to be zero. A similar construction based on the Lie bracket is used by Tax et al. in [12] to estimate sheet probability in the brain from diffusion data. We define an energy based on the $L2$ norm of the Lie bracket in a local neighbourhood $N(p)$ of point $p$. Then, given **a**, we estimate the vector field **b** that minimizes this energy at every point in space. In order to obtain a meaningful non-degenerate solution to our problem, we constrain the solution space to vectors **b** in the plane perpendicular to **a**. In the following section we present the details of our method.

## 3   Sheet Estimation by Lie Bracket Energy Minimization

Let **u** be the unit norm vector field derived from the principal eigenvector of a cardiac DTI reconstruction. We assume the principal eigenvector field **u** to be aligned with the local direction of myofibers. The second and third eigenvectors can be associated with similar eigenvalues (see Fig. 1) and thus may not be reliable for sheet estimation. We use the following alternate strategy to recover the sheet direction, **v**, using only the principal eigenvector **u**. Let **v**$(\theta)$ be a vector field in the local plane perpendicular to the vector field **u**. Our objective is to estimate the field $\hat{\mathbf{v}}(\theta)$, which together with **u** spans a local sheet, by minimizing a Lie bracket based energy $E_p(\theta)$ which at point $p$ is defined as follows:

$$E_p(\theta) = \sum_{q \in N(p)} L_q(\theta)^T L_q(\theta), \tag{3}$$

where $L_q(\theta) = J_{\mathbf{v}_q(\theta)}\mathbf{u} - J_{\mathbf{u}_q(\theta)}\mathbf{v}$ is the Lie bracket at the point $q$ and $\mathbf{v}_q(\theta)$ and $\mathbf{u}_q$ are the values of vector fields **v** and **u** respectively at the point $q$. Note that $\mathbf{v}_p$ is a function of a single parameter $\theta$ - the orientation of the vector in the plane perpendicular to the fixed field $\mathbf{u}_q$. We shall not express $\mathbf{v}_p$ in terms of $\theta$ since this parameter is not used explicitly in our algorithm.

In order to minimize the energy $E(\theta)$ we use gradient descent with respect to $\theta$ pointwise. Rather than using $\theta$ explicitly we directly update the vector field **v** using the following update equation:

$$\mathbf{v}_p^{t+1}(\theta) = \mathbf{v}_p^t(\theta - \eta \frac{\partial E(\theta)}{\partial \theta}).$$

The update is performed by rotating $\mathbf{v}_p$ about the axis $\mathbf{u}_p$ by an angle $\eta \dfrac{\partial E(\theta)}{\partial \theta}$. This method of update allows us to not have to write **v** explicitly in terms of $\theta$ and also ensures that the vector field **v** stays in the plane perpendicular to **u** after the update.

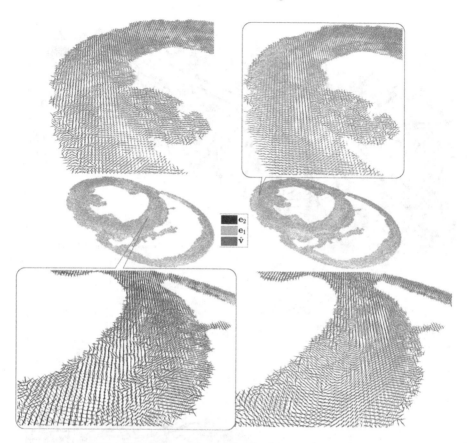

**Fig. 2.** (Best viewed by zooming in on the electronic version.) Using a sample slice from a canine heart (middle row) we focus on regions in the left ventricular wall (top row) and the septum (bottom row). In all panels the principal eigenvector direction $\mathbf{e}_1$ is shown in green. In the panels on the left the tensor eigenvector $\mathbf{e}_2$ is shown in blue. For comparison, in the panels on the right our estimated sheet direction vector $\hat{\mathbf{v}}$ is shown in red. (Color figure online)

## 4 Results and Discussion

We tested our algorithm on ex vivo DTI data of 8 rat hearts and 8 canine hearts from the open access repository in [11], with an initial step size of $\eta = 0.2$ and an exponential decay with decay rate of 0.9 and decay every 20 steps. Each rat heart dataset is $64 \times 64 \times 128$, with a voxel dimension of $0.25 \times 0.25 \times 0.25\,\mathrm{mm}^3$. Each canine dataset is $300 \times 300 \times 333$, with a voxel dimension of $0.3125 \times 0.3125 \times 0.8\,\mathrm{mm}^3$. The canine specimens are therefore 57 times larger than the rat specimens in terms of the number of voxels. In practice our algorithm converges relatively fast, with the gradient descent loss plateauing after about 150 iterations. We ran each specimen, from both the rat and canine datasets,

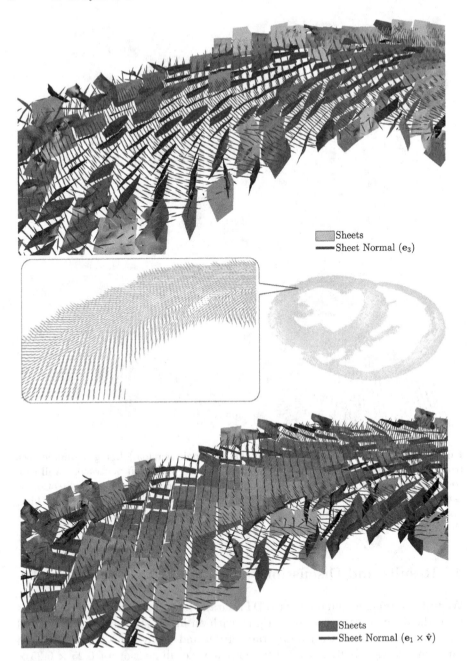

**Fig. 3.** (Best viewed by zooming in on the electronic version.) The top row shows the sheets spanned by $\mathbf{e}_1$ and $\mathbf{e}_2$ in cyan, with the corresponding normal vector $\mathbf{e}_3$ (blue). The bottom row shows our proposed sheets, spanned by $\mathbf{e}_1$ and $\hat{\mathbf{v}}$, in blue, along with our sheet normal vector $\mathbf{n} = \mathbf{e}_1 \times \hat{\mathbf{v}}$ (red). The middle row shows the zoomed-in region in the left ventricular wall from a short axis slice of the canine heart being analyzed, with the principal eigenvector $\mathbf{e}_1$ shown in green. (Color figure online)

for 200 iterations. Our Matlab implementation, which has not been optimized, took about 50 min per rat heart and around a day for each canine heart on an Intel Core i9-7900x machine. We note that since our energy is local, it would be possible to get a significant boost in speed by implementing the algorithm on a GPU, thus making the approach quite practical.

Given the noisy nature of DTI data, in our implementation we estimate the partial derivatives using the minimum of central, forward and backward finite differences. Using the minimum of these three finite differences helps smooth the derivatives while preserving local structure in the presence of noise. Further, we also correct for any random 180° flips present in the raw eigenvector directions due to the directionless nature of DTI tensor measurements.

Figure 2 compares our estimated sheet direction vector $\hat{\mathbf{v}}$ (red) with the commonly used second eigenvector $\mathbf{e}_2$ (blue) of the DTI tensor data. For clarity, the figures show a single short axis slice around the middle of a canine heart. In all panels the principal eigenvector $\mathbf{e}_1$ is shown in green. Observe that in the lower right panel the estimated vector field $\hat{\mathbf{v}}$ (red) is relatively smooth, resembling the nature of $\mathbf{e}_1$ (green), as it should. In contrast, in the lower left panel the second eigenvector $\mathbf{e}_2$ (blue) is considerably noisier, even in smooth regimes of $\mathbf{e}_1$. As such, the sheet structure estimated using $\mathbf{e}_1$ and $\mathbf{e}_2$ would be noisy as well.

In Fig. 3 we use the estimated vector field $\hat{\mathbf{v}}$ to draw sheets spanned by the vector fields $\mathbf{e}_1$ and $\hat{\mathbf{v}}$, and compare them with the sheets spanned by the vector fields $\mathbf{e}_1$ and $\mathbf{e}_2$. Local sheets are drawn by fitting a surface to points estimated by following streamlines along $\mathbf{e}_1$ and $\hat{\mathbf{v}}$. Specifically, we start at a grid point, say $\mathbf{r}$, and place it in a queue. We iteratively dequeue a point, follow the sheet by moving in the directions of $\pm\mathbf{e}_1$ and $\pm\hat{\mathbf{v}}$ at the point, and enqueue each neighbouring point. In our implementation we use bilinear interpolation to estimate the value of vector fields at non-grid points and we fit a sheet once 40 such points have been obtained.

In fact, even in the relatively smooth region shown in the top half of Fig. 2, we observe in Fig. 3 (top row) that the estimated sheets using $\mathbf{e}_1$ and $\mathbf{e}_2$ (cyan) are all oriented in a radial direction despite the varying myofiber orientation in the middle and border regions of the heart wall. The principal direction of fibers (shown in green in the middle row) are in the circumferential direction. However, the third eigenvector $\mathbf{e}_3$ (blue), which is normal to the sheets in cyan, does not follow this orientation. Our estimated normals $\mathbf{n} = \mathbf{e}_1 \times \hat{\mathbf{v}}$ (red), shown in the bottom row, are oriented in the long axis direction in the middle of the wall and in the radial direction closer to the inner and outer walls. The sheets estimated using our computed normals better follow the local orientation $\mathbf{e}_1$ of the fibers.

For a more quantitative assessment of our approach we use the holonomicity of the sheet normal vector field $\mathbf{e}_1 \times \hat{\mathbf{v}} = \mathbf{n}$ as a measure of the quality of our estimated sheet direction $\hat{\mathbf{v}}$. The holonomicity $\rho$ of a vector field $\mathbf{n}$ is given by

$$\rho = \langle \mathbf{n}, \operatorname{curl} \mathbf{n} \rangle .$$

The holonomicity $\rho$ is zero everywhere for a vector field if and only if there exists a family of surfaces normal to that field [1]. Therefore, $\rho$ provides a measure of

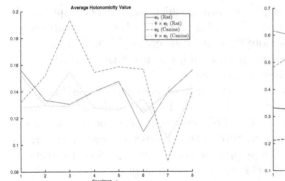 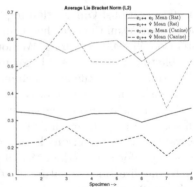

**Fig. 4.** On the left we plot the average holonomicity value for the normal to local sheets, in red for $\mathbf{e}_3$, and in cyan for $\mathbf{e}_1 \times \hat{\mathbf{v}}$, the normal estimated using our method. The right panel shows the average $L2$ norm of the Lie bracket, in red for eigenvectors $\mathbf{e}_1$ and $\mathbf{e}_2$, and in black between $\mathbf{e}_1$ and our estimated vector field $\hat{\mathbf{v}}$. (Color figure online)

the degree to which it is not possible to have a surface normal to the field $\mathbf{n}$. A vector field normal to sheet-like surfaces is expected to have a value close to zero. In Fig. 4 we compare the average holonomicity value of the normal vector field $\mathbf{n}$, derived from our estimated field $\hat{\mathbf{v}}$, with the normal vector field of the tensor direction $\mathbf{e}_3$. The average holonomicity value for the estimated normal $\mathbf{n}$ is smaller than that of $\mathbf{e}_3$ for the canine dataset and is close to that of $\mathbf{e}_3$ for the rat dataset. The lower value of holonomicity of $\mathbf{e}_3$ for the rat dataset may be due to over-smoothing of $\mathbf{e}_3$. The right plot in Fig. 4 compares the average $L2$ norm of the Lie bracket between $\mathbf{e}_1$ and our estimated sheet direction $\hat{\mathbf{v}}$ with that between $\mathbf{e}_1$ and $\mathbf{e}_2$ In other words we compare $[\mathbf{e}_1, \mathbf{e}_2]$ with $[\mathbf{e}_1, \hat{\mathbf{v}}]$. Our estimated sheets result in a much smaller Lie bracket $L2$ norm, which is expected since we explicitly minimize an energy which is a function of this norm.

We note that our method implicitly ignores noise in the measurement of $\mathbf{e}_1$ and also treats the principal eigenvector as a reliable estimator of local myofiber orientation. This assumption may lead to unreliable sheet estimates in regions with noisy DTI measurements, or in regions such as the apex of the heart, where fibers turn rapidly.

## 5    Conclusion

We have presented a method for estimating a frame field spanning a sheet locally, from a single vector field. The method minimizes an energy based on the Lie bracket, by using gradient descent. We have applied our algorithm to find sheets in ex vivo cardiac DTI data. While the current paper focuses on cardiac data, the method is quite general and may find use in other applications as well. It is also possible to minimize the holonomicity directly, a direction that we shall explore in future work. Ultimately, a gold standard for validating our sheet estimates would be given by reconstructions of the sheets revealed by histology. Reconstructing sheet structure from histological sections and comparing such reconstructions with DTI measurements are worthwhile but challenging goals, which are beyond the scope of the present paper.

**Acknowledgements.** We thank the reviewers for many insightful comments and suggestions. We are grateful to the Natural Sciences and Engineering Research Council of Canada for research funding.

# References

1. Aminov, Y.: Geometry of Vector Fields. CRC Press, Boca Raton (2000)
2. Dou, J., Tseng, W.Y.I., Reese, T.G., Wedeen, V.J.: Combined diffusion and strain MRI reveals structure and function of human myocardial laminar sheets in vivo. Magn. Reson. Med. **50**(1), 107–113 (2003)
3. Eriksson, T.S., Prassl, A., Plank, G., Holzapfel, G.A.: Influence of myocardial fiber/sheet orientations on left ventricular mechanical contraction. Math. Mech. Solids **18**(6), 592–606 (2013)
4. Gilbert, S.H., et al.: Visualization and quantification of whole rat heart laminar structure using high-spatial resolution contrast-enhanced MRI. Am. J. Phys.-Heart Circ. Physiol. **302**(1), H287–H298 (2011)
5. Helm, P., Beg, M.F., Miller, M.I., Winslow, R.L.: Measuring and mapping cardiac fiber and laminar architecture using diffusion tensor MR imaging. Ann. NY Acad. Sci. **1047**(1), 296–307 (2005)
6. LeGrice, I.J., Smaill, B.H., Chai, L.Z., Edgar, S.G., Gavin, J.B., Hunter, P.J.: Laminar structure of the heart: ventricular myocyte arrangement and connective tissue architecture in the dog. Am. J. Physiol. **269**(2), H571–82 (1995)
7. Lombaert, H., et al.: Human atlas of the cardiac fiber architecture: study on a healthy population. IEEE Trans. Med. Imaging **31**(7), 1436–1447 (2012)
8. Nash, M.P., Hunter, P.J.: Computational mechanics of the heart. J. Elast. Phys. Sci. Solids **61**(1–3), 113–141 (2000)
9. Peyrat, J.M., et al.: A computational framework for the statistical analysis of cardiac diffusion tensors: application to a small database of canine hearts. IEEE Trans. Med. Imaging **26**(11), 1500–1514 (2007). https://doi.org/10.1109/TMI.2007.907286
10. Rohmer, D., Sitek, A., Gullberg, G.T.: Reconstruction and visualization of fiber and laminar structure in the normal human heart from ex vivo diffusion tensor magnetic resonance imaging (DTMRI) data. Invest. Radiol. **42**(11), 777–789 (2007)
11. Savadjiev, P., Strijkers, G.J., Bakermans, A.J., Piuze, E., Zucker, S.W., Siddiqi, K.: Heart wall myofibers are arranged in minimal surfaces to optimize organ function. Proc. Natl. Acad. Sci. **109**(24), 9248–9253 (2012)
12. Tax, C.M., et al.: Sheet probability index (SPI): characterizing the geometrical organization of the white matter with diffusion MRI. NeuroImage **142**, 260–279 (2016)
13. Young, R.J., Panfilov, A.V.: Anisotropy of wave propagation in the heart can be modeled by a riemannian electrophysiological metric. Proc. Natl. Acad. Sci. **107**(34), 15063–15068 (2010)

# Automated Motion Correction and 3D Vessel Centerlines Reconstruction from Non-simultaneous Angiographic Projections

Abhirup Banerjee[2,3](✉) [iD], Rajesh K. Kharbanda[2], Robin P. Choudhury[1,2], and Vicente Grau[3] [iD]

[1] Oxford Acute Vascular Imaging Centre, Oxford, UK
[2] Radcliffe Department of Medicine, Division of Cardiovascular Medicine, University of Oxford, Oxford, UK
{abhirup.banerjee,robin.choudhury}@cardiov.ox.ac.uk
[3] Department of Engineering Science, Institute of Biomedical Engineering, University of Oxford, Oxford, UK
vicente.grau@eng.ox.ac.uk

**Abstract.** Automated estimation of 3D centerlines is necessary to transform angiographic projections into accurate 3D reconstructions. Although several methods exist for 3D centerline reconstruction, most of them are sensitive to the motion in coronary arteries when images are acquired by a single-plane rotational X-ray system. The objective of the proposed method is to rectify the motion-related deformations in coronary vessels from 2D projections and subsequently achieve an optimal 3D centerline reconstruction. Rigid motion in arteries is removed by estimating the optimal rigid transformation from all projection planes. The remaining non-rigid motion at end-diastole is modelled by a radial basis function based warping of 2D centerlines. Point correspondences are then generated from all projection planes by least squares matching. The final 3D centerlines are obtained by 3D non-uniform rational basis splines fitting over generated point correspondences. Experimental analysis over 20 coronary vessel trees (12 right coronary artery: RCA and 8 left coronary artery: LCA) demonstrates that the rigid transformation is able to reduce the coronary vessel movements to 0.72 mm average, while the final 3D centerline reconstruction achieves an average rms error of 0.31 mm, when backprojected on angiographic planes.

**Keywords:** 3D centerline reconstruction · Coronary tree · Angiograms · Motion correction · NURBS

## 1   Introduction

Invasive coronary X-ray arteriography (ICA) is one of the best diagnostic tools for the assessment of coronary arterial diseases, due to its high temporal and

© Springer Nature Switzerland AG 2019
M. Pop et al. (Eds.): STACOM 2018, LNCS 11395, pp. 12–20, 2019.
https://doi.org/10.1007/978-3-030-12029-0_2

spatial resolution [11]. However, the limitation of depicting only 2D projection images of the complex 3D anatomy makes it susceptible to artifacts (due to vessel overlap and foreshortening) and affects the accuracy of the estimation of lesion severity and stent size (due to subjective interpretation) [7]. To overcome this inherent limitation of ICA, several groups have attempted 3D reconstruction of coronary arterial (CA) trees from a limited number of 2D projections [4]. A major component of the 3D arterial tree reconstruction is the accurate estimation of the 3D vessel centerline that uses the geometrical and topological information from 2D image projections, to compute the geometry of the 3D CA tree.

One of the earliest 3D centerline reconstruction methods introduced an epipolar line based technique [9]. For each point along a vessel in the first image plane, an epipolar straight line was generated in the second plane and its intersection with the vessel axis generated the point correspondence. Despite being susceptible to the distortions due to cardiac and breathing movements, this method was applied widely in several CA tree reconstructions. Ding and Friedman [6] incorporated the method for quantification of arterial motion from biplane coronary cineangiograms. Chen and Carroll [5] determined the transformation in imaging geometry before applying the same centerline reconstruction. Their experimental analysis over 20 LCA and 20 RCA cases by comparing the projection on a third plane generated the average rms errors of 3.09 mm and 3.13 mm, respectively.

Since reconstructions using two projections have not been shown as sufficiently reliable in clinical studies up to now, Blondel et al. [1] projected the reconstructed 3D vessel points from two angiograms on the third plane to generate a score based on multiscale analysis. The final correspondence was generated by minimizing a semi-local energy function. Cañero et al. [3] calibrated the system for geometrical distortions before applying the 3D centerline reconstruction using a biplane snakes model. Along with the epipolar constraint, Mourgues et al. [12] introduced a cost on the distance to epipolar lines and a penalty to constraint the match of all bifurcations to define point correspondences between two projections. This reconstruction method was later integrated with 3D motion tracking [13] and prospective motion correction [14] in biplane angiograms.

Recently, Galassi et al. [8] have developed a nonuniform rational basis splines (NURBS) based CA reconstruction method where the 3D centreline is reconstructed as the intersection of surfaces from corresponding branches. The method identified corresponding points sampled at uniform distances on the projection lines. Another NURBS-based 3D CA tree reconstruction method is proposed in [15], where the point correspondences between two angiographic projections are identified based on a cost matrix and the geometry correction using calibration has generated average mse 2.532–3.259 mm$^2$ in clinical data.

Most of the existing reconstruction methods were performed on end-diastole frames of angiograms, where cardiac motion was assumed to not exist between projections. However, in a single-plane rotational X-ray system, this assumption does not hold for non-simultaneous acquisition. Single-plane systems have some advantages over biplane ones, such as their lower cost and the possibility of acquiring any number of projections. To remove the effects of respiration and patient related movements, in previous works the cardiac images for reconstruction were mostly acquired by breath-holding and ensuring no patient movement [1,3,15]. However, breath-holding does not necessarily ensure the generation of images at the same respiratory cycle and often the patients are not in a state to follow the breath-hold protocol. Also, the assumption of no patient movement is often violated due to invasive nature of the procedure.

The purpose of the proposed research work is to generate 3D vessel centerlines from multiple angiographic projections without the need for any assumptions during image acquisition. The proposed method initially estimates optimal rigid transformations from all projection planes to remove rigid motion due to respiration and patient movements. The remaining non-rigid movements at end-diastole, mostly due to cardiac motion, are modelled by a radial basis function based warping of the 2D centerlines. Pointwise correspondences are then identified by least squares matching from all projection planes. Finally, the 3D centerlines of all coronary vessels are obtained by 3D non-uniform rational basis splines (NURBS) fitting over generated point correspondences. The method is qualitatively and quantitatively evaluated on 12 RCA and 8 LCA trees from clinical datasets.

## 2    Proposed Method

### 2.1    Rigid Motion Correction

One of the main problems with 3D reconstruction of CA tree is that acquired angiogram images from different views do not correspond to the same geometry. Even after selecting the end-diastolic frames for non-simultaneous image acquisition, the discrete selection of image frame from end-diastolic cardiac phase retains certain amount of temporal misalignment, and thus non-rigid deformations in the coronary vessels. Respiratory movements of the lungs cause rigid movement in the heart and, hence, in the vessels. Patient or device-related movements during or between image acquisitions also cause similar rigid movement.

In order to remove the effects of rigid movement due primarily to respiration and patient or device related motion, the proposed algorithm aims to estimate the optimal rigid transformations of all projection planes that minimize the rigid movement of the heart during imaging. The proposed technique models this motion by estimating the transformed location of the projection planes that optimizes the match between projections. The proposed rigid motion correction procedure requires the selection of a few (preferably 4–6) corresponding landmark points from all projection planes. In our current implementation, the bifurcation points of vessels are manually selected as landmark points.

Let us assume the motion corrected 3D locations of $r$ landmark points are given by $B_i, i = 1, \cdots, r$ and the true location of landmark $B_i$ on the $j$th projection plane is $b_{ij}, j = 1, \cdots, s$. Also, assume the location of X-ray source, origin, and normal of projection plane are defined as $F_j, M_j,$ and $n_j, j = 1, \cdots, s$, respectively. Hence, the objective is to estimate the optimal rigid transformation (translation $t_j$ and rotation $R_j, j = 1, \cdots, s$) from each plane, so that the

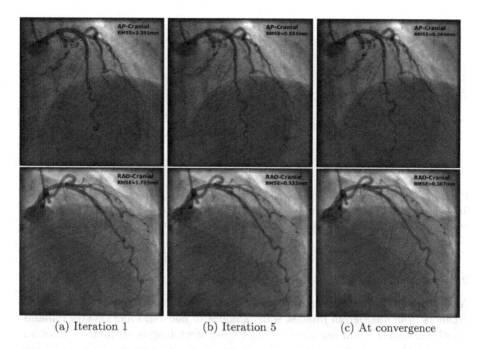

(a) Iteration 1            (b) Iteration 5            (c) At convergence

**Fig. 1.** Rigid-motion correction on two LCA image projections. Red and green circles in the images mark, respectively, the rigid-motion corrected and true landmark locations. (Color figure online)

projections of $B_i, i = 1, \cdots, r$ on each of the transformed angiographic planes match with the corresponding landmarks in 2D image planes, that is,

$$\min_{t_j, R_j} \sum_{i=1}^{r} \|B_{i,j'} - (t_j + R_j * b_{ij})\|^2 \quad \forall j = 1, \cdots, s. \tag{1}$$

Here, $j'$ denotes the $j$th plane transformed by $(t_j, R_j)$ and $B_{i,j'} = B_i - (n_{j'} \cdot (B_i - M_{j'})) * n_{j'}$ denotes the projection of $B_i$ on plane $j'$. The optimization in (1) is solved using Horn's quaternion-based method [10]. Now, since the motion corrected 3D locations $B_i$ are not known, these 3D landmark locations are estimated as the point of intersection of 3D straight lines $\overrightarrow{F_j b_{ij}}, j = 1, \cdots, s$, i.e.,

$$B_i = \arg \min_{p} \sum_{j} D(p; F_j, b_{ij}), \tag{2}$$

where $D(p; F_j, b_{ij}) = \|(b_{ij} - p) - ((b_{ij} - p)^T (b_{ij} - F_j))(b_{ij} - F_j)\|^2$ denotes the distance of the point $p$ to $\overrightarrow{F_j b_{ij}}$. The minimization is solved iteratively, where, in each iteration, the 3D landmark points are generated using (2) and then projected on angiographic planes to estimate the optimal rigid transformation using (1). The algorithm usually requires 8–20 iterations to converge.

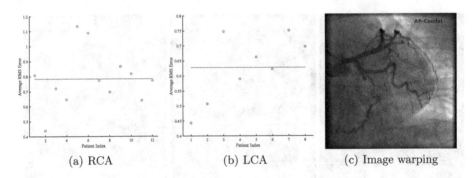

(a) RCA                    (b) LCA                    (c) Image warping

**Fig. 2.** (a), (b): Evaluating rigid-motion correction: red circles represent average RMSE for each patient, while blue line represents the average over all patients. (c) Modelling non-rigid motion (blue: original centerline, red: warped motion-corrected centerline). (Color figure online)

### 2.2   Correction of Non-rigid Motion

Although the proposed rigid motion correction can remove most of the motion artifacts from angiogram images, some amount of non-rigid motion still remains mostly around the mid-region of vessels (vessel stretching or shrinking). In order to model this non-rigid motion, the proposed work employs the radial basis function based image warping method, proposed by Bookstein [2].

Let $p_i = (x_i, y_i)$ be the true 2D coordinate of $b_{ij}$ in the $j$th projection plane and $p'_i = (x'_i, y'_i)$ be the rigid-motion corrected projection of $B_i$ in the same plane. To model the difference between $p_i$ and $p'_i$ due to non-rigid motion in arteries, we define, $K = ((k_{mn}))$ where $k_{mn} = \exp(-\frac{\|z_m - z_n\|^2}{r^2})$, $z_m = \|p'_m - p_m\|$, $P = [1, \underset{\sim}{p}]$, and $L = \begin{bmatrix} K & P \\ P^T & 0 \end{bmatrix}$. $W = L^{-1} * \underset{\sim}{p}'$ is called the coefficient matrix. Hence, the warped 2D centerline, modelling non-rigid motion at $j$th projection plane, is given by

$$c'_j = L^\star * W, \tag{3}$$

where $c_j$ is the rigid-motion corrected 2D centerline, $L^\star = [K', 1, c_j]$, $K' = ((k'_{mn}))$, $k'_{mn} = \exp(-\frac{\|z'_m - z'_n\|^2}{r^2})$, and $z'_m = \|c_j - p'_m\|$.

### 2.3   3D Centerline Reconstruction

After correcting rigid and non-rigid movement in the coronary vessels, point correspondences among 2D centerlines of projected planes are identified. From a 2D centerline of any of the projected planes, discrete equidistant image points (at least 250–300) are automatically selected (say, $c_{1,i}, i = 1, \cdots, n$). The corresponding locations in the remaining image planes are estimated from the optimization problem:

$$\min_{c_{j,i}; j=2,\cdots,s} \sum_{j=1}^{s} D(C_i; F_j, c_{j,i}) \quad \forall i = 1, \cdots, n \tag{4}$$

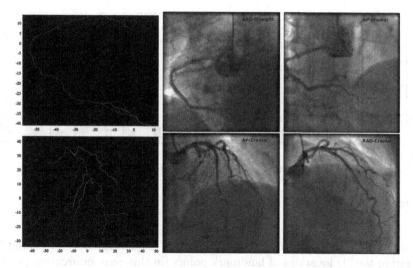

**Fig. 3.** From left to right: 3D centerline and back-projection of centerline on original image planes (top: RCA, bottom: LAD).

where $C_i = \arg\min_p \sum_j D(p; F_j, c_{j,i})$ and $D(\cdot)$ follows the same definition as (2). $C_i, i = 1, \cdots, n$ are estimated using the method of least squares matching, which minimizes the sum of perpendicular distances to the lines (if they do not intersect to a unique point).

Equation (4) not only identifies the point correspondences $c_{j,.}$ among 2D centerlines of all projection planes, it simultaneously generates the 3D vessel points $C_i$. In order to produce the final 3D centerline, the 3D non-uniform rational basis splines (NURBS) function is fitted through the vessel points $C_i$, as follows:

$$\gamma(u) = \sum_{i=1}^{n} B_{i,p}(u)C_i, \quad u \in [0, 1], \tag{5}$$

where $p$ is the degree of the curve and $B_{i,p}(\cdot)$ is a rational basis function of $C_i$ of $p$th degree.

## 3 Experimental Analysis

The performance of the proposed technique is evaluated on patients enrolled in clinical studies. 12 RCA and 8 LCA images are involved in the analysis. The number of angiographic projections varied between 2 (mostly for RCA) to 5.

### 3.1 Motion Correction

The qualitative performance of rigid-motion correction on two LCA image projections is depicted in Fig. 1. The algorithm usually converges in 8–12 iterations

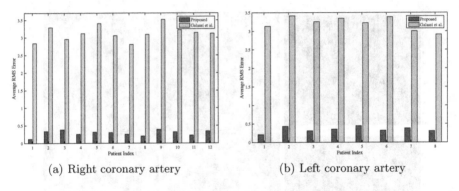

(a) Right coronary artery          (b) Left coronary artery

**Fig. 4.** Evaluating the 3D centerline reconstruction on RCA and LCA images.

(maximum 20). The performance of the method is quantitatively evaluated by comparing the 2D locations of landmark points (in this case, bifurcation points) in each plane with the projected locations of motion-corrected 3D landmark points. For quantitative validation, we calculated the average rms error, which is presented in Fig. 2, separately for RCA and LCA. The average rms error for 12 RCA cases is 0.784 mm, while the same for 8 LCA cases is 0.628 mm. Together, the average rmse for all 20 cases is 0.7216 mm. After the rigid motion correction, the 2D centerlines of each projection plane are warped to correct the remaining non-rigid deformation, qualitatively depicted in Fig. 2(c).

### 3.2  3D Centerline Reconstruction

The performance of the 3D centerline reconstruction is evaluated by comparing the projected 3D centerlines on each plane with the original 2D centerlines (Fig. 3). The performance is quantitatively measured using the average rms error and presented in Fig. 4. From the results, it is observed that the proposed 3D centerline reconstruction produces average rms value 0.286 mm for 12 RCA cases and 0.346 mm for 8 LCA cases. The reconstruction performance is compared with a recently developed 3D CA reconstruction algorithm by Galassi et al. [8] (Fig. 4), showing a substantial reduction in matching error. We performed non-parametric Wilcoxon signed rank and parametric paired-$t$ tests, both showing the improvement to be statistically significant (p-value = 2.4e−04 and 4.4e−15).

## 4  Discussion and Conclusions

The objective of the proposed research work is to develop a new 3D centerline reconstruction algorithm, maintaining the standard clinical image acquisition procedure, without the need for any additional constraint to reduce motion between acquisitions. The proposed approach aims at removing the effects of both rigid and non-rigid motion, so that the point correspondences can be identified for 3D centerline generation. Experimental analysis has showed that the proposed algorithm reduces the rigid motion effects to average 0.72 mm. The performance of the final 3D centerline reconstruction shows an average rms error of 0.31 mm over 20 cases, significantly lower compared to a recent 3D reconstruction algorithm [8]. Although the proposed algorithm provides satisfactory reconstruction performance, qualitatively as well as quantitatively, at proximal and distal areas of vessels, there is scope for improvement around the middle sections of vessels, where non-rigid motions have a high impact. The future work will aim to develop new algorithms that can improve this limitation. The proposed approach is completely automated, except for the manual selection of landmark points for rigid motion correction. These landmarks were selected to correspond to clearly visible structures in the images to facilitate accurate selection. While any obvious errors arising from possible landmark inaccuracies were not detected during the analysis, a complete automation would be desirable to facilitate the clinical application of the technique.

**Acknowledgments.** This work was supported in part by a British Heart Foundation (BHF) Project Grant. The authors acknowledge the use of services and facilities of the Oxford Acute Vascular Imaging Centre (AVIC) and the Institute of Biomedical Engineering (IBME), University of Oxford.

# References

1. Blondel, C., Vaillant, R., Devernay, F., Malandain, G., Ayache, N.: Automatic trinocular 3D reconstruction of coronary artery centerlines from rotational X-ray angiography. In: Proceedings of the 16th International Congress and Exhibition on Computer Assisted Radiology and Surgery, pp. 832–837 (2002)
2. Bookstein, F.L.: Principal warps: thin plate splines and the decomposition of deformations. IEEE Trans. Pattern Anal. Mach. Intell. **11**(6), 567–585 (1989)
3. Cañero, C., Vilariño, F., Mauri, J., Radeva, P.: Predictive (un)distortion model and 3-D reconstruction by biplane snakes. IEEE Trans. Med. Imaging **21**(9), 1188–1201 (2002)
4. Çimen, S., Gooya, A., Grass, M., Frangi, A.F.: Reconstruction of coronary arteries from X-ray angiography: a review. Med. Image Anal. **32**, 46–68 (2016)
5. Chen, S.Y.J., Carroll, J.D.: 3-D reconstruction of coronary arterial tree to optimize angiographic visualization. IEEE Trans. Med. Imaging **19**(4), 318–336 (2000)
6. Ding, Z., Friedman, M.H.: Quantification of 3-D coronary arterial motion using clinical biplane cineangiograms. Int. J. Card. Imaging **16**(5), 331–346 (2000)
7. Eng, M.H., Hudson, P.A., Klein, A.J., Chen, S.J., et al.: Impact of three dimensional in-room imaging (3DCA) in the facilitation of percutaneous coronary interventions. J. Cardiol. Vasc. Med. **1**, 1–5 (2013)
8. Galassi, F., Alkhalil, M., Lee, R., Martindale, P., et al.: 3D reconstruction of coronary arteries from 2D angiographic projections using non-uniform rational basis splines (NURBS) for accurate modelling of coronary stenoses. Plos One **13**(1), 1–23 (2018)
9. Guggenheim, N., Doriot, P.-A., Dorsaz, P.-A., Descouts, P., Rutishauser, W.: Spatial reconstruction of coronary arteries from angiographic images. Phys. Med. Biol. **36**(1), 99–110 (1991)

10. Horn, B.K.P.: Closed-form solution of absolute orientation using unit quaternions. J. Opt. Soc. Am. A **4**(4), 629–642 (1987)
11. Mark, D.B., Berman, D.S., Budoff, M.J., Carr, J.J., et al.: Expert consensus document on coronary computed tomographic angiography. J. Am. Coll. Cardiol. **55**(23), 2663–2699 (2010)
12. Mourgues, F., Devernay, F., Malandain, G., Coste-Manière, E.: 3D+t modeling of coronary artery tree from standard non simultaneous angiograms. In: Proceedings of the 4th International Conference on Medical Image Computing and Computer-Assisted Intervention, pp. 1320–1322 (2001)
13. Shechter, G., Devernay, F., Coste-Manière, E., Quyyumi, A., McVeigh, E.R.: Three-dimensional motion tracking of coronary arteries in biplane cineangiograms. IEEE Trans. Med. Imaging **22**(4), 493–503 (2003)
14. Shechter, G., Shechter, B., Resar, J.R., Beyar, R.: Prospective motion correction of X-ray images for coronary interventions. IEEE Trans. Med. Imaging **24**(4), 441–450 (2005)
15. Vukicevic, A.M., Çimen, S., Jagic, N., Jovicic, G., Frangi, A.F., Filipovic, N.: Three-dimensional reconstruction and NURBS-based structured meshing of coronary arteries from the conventional X-ray angiography projection images. Sci. Rep. **8**, 1711 (2018)

# Left Ventricle Segmentation and Quantification from Cardiac Cine MR Images via Multi-task Learning

Shusil Dangi[1(✉)], Ziv Yaniv[2,3], and Cristian A. Linte[1,4]

[1] Center for Imaging Science, Rochester Institute of Technology, Rochester, NY, USA
{sxd7257,calbme}@rit.edu
[2] TAJ Technologies Inc., Bloomington, MN, USA
[3] National Library of Medicine, National Institutes of Health, Bethesda, MD, USA
[4] Biomedical Engineering, Rochester Institute of Technology, Rochester, NY, USA

**Abstract.** Segmentation of the left ventricle and quantification of various cardiac contractile functions is crucial for the timely diagnosis and treatment of cardiovascular diseases. Traditionally, the two tasks have been tackled independently. Here we propose a convolutional neural network based multi-task learning approach to perform both tasks simultaneously, such that, the network learns better representation of the data with improved generalization performance. Probabilistic formulation of the problem enables learning the task uncertainties during the training, which are used to automatically compute the weights for the tasks. We performed a five fold cross-validation of the myocardium segmentation obtained from the proposed multi-task network on 97 patient 4-dimensional cardiac cine-MRI datasets available through the STA-COM LV segmentation challenge against the provided gold-standard myocardium segmentation, obtaining a Dice overlap of $0.849 \pm 0.036$ and mean surface distance of $0.274 \pm 0.083$ mm, while simultaneously estimating the myocardial area with mean absolute difference error of $205 \pm 198$ mm$^2$.

## 1 Introduction

Magnetic Resonance Imaging (MRI) is the preferred imaging modality for non-invasive assessment of cardiac performance, thanks to its lack of ionizing radiation, good soft tissue contrast, and high image quality. Cardiac contractile function parameters such as systolic and diastolic volumes, ejection fraction, and myocardium mass are good indicators of cardiac health, representing reliable diagnostic value. Segmentation of the left ventricle (LV) allows us to compute these cardiac parameters, and also to generate high quality anatomical models for surgical planning, guidance, and regional analysis of the heart. Although manual delineation of the ventricle is considered as the gold-standard, it is time consuming and highly susceptible to inter- and intra-observer variability. Hence, there is a need for fast, robust, and accurate semi- or fully-automatic segmentation algorithms.

© Springer Nature Switzerland AG 2019
M. Pop et al. (Eds.): STACOM 2018, LNCS 11395, pp. 21–31, 2019.
https://doi.org/10.1007/978-3-030-12029-0_3

Cardiac MR image segmentation techniques can be broadly classified into: (1) no-prior based methods, such as thresholding, edge-detection and linking, and region growing; (2) weak-prior based methods, such as active contours (snakes), level-set, and graph-theoretical models; (3) strong-prior based methods, such as active shape and appearance models, and atlas-based models; and (4) machine learning based methods, such as per pixel classification and convolutional neural network (CNN) based models. A comprehensive review of various cardiac MR image segmentation techniques can be found in [1].

Recent success of deep learning techniques [2] in high level computer vision, speech recognition, and natural language processing applications has motivated their use in medical image analysis. Although the early adoption of deep learning in medical image analysis encountered various challenges, such as the limited availability of medical imaging data and associated costly manual annotation, those challenges were circumvented by patch-based training, data augmentation, and transfer learning techniques [3,4].

Long *et al.* [5] were the first to propose a fully convolutional network (FCN) for semantic image segmentation by adapting the contemporary classification networks fine-tuned for the segmentation task, obtaining state-of-the-art results. Several modifications to the FCN architecture and various post-processing schemes have been proposed to improve the semantic segmentation results as summarized in [6]. Notably, the U-Net architecture [7] with data augmentation has been very successful in medical image segmentation.

While segmentation indirectly enables the computation of various cardiac indices, direct estimation of these quantities from low-dimensional representation of the image have also been proposed in the literature [8–10]. However, these methods are less interpretable and the correctness of the produced output is often unverifiable, potentially limiting their clinical adoption.

Here we propose a CNN based multi-task learning approach to perform both LV segmentation and cardiac indices estimation simultaneously, such that these related tasks regularize the network, hence improving the network generalization performance. Furthermore, our method increases the interpretablity of the output cardiac indices, as the clinicians can infer its correctness based on the quality of produced segmentation result.

## 2    Methodology

Traditionally, the segmentation of the LV and quantification of the cardiac indices have been performed independently. However, due to a close relation between the two tasks, we identified that learning a CNN model to perform both tasks simultaneously is beneficial in two ways: (1) it forces the network to learn features important for both tasks, hence, reducing the chances of over-fitting to a specific task, improving generalization; (2) the segmentation results can be used as a proxy to identify the reliability of the obtained cardiac indices, and also to perform regional cardiac analysis and surgical planning.

## 2.1   Data Preprocessing and Augmentation

This study employed 97 de-identified cardiac MRI image datasets from patients suffering from myocardial infarction and impaired LV contraction available as a part of the STACOM Cardiac Atlas Segmentation Challenge project [11,12] database[1]. Cine-MRI images in short-axis and long-axis views are available for each case. The semi-automated myocardium segmentation provided with the dataset served as gold-standard for assessing the proposed segmentation technique. The dataset was divided into 80% training and 20% testing for five-fold cross-validation.

The physical pixel spacing in the short-axis images ranged from 0.7031 to 2.0833 mm. We used SimpleITK [13] to resample all images to the most common spacing of 1.5625 mm along both x- and y-axis. The resampled images were center cropped or zero padded to a common resolution of $192 \times 192$ pixels. We applied two transformations, obtained from the combination of random rotation and translation (by maximum of half the image size along x- and y-axis), to each training image for data augmentation.

## 2.2   Multi-task Learning Using Uncertainty to Weigh Losses

We estimate the task-dependent uncertainty [14] for both myocardium segmentation and myocardium area regression via probabilistic modeling. The weights for each task are determined automatically based on the task uncertainties learned during the training [15].

For a neural network with weights $\mathbf{W}$, let $\mathbf{f}^{\mathbf{W}}(\mathbf{x})$ be the output for the corresponding input $\mathbf{x}$. We model the likelihood for segmentation task as the squashed and scaled version of the model output through a softmax function:

$$p(\mathbf{y}|\mathbf{f}^{\mathbf{W}}(\mathbf{x}), \sigma) = \text{Softmax}\left(\frac{1}{\sigma^2}\mathbf{f}^{\mathbf{W}}(\mathbf{x})\right) \tag{1}$$

where, $\sigma$ is a positive scalar, equivalent to the *temperature*, for the defined Gibbs/Boltzmann distribution. The magnitude of $\sigma$ determines the *uniformity* of the discrete distribution, and hence relates to the uncertainty of the prediction. The log-likelihood for the segmentation task can be written as:

$$\log p(\mathbf{y} = c|\mathbf{f}^{\mathbf{W}}(\mathbf{x}), \sigma) = \frac{1}{\sigma^2}f_c^{\mathbf{W}}(\mathbf{x}) - \log \sum_{c'} \exp\left(\frac{1}{\sigma^2}f_{c'}^{\mathbf{W}}(\mathbf{x})\right) \tag{2}$$

where $f_c^{\mathbf{W}}(\mathbf{x})$ is the $c$'th element of the vector $\mathbf{f}^{\mathbf{W}}(\mathbf{x})$.

Similarly, for the regression task we define our likelihood as a Lapacian distribution with its mean given by the neural network output:

$$p(\mathbf{y}|\mathbf{f}^{\mathbf{W}}(\mathbf{x}), \sigma) = \frac{1}{2\sigma}\exp\left(-\frac{|\mathbf{y} - \mathbf{f}^{\mathbf{W}}(\mathbf{x})|}{\sigma}\right) \tag{3}$$

---

[1] http://www.cardiacatlas.org.

The log-likelihood for regression task can be written as:

$$\log p(\mathbf{y}|\mathbf{f}^{\mathbf{W}}(\mathbf{x}), \sigma) \propto -\frac{1}{\sigma}|\mathbf{y} - \mathbf{f}^{\mathbf{W}}(\mathbf{x})| - \log\sigma \tag{4}$$

where $\sigma$ is the neural networks observation noise parameter—capturing the noise in the output.

For a network with two outputs: continuous output $\mathbf{y}_1$ modeled with a Laplacian likelihood, and a discrete output $\mathbf{y}_2$ modeled with a softmax likelihood, the joint loss $\mathcal{L}(\mathbf{W}, \sigma_1, \sigma_2)$ is given by:

$$
\begin{aligned}
\mathcal{L}(\mathbf{W}, \sigma_1, \sigma_2) &= -\log p(\mathbf{y}_1, \mathbf{y}_2 = c|\mathbf{f}^{\mathbf{W}}(\mathbf{x})) \\
&= -\log \left( p(\mathbf{y}_1|\mathbf{f}^{\mathbf{W}}(\mathbf{x}), \sigma_1) \cdot p(\mathbf{y}_2 = c|\mathbf{f}^{\mathbf{W}}(\mathbf{x}), \sigma_2) \right) \\
&\approx \frac{1}{\sigma_1}\mathcal{L}_1(\mathbf{W}) + \frac{1}{\sigma_2^2}\mathcal{L}_2(\mathbf{W}) + \log\sigma_1 + \log\sigma_2
\end{aligned}
\tag{5}
$$

where $\mathcal{L}_1(\mathbf{W}) = |\mathbf{y}_1 - \mathbf{f}^{\mathbf{W}}(\mathbf{x})|$ is the mean absolute distance (MAD) loss of $\mathbf{y}_1$ and $\mathcal{L}_2(\mathbf{W}) = -\log \text{Softmax}(\mathbf{y}_2, \mathbf{f}^{\mathbf{W}}(\mathbf{x}))$ is the cross-entropy loss of $\mathbf{y}_2$. To arrive at (5), the two tasks are assumed independent and simplifying assumptions have been made for the softmax likelihood, resulting in a simple optimization objective with improved empirical results [15]. During the training, the joint likelihood loss $\mathcal{L}(\mathbf{W}, \sigma_1, \sigma_2)$ is optimized with respect to $\mathbf{W}$ as well as $\sigma_1$, $\sigma_2$.

As observed in (5), the uncertainties ($\sigma_1$, $\sigma_2$) learned during the training are weighting the losses for individual tasks, such that, the task with higher uncertainty is weighted less and vice versa. Furthermore, the uncertainties can't become too large, as they are penalized by the last two terms in (5).

## 2.3   Network Architecture

In this work, we adapt the U-Net architecture [7], highly successful in medical image segmentation, to perform an additional task of myocardium area estimation as shown in Fig. 1. The segmentation and regression paths are split at the final up-sampling and concatenation layer. The final feature map in the segmentation path is passed through a sigmoid layer to obtain a per-pixel image segmentation. Similarly, the regression output is obtained by down-sampling the final feature map in the regression path by $1/4^{th}$ of its size and passing it through a fully-connected layer. The logarithm of the task uncertainties ($\log\sigma_1$, $\log\sigma_2$) added as the network parameters are used to construct the loss function (5), and are learned during the training. Note that we train the network to predict the log uncertainty $s = \log(\sigma)$ due to its numerical stability and the positivity constraint imposed on the computed uncertainty via exponentiation, $\sigma = \exp(s)$.

# 3   Results

The network was initialized with the *Kaiming uniform* [16] initializer and trained for 50 epochs using *RMS prop* optimizer with a learning rate of 0.001 (decayed by 0.95 every epoch) in PyTorch[2]. The best performing network, in terms of the Dice overlap between the obtained and gold-standard segmentation, in the test set, was saved and used for evaluation.

The network training required 9 min per epoch on average using a 12GB Nvidia Titan Xp GPU. It takes 0.663 milliseconds on average to process a slice during testing. The log uncertainties learned for the segmentation and regression tasks during training are $-3.9$ and $3.45$, respectively, which correspond to weighting the cross-entropy and mean absolute difference (MAD) loss by a ratio of 1556:1. Note that the scale for cross-entropy loss is $10^{-2}$, whereas that for MAD loss is $10^{2}$.

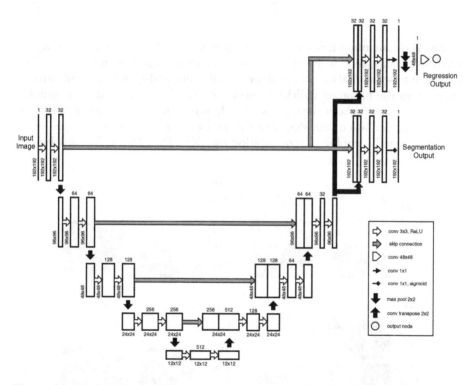

**Fig. 1.** Modified U-Net architecture for multi-task learning. The segmentation and regression tasks are split at the final up-sampling and concatenation layer. The final feature map in the segmentation path is passed through a sigmoid layer to obtain a per-pixel image segmentation. Similarly, the final feature map in the regression path is down-sampled (by max-pooling) to $1/4^{th}$ of its size and fed to a fully-connected layer to generate a single regression output. The logarithm of the task uncertainties ($\log\sigma_1$, $\log\sigma_2$) are set as network parameters and are encoded in the loss function (5), hence learned during the training.

---

[2] https://github.com/pytorch/pytorch.

**Table 1.** Evaluation of the segmentation results obtained from the baseline U-Net (UNet) architecture and the proposed multi-task network (MTN) against the provided gold-standard myocardium segmentation using—Dice Index, Jaccard Index, Mean Surface Distance, and Hausdorff Distance.

| Assessment metric | End-diastole | | End-systole | | All phases | |
|---|---|---|---|---|---|---|
| | UNet | **MTN** | UNet | **MTN** | UNet | **MTN** |
| Dice Index | 0.836 ± 0.036 | **0.837 ±** 0.038 | **0.850 ±** 0.033 | 0.849 ± 0.036 | 0.847 ± 0.035 | **0.849 ±** 0.036 |
| Jaccard Index | 0.719 ± 0.052 | **0.721 ±** 0.054 | **0.740 ±** 0.048 | 0.739 ± 0.053 | 0.736 ± 0.050 | **0.739 ±** 0.053 |
| Mean surface distance (mm) | 0.318 ± 0.089 | **0.286 ±** 0.087 | 0.299 ± 0.095 | **0.274±** 0.090 | 0.305 ± 0.088 | **0.274±** 0.083 |
| Hausdorff distance (mm) | 13.582± 4.337 | **13.364±** 4.108 | **13.083±** 3.630 | 13.355± 3.861 | **13.211±** 4.212 | 13.233± 3.810 |

The 2D segmentation results are stacked to form a 3D volume, and the largest connected component is selected as the final myocardium segmentation. The myocardium segmentation obtained for end-diastole, end-systole, and all cardiac phases from the proposed multi-task network (MTN) and from the baseline U-Net architecture (without the regression task) are both assessed against the gold-standard segmentation provided with the dataset as part of the challenge, using four traditionally employed segmentation metrics—Dice Index, Jaccard Index,

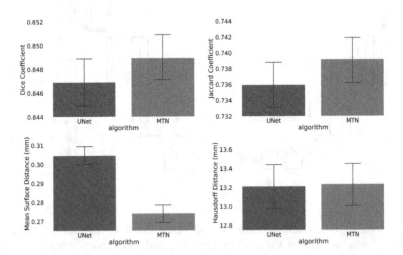

**Fig. 2.** Mean and 99% confidence interval for (a) dice coefficient, (b) Jaccard coefficient, (c) Mean surface distance (mm), and (d) Hausdorff distance (mm), for baseline U-Net and the proposed MTN architecture across all cardiac phases. Confidence interval is obtained based on 1000 bootstrap re-sampling with replacement for 2191 test volumes across five-fold cross-validation.

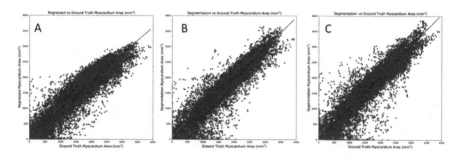

**Fig. 3.** The myocardium area computed from (A) regression path of the proposed multi-task network, (B) segmentation obtained from the proposed multi-task network, (C) segmentation obtained from the baseline U-Net model, plotted against the corresponding myocardium area obtained from the provided gold-standard segmentation for all cardiac phases. The best fit line is shown in each plot. The correlation coefficients for A, B, and C are 0.9466, 0.9565, 0.9518, respectively

Mean surface distance (MSD), and Hausdorff distance (HD)—summarized in Table 1. Note that the myocardium dice coefficient is higher for end-systole phase where the myocardium is thickest.

The Kolmogorov-Smirnov test shows that the difference in distributions for Dice, Jaccard and MSD metrics between the proposed multi-task network and baseline U-Net architecture are statistically significant with p-values: $2.156e^{-4}$, $2.156e^{-4}$, and $6.950e^{-34}$, respectively. However, since the segmentation is evaluated on a large sample of 2191 volumes across five-fold cross validation, the p-values quickly go to zero even for slight difference in distributions being

**Table 2.** Mean absolute difference (MAD), in $mm^2$, between the myocardium area obtained from the provided gold-standard segmentation and the results computed from: (a) the regression path of the proposed multi-task network, (b) segmentation obtained from the proposed multi-task network, and (c) segmentation obtained from the baseline U-Net model, for end-diastole, end-systole, and all cardiac phases, sub-divided into apical, mid, and basal regions of the heart.

| Cardiac regions | End-diastole | | | End-systole | | | All phases | | |
|---|---|---|---|---|---|---|---|---|---|
| | Reg-MTN | Seg-MTN | Seg-UNet | Reg-MTN | Seg-MTN | Seg-UNet | Reg-MTN | Seg-MTN | Seg-UNet |
| All | 201 ± 199 | **174 ±** 209 | 203 ± 221 | 211 ± 209 | **173 ±** 203 | 187 ± 204 | 206 ± 198 | **170 ±** 199 | 193 ± 208 |
| Apical | **185 ±** 180 | 187 ± 204 | 194 ± 186 | 193 ± 199 | 190 ± 226 | **185 ±** 185 | 184 ± 183 | **181 ±** 210 | 187 ± 189 |
| Mid | 190 ± 172 | **141 ±** 132 | 179 ± 142 | 228 ± 194 | **160 ±** 151 | 174 ± 135 | 212 ± 178 | **149 ±** 132 | 176 ± 141 |
| Basal | **250 ±** 269 | 252 ± 331 | 282 ± 368 | 193 ± 241 | **186 ±** 267 | 216 ± 312 | 213 ± 248 | **210 ±** 289 | 237 ± 319 |

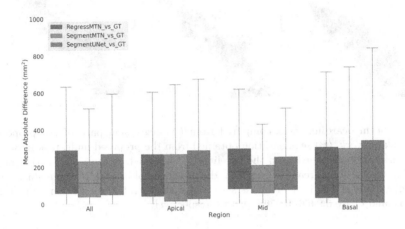

(a) Box-plot for the mean absolute difference (MAD).

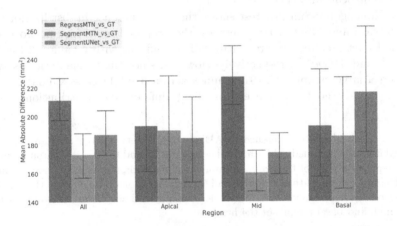

(b) Mean and 99% confidence interval for the mean absolute difference (MAD).

**Fig. 4.** (a) Box-plot (outliers removed for clarity) and (b) Mean and 99% confidence interval, for the mean absolute difference (MAD) between the myocardium area obtained from the provided gold-standard segmentation and the results obtained from: (1) the regression path of the proposed multi-task network, (2) segmentation obtained from the proposed multi-task network, and (3) segmentation obtained from the baseline U-Net model. Confidence intervals are obtained based on 1000 bootstrap re-sampling with replacement.

compared, representing no practical significance [17]. Hence, we computed the 99% confidence interval for the mean value of each segmentation metric based on 1000 bootstrap re-sampling with replacement, as shown in Fig. 2. As evident from Fig. 2, Dice, Jaccard and HD metrics are statistically similar, whereas the reduction in MSD for the proposed multi-task network compared to the baseline U-Net architecture is statistically significant.

In addition to obtaining the myocardium area from the regression path of the proposed network, it can also be computed indirectly from the obtained myocardium segmentation. Hence, we compute and evaluate the myocardium area estimated from three different sources: (a) regression path of the MTN, (b) segmentation obtained from the MTN, and (c) segmentation obtained from the baseline U-Net model. Figure 3 shows the myocardium area obtained from these three methods for all phases of the cardiac cycle plotted against the ground-truth myocardium area estimated from the gold-standard myocardium segmentation provided as part of the challenge data. We can observe a linear relationship between the computed and gold-standard myocardium areas, and the corresponding correlation coefficients for the methods (a), (b), and (c) are 0.9466, 0.9565, 0.9518, respectively.

Further, we computed the MAD between the ground-truth myocardium area and the area estimated by each of the three methods for end-diastole, end-systole, and all cardiac phases (for 26664 slices across five-fold cross validation). For the regional analysis, slices in the ground-truth segmentation after excluding two apical and two basal slices are considered as mid-slices. Table 2 summarizes the mean and standard deviation for the computed MADs. Box-plots (outliers removed for clarity) comparing the three methods for different regions of the heart throughout the cardiac cycle are shown in Fig. 4a. The MAD in myocardium area estimation of $206 \pm 198\,\text{mm}^2$ obtained from the regression output of the proposed method is similar to the results presented in [9]: $223 \pm 193\,\text{mm}^2$, while acknowledging the limitation that the study in [9] was conducted on a different dataset than our study. Moreover, while the regression output of the proposed network yields good estimate of the myocardial area, the box-plot in Fig. 4a suggests that even further improved myocardial area estimates can be obtained from a segmentation based method, provided that the quality of the segmentation is good.

Lastly, we computed the 99% confidence interval for the mean value of myocardium area MAD based on 1000 bootstrap re-sampling with replacement, as shown in Fig. 4b. This confirms that the myocardium area estimated from the segmentation output of the proposed multi-task network is significantly better than that obtained from the regression output, however, there is no statistical significance between other methods. Furthermore, we can observe the variability in MAD is highest for the basal slices, followed by apical and mid slices.

## 4  Discussion, Conclusion, and Future Work

We presented a multi-task learning approach to simultaneously segment and quantify myocardial area. We adapt the U-Net architecture, highly successful in medical image segmentation, to perform an additional regression task. The best location to incorporate the regression path into the network is a hyper-parameter, tuned empirically. We found that adding the regression path in the bottleneck or intermediate decoder layers is detrimental for the segmentation performance of the network, likely due to high influence of the skip connections in the U-Net architecture.

Myocardium area estimates obtained from the regression path of the proposed network are similar to the direct estimation-based results found in the literature. However, our experiments suggest that segmentation-based myocardium area estimation is superior to that obtained from a direct estimation-based method. Lastly, the myocardium segmentation obtained from our method is at least as good as the segmentation obtained from the baseline U-Net model.

To test the generalization performance of the proposed multi-task network, we plan to evaluate the network performance using a lower number of training images. Similarly, we plan to extend this work to segment left ventricle myocardium, blood-pool, and right ventricle, and regress their corresponding areas using the Automated Cardiac Diagnosis Challenge (ACDC)[3] 2017 dataset.

**Acknowledgements.** Research reported in this publication was supported by the National Institute of General Medical Sciences of the National Institutes of Health under Award No. R35GM128877 and by the Office of Advanced Cyberinfrastructure of the National Science Foundation under Award No. 1808530. Ziv Yaniv's work was supported by the Intramural Research Program of the U.S. National Institutes of Health, National Library of Medicine.

## References

1. Petitjean, C., Dacher, J.N.: A review of segmentation methods in short axis cardiac MR images. Med. Image Anal. **15**(2), 169–184 (2011)
2. LeCun, Y., Bengio, Y., Hinton, G.: Deep learning. Nature **521**(7553), 436–444 (2015)
3. Shen, D., Wu, G., Suk, H.I.: Deep learning in medical image analysis. Ann. Rev. Biomed. Eng. **19**, 221–248 (2017)
4. Litjens, G., et al.: A survey on deep learning in medical image analysis. Med. Image Anal. **42**, 60–88 (2017)
5. Long, J., Shelhamer, E., Darrell, T.: Fully convolutional networks for semantic segmentation. In: The IEEE Conference on CVPR, June 2015
6. Garcia-Garcia, A., Orts-Escolano, S., Oprea, S., Villena-Martinez, V., Rodríguez, J.G.: A review on deep learning techniques applied to semantic segmentation. CoRR abs/1704.06857 (2017)
7. Ronneberger, O., Fischer, P., Brox, T.: U-net: Convolutional networks for biomedical image segmentation. CoRR abs/1505.04597 (2015)

---

[3] https://www.creatis.insa-lyon.fr/Challenge/acdc/.

8. Zhen, X., Islam, A., Bhaduri, M., Chan, I., Li, S.: Direct and simultaneous four-chamber volume estimation by multi-output regression. In: Navab, N., Hornegger, J., Wells, W.M., Frangi, A.F. (eds.) MICCAI 2015. LNCS, vol. 9349, pp. 669–676. Springer, Cham (2015). https://doi.org/10.1007/978-3-319-24553-9_82
9. Xue, W., Islam, A., Bhaduri, M., Li, S.: Direct multitype cardiac indices estimation via joint representation and regression learning. IEEE Trans. Med. Imaging 36(10), 2057–2067 (2017)
10. Xue, W., Brahm, G., Pandey, S., Leung, S., Li, S.: Full left ventricle quantification via deep multitask relationships learning. Med. Image Anal. 43, 54–65 (2018)
11. Fonseca, C.G., et al.: The cardiac atlas project - an imaging database for computational modeling and statistical atlases of the heart. Bioinformatics 27(16), 2288–2295 (2011)
12. Suinesiaputra, A., et al.: A collaborative resource to build consensus for automated left ventricular segmentation of cardiac MR images. Med. Image Anal. 18(1), 50–62 (2014)
13. Yaniv, Z., Lowekamp, B.C., Johnson, H.J., Beare, R.: SimpleITK image-analysis notebooks: a collaborative environment for education and reproducible research. J. Digit. Imaging 31, 290–303 (2017)
14. Kendall, A., Gal, Y.: What uncertainties do we need in Bayesian deep learning for computer vision? In: Guyon, I., et al. (eds.) Advances in Neural Information Processing Systems 30, pp. 5574–5584. Curran Associates, Inc. (2017)
15. Kendall, A., Gal, Y., Cipolla, R.: Multi-task learning using uncertainty to weigh losses for scene geometry and semantics. CoRR abs/1705.07115 (2017)
16. He, K., Zhang, X., Ren, S., Sun, J.: Delving deep into rectifiers: surpassing human-level performance on imagenet classification. In: Proceedings of the IEEE International Conference on Computer Vision, pp. 1026–1034 (2015)
17. Lin, M., Lucas Jr., H.C., Shmueli, G.: Research commentary - too big to fail: large samples and the p-value problem. Inf. Syst. Res. 24(4), 906–917 (2013)

# Statistical Shape Clustering of Left Atrial Appendages

Jakob M. Slipsager[1]([✉]), Kristine A. Juhl[1], Per E. Sigvardsen[2],
Klaus F. Kofoed[2], Ole De Backer[2], Andy L. Olivares[3], Oscar Camara[3],
and Rasmus R. Paulsen[1]

[1] DTU Compute, Technical University of Denmark,
Kongens Lyngby, Denmark
jmsl@dtu.dk
[2] Department of Cardiology, Rigshospitalet, University of Copenhagen,
Copenhagen, Denmark
[3] Physense, Department of Information and Communication Technologies,
Universitat Pompeu Fabra,
Barcelona, Spain

**Abstract.** Fifteen percent of all strokes are caused by emboli formed in
the left atrium (LA) in case of atrial fibrillation (AF). The most common
site of thrombus formation is inside the left atrial appendage (LAA). The
LAA is accounting for 70% to 90% of the thrombi formed in the LA in
patients with non-valvular AF. Studies have shown there is a correlation between the LAA morphology and risk of ischemic stroke; Chicken
Wing and Cauliflower LAA shapes are associated with lower and higher
risk, respectively. These two LAA shape categories come from a popular
classification in the medical domain, but it is subjective and based on
qualitative shape parameters. In this paper, we describe a full framework for shape analysis and clustering of the LAA. Initially, we build
a point distribution model to quantitatively describe the LAA shape
variation based on 103 LAA surfaces segmented and reconstructed from
multidetector computed tomography volumes. We are successfully able to
determine point correspondence between LAA surfaces, by non-rigid volumetric registration of signed distance fields. To validate if LAA shapes
are clustered, we employ an unsupervised clustering on the shape models
parameters to estimate the natural number of clusters in our training set,
where the number of shape clusters is estimated by validating the test
log-likelihood of several Gaussian mixture models using two level cross-
validation. We found that the LAAs surfaces basically formed two shape
clusters broadly corresponding to the Chicken wing and non-Chicken
Wing morphologies, which fits well with clinical knowledge.

**Keywords:** Left atrial appendage · Point distribution models ·
Clustering · Gaussian mixture models

© Springer Nature Switzerland AG 2019
M. Pop et al. (Eds.): STACOM 2018, LNCS 11395, pp. 32–39, 2019.
https://doi.org/10.1007/978-3-030-12029-0_4

# 1   Introduction

Atrial fibrillation (AF) causes a 5-fold increase in risk of ischemic stroke, being the cause for approximately 15% all strokes in the United States [7]. Around 70% to 90% of the cases, thrombi are formed inside the left atrial appendage (LAA) in patients with non-valvular AF [17]. The LAA is a complex tubular structure, with a high inter-patient variability, originating from the left atrium (LA). Studies have shown there is a correlation between LAA morphology and risk of ischemic stroke [4,8]. Di Biase et al. [4] reported that the popular named Chicken Wing morphology is associated with lower risk of stroke compared to non-Chicken Wing morphology. Several studies have focused on describing the varying LAA morphology, where the morphology is described by the LAA length, width, orfice/ostium size, and number of lobes. In a study based on 220 LAA obtained from necropsy studies, Ernt et al. [5] reported variation in LAA volumes ranging from 770 to 19,270 $mm^3$, minor orifice diameters ranging from 5 to 27 mm, major orifice diameters between 10 and 40 mm, and LAA lengths ranging between 16 and 51 mm.

The aim of this work was to quantitatively describe the LAA shape variation and clustering using a statistical shape model. We trained a point distribution model (PMD) based on LAA surfaces reconstructed from multidetector computed tomography (CT) images and later combined the trained PMD together with unsupervised clustering methods to examine the natural clustering of the LAA shapes.

# 2   Data and Preprocessing

The LAA surfaces are reconstructed from CT images, provided by the Department of Radiology, Rigshospitalet, University of Copenhagen. The data are acquired as part of the Copenhagen General Population Study [12], where participants are offered a research cardiac computed tomography angiography (CCTA) examination [6]. Participants are excluded from the examination if they, among other things, suffer from AF. The CCTA examinations are performed on a 320 detector CT scanner (Aquilion One, Toshiba, Medical Systems), with the scanner settings: Gantry rotation time 350 ms, detector collimation $0.5 \times 320$, X-ray tube voltage 100–120 kV, and X-ray tube current 280–500 mA. The acquired CT images have a matrix size $512 \times 512 \times 560$ and a voxel size $0.5 \times 0.5 \times 0.25$ mm.

One hundred and five CT images with high contrast are randomly selected from the database (see Fig. 1a for an example of CT image). The raw CT-volumes are manually cropped, using Osirix, to only contain the tracer-enhanced regions with the LAA. After cropping, CT-volumes are blurred with a Gaussian filter kernel with standard deviation at 0.5 mm and the iso-surfaces of the inner part of the LAA is computed using the Marching Cubes algorithm [11] with a manually set iso-level in the range 150–250 Hounsfield Units. The selected iso-surface level varies, due to variations of the amount of tracer in the LAA. Image blurring and surface reconstruction are conducted using 3D Slicer [1]. A reconstructed LAA surface from the example CT image is shown in Fig. 1b.

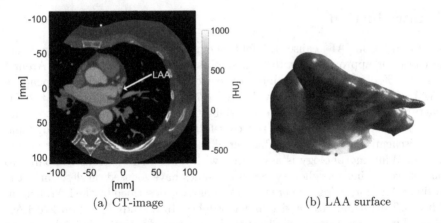

(a) CT-image                                    (b) LAA surface

**Fig. 1.** (a) Slice from raw cardiac computed tomography (CT) image. (b) Left atrial appendage (LAA) surface reconstructed from the CT image shown in (a), where the red marks are manually placed landmarks. HU: Hounsfield Units (Color figure online)

## 3   Methods

The first goal of this work is to build a statistical shape model [3] to quantitatively describe the shape variation of the LAA. This model is created from a training set containing $N$ Procrustes-aligned shapes; shapes in the training set are represented as a series of corresponding points.

### 3.1   Point Correspondence

Point correspondences between LAA surfaces are determined by registering a source surface $\mathcal{S}$ to each target surface $\mathcal{T}$ in the training set, such that each vertex is positioned on the same anatomical structures in both $\mathcal{S}$ and $\mathcal{T}$. Initially, $\mathcal{S}$ is aligned to $\mathcal{T}$ with a similarity transform by registration of four manually placed landmarks equally distributed in the LAA orifice (two out of the four landmarks are visible as the red marks in Fig. 1b). Furthermore, the registration is fine-tuned by an iterative close point (ICP) alignment [16]. The aligned source is now denoted $\mathcal{S}_{ICP}$. The surface registration of $\mathcal{S}_{ICP}$ and $\mathcal{T}$ is performed using a non-rigid volumetric registration algorithm. To be able to use the volumetric registration algorithm, $\mathcal{S}_{ICP}$ and $\mathcal{T}$ must be represented as volumes. We represent $\mathcal{S}_{ICP}$ and $\mathcal{T}$, as signed distance fields (SDF), where each voxel value in the SDF is equal to the signed Euclidean distance to the surface [13,15].

The non-rigid volumetric registration is conducted by solving the optimization given by:

$$\hat{\mathbf{T}}_\mu = \arg\min_{\mathbf{T}_\mu} \left( \mathcal{C} \left( \mathbf{T}_\mu ; I_F, I_M \right) \right) \tag{1}$$

Here $\mathcal{C}$ is a cost-function, $I_F$ is the fixed volume and $I_M$ is the moving volume, where $I_F$ and $I_M$ are the SDF representation of $\mathcal{S}_{ICP}$ and $\mathcal{T}$ respectively. $\mathbf{T}_\mu$ is the non-rigid volumetric transformation that transform $I_M$ to $I_F$.

The transformation is parameterised by a parameter-vector $\boldsymbol{\mu}$. In this work, we use a multi-level B-Spline transformation with five resolution levels. The cost function we are going to minimize is described by:

$$C = \omega_1 MSD\left(\mu; I_F, I_M\right) + \omega_2 \mathcal{P}_{CP}(\mathbf{x}, \mathbf{y}) + \omega_3 \mathcal{P}_{BE}\left(\mu\right), \qquad (2)$$

where $MSD$ is the mean squared voxel value difference similarity measure, $\mathcal{P}_{CP}(\mathbf{x}, \mathbf{y})$ is penalizing large distances between landmarks and $\mathcal{P}_{BE}\left(\mu\right)$ is the bending energy penalty term. The weights: $\omega_1 = 1$, $\omega_2 = 0.15$ and $\omega_3 = 2$ are optimized using a grid search. The optimal transformation parameters are found using adaptive stochastic gradient descent [9] as optimizer, with 2048 random samples per iteration for a maximum of 500 iterations as implemented in the elastix library [10]. The estimated transformation determined between $I_F$ and $I_M$ is applied to $\mathcal{S}_{ICP}$ and the transformed surface is $\mathcal{S}_T$.

Since the volumetric registration is conducted on the SDF, it is not guaranteed that the zero level iso-surfaces fits perfect after the registration. This problem is solved using an approach originally described in [14], where vertices in $\mathcal{S}_T$ are propagated to $\mathcal{T}$ using Markov Random Field regularization of the correspondence vector field. After the vertices in $\mathcal{S}_T$ are propagated to $\mathcal{T}$ we have obtained a point correspondence surface $\mathcal{S}_{COR}$, where each vertex corresponds to a vertex in $\mathcal{T}$. The set of surfaces with point correspondence is used to construct a point distribution model using Procrustes alignment and principal component analysis (PCA) as described in [3].

## 3.2  Shape Clustering

To examine the natural shape clusters formed by our data set, we use the trained point distribution model to represent the surfaces by their PCA loadings and use the loadings to identify shape clusters. The PCA loadings $\mathbf{b}$ of a given surface is determined by:

$$\mathbf{b} = \mathbf{P}(\mathbf{x}' - \bar{\mathbf{x}}), \qquad (3)$$

where $\mathbf{x}'$ is the input surface, $\bar{\mathbf{x}}$ is the Procrustes average shape of the $N$ aligned $\mathcal{S}_{COR}$ and $\mathbf{P}$ is the set of the $t$ first eigenvectors. We use the PCA loadings to estimate the natural number of shape clusters, by examining the log-likelihood (LLH) computed from multivariate Gaussian mixture models (GMM) fitted to the loadings. The probability density function of a GMM can be written as:

$$p(\mathbf{x}) = \sum_{i=k}^{K} \pi_k \mathcal{N}\left(\mathbf{x}|\boldsymbol{\mu}_k, \boldsymbol{\Sigma}_k\right), \qquad (4)$$

where $\mathbf{x}$ is the loadings, $\pi_k$ is the mixing coefficient, $K$ is the number of mixture components and $\mathcal{N}\left(\mathbf{x}|\boldsymbol{\mu}_k, \boldsymbol{\Sigma}_k\right)$ is the multivariate Gaussian distribution with mean $\boldsymbol{\mu}_k$ and covariance matrix $\boldsymbol{\Sigma}_k$. From Eq. (4) the LLH function is given by [2]:

$$p(\mathbf{x}|\boldsymbol{\pi}, \boldsymbol{\mu}, \boldsymbol{\Sigma}) = \sum_{i=1}^{N} \ln\left(\sum_{k=1}^{K} \pi_k \mathcal{N}(\mathbf{x}_i|\boldsymbol{\mu}_k, \boldsymbol{\Sigma}_k)\right) \qquad (5)$$

In order to avoid over-fitting, the number of shape clusters is determined by using two level cross-validation. The first level performs leave-one-out cross-validation. Here the data are divided into $N-1$ training shapes and one test shape. The training set is used to train a GMM with $K$ mixture components, while the test shape is used to validate the trained GMM, using the LLH as quality metric. This procedure is repeated until all $N$ shapes have been used as the test shape, after which the mean test LLH is computed based on the $N$ test LHH. The second cross-validation level iterates through $K = 1 \ldots 10$ mixture components, where the first level is conducted for every $K$. The number of shape clusters is equal to the number of mixture components, which results in the highest mean test LLH.

In order to identify shape appearance of the natural formed clusters, a new GMM is trained on the entire data set, where the number of mixture components is equal to the number of estimated shape clusters. We can now use the model to randomly sample PCA loadings within the different shape clusters and generate synthetic shapes base on the loadings by:

$$\mathbf{x} = \bar{\mathbf{x}} + \mathbf{P}\mathbf{b} \tag{6}$$

The synthetic shapes can be visualized to identify the different shape appearance of each cluster.

The GMMs are fitted to the training data by estimating a set of model parameters: $\pi$, $\mu$, and $\Sigma$, that maximize the LLH function. In this work, we estimate the parameters by the Expectation Maximization algorithm, with 100 random initialization and use the set of model parameters with highest training LLH.

## 4    Results

The point correspondence framework is applied to our 105 reconstructed LAA surfaces. We use the template surface shown in Fig. 2a as source. The template is the average shape of $N$ Procrustes aligned $\mathcal{S}_{COR}$. The set of $\mathcal{S}_{COR}$ is computed as an initial registration of $\mathcal{S}$ and $\mathcal{T}$, where $\mathcal{S}$ is selected randomly from the pool of LAA surfaces.

We are able to determine point correspondences of the majority of the target surfaces (103 out of 105), with a median root mean square distance (RMS) between $\mathcal{T}$ and $\mathcal{S}_{COR}$ at 0.6 mm and a 75th percentile at 0.9 mm. The surfaces with RMS equal to the median and 75th percentile are shown in Fig. 2b and c, respectively. The figure shows $\mathcal{T}$, where the color scale indicate the distance between $\mathcal{T}$ and $\mathcal{S}_{COR}$. It is seen that $\mathcal{S}_{COR}$ matches $\mathcal{T}$ in most of the surface. It is also seen that the point correspondence framework are not able to find point correspondences in the most distal lobes of the LAA. A visual analysis of all $\mathcal{S}_{COR}$ shows that two of the surfaces have poor point correspondence and are therefore excluded from the training set, leaving 103 surfaces for the rest of the analysis.

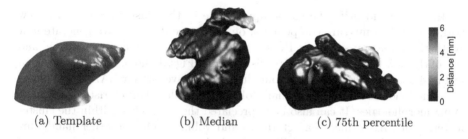

(a) Template           (b) Median           (c) 75th percentile

**Fig. 2.** (a): Template surface determined as the average shape of the $N$ Procrustes-aligned point correspondence surfaces ($\mathcal{S}_{COR}$). (b) and (c) target surface, where the color scale indicate the distance to the $\mathcal{S}_{COR}$. (Color figure online)

The point distribution model is trained on the 103 Procrustes-aligned $\mathcal{S}_{COR}$ and we choose to represent the shapes using their first five PCA loadings. The first five PCA loadings are used, since the remaining 98 PCA loadings each describes only a small fraction (less than 5 %) of the total shape variation in the studied data. Ten GMMs, with $K = 1 \ldots 10$ mixture components, are trained on the PCA loadings and the test LLH is computed from each GMM using cross-validation. The mean test LLH and mean train LLH are shown in Fig. 3 for each validated GMM. It can be seen that, according to the LLH test, a GMM with two mixture components gets the best validation performance. This means that the studied dataset of LAA most likely form two different shape clusters.

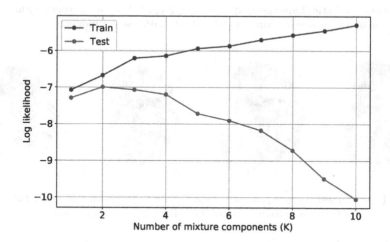

**Fig. 3.** Training and testing log likelihood (LLH) after training Gaussian mixture models (GMMs) on 103 PCA loadings using two level leave-one-out cross-validation.

In order to identify the shape appearance of the clusters, we train a new GMM, with two mixture components, on the entire data set. We generate four synthetic shapes by sampling PCA loadings from mixture component one and two of the new GMM, which can be visualized in Figs. 4 and 5. It can be observed in Fig. 4 that surfaces sampled from cluster one have similar LAA morphology, with an obvious bend in the primary lobe, a particular characteristic of Chicken Wing morphologies. It can also be appreciated the variability within the cluster in terms of LAA orifice characteristics and volumes. On the other hand, surfaces samples from cluster two, illustrated in Fig. 5, do not present a bending of the primary lobe, but a wider one with several secondary lobes. These particular characteristics are typical of non-Chicken Wing LAA morphologies such as Cauliflower ones.

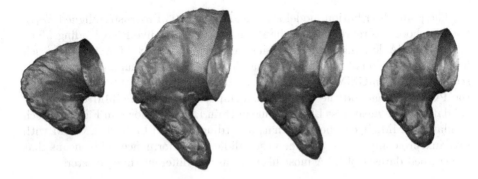

**Fig. 4.** Four synthetic shapes generated by sampling PCA loadings from mixture component one of a Gaussian mixture model (GMM) with two components.

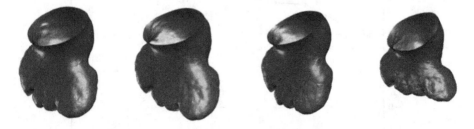

**Fig. 5.** Four synthetic shapes generated by sampling PCA loadings from mixture component two of a Gaussian mixture model (GMM) with two components.

## 5    Conclusion

In this work we have presented a full framework for the extraction and quantification of shape clusters of left atrial appendages and demonstrated that the two primary shape clusters broadly correspond to the main LAA morphological categories in standard clinical classification, Chicken Wing and non-Chicken Wing LAA shapes. The framework enables future statistical inference on the relation between LAA shape characteristics and stroke risk.

# References

1. 3D Slicer. https://www.slicer.org/
2. Bishop, C.M.: Pattern Recognition and Machine Learning, 1st edn. Springer, New York (2006)
3. Cootes, T.F., Taylor, C.J., Cooper, D.H., Graham, J.: Active shape models-their training and application. Comput. Vis. Image Underst. **61**(1), 38–59 (1995)
4. Di Biase, L., et al.: Does the left atrial appendage morphology correlate with the risk of stroke in patients with atrial fibrillation?: results from a multicenter study. J. Am. Coll. Cardiol. **60**(6), 531–538 (2012)
5. Ernst, G., et al.: Morphology of the left atrial appendage. Anat. Rec. **242**(4), 553–561 (1995)
6. Fuchs, A., et al.: Normal values of left ventricular mass and cardiac chamber volumes assessed by 320-detector computed tomography angiography in the copenhagen general population study. Eur. Heart J.-Cardiovasc. Imaging **17**(9), 1009–1017 (2016)
7. Go, A.S., et al.: Prevalence of diagnosed atrial fibrillation in adults: national implications for rhythm management and stroke prevention: the anticoagulation and risk factors in atrial fibrillation (ATRIA) study. JAMA **285**(18), 2370–2375 (2001)
8. Khurram, I.M., et al.: Relationship between left atrial appendage morphology and stroke in patients with atrial fibrillation. Heart Rhythm **10**(12), 1843–1849 (2013)
9. Klein, S., Pluim, J.P., Staring, M., Viergever, M.A.: Adaptive stochastic gradient descent optimisation for image registration. Int. J. Comput. Vis. **81**(3), 227 (2009)
10. Klein, S., Staring, M., Murphy, K., Viergever, M.A., Pluim, J.P.: Elastix: a toolbox for intensity-based medical image registration. IEEE Trans. Med. Imaging **29**(1), 196–205 (2010)
11. Lorensen, W.E., Cline, H.E.: Marching cubes: a high resolution 3D surface construction algorithm. ACM SIGGRAPH Comput. Graph. **21**, 163–169 (1987)
12. Nordestgaard, B.G., et al.: The effect of elevated body mass index on ischemic heart disease risk: causal estimates from a mendelian randomisation approach. PLoS Med. **9**(5), e1001212 (2012)
13. Paulsen, R.R., Baerentzen, J.A., Larsen, R.: Markov random field surface reconstruction. IEEE Trans. Vis. Comput. Graph. **16**(4), 636–646 (2010)
14. Paulsen, R.R., Hilger, K.B.: Shape modelling using markov random field restoration of point correspondences. In: Taylor, C., Noble, J.A. (eds.) IPMI 2003. LNCS, vol. 2732, pp. 1–12. Springer, Heidelberg (2003). https://doi.org/10.1007/978-3-540-45087-0_1
15. Paulsen, R.R., Marstal, K.K., Laugesen, S., Harder, S.: Creating ultra dense point correspondence over the entire human head. In: Sharma, P., Bianchi, F.M. (eds.) SCIA 2017. LNCS, vol. 10270, pp. 438–447. Springer, Cham (2017). https://doi.org/10.1007/978-3-319-59129-2_37
16. Rusinkiewicz, S., Levoy, M.: Efficient variants of the ICP algorithm. In: 2001 Proceedings of the Third International Conference on 3-D Digital Imaging and Modeling, pp. 145–152. IEEE (2001)
17. Wunderlich, N.C., Beigel, R., Swaans, M.J., Ho, S.Y., Siegel, R.J.: Percutaneous interventions for left atrial appendage exclusion: options, assessment, and imaging using 2D and 3D echocardiography. JACC Cardiovasc. Imaging **8**(4), 472–488 (2015)

# Deep Learning Segmentation of the Left Ventricle in Structural CMR: Towards a Fully Automatic Multi-scan Analysis

Hakim Fadil, John J. Totman, and Stephanie Marchesseau[✉]

Clinical Imaging Research Centre, A*STAR-NUS, Singapore, Singapore
stephanie.marchesseau@circ.a-star.edu.sg

**Abstract.** In the past three years, with the novel use of artificial intelligence for medical image analysis, many authors have focused their efforts on defining automatically the ventricular contours in cardiac Cine MRI. The accuracy reached by deep learning methods is now high enough for routine clinical use. However, integration of other cardiac MR sequences that are routinely acquired along with the functional Cine MR has not been investigated. Namely, T1 maps are static T1-based images that encode in each pixel the T1 relaxation time of the tissue, enabling the definition of local and diffuse fibrosis; T2 maps are static T2-based images that highlight excess water (edema) within the muscle; Late Gadolinium Enhancement (LGE) images are acquired 10 min after injection of a contrast agent that will linger in infarct areas. These sequences are acquired in short-axis plane similar to the 2D Cine MRI, and therefore contain similar anatomical features. In this paper we focus on segmenting the left ventricle in these structural images for further physiological quantification. We first evaluate the use of transfer learning from a model trained on Cine data to analyze these short-axis structural sequences. We also develop an automatic slice selection method to avoid over-segmentation which can be critical in scar/fibrosis/edema delineation. We report good accuracy with dice scores around 0.9 for T1 and T2 maps and correlation of the physiological parameters above 0.9 using only 40 scans and executed in less than 15 s on CPU.

## 1 Introduction

Cardiac imaging is an active research field with increasingly new imaging techniques being proposed to better understand, diagnose, treat and anticipate the evolution of cardiac diseases, especially after myocardial infarction. Manual analysis of these images made of thousands of slices is extremely tedious, and more importantly leads to high variability between observers which impacts diagnosis as well as research methodology (e.g. increased recruitment is needed to account for measurement variability). Hence, automatic and reproducible pipelines to process all imaging data are strongly needed. With the emergence of high computational power and the availability of massive labeled image data, recent advances in deep neural networks have made fully automatic cardiac MRI segmentation possible.

© Springer Nature Switzerland AG 2019
M. Pop et al. (Eds.): STACOM 2018, LNCS 11395, pp. 40–48, 2019.
https://doi.org/10.1007/978-3-030-12029-0_5

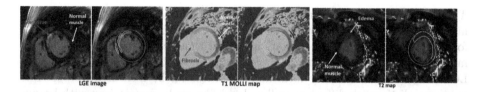

**Fig. 1.** Illustration of the three studied structural MRI sequences for segmentation of the scar tissue (in LGE), fibrosis (in T1 maps) and edema (in T2 maps). (Color figure online)

## 1.1  Deep Learning for Cardiac Cine MRI

Last year's Automated Cardiac Diagnosis Challenge [10] allowed the comparison of several deep learning models for the segmentation of the left and right ventricles from Cine MRI. Very high accuracies were reached, for instance by models derived from the DenseNets model [6] which connects each layer to every other layer; or the U-Net architecture [1,4,11] made of a series of symmetric downsampling convolutional layers followed by upsampling convolutional layers and skip connections for each resolution.

A modified version of the 2D U-Net was optimized for cardiac segmentation [1] in which the authors reduced the up-sampling path complexity by setting the number of filters in the up-convolution layers to the number of classes. Since this model gives very accurate results on our Cine data, we based our implementation on this modified 2D U-Net model.

## 1.2  Structural MRI Sequences

In addition to Cine MRI, which allows the evaluation of functional parameters, most MRI scans also include structural images. In this paper we will focus on three of the available sequences that are routinely used in cardiac scans.

**Late Gadolinium Enhancement** - Acquired in most imaging centers, short-axis Late Gadolinium Enhancement (LGE) images (also known as delayed enhancement images) are T1-weighted images acquired 10–15 min after injection of a contrast agent (gadolinium) that washes-out more slowly in infarct tissue than healthy tissue, resulting in areas of high intensities in the presence of scarring (see example Fig. 1 (left)). Despite many studies, no consensus has yet been reached to determine the best segmentation method to extract scar tissue once the myocardium is segmented [8]. On the other hand, only very few papers focus their energy on segmenting the myocardium in these sequences, using either deformable models [2], or Cine to LGE registration [12]. As one can notice on Fig. 1 (left), the intensity difference between the blood pool and the area of interest is rather low, leading to high difficulty in distinguishing accurately between endocardium and scar, making automatic segmentation of the myocardium in LGE images for scar quantification an open challenge.

**T1 maps -** Short-axis T1 maps are routinely acquired since the development of fast acquisition methods such as the Modified Look-Locker imaging (MOLLI), for its ability to measure T1 relaxation times without contrast. More specifically, myocardial T1 relaxation time is prolonged in most forms of pathology. Unlike LGE images for which only local abnormalities can be measured, T1 MOLLI maps can distinguish diffuse as well as local fibrosis by comparison with known reference T1 values. An example of local fibrosis definition (in yellow) on T1 MOLLI map is shown Fig. 1 (middle). In order to assess diffuse fibrosis, the T1 value of the entire myocardium must be compared to healthy muscle, therefore in T1 maps the accuracy of the LV segmentation is crucial for diffuse fibrosis quantification as well as further thresholding for local fibrosis quantification.

**T2 maps -** Similarly, T2 maps are acquired to measure myocardial edema, which manifests as free water content with prolonged T2 relaxation times. Differentiating edema from infarction has been shown to be useful to characterize myocardial salvage by comparison with LGE, and hence quantify the success of coronary revascularization surgery. Prolonged T2 value is also an indicator of diffuse pathology such as cardiomyopathy and is therefore a key biomarker to measure. An example of T2 maps is presented Fig. 1 (right) where a zone with edema is delineated in yellow.

In this paper, we propose to automatically segment the endocardium and epicardium contours on the three structural images described above. To the best of our knowledge, this is the first attempt to automatically segment the left ventricle in T1 and T2 maps. Since a large amount of Cine data is available for training compared to available structural data, we will test a transfer learning approach from the Cine model as well as a full U-Net architecture on our dataset made of 40 patients.

## 2   Methods

### 2.1   Available Data: Pre and Post-processing

The T1 and T2 maps being relatively new sequences, and varying from scanner to scanner, only 40 patient scans were available at the time of this study. The same number of LGE data was used for fair comparison. The train/validation partition was set to 80/20% for each experiment without data augmentation. All images were acquired post-myocardial infarction on Siemens scanners (3T mMR and 3T Prisma). To account for various image sizes and resolutions, all 2D slices were rescaled to $212 \times 212$ pixels of 1.37 mm × 1.37 mm resolution, and intensities were normalized. Softmax predictions were rescaled to the original resolution and size, and 3D masks were created choosing the labels with the highest score. Finally, only large connected components were kept, and a convex hull was defined to smooth the endocardium and epicardium contours. Since no consensus exist on the best way to define scar, fibrosis or edema once the left ventricle contours exist, we decided to use 2-dimensional Otsu automatic threshold [5] to estimate the amount of damaged tissue for both the ground truth contours (obtained manually) and the predicted contours, instead of relying on a manual ground truth.

## 2.2   Transfer Learning from U-Net vs Full U-Net Model

Transfer learning is based on the hypothesis that in a deep neural network, the first layers extract global features that are similar among many types of images, hence do not require to be retrained for every set of similar images. In our case, all images are short-axis slices showing similar shapes for which only the contrast between the various structures changes. We therefore hypothesize that training only the last few layers will allow a smaller database to be used for training while giving accuracy similar to a full U-Net model. A transfer model also has the advantage of requiring less computational memory and time and can therefore be trained on CPU. To apply transfer learning, we base our code on the open-source 2D U-Net python code [1]. We apply our pretrained Cine model up to the last 3 layers to create the input image and define a small model (named *3 Layers*) only made of the last 3 layers of U-Net (3 convolution blocks with 64, 64 and 3 filters). This model as well as the full U-Net model (named *Full Model*) are optimized to maximize the foreground dice (combining endocardium and myocardium contours) with Adam optimizer (learning rate of 0.01, $\beta_1 = 0.9$, $\beta_2 = 0.999$, batch size = 5). The training of the *3 Layers* and *Full Model* took respectively 5 h and 20 h for 40K steps on a GPU NVIDIA K40. The volume segmentation takes less than 14 s on an Intel Xeon E5-1650 CPU.

## 2.3   Automatic Image Selection

Imbalance between foreground and background data led to over-segmentation of basal and apical slices that should not be included in the segmentation (do not contain myocardium), as previously noticed in right ventricle segmentation using the full 2D U-Net on Cine MRI. To cope with this issue, we decided to independently train a binary classification network that automatically selects for each scan the slices that include the myocardium. We designed a neural network (see Fig. 2) made of 7 convolutional blocks, a flatten layer and two fully connected blocks, inspired by the AlexNet [7] architecture. Each convolution block consists of a $3 \times 3$ convolution without padding, followed by a $2 \times 2$ max pooling layer and a ReLU activation. The model was optimized to fit weighted cross-entropy with a weight of 0.2 to decrease false negative rate since removing a slice to segment is more critical than over-segmenting. The decision to remove a slice was made if the probability to remove it was 10 times higher than the probability to keep it. We used Adam optimizer (learning rate of 0.0001, $\beta_1 = 0.9$, $\beta_2 = 0.999$, batch size = 16) and dropout layers after each block, to regularize the model that initially showed high variance, with the keep prob adjusted for each sequence from 0.7 to 0.9. The model was trained in approximately 2000 steps which took 20 min on CPU. The classification of each scan took less than 1 s on CPU. Each 3D scan is therefore first segmented through the U-Net model and then over-segmentation are deleted using the results of this selection network applied on the original scan.

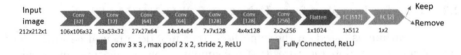

**Fig. 2.** Representation of the designed CNN, inspired by AlexNet [7] architecture. Each blue arrow represents a convolution layer made of a 3 × 3 convolution, a 2 × 2 max pooling and a ReLU activation. The number of filters is written in each arrow. A flatten layer and two fully connected layers terminate the network. Resulting image sizes are written below each block. (Color figure online)

# 3   Results

## 3.1   Accuracy of Image Selection for Structural Images

Table 1 reports the sensitivity (true positive rate, where positive means keep a slice), specificity (true negative rate, where negative means remove a slice) and total accuracy for the training and validation sets. Dropout coefficients were adjusted to each sequence to balance the high variance problem we observed in our first experiments. Notice the very high sensitivity rates in the validation set that guarantees that all slices to be segmented will be rightly kept. The model also manages to remove between 78% (for T2 maps) and 91% (for T1 maps) of all unwanted slices.

**Table 1.** Sensitivity, specificity and accuracy of the designed method to select the slices to keep for further segmentation. KP: keep prob in the dropout regularization

|  | Training (n = 384 slices) | | | Validation (n = 81 slices) | | |
|---|---|---|---|---|---|---|
|  | Sensitivity | Specificity | Accuracy | Sensitivity | Specificity | Accuracy |
| LGE (KP = 0.7) | 98% | 74% | 91% | 100% | 85% | 96% |
| T1 maps (KP = 0.9) | 100% | 100% | 100% | 97% | 91% | 94% |
| T2 maps (KP = 0.8) | 89% | 76% | 86% | 100% | 78% | 91% |

## 3.2   Comparison of the Methods

Using the image selection results, corrections were applied (slices were removed according to the classification) after prediction, for the *3 Layers* and the *Full Model* leading to a comparison of four methods. Results were measured in terms of dice scores for the endocardium and the myocardium masks. Additionally, using the Otsu thresholding to define damaged zones, Pearson's correlation (R), bias and reproducibility error (RPC) were assessed for the scar size. For T1 and T2 maps, since the actual relaxation times are also predictors of cardiac dysfunction, the mean value within the myocardium was also measured and a Pearson's correlation reported.

**Table 2.** Segmentation accuracy and statistical results for the estimation of infarct size using the various proposed methods on LGE images. Bold values represent best values in the validation set.

| | Training(n = 32) | | | | | Validation (n=8) | | | | |
|---|---|---|---|---|---|---|---|---|---|---|
| | Dice | | Scar %LVM | | | Dice | | Scar %LVM | | |
| | Endo | Myo | R | Bias | RPC | Endo | Myo | R | Bias | RPC |
| 3 Layers | 0.93 | 0.84 | 0.97 | 1.96 | 3.63 | 0.83 | 0.68 | 0.94 | 0.09 | 5.58 |
| Corrected 3L | 0.95 | 0.85 | 0.96 | 2.11 | 4.38 | 0.87 | **0.69** | **0.95** | 0.16 | **4.92** |
| Full Model | 0.84 | 0.88 | 0.97 | −0.82 | 3.85 | 0.78 | 0.64 | 0.91 | −1.11 | 7.34 |
| Corrected FM | 0.96 | 0.96 | 0.99 | −0.03 | 2.06 | **0.89** | 0.66 | 0.94 | **0.05** | 5.67 |

**LGE Image Segmentation.** As shown in Table 2, training accuracy was extremely high for the *3 Layers* (with or without correction) and the *Corrected Full Model*, which proves the capacity of these models to estimate myocardium contours in LGE images despite the close signal intensity values between blood pool and scar tissue. However, the validation set did not reach such performance in terms of dice scores despite a good correlation in the scar size. Fair and poor results are mainly due to a wrong endocardium segmentation that includes the infarct zone as shown Fig. 4 (left). Since scar appearance in MRI varies considerably in shape, location and intensity from patient to patient, and depends on the contrast injected and the acquisition method, we believe a larger database is mandatory to reduce the high variance we obtained.

**T1 Maps Segmentation.** As illustrated in Fig. 3 (left), best dice scores and correlations are reported with the *Corrected Full Model* (dice = 0.96 for endocardium, dice = 0.92 for myocardium, R = 0.996 for Mean T1 and R = 0.983 for fibrosis size) while the *3 Layers* model seems unreliable for fibrosis detection. Figure 4 (middle) gives three example segmentations of local fibrosis and

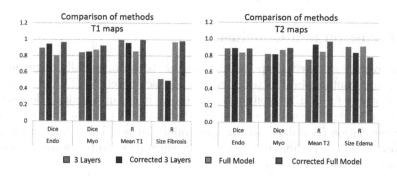

**Fig. 3.** Segmentation accuracy and correlation of T1 and T2 measures using the various proposed methods for the validation set made of 8 patient scans.

endocardium in T1 maps. Fair results correspond to slight underestimation of the fibrosis due to over-segmentation of the endocardium. Poor results arise in cases of important artifacts (probably due to breathing motion that perturbed T1 reconstruction).

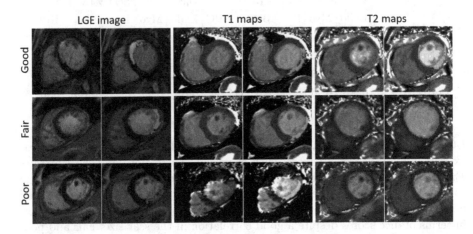

**Fig. 4.** Example endocardium (green) and scar, fibrosis or edema (red) segmentations obtained using the *Corrected Full Model* on validation data. Good, median and bad results are shown for LGE, T1 and T2 images. (Color figure online)

**T2 Maps Segmentation.** As shown Fig. 3 (right), the *Corrected 3 Layers* showed results similar to the *Corrected Full Model* in terms of dice scores for the endocardium (0.889 and 0.885) and correlation for Mean T2 values (0.937 and 0.977 respectively). Surprisingly, the edema size showed better correlation without the added correction (R = 0.919). Figure 4 (right) illustrates examples of endocardium and edema delineations showing that poor results were due to wrong segmentation of non-existing edema, while fair results, similarly to T1 and LGE, were due to over-segmentation of the endocardium.

## 4    Discussion and Conclusions

In this paper, we used deep neural network to automatically segment the myocardium contours in structural images, very accurately for T1 and T2 maps, and reasonably for LGE. Despite a fair dice score in the myocardium delineation of the LGE images, the scar size was well estimated with an error around 5% and a high correlation (R = 0.94) on our small validation set. These correlations and error levels are comparable to reported reproducibility errors obtained between two manual segmentations [3].

We evaluated the use of transfer learning to segment the structural maps since these maps share features with Cine data for which our pretrained model reach high performance. Results were extremely comparable to the full U-Net model in terms of dice scores or correlations, apart from the estimation of the fibrosis size. It can therefore be a good alternative when working with restricted memory or computation power.

Better results were achieved after correction using our proposed CNN image selection model. High accuracies were reached, preventing most over-segmentation without deleting required slices. Cine MRI segmentation also suffers from this type of over-segmentation as the definition of the basal slice varies among observers. Recent active shape algorithms have been proposed to automatically define the basal slice [9] but once again deep learning methods, such as the one proposed here, showed more promising results. For a fully automatic cardiac analysis pipeline, we will extend the proposed slice selection neural network to our Cine data in combination with the current U-Net model that we are routinely using.

Future work will extend the current database and validate the robustness of the proposed pipelines. Similar concepts will also be applied to other structural images such as T2*, T2 black blood, post-contrast T1 maps or even ECV maps. Estimating the scar/fibrosis/edema zones directly from a deep learning network would also be appreciated once a reliable ground truth is given, or a consensus among experts has been reached.

After further validation, these models and the Cine model will be incorporated into our in-house cardiac analysis software. The amount of post-prediction corrections required will be measured in order to obtain confidence intervals for a fully automatic multi-scan cardiac MRI analysis pipeline in the near future.

## References

1. Baumgartner, C.F., Koch, L.M., Pollefeys, M., Konukoglu, E.: An exploration of 2D and 3D deep learning techniques for cardiac MR image segmentation. In: Pop, M., et al. (eds.) STACOM 2017. LNCS, vol. 10663, pp. 111–119. Springer, Cham (2018). https://doi.org/10.1007/978-3-319-75541-0_12
2. Ciofolo, C., Fradkin, M., Mory, B., Hautvast, G., Breeuwer, M.: Automatic myocardium segmentation in late-enhancement MRI. In: ISBI 2008, pp. 225–228. IEEE (2008)
3. Flett, A.S., et al.: Evaluation of techniques for the quantification of myocardial scar of differing etiology using cardiac magnetic resonance. JACC Cardiovasc. Imaging. 4(2), 150–156 (2011)
4. Isensee, F., Jaeger, P.F., Full, P.M., Wolf, I., Engelhardt, S., Maier-Hein, K.H.: Automatic cardiac disease assessment on cine-MRI via time-series segmentation and domain specific features. In: Pop, M., et al. (eds.) STACOM 2017. LNCS, vol. 10663, pp. 120–129. Springer, Cham (2018). https://doi.org/10.1007/978-3-319-75541-0_13
5. Liu, J., Li, W., Tian, Y.: Automatic thresholding of gray-level pictures using two-dimension Otsu method. In: International Conference on Circuits and Systems, pp. 325–327. IEEE (1991)
6. Khened, M., Alex, V., Krishnamurthi, G.: Densely connected fully convolutional network for short-axis cardiac cine MR image segmentation and heart diagnosis using random forest. In: Pop, M., et al. (eds.) STACOM 2017. LNCS, vol. 10663, pp. 140–151. Springer, Cham (2018). https://doi.org/10.1007/978-3-319-75541-0_15

7. Krizhevsky, A., Sutskever, I., Hinton, G.E.: ImageNet classification with deep convolutional neural networks. In: Advances in Neural Information Processing Systems, pp. 1097–1105 (2012)
8. McAlindon, E., Pufulete, M., Lawton, C., Angelini, G.D., Bucciarelli-Ducci, C.: Quantification of infarct size and myocardium at risk: evaluation of different techniques and its implications. EHJ-CVI **16**(7), 738–746 (2015)
9. Paknezhad, M., Marchesseau, S., Brown, M.S.: Automatic basal slice detection for cardiac analysis. J. Med. Imaging **3**(3) (2016). https://doi.org/10.1117/1.JMI.3.3.034004
10. Pop, M., Sermesant, M., Jodoin, P.-M., Lalande, A., Zhuang, X., Yang, G., Young, A., Bernard, O. (eds.): STACOM 2017. LNCS, vol. 10663. Springer, Cham (2018). https://doi.org/10.1007/978-3-319-75541-0
11. Ronneberger, O., Fischer, P., Brox, T.: U-Net: convolutional networks for biomedical image segmentation. In: Navab, N., Hornegger, J., Wells, W.M., Frangi, A.F. (eds.) MICCAI 2015. LNCS, vol. 9351, pp. 234–241. Springer, Cham (2015). https://doi.org/10.1007/978-3-319-24574-4_28
12. Wei, D., Sun, Y., Chai, P., Low, A., Ong, S.H.: Myocardial segmentation of late Gadolinium enhanced MR images by propagation of contours from cine MR images. In: Fichtinger, G., Martel, A., Peters, T. (eds.) MICCAI 2011. LNCS, vol. 6893, pp. 428–435. Springer, Heidelberg (2011). https://doi.org/10.1007/978-3-642-23626-6_53

# Cine and Multicontrast Late Enhanced MRI Registration for 3D Heart Model Construction

Fumin Guo[1,2(✉)], Mengyuan Li[3], Matthew Ng[1,2], Graham Wright[1,2], and Mihaela Pop[1,2]

[1] Sunnybrook Research Institute, University of Toronto, Toronto, Canada
fumin.guo@sri.utoronto.ca
[2] Department of Medical Biophysics, University of Toronto, Toronto, Canada
[3] Department of Biomedical Engineering, University of Toronto, Toronto, Canada

**Abstract.** Cardiac MR imaging using multicontrast late enhancement (MCLE) acquisition provides a way to identify myocardium infarct scar and arrhythmia foci in the peri-infarct. In image-guided RF ablations of ventricular arrhythmia and computational modeling of cardiac function, construction of a 3D heart model is required but this is hampered by the challenges in myocardium segmentation and slice misalignment in MCLE images. Here we developed an approach for cine and MCLE registration, and MCLE scar-cine myocardium label fusion to build high-fidelity 3D heart models. MCLE-cine image alignment was initialized using a block-matching-based rigid registration approach followed by a deformable registration refinement step. The deformable registration approach employed a self similarity context descriptor for image similarity measurements, optical flow as a transformation model and convex optimization to derive the optimal solution. We applied the developed approach to a preclinical dataset of 10 pigs with myocardium infarction and evaluated the registration accuracy by comparing cine and MCLE myocardium masks using Dice-similarity-coefficient (DSC) and average symmetric surface distance (ASSD). For 10 pigs, we achieved a mean DSC of $80.4 \pm 7.8\%$ and ASSD of $1.28 \pm 0.47$ mm for myocardium with a mean runtime of 1.5 min for each dataset. These results suggest that the developed approach provide the registration accuracy and computational efficiency that may be suitable for clinical applications of cardiac MRI that involve a cine and MCLE MRI registration component.

**Keywords:** Multicontrast late enhancement MRI ·
Myocardium infarct tissue characterization · Image registration ·
Convex optimization

## 1 Introduction

Ventricular arrhythmia associated with myocardial infarction is a leading cause of death. Currently, arrhythmogenic foci are identified using surfacic and invasive electrophysiological mapping. Cardiac MRI using contrast-enhanced acquisitions, e.g., late gadolinium enhancement (LGE), provides a way to non-invasively

© Springer Nature Switzerland AG 2019
M. Pop et al. (Eds.): STACOM 2018, LNCS 11395, pp. 49–57, 2019.
https://doi.org/10.1007/978-3-030-12029-0_6

identify the arrhythmogenic foci hidden deep in the myocardium. In image-guided RF ablations of ventricular arrhythmogenic foci and computational modeling of cardiac function, construction of a 3D heart model that depicts the extent, location and transmurality of myocardial infarct within the myocardium is urgently required. However, this is challenging because of the difficulties in myocardium segmentation and slice misalignment in late enhancement MR images. Cardiac MRI using cine acquisitions provide excellent visualization of myocardium and cine-myocardium may be employed to facilitate a 3D heart model construction but this requires cine-LGE MRI registration as a first step.

Limited efforts have been dedicated to cine-LGE MRI registration and most of these efforts aimed to propagate cine myocardium contours to LGE images for myocardium infarct scar heterogeneity characterization. For example, Chenoune et al. [2] initialized 3D cine and LGE MR alignment using image orientation and position information, and the rough alignments were refined using a rigid registration approach that employed normalized mutual information metrics and a Powell's optimization scheme. Wei et al. [12] registered cine to LGE images using an enhanced cross correlation-based constrained affine registration algorithm followed by a B-spline free-form deformable registration using pattern intensity. Tao et al. [11] performed global affine registration using Elastix toolbox and then locally refined cine-LGE alignments by maximizing the correlation between cine contour maps and LGE images.

Multicontrast late enhancement (MCLE) MRI represents a new method that also permits myocardium infarct tissue characterization and has demonstrated advantages over conventional LGE MRI [3]. Previous studies have shown that MCLE provides better SNR, higher contrast for improved myocardium infarct visualization and scar characterization. However, there are still known issues associated with myocardium segmentation and slice misalignment that hamper 3D MCLE-based heart model construction. In this work, we provided an approach for cine and MCLE cardiac MRI registration and cine myocardium-MCLE scar label fusion for 3D heart model construction. As shown in Fig. 1, our developed approach employed a rigid registration step to initialize the cine-MCLE alignment followed by a deformable registration component to refine the registration results. The resulting deformation field was used to transform MCLE-scar and cine-myocardium masks for label fusion and 3D heart model construction.

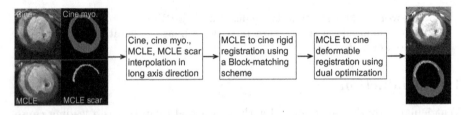

**Fig. 1.** Cine-MCLE registration and cine myocardium-MCLE scar fusion workflow.

# 2   Methods

## 2.1   Cine-MCLE Rigid Registration: A Block Matching Scheme

Rigid or affine registrations are usually performed to initialize the rough alignment of images prior to finer deformable registrations. Most of the rigid and affine registration methods involve finding a global relationship or correspondence of the features (i.e., image signal intensities, landmarks) between two images and maximizing the similarity between these features. Block matching represents a popular technique and has been widely used for rigid (and affine) and non-rigid image registration. Instead of measuring the global similarity between image features with respect to the global transformation parameters, block matching exploits local similarity measurements of image features and generates a local transformation, from which a global transformation is derived [9].

In block matching-based rigid registration algorithms, the basic concept is to divide the images into blocks and to find the block-wise correspondences. The $i^{th}$ subimage or block $B_m^i(x)$ in the moving image $I_m(x)$ is moved within a search window and the similarity between block $B_m^i(x)$ and the block in the fixed image $I_f(x)$ with coincident positions as $B_m^i(x)$ is measured, i.e., using sum of absolute differences, cross correlation or mutual information. The block $\hat{B}_f^i(x)$ that best matches $B_m^i(x)$ is determined under the matching criterion, resulting in a correspondence between the centre of $B_m^i(x)$ and the centre of $\hat{B}_f^i(x)$. Depending on applications, the computational efficiency of the block-to-block similarity measurements may be improved in a few ways, i.e., subsampling the search window and features in the block for similarity measurements. To facilitate robust estimation of the global transformation matrix $T$, outlier correspondences (bad block match) were removed and the remainder were used for a least trimmed squared regression. The correspondences with low regression errors (top 50%) were used to re-estimate the global transformation $T$ and this process was iterated until convergence.

## 2.2   Cine-MCLE Deformable Registration Using Dual Optimization

From the perspective of energy minimization, deformable registration aims to estimate plausible deformation filed $\phi(x)$ to align two images or to maximize the similarity between a fixed image $I_f(x)$ and a moving image $I_m(x)$, $x \in \Omega$. In cardiac cine and MCLE images, we noticed that the two images are not directly comparable because of different contrast mechanisms [4]. Here we employed the self similarity context ($SSC$) [7] instead of the original image signal intensities for robust similarity measurements of cine and MCLE images. For $\forall x \in \Omega$, $SSC(x)$ defines an eight-element vector each measuring the exponent of the negative absolute differences of image signal intensities between $x$ and its eight-neighbour. Therefore, the similarity between $I_f(x)$ and $I_m(x)$ can be measured by computing the sum of absolute differences between $SSC_f(x)$ and $SSC_m(x)$ under the deformation field $\phi(x)$. In addition, the desired "plausible" deformation field was constrained by regularizing the smoothness of $\phi(x)$ using total variation, i.e., $\int_\Omega |\nabla\phi(x)|\,dx$. To this end, the deformable registration problem can be formulated by minimizing the energy function $E(\phi)$ (subject to some given deformation $\Phi(x)$ if there is) as follows:

$$E(\phi) = \int_\Omega \left| SSC_f(x) - SSC_m(x + \Phi + \phi) \right| dx + \alpha \sum_{d=1}^{3} \int_\Omega |\nabla(\Phi_d + \phi_d)|\,dx, \quad (1)$$

where the first term measures the similarity between $I_f(x)$ and $I_m(x)$ using the respective $SSC$ under the deformation field $\phi(x)$ (with given $\Phi(x)$ if there is), and the second term evaluates the smoothness of the deformation field.

Direct minimization of $E(\phi)$ (1) is challenging because of the nonlinearity and nonsmoothness of the energy function. Here we employed optical flow technique [8] to decompose the nonlinear similarity measurement term following the fact that $f(x+t) \approx f(x) + \nabla f(x) \cdot t$, where $t$ is the small displacement field between $f(x)$ and $f(x+t)$, and can be obtained by estimating a series of incremental deformation field $t'(x)$, i.e., $t(x) = \sum_i t_i'(x)$ [6]. In addition, the nonsmooth absolute function terms can be smoothed by introducing additional function, i.e., $|g(x)| = \max_{|h(x)| \leq 1} h(x) \cdot g(x)$. Furthermore, the total variation function can be formulated as $\int |\nabla q(x)|\,dx = \max_{|p(x)| \leq 1} \int divp(x)q(x)dx$ (see [5,6] for detailed analyses). Therefore, we can equivalently rewrite the complicated minimization problem Eq. (1) as follows:

$$\max_{|p| \leq 1, |q| \leq \alpha} \min_{\phi} E(\phi; p, q) := \int_\Omega (p \cdot S_0 + \sum_{d=1}^{3} \Phi_d \cdot \operatorname{div} q_d)dx$$

$$+ \sum_{d=1}^{3} \int_\Omega \phi_d \cdot (\operatorname{div} q_d - p \cdot \partial_d S)dx, \quad (2)$$

where $S_0 = SSC_f(x) - SSC_m(x + \Phi)$ and $S = SSC_m(x + \Phi)$. Clearly, minimization of Eq. (2) over free variable $\phi_d(x)$ requires vanishing of $(\operatorname{div} q_d - p \cdot \partial_d S)$, i.e., $(\operatorname{div} q_d - p \cdot \partial_d S) = 0, d \in \{1, 2, 3\}$. Alternatively, the deformation field $\phi_d(x)$ just acted as the multiplier function of the respective constraints $(\operatorname{div} q_d - p \cdot \partial_d S) = 0$, $d \in \{1, 2, 3\}$, on top of the maximization component. Therefore, we developed an augmented Lagrangian algorithm based on convex optimization theories [1] to derive the optimal $\phi_d(x)$ as follows:

$$\max_{|p| \leq 1, |q| \leq \alpha} \min_{\phi} E(\phi; p, q) - \frac{c}{2} \sum_{d=1}^{3} \| \operatorname{div} q_d - p \cdot \partial_d S \|^2, \quad (3)$$

where $c > 0$ is a scale.

Clearly, the dual relationship between Eqs. (2) and (1) indicates that optimization of the convex problem Eq. (2) solves the original registration problem Eq. (1) equivalently but demonstrated greater simplicity in numerics than the original non-linear non-convex formulation (1). In addition, the point-wise iterative implementation of the registration algorithm Eq. (3) (see [6] for the details) leads to high performance parallel implementation on a GPU for speed-up.

# 3   Experiments and Results

## 3.1   Animal Preparation and MRI Acquisition

Myocardial infarction was generated in 10 pigs (weighing 20–25 kg) by occluding either the left anterior descending artery (LAD) or the left circumflex (LCX) artery. We inflated a balloon catheter in the LAD or LCX for ~90 min followed by reperfusion to create heterogeneous infarcts as previous described [10]. The animal study was approved by our institution and the animals were allowed to heal for ~5 weeks prior to MR imaging.

MRI was performed at 1.5 T with a GE Signa Excite Scanner. Animals were sedated and the respiration was controlled using a mechanical ventilator. For cine MRI, 2D short-axis slices were acquired using a conventional segmented steady state pulse sequence (SSFP, bandwidth = 125 KHz, number of phases = 20, views per segment = 16, TR/TE/flip angle = 3.7 ms/1.6 ms/45°, field of view = $23 \times 23 \, \text{cm}^2$, image matrix = 256 × 256, NEX = 1, number of slices = 9–18, slice thickness = 5 mm). MCLE images were obtained ~15 min after a bolus injection of 0.2 mmol/kg Gd-DPTA (Gd-DPTA Magnevist; Berlex Inc., Wayne, NJ, USA). 2D short-axis images were acquired using an inversion recovery-prepared b-SSFP sequence at different inversion times ranging from 175 to 250 ms (bandwidth = 125 KHz, number of phases = 20, views per segment = 16, TR/TE/flip angle = 5.5 ms/1.9 ms/45°, field of view = $25.6 \times 25.6 \, \text{cm}^2$, image matrix = 256 × 256, slice thickness = 5 mm).

## 3.2   Algorithm Implementation

MCLE scar heterogeneity was segmented slice-by-slice using a fuzzy-logic clustering approach [3]. For each dataset, one out of the 6–8 images at diastolic phases with minimal cardiac motion was extracted for each slice and the extracted phases for all the slices were stacked into a 3D MCLE image $I_m(x)$. Similarly, 1 out of the 20 phases of each cine slice that matches the shape of the heart in MCLE was selected and all the selected phases were stacked into a 3D cine image $I_f(x)$. Prior to algorithm registration, 3D cine and MCLE images were interpolated to isotropic voxel size of respective in-plane voxel width (~0.9 mm for cine and ~1 mm for MCLE). Both the rigid and deformable registrations were implemented in a coarse-to-fine manner (3 levels, scaling factors = {4, 2, 1}), whereby large displacements field was estimated in lower levels and smaller displacements field in higher levels. The coarse-to-fine implementation provides both registration accuracy and computational efficiency. For the deformable registration, the given (initial) deformation field $\Phi(x)$ in Eq. (1) at the lowest level (scaling factor = 4) was set to $\mathbf{0}$ while for the other levels $\Phi(x)$ was initialized using the final deformation from the preceding levels. The MCLE scar (infarct core ∪ gray zone) were deformed and fused with cine myocardium masks.

### 3.3   Validation

Cine and MCLE image registration accuracy was evaluated by comparing cine and MCLE myocardium masks. In particular, a single observer manually segmented the myocardium in cine slice-by-slice using ITK-SNAP and the segmentation results were approved by an experienced observer with more than 10 years' experience in cardiac MRI segmentation. MCLE images were segmented slice-by-slice using a fuzzy-logic clustering approach [3]. For each slice, 6–8 images at diastolic phases with minimal cardiac motion were used for exponential fitting to derive a T1* and steady-state (SS) signal intensity value for each pixel. The generated T1*-SS maps were classified into: infarct core (IC), gray zone (GZ), healthy myocardium (HM) and blood (B). MCLE myocardium masks were obtained by combining the IC, GZ and HM sub-regions. The derived deformation field was used to warp the MCLE myocardium masks. Cine-MCLE registration accuracy was quantified using Dice similarity coefficient ($DSC$) and average symmetric surface distance ($ASSD$) of cine and MCLE manual myocardium masks. Let $R_c$ and $R_m$ be the cine and deformed MCLE myocardium mask, respectively. $DSC$ is calculated as: $\frac{2(R_c \cap R_m)}{R_c + R_m} \cdot 100\%$. $ASSD$ is given as: $\frac{1}{2}\{\frac{1}{|\partial R_c|} \sum_{p \in \partial R_c} d(p, \partial R_m) + \frac{1}{|\partial R_m|} \sum_{p \in \partial R_m} d(p, \partial R_c)\}$, where $\sum_{p \in \partial R_c} d(p, \partial R_m)$ represents the summation of the minimal distance $d$ from the points $p$ in surface $\partial R_c$ to surface $\partial R_m$ and $|\partial R_c|$ the total number of points in surface $\partial R_c$. In addition, we reported the runtime to evaluate the computational efficiency of our approach.

### 3.4   Results

Figure 2 illustrates representative cine-MCLE image registration results. As shown in Table 1, we achieved a mean DSC of $80.4 \pm 7.8\%$ (range: 62.3%–86.7%) and ASSD of $1.28 \pm 0.47$ mm (range: 0.82 mm–2.25 mm) by comparing the cine and deformed MCLE myocardium masks in 10 pigs. Figure 3 shows example fused MCLE scar (IC in green and GZ in yellow) and cine myocardium (HM in blue).

All image analysis was performed in 3D space on a Linux Desktop (Ubuntu 14.04, 16G RAM, Inter(R) i7-3770, 3.4 GHz) with a NVIDIA graphics processing unit (GeForce GTX TITAN BLACK). Given pre-segmented MCLE scar geometry, the developed fully automated registration approach required ~45 s for the rigid registration, ~50 s for the deformable registration refinement, resulting in an overall runtime of ~1.5 min for each pig.

**Table 1.** Cine and MCLE registration accuracy measurements. (n = 10)

| Pig ID | 1 | 2 | 3 | 4 | 5 | 6 | 7 | 8 | 9 | 10 | All |
|---|---|---|---|---|---|---|---|---|---|---|---|
| DSC (%) | 86.7 | 86.6 | 76.2 | 84.8 | 84.0 | 84.5 | 83.5 | 62.3 | 72.4 | 82.6 | $80.4 \pm 7.8$ |
| ASSD (mm) | 0.82 | 0.84 | 1.85 | 1.27 | 0.93 | 1.00 | 1.26 | 2.25 | 1.53 | 1.05 | $1.28 \pm 0.47$ |

MCLE      Cine      Overlay      MCLE      Cine      Overlay

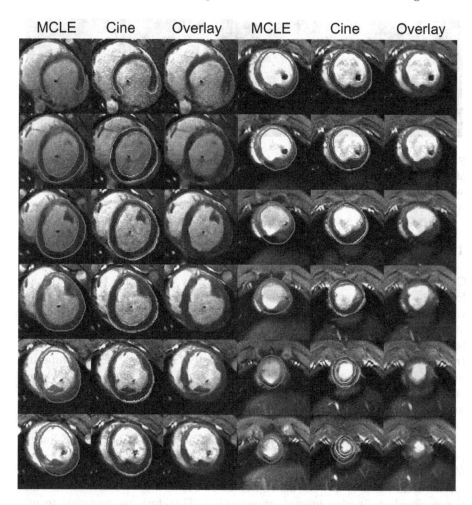

**Fig. 2.** Representative MCLE and cine MRI registration results. **First** and **Fourth** columns: MCLE images and myocardium segmentation (purple contours); **Second** and **Fifth** columns: cine images, cine (green) and deformed MCLE (purple) myocardium contours; **Third** and **Sixth** columns: cine (green) and registered MCLE (purple) image overlay (green or purple shows the differences and gray represents match of the two images.) Basal to apex slices are shown from upper left to lower right. (Color figure online)

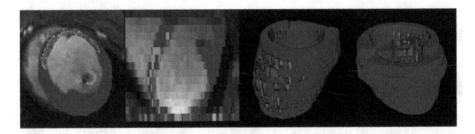

**Fig. 3.** A representative heart model built from cine-MCLE MRI registration and MCLE scar-cine myocardium fusion. Short axis, long axis and 3D view of the heart models are shown from left to right. (Blue: Cine myocardium, Green: MCLE infarct core, Yellow: MCLE gray zone.) (Color figure online)

## 4   Discussion and Conclusions

Cine and MCLE MRI registration provides a way to combine the information from the two imaging methods and holds great promise for improved cardio-vascular disease care. In this preclinical study, we developed an approach for automated cine and MCLE image registration and MCLE scar-cine myocardium label fusion. For a group of 10 pigs, we demonstrated cine and MCLE MRI registration accuracy of $80.4 \pm 7.8\%$ for DSC, $1.28 \pm 0.47$ mm for ASSD by comparing manual myocardium masks from the two images. The resulting MCLE-scar and cine-myocardium mask fusion provides a way to build high fidelity 3D heart models for image-guided RF ablation therapies and computational modeling of cardiac electro-mechanical function.

Lastly, we acknowledge a number of limitations of this study. MCLE images provide visualization of myocardium but the image quality is generally poor for myocardium segmentation. In parallel, we observed non-smooth MCLE myocardium surfaces due to the displacement between consecutive slices while the myocardium in cine images are smoother. Therefore, we proposed to utilize the smooth myocardium surfaces provided by cine images through MCLE-cine registration. We note that in most cases the myocardium surfaces in cine images are smooth but in some situations the cine myocardium masks present discontinuities. In these situations, the non-smooth myocardium masks may be improved by re-aligning the cine slices prior to MCLE-cine registration, i.e., using the block-matching-based rigid registration approach [9]. Regarding the deformable registration method, our approach required point-wise similarity measurements while other metrics, including global mutual information/cross correlation, also demonstrate promising performance and are widely used in various registration tasks. Unfortunately, these global similarity metrics are not directly amenable to our registration framework further investigations, i.e., relaxing these global measure to point-wise measure are required. Currently, both the cine and MCLE images were acquired in a segment-by-segment and slice-by-slice manner under ECG-gating. However, it is still challenging to ensure the alignment within and between slices due to imperfectness of ECG gating, leading to jagged myocardium contours along the long-axis direction. Recent developments in fast MR image acquisition methods including compressed sensing and parallel imaging and the combination provide a way for rapid volumetric image acquisition. We think these techniques may mitigate the within and between slice misalignment issue and we are planning to implement fast cine and MCLE volumetric image acquisitions in the future.

# References

1. Bertsekas, D.P.: Nonlinear Programming. Athena Scientific, Belmont (1999)
2. Chenoune, Y., et al.: Rigid registration of delayed-enhancement and cine cardiac MR images using 3D normalized mutual information. In: Computing in Cardiology, pp. 161–164. IEEE (2010)
3. Detsky, J.S., Paul, G., Dick, A.J., Wright, G.A.: Reproducible classification of infarct heterogeneity using fuzzy clustering on multicontrast delayed enhancement magnetic resonance images. IEEE Trans. Med. Imaging 28(10), 1606–1614 (2009)
4. Detsky, J., Stainsby, J., Vijayaraghavan, R., Graham, J., Dick, A., Wright, G.: Inversion-recovery-prepared SSFP for cardiac-phase-resolved delayed-enhancement MRI. Magn. Reson. Med. 58(2), 365–372 (2007)
5. Guo, F., Capaldi, D.P., Di Cesare, R., Fenster, A., Parraga, G.: Registration pipeline for pulmonary free-breathing 1 H MRI ventilation measurements. In: Medical Imaging 2017: Biomedical Applications in Molecular, Structural, and Functional Imaging, vol. 10137, p. 101370A. International Society for Optics and Photonics (2017)
6. Guo, F., et al.: Thoracic CT-MRI coregistration for regional pulmonary structure-function measurements of obstructive lung disease. Med. Phys. 44(5), 1718–1733 (2017)
7. Heinrich, M.P., Jenkinson, M., Papież, B.W., Brady, S.M., Schnabel, J.A.: Towards realtime multimodal fusion for image-guided interventions using self-similarities. In: Mori, K., Sakuma, I., Sato, Y., Barillot, C., Navab, N. (eds.) MICCAI 2013. LNCS, vol. 8149, pp. 187–194. Springer, Heidelberg (2013). https://doi.org/10.1007/978-3-642-40811-3_24
8. Lucas, B.D., Kanade, T., et al.: An iterative image registration technique with an application to stereo vision (1981)
9. Ourselin, S., Roche, A., Prima, S., Ayache, N.: Block matching: a general framework to improve robustness of rigid registration of medical images. In: Delp, S.L., DiGoia, A.M., Jaramaz, B. (eds.) MICCAI 2000. LNCS, vol. 1935, pp. 557–566. Springer, Heidelberg (2000). https://doi.org/10.1007/978-3-540-40899-4_57
10. Pop, M., et al.: Quantification of fibrosis in infarcted swine hearts by ex vivo late gadolinium-enhancement and diffusion-weighted MRI methods. Phys. Med. Biol. 58(15), 5009 (2013)
11. Tao, Q., Piers, S.R., Lamb, H.J., van der Geest, R.J.: Automated left ventricle segmentation in late gadolinium-enhanced mri for objective myocardial scar assessment. J. Magn. Reson. Imaging 42(2), 390–399 (2015)
12. Wei, D., Sun, Y., Chai, P., Low, A., Ong, S.H.: Myocardial segmentation of late gadolinium enhanced MR images by propagation of contours from cine MR images. In: Fichtinger, G., Martel, A., Peters, T. (eds.) MICCAI 2011. LNCS, vol. 6893, pp. 428–435. Springer, Heidelberg (2011). https://doi.org/10.1007/978-3-642-23626-6_53

# Joint Analysis of Personalized In-Silico Haemodynamics and Shape Descriptors of the Left Atrial Appendage

Jordi Mill[1], Andy L. Olivares[1], Etelvino Silva[2], Ibai Genua[1],
Alvaro Fernandez[1], Ainhoa Aguado[1], Marta Nuñez-Garcia[1],
Tom de Potter[2], Xavier Freixa[3], and Oscar Camara[1(✉)]

[1] DTIC (Department of Information and Communication Technologies),
Universitat Pompeu Fabra, Barcelona, Spain
oscar.camara@upf.edu
[2] Arrhythmia Unit, Department of Cardiology, Cardiovascular Center,
OLV Hospital, Aalst, Belgium
[3] Department of Cardiology, Hospital Clinic de Barcelona,
Universitat de Barcelona, Barcelona, Spain

**Abstract.** The left atrial appendage (LAA) is a complex and heterogeneous bulge structure of the left atrium (LA). It is known that, in atrial fibrillation (AF) patients, around 70% to 90% of the thrombi are formed there. However, the exact mechanism of the process of thrombus formation and the role of the LAA in that process are not fully understood yet. The main goal of this work is to perform patient-specific haemodynamics simulations of the LA and LAA and jointly analyse the resulting blood flow parameters with morphological descriptors of these structures in relation with the risk of thrombus formation. Some LAA morphological descriptors such as ostium characteristics and pulmonary configuration were found to influence LAA blood flow patterns. These findings improve our knowledge on the required conditions for thrombus formation in the LAA.

**Keywords:** Computational fluid dynamics · Left atrial appendage ·
Blood flow patterns · Thrombus formation

## 1 Introduction

Atrial fibrillation (AF) is an arrhythmia produced by a re-entry and/or rapid focal ectopic firing that affects 1 to 2% of the population, and about 8% of the individuals over 80 years of age [1]. It is known that AF is the most common cardiac abnormality associated with ischemic strokes [2]. Chronic atrial fibrillation causes rigidity on left atrial (LA) walls, preventing its proper contraction and altering the normal left atrial haemodynamics. This scenario could lead to blood stagnation in the left atrial appendage (LAA), increasing the risk of thrombus formation [3]. In fact, 25% of embolic strokes are due to AF, mainly caused by the formation of thrombus; a high percentage, around 70% to 90%, of thrombi occurring during AF derive from the LAA [4]. The LAA is a little portion of cardiac anatomy structure that develops during the third week of gestation. It has a wide shape spectrum among individuals and its

© Springer Nature Switzerland AG 2019
M. Pop et al. (Eds.): STACOM 2018, LNCS 11395, pp. 58–66, 2019.
https://doi.org/10.1007/978-3-030-12029-0_7

classification depends on the number of lobes, the length or the angle between the LAA and its ostium. If the thrombus leaves the LAA and is released onto the circulatory system, it can travel to the brain and cause an embolic stroke.

Despite of the high incidence of embolic strokes on AF patients and the known role of the LAA, the relationship among LA, LAA and thrombus is not clear yet. The exact dependence between morphological parameters and LA haemodynamics on patient-specific geometries needs to be established. Recent studies claimed that some LAA morphologies (e.g. non chicken-wing) were more likely to produce thrombi [5]. Others used indices such as the ostium diameter, the number of lobes or other anatomical parameters that are difficult to reproduce or reach a consensus [4, 6]. The influence of the haemodynamics stimuli in the LAA on the thrombi formation process has been studied by many researchers [7]. Low flow velocities in the LAA are associated with a higher risk of stroke among AF patients. However, the flow is assessed from Trans-esophageal Echocardiographic (TEE) images, only providing partial flow data on a plane or a given point of view, which is insufficient. Novel technics such as 3D echography and 4D PC-MRI can provide information in terms of blood velocity and thrombus localization, but not all the hospitals have access to these and their resolution is not good enough to produce a subsequent quantitative haemodynamic analysis such as the presented in this manuscript.

Considerable progress has been achieved on the development of computational technologies and computational fluid dynamics (CFD) methods. Such methods allow the application of image-based CFD simulations to understand the effects of haemo-dynamics in vascular pathologies and its pathophysiology, which is known to be related with thrombus formation [8]. Some researchers have already used CFD to study the LAA, assessing 3D blood flow patterns in different LA/LAA configurations [9–11]. However, these studies either used synthetic models or not jointly analyzing blood flow simulations with morphological descriptors.

The main goal of this work was to perform a joint analysis of morphological and in silico haemodynamics descriptors on clinical data from four AF patients. Blood flow simulations were personalized with patient-specific pressure data and validated with Doppler ultrasound images.

## 2   Material and Methods

### 2.1   Patient Data and Model Reconstruction

Imaging data from four patients was processed in this work to build the anatomical reference for the simulations: three 3D Rotational Angiography (3DRA) images were provided by OLV Hospital, Aalst, Belgium; the fourth dataset was a Computational Tomography (CT) scan from Hospital Clinic de Barcelona, Spain. The data was anonymized and a written consent was obtained from all patients. Regarding patients from Belgium, the following data were also provided: the pressure in the mitral valve, the atrium, the pulmonary veins and in the LAA. This anonymous information is valuable because it provides data for validation and verification. Moreover, full echocardiography datasets including multiple chamber views and Doppler acquisitions were provided for all cases.

The anatomical structure of the left atrium was semi-automatically extracted from each medical image using Seg3D: Volumetric Image Segmentation and Visualization[1]. The output segmentation was transformed into a surface mesh of triangular elements with the classical Marching Cubes algorithm. Subsequently, a Taubin filter with the threshold set to 0.7 was applied to smooth the surface mesh using Meshlab[2].

Due to the surgery and data acquisition process, a catheter was needed during the intervention, which appears in the 3DRA images. Furthermore, the direction and shape of the veins were too irregular from the segmentations. In addition, the resolution was still too low to allow a good jet flow development in the simulations as well as to extract the cardiac valve accurately enough (mitral valve in the case of the left atria). For all the reasons mentioned above, the surface mesh was refined using Meshlab and MeshMixer[3], as can be seen in Fig. 1. The next step was to generate a tetrahedral volumetric mesh from the surface mesh, which was obtained with Gmsh[4].

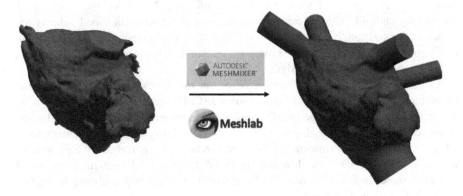

**Fig. 1.** An example of computational mesh building from patient-specific geometries (Patient 1).

## 2.2    Computational Fluid Dynamics Model

For the four processed cases a second order implicit unsteady formulation was used for the solution of the momentum equations in conjunction with a standard partial discretization for the pressure (under-relaxation factors set by default to 0.3 and 0.7 for the pressure and momentum, respectively). The equations were solved using finite volume method (FVM) in Ansys Fluent 18.2 (academic research CFD solver)[5]. Blood flow in the LA was modeled using the incompressible Navier Stokes and continuity equations. Absolute residuals of mass and momentum conservation equations lower than 0.005

---

[1] http://www.seg3d.org.

[2] http://www.meshlab.net/.

[3] http://www.meshmixer.com.

[4] http://gmsh.info/.

[5] https://www.ansys.com/products/fluids/ansys-fluent.

were absolute convergence criteria. Blood was modeled as an incompressible New-tonian fluid with a density value of 1060 kg/m$^3$ [10]. The dynamic viscosity of blood in large vessels and the heart was set to 0.0035 Pa · s [10]. Simulations were run using a laminar flow hypothesis under isothermal and no-gravitational effects. All walls were simulated rigid and with no-slip conditions and therefore replicating the arguably worst-case scenario in AF, without atrial contraction.

The differential equations were solved using a time-step of 0.01 s during one beat, that means 105 steps overall, where the ventricular systole phase lasts 0.4 s and the ventricular diastole lasts 0.65 s. Following the clinical observations of Fernandez-Perez et al. [12] an equal time-velocity function was applied at the four pulmonary veins for all the four patients. At the mitral valve a pressure of 8 mmHg was imposed during the ventricular diastolic phase. In ventricular systole, the MV is closed and was simulated as a wall boundary. Subsequently, the model of each case was imported to Ansys Fluent.

## 2.3  Morphological and Haemodynamics Descriptors

The morphological indices, extracted using MATLAB R2017b (Mathworks, MS, USA)[6] were the following: (1) mean ostium diameter, being the average between the maximum and minimum diameter; (2) ostium maximum diameter; (3) ostium perimeter; (4) ostium area; (5) volume of the LAA; and (6) the centreline length, being the distance between the centre of the ostium and the furthest point of the anatomy using a heat equation. Therefore, a gradual distribution is obtained and to find the farthest point of the geometry. From that point, the other points can be calculated using a marching algorithm and generating the centerline.

Flow simulations were assessed at the pulmonary veins (PV), mitral valve (MV), the middle of the left atrium and the LAA. Furthermore, wall shear stress (WSS), defined as the frictional force exerted by the flowing blood tangentially to the wall, was also estimated from the flow simulations. This parameter has been strongly associated with endothelial lesions and thrombus formation [13]. From the WSS, some parameters that have also been related to the risk of thrombus formation were calculated: Time Averaged Wall Shear Stress (TAWSS), the Oscillatory Shear Index (OSI), the Endothelial Cell Activation Potential (ECAP). Theoretically, regions with low TAWSS and high OSI are more prone to develop atherosclerosis [6]. In an attempt to localize these regions, Di Achille et al. [14] proposed using the ratio of these two indices to characterize the degree of susceptibility of the area to generate a thrombus, also known as ECAP. Whether the ECAP values are high, then that area has more potential to be susceptible to generate a thrombus.

---

[6] https://es.mathworks.com/products/matlab.html.

## 3  Results

This work has presented a methodology to assess the probability to generate a thrombus at LAA in a non-invasive and patient-specific manner from medical images and computational simulations. In order to validate this workflow and its results, personalized data for each patient such as ultrasound images and pressure profiles were collected. According to literature, patients with atrial fibrillation do not have the A curve that can be seen in healthy patients [15]. Figure 2(a) and (c) displays the Doppler ultrasound image from Patient 4 and its respective blood flow simulation at the MV. As shown, there is no A wave in both, simulation and ultrasound imaging, only the E wave is present.

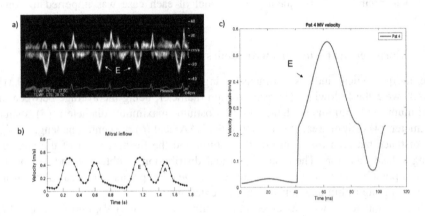

**Fig. 2.**  (a) Doppler ultrasound image from Patient 4; (b) Mitral Valve (MV) velocity curve from the literature [15]; (c) Mitral valve velocity curve from our blood flow simulations.

### 3.1  Morphological Descriptors

The morphological parameters from the LAA of the four patients under study are displayed in Table 1. The results among the four patients are relatively similar except Patient 4, who seems to have a larger LAA. However, the longest centreline length corresponds to Patient 1 since it has more secondary lobes and bulges than Patient 4

**Table 1.**  Morphological parameters for each LAA

|  | Mean diam. (mm) | Max. diam. (mm) | Ostium perimeter (mm) | Ostium area (mm$^2$) | LAA volume (mL) | LA volume (mL) | Centreline length (mm) |
|---|---|---|---|---|---|---|---|
| Patient 1 | 22.53 | 26.58 | 72.07 | 384.93 | 6.86 | 188.51 | 42.76 |
| Patient 2 | 23.08 | 26.40 | 76.26 | 422.95 | 6.06 | 181.07 | 27.56 |
| Patient 3 | 25.97 | 32.56 | 82.40 | 499.63 | 7.02 | 137.59 | 31.16 |
| Patient 4 | 31.45 | 35.43 | 97.40 | 762.00 | 10.16 | 303.56 | 30.11 |

(see Fig. 3). As has been mentioned previously, these parameters are related with the risk of thrombus formation [10, 16]. The higher are these morphological values, the higher is the probability to generate a thrombus. Therefore, Patient 1 and Patient 4 would be the ones with more risk according to the morphological parameters.

## 3.2   Haemodynamic Descriptors

Streamlines are used to visualize which flow trajectories are following the particles and, in our study, if these are going into the LAA. It can also help to find the origin of these particles and figure out which pulmonary veins contribute to flow entering the LAA, depending on PV number, orientation and spatial distribution. Figure 3 shows the same time step from diastole in all cases. The left superior pulmonary vein always provides blood flow to the LAA. This makes sense since it is the closest to the LAA; in some cases, just below it. However, the other PV that provide flow to the LAA vary

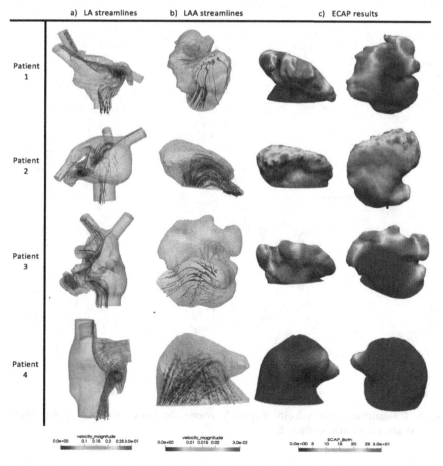

**Fig. 3.** (a) Streamlines of blood flow in the left atria entering the left atrial appendage at second 0.6 (diastole); (b) Flow streamlines in the LAA; (c), (d) frontal and posterior views, respectively, of Endothelial Cell Activation Potential (ECAP) maps in the LAA.

depending on their orientation, without a common pattern among the studied patients. Moreover, the combination of both, the PV configuration and LAA morphology, especially ostium characteristics, determines the amount of flow reaching the distant parts of the LAA (see Fig. 3).

The ECAP results presented in Fig. 3 show that Patient 4 had the lowest ECAP values, although it had a high probability of developing thrombus according to the morphological parameters. On the other hand, Patient 1 and 3 reached high ECAP values on some secondary lobes. Patient 2 presented a high probability to generate thrombi in almost all the parts of its LAA. Nevertheless, Patient 2 ECAP values could be explained by the lack of flow entering the LAA over the whole the cardiac cycle, especially during the systole. If velocity values inside the LAA are very low, large values of ECAP are reached.

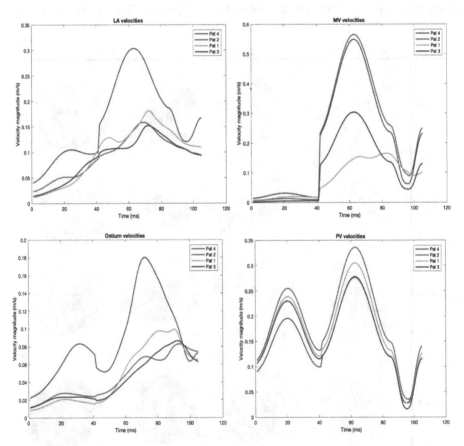

**Fig. 4.** Comparison of the velocity magnitude among the cases with LAA in the different sampling areas (PV, MV, Ostium, LA).

For all the patients, a velocity average at different parts of the left atrium was carried out: (1) at the middle of the LA; (2) at the MV; (3) at the end of the LSPV and LAA's ostium (see Fig. 4). From 0 ms to 40 ms corresponds to systole phase, and from 60 ms to 105 ms the diastole. Since during the systole the MV is closed, the velocities are lower at the beginning of the cycle and at the MV are almost zero.

## 4 Conclusions

There are studies aimed at determining the probability to generate thrombi solely based on morphological parameters of the LAA [16]. Such an approach simplifies the study excessively, in particular due to the absence of robust ways to fully characterize LAA morphology [17]. The work described in this manuscript presents a workflow to assess patient haemodynamics in the LA/LAA and its risk to develop thrombi with blood flow simulations, which provide complementary information to morphological parameters.

The four studied patients had similar morphological parameters but their ECAP distribution were very different. Furthermore, Patient 4, being the one with the highest probability of developing thrombi according to morphological parameters (maximum ostium diameter and larger volume), was the one with the lowest ECAP. However, these differences might be due to the different imaging system used in Patient 4 with respect to the other patients. On the other hand, Patient 2 presented the highest like-lihood to produce thrombi according to ECAP distribution, likely due to the low blood flow velocities entering the LAA during systole.

Overall, differences in ECAP distribution in the analyzed patients should be related to their LA/LAA anatomies since boundary conditions were the same in the simulations. Therefore, it is clear that LAA morphological parameters play a critical role in thrombus formation but other factors need to be considered. For instance, we have observed that LAA flow is highly dependent on how PV are localized in the LA, allowing to produce complex flow patterns in the cavity. In consequence, both morphological and haemodynamics parameters should be jointly analyzed for a more complete understanding of the mechanisms inducing thrombus formation. Future work would focus on running the whole modelling pipeline in a large number of cases and perform advanced statistical studies to identify which parameters are different in AF patients with and without a history of thromboembolic events.

**Acknowledgments.** This work was partially supported by COMPILAAO Retos I+D project (DPI2015-71640-R), funded by the Spanish Government of Economy and Competitiveness.

## References

1. Pellman, J., Sheikh, F., Diego, S., Jolla, L.: Atrial fibrillation: mechanisms, therapeutics, and future directions. Compr. Physiol. **5**(2), 649–665 (2017)
2. Chugh, S.S., Havmoeller, R., Narayanan, K., Kim, Y., McAnulty Jr., J.H.M., Zheng, Z.: Worldwide epidemiology of atrial fibrillation: a global burden of disease 2010 study. Circulation **129**(8), 837–847 (2014)

3. Fatkin, D., Kelly, R.P., Feneley, M.P.: Relations between left atrial appendage blood flow velocity, spontaneous echocardiographic contrast and thromboembolic risk in vivo. J. Am. Coll. Cardiol. **23**(4), 961–969 (1994)

4. Lee, J.M., et al.: Why is left atrial appendage morphology related to strokes? An analysis of the flow velocity and orifice size of the left atrial appendage. J. Cardiovasc. Electrophysiol. **26**(9), 922–927 (2015)

5. Beigel, R., Wunderlich, N.C., Ho, S.Y., Arsanjani, R., Siegel, R.J.: The left atrial appendage: anatomy, function, and noninvasive evaluation. JACC Cardiovasc. Imaging **7**(12), 1251–1265 (2014)

6. Di Biase, L., et al.: Does the left atrial appendage morphology correlate with the risk of stroke in patients with atrial fibrillation? Results from a multicenter study. J. Am. Coll. Cardiol. **60**(6), 531–538 (2012)

7. Wootton, D.M., Ku, D.N.: Fluid mechanics of vascular systems, diseases, and thrombosis. Annu. Rev. Biomed. Eng. **1**(1), 299–329 (1999)

8. Nedios, S., et al.: Left atrial appendage morphology and thromboembolic risk after catheter ablation for atrial fibrillation. Heart Rhythm **11**(12), 2239–2246 (2014)

9. Otani, T., Al-Issa, A., Pourmorteza, A., McVeigh, E.R., Wada, S., Ashikaga, H.: A computational framework for personalized blood flow analysis in the human left atrium. Ann. Biomed. Eng. **44**(11), 3284–3294 (2016)

10. Guadalupe, G.I., et al.: Sensitivity analysis of geometrical parameters to study haemodynamics and thrombus formation in the left atrial appendage. Int. J. Numer. Method Biomed. Eng. **34**(8), 122–140 (2018)

11. Olivares, A.L., et al.: In silico analysis of haemodynamics in patient-specific left atria with different appendage morphologies. In: Pop, M., Wright, G.A. (eds.) FIMH 2017. LNCS, vol. 10263, pp. 412–420. Springer, Cham (2017). https://doi.org/10.1007/978-3-319-59448-4_39

12. Fernandez-Perez, G.C., Duarte, R., la de Calle, M.C., Calatayud, J., Sanchez Gonzalez, J.: Analysis of left ventricular diastolic function using magnetic resonance imaging. Radiologia **54**(4), 295–305 (2012)

13. Meng, H., Tutino, V.M., Xiang, J., Siddiqui, A.: High WSS or Low WSS? Complex interactions of hemodynamics with intracranial aneurysm initiation, growth, and rupture: toward a unifying hypothesis. Am. J. Neuroradiol. **35**(7), 1254–1262 (2014)

14. Di Achille, P., Tellides, G., Figueroa, C.A., Humphrey, J.D.: A haemodynamic predictor of intraluminal thrombus formation in abdominal aortic aneurysms. Proc. R. Soc. A Math. Phys. Eng. Sci. **470**(2172) (2014). https://doi.org/10.1098/rspa.2014.0163

15. Gautam, S., John, R.M.: Interatrial electrical dissociation after catheter-based ablation for atrial fibrillation and flutter. Circ. Arrhythmia Electrophysiol. **4**(4), 26–29 (2011)

16. Yamamoto, M., et al.: Complex left atrial appendage morphology and left atrial appendage thrombus formation in patients with atrial fibrillation. Circ. Cardiovasc. Imaging **7**(2), 337–343 (2014)

17. Khurram, I.M., et al.: Relationship between left atrial appendage morphology and stroke in patients with atrial fibrillation. Heart Rhythm **10**(12), 1843–1849 (2013)

# How Accurately Does Transesophageal Echocardiography Identify the Mitral Valve?

Claire Vannelli[1,3](✉), Wenyao Xia[2,3], John Moore[3], and Terry Peters[1,2,3]

[1] School of Biomedical Engineering, Western University, London, ON, Canada
cvannel@uwo.ca
[2] Department of Medical Biophysics, Western University, London, ON, Canada
[3] Robarts Research Institute, London, ON N6A5B7, Canada

**Abstract.** Mitral Valve Disease (MVD) describes a variety of pathologies that cause regurgitation of blood during the systolic phase of the cardiac cycle. Decisions in valvular disease management rely heavily on non-invasive imaging. Transesophageal echocardiography (TEE) is widely recognized as the key evaluation technique where backflow of high velocity blood can be visualized under Doppler. However, the heavy reliance on TEE segmentation for diagnosis and modelling necessitates an evaluation of the accuracy of this oft-used mitral valve imaging modality. In this pilot study, we acquire simultaneous CT and TEE images of both a silicone mitral valve phantom and an iodine-stained bovine mitral valve. We propose a pipeline to use CT as ground truth to study the relationship between TEE intensities and the underlying valve morphology. Preliminary results demonstrate that even with an optimized threshold selection based solely on TEE pixel intensities, only 40% of pixels are correctly classified as part of the valve. In addition, we have shown that emphasizing the center line rather than the boundaries of the high intensity regions in TEE provides a better representation and segmentation of the valve morphology. The root mean squared distance between the TEE and CT ground truth is 1.80 mm with intensity-based segmentation and improves to 0.81 mm when comparing the center line extracted from the segmented volumes.

**Keywords:** Transesophageal echocardiography · Mitral valve disease · Ground truth phantom validation

## 1 Introduction

Due to the significant increase in life expectancy over the past century, valvular heart disease has come to the forefront of challenges facing modern medicine [1]. Mitral valve disease (MVD) classifies a range of different pathologies that cause regurgitation of blood into the left atrium during ventricular contraction. Mitral valve regurgitation (MVR) can occur with both prolapsed and stenotic valves,

© Springer Nature Switzerland AG 2019
M. Pop et al. (Eds.): STACOM 2018, LNCS 11395, pp. 67–76, 2019.
https://doi.org/10.1007/978-3-030-12029-0_8

as well as flail leaflets due to ruptured chordae tendineae [2]. Transesophageal echocardiography (TEE) is currently the state-of-the-art technique for diagnosis and evaluation of MVD. A diseased valve presents with regurgitation of high velocity blood which can be visualized with the TEE probe under Doppler [3]. However, there are limitations to using TEE for diagnosis and intervention of regurgitant mitral valves. A study of 242 patients who underwent TEE-guided mitral valve repair for degenerative MVD found the rate of recurrence of regurgitation to be 5% after 1 month and 40% after 4 years [4].

Segmentation of 3D TEE images of the mitral valve has important applications in pre-operative modelling for surgical planning and quantitative analysis of valve function [5]. Clinicians can calculate valve surface area, commissure-commissure lengths and leaflet area to identify various pathological features, evaluate MVR severity and inform subsequent interventions and treatment. Measurements made to evaluate lengths for neo-chordae, valve replacements and other volumetric data rely on intensities in TEE images as indications of valve and papillary muscle tissue [6]. However, there are multiple pervading issues with the TEE datasets such as signal dropout and high-intensity signal artifacts that can be misconstrued as leaflets or other anatomical structures [7]. Both modelling and quantitative assessment are essential steps in the evaluation of patient-specific geometry, pathophysiologic mechanisms and abnormalities of the mitral valve.

The major challenge with 3D TEE segmentation is that a limited number of methods exist that can accurately and robustly isolate the mitral valve. Several anatomical details of the valve introduce nontrivial challenges to its segmentation. First, there is no intensity-based boundary between the leaflets and the adjacent heart tissue, making it difficult to identify the valve complex solely with image intensity information. As well, during diastole, the posterior leaflet is often pressed against the left ventricular wall, and the resulting echo image is characterized by signal dropout. In addition, there is no intensity-based distinction between the two leaflets which form the coaptation zone during systole [8]. The majority of current ultrasound segmentation techniques involve and rely on intensity thresholds. Methods include intensity-based level sets, active contours graph cuts and thresholding to segment the valve [9]. However, it remains ambiguous whether regions of high intensity correspond to presence of mitral valve or papillary muscle tissue. As demonstrated in Fig. 1, and an intensity-based approach cannot always accurately delineate the valve. Particularly along the appearance of the tissue edge, there is the option to over- or under-estimate the leaflet with intensity-based segmentation.

The main contribution of this work is its analysis of the relationship between TEE intensities and underlying mitral valve tissue. With comparison to a CT ground truth, we have quantified the observed overestimation of valve structure in TEE [10]. This will inform intensity-based techniques for valve segmentation used in diagnosis, quantitative assessment and patient-specific modelling of mitral valve regurgitation. Further, we have explored and demonstrated the efficacy of the center line method for improved valve segmentation.

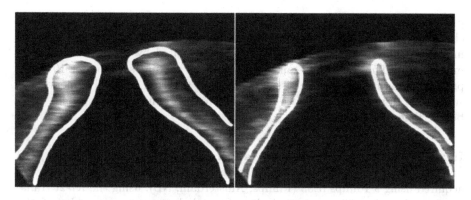

**Fig. 1.** A 2D slice of a 3D TEE volume that demonstrates the two different ways to estimate the boundary (depicted in red). If is unclear if the edges of the bright intensity regions are due to echoes or represent underlying valve tissue (Color figure online)

## 2   Materials and Methods

### 2.1   Ground Truth Creation

**Motorized Valve Manipulator.** We constructed a motorized system that provides repeatable, ECG-gated, realistic valve movements over a complete cardiac cycle. The setup consists of two Arduino-controlled (https://www.arduino.cc) servomotors (HiTEC, https://www.hitecrcd.com, HS-625MG), strategically positioned pulleys for adjusting the papillary muscles, and chordae to open the valve leaflets. Servomotors were selected to ensure precise control of the quasi-static valve model by modulating their angular position. The system mimics the full range of cardiac motion from diastole to systole and arrests the phantom at any point in the cardiac cycle. For dynamic scans, the motors operate continuously to mimic valve movement at either 30, 60 or 90 bpm. This provided us with phantoms for TEE and CT image collection of both static and dynamic scenarios. The servomotors were affixed to the top of a plastic box with 3D-printed interfacing components. Similarly, the pulleys were 3D-printed and placed around the edge of the box (Fig. 2, left center).

**Silicone and Bovine Phantom Preparation.** Two types of dynamic mitral valve phantoms were designed in our study. The first was a proof-of-concept, silicone valve based on a patient-specific, 3D-printed mold, chordae and adjustable papillary muscles supported by a flange angled to mimic a patient lying supine (Fig. 2, top left). Strings acting as chordae were cured within the silicone leaflets and secured to the end of the servomotors. As the motorized valve manipulator operates, the string-chordae were adjusted to simulate a realistic systolic-to-diastolic valve motion pattern. The silicone model had the advantage of an indefinite shelf life, and excellent CT image quality. For improved ultrasound imaging, and more realistic valvular anatomy, we have also constructed a bovine phantom (Fig. 2, top right). For this model, the valve, chordae and papillary muscles were dissected and sutured into a silicone flange, and strings were sutured into the papillary muscles, so they could be actuated by the motorized valve manipulator. The valve was stained with a Lugol's iodine solution for 10 hours prior to scanning to increase tissue contrast in CT [10]. The valve model was submerged in water to introduce hydrodynamics and ensure compatibility with both CT and TEE imaging.

**Image Acquisition.** The data were acquired with simultaneous CT and TEE imaging of both the silicone and bovine valves. All of the 3D TEE scans were acquired using a Philips iE33 (Philips, Amsterdam, NL), using the X7-2t transesophageal probe. Three strategically placed holes on opposite sides of the annulus served as fiducials that were visible in both CT and TEE for subsequent landmark-based registration. The resulting dynamic image used for post-processing is an average over multiple TEE acquisitions. To create the static ground truth, the CT images were acquired in an O-arm (Medtronic, Minnesota, MN) at resolution of $512 \times 512 \times 120$ and the pixel spacing of $0.415 \times 0.415 \times 0.833$ mm. To obtain the dynamic ground truth, the 4D-CT scans were performed on the GE Revolution Discovery multi-slice scanner (GE Medical Systems, Waukesha, WI) in cine mode. These data were acquired with retrospective gating and the scan parameters were set as follows: 0.28 s gantry rotation, 2.2 s length of scan and 0.625 mm slice thickness. Each image reconstruction used data acquired from a full $360°$ rotation of the gantry.

**Image Processing.** After acquisition, specific images were selected for further analysis based on visual inspection. Due to the poor image quality of TEE images during systole, this primarily consisted of selecting 8 CT-TEE cardiac phases from end-diastole. In addition, we chose to use images with the motorized valve manipulator operating at 30 bpm to avoid duplicate structure artifacts where a valve leaflet appears twice in one reconstructed slice in 4DCT. Once selected and sorted into image pairs, corresponding time frames were registered. The CT-TEE registration process was not trivial; to ensure a high degree of accuracy, most of the registration was performed manually. First, we placed landmarks in the centers of the holes visible in both CT and TEE, and the landmark-based registration was performed to initialize the CT-TEE registration. We then employed the BRAINS registration module in Slicer 4.6.2 (https://www.slicer.org/) to improve the alignment. We implemented both Mutual Information (MI) and Normalized Cross Correlation (NCC) as our registration metrics, with NCC providing the more accurate alignment based on visual inspection. After automatic registration using NCC, the alignment was manually fine-tuned to further improve the registration accuracy. The aligned TEE and CT images were then apodized manually to focus on the valve region, excluding the supporting apparati. Finally, we applied the continuous max flow [11] algorithm to segment the processed CT images for subsequent use as ground truth. The workflow of the proposed ground truth creation pipeline is represented in Fig. 2.

## 2.2   Image Analysis

With the ground-truth correspondence and registration established, we then completed an intensity-based analysis of the static silicone (1 image pair), dynamic silicone (8 image pairs) and dynamic bovine (8 image pairs) CT and TEE image pairs.

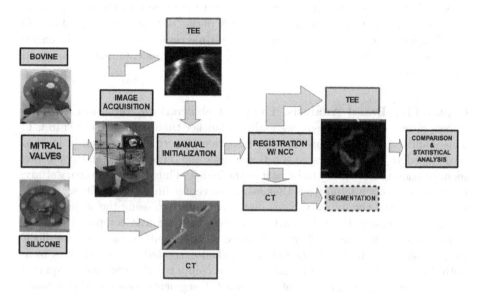

**Fig. 2.** Pipeline for image acquisition, image processing and statistical analysis

**Pixel Based Analysis.** First, we directly studied the relationship between TEE image pixel intensity and the likelihood that a pixel belongs to the leaflet. Thus, for each discretized pixel intensity $i$ ranging from 1 to 255, we computed the conditional probability of a pixel belonging to the valve region given its intensity:

$$P(i) = P(x \in L | I(x) = i) = \frac{P(x \in L \cap I(x) = i)}{P(I(x) = i)}, \qquad (1)$$

where $P(I(x) = i)$ represents the probability that a pixel in TEE images has intensity $i$, $P(x \in L)$ represents the probability that a pixel at location $x$ belongs to the leaflet, and $P(x \in L \cap I(x) = i)$ represents the probability of the joint event.

**Segmentation Based Analysis.** To further analyze the intensity correspondence of the TEE intensities and valve tissue, we also studied the performance of an advanced and commonly-used intensity-based segmentation approach. Since we are interested in correctly predicting the location of the leaflet based on intensity, we used the positive predictive value (PPV) as our measurement of segmentation accuracy. This accounts for the fact that the majority of the TEE image is background, while we are concerned with correctly identifying valve regions. In the context of ROC, the PPV is given as:

$$PPV = \frac{TP}{TP + FP},$$ (2)

where TP represents the true positive rate and FP represents the false positive rate. Using this metric, we examined the PPV of the 3D Continuous Maxflow (3D CMF) algorithms. We ran the algorithm multiple times with different intensity threshold values for a thorough analysis of any improvements due to threshold optimization.

**Center-Line Based Analysis.** From past observations, we noticed that the leaflets shown on TEE images were always thicker than those presented in CT. This could be an indication that the classification errors were mainly located near the edges of the TEE segmentations, while the center of the intensity-based segmentation is more likely to be part of the leaflet. Thus, as a preliminary study, we wanted to test a hypothesis that using the center line of the TEE segmentation to represent the mitral valve is more accurate than using the segmentation surface to represent the valve surface. To test this, we first computed the segmentations of the static silicone, dynamic silicone and dynamic bovine TEE images using the 3D CMF algorithm and then extracted the center lines from both TEE and CT segmentations. Then we compared the root mean squared distances between the surfaces of TEE and CT segmentations and the distance between TEE and CT center lines.

## 3    Results

In this study, we created a dynamic silicone phantom and dynamic bovine phantom, which allowed us to image them in both static and dynamic states. Following the registration scheme shown in Fig. 2, we created three sets of CT ground truth images for static silicone TEE, dynamic silicone TEE, and dynamic silicone TEE. For each dataset, we performed the pixel- and segmentation-based analyses as described in Sect. 2.2. And finally, we performed a center line analysis using the static silicone dataset, as a pilot study, to prove the validity of our hypothesis given in Sect. 2.2 Center-Line based analysis.

### 3.1    Pixel Based Analysis

For each dataset, we computed the conditional probability of a pixel belonging to the valve region given its intensity following Eq. (1). The relationships between the likelihood and the TEE image intensity are shown in Fig. 3. Generally, the results indicate that even with an optimized naive based threshold segmentation, only 40% of pixels are correctly classified as part of the valve.

The results shown in Fig. 3 indicate the likelihood of any individual pixel intensity value corresponding to part of the valve. In Fig. 3(a), we see that for the static silicone valve, there is a maximum 42% likelihood that any pixel would be classified as valve tissue, given its TEE intensity. From Fig. 3(b), the maximum likelihood of intensity-based valve classification for the dynamic silicone valve is 35%. For dynamic bovine, the maximum likelihood given its intensity is below 51%, as shown in Fig. 3(c). As a result, our data suggests that classification based on intensity alone may provide an excess of false positives.

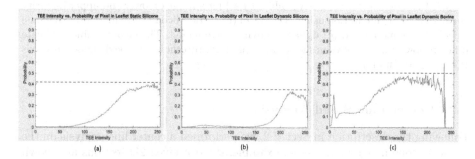

**Fig. 3.** Plots of probability of each individual pixel intensity value in TEE images to correctly classify the pixel as part of the valve leaflet for (a) static silicone valve model (b) dynamic silicone valve model and (c) dynamic bovine valve model

## 3.2  Segmentation Based Analysis

After the pixel-based analysis was completed, we performed a 3D Continuous Max Flow (CMF) segmentation [11] on all three image pairs (static silicone, dynamic silicone and dynamic bovine). CMF is an advanced, intensity-based segmentation technique that needs both a lower and upper threshold to indicate the expected intensity value of the background and foreground, respectively.

The results from the 3D CMF segmentations are depicted in Fig. 4. The multiple coloured lines represent the lower threshold, and the x-axis marks the upper threshold values. The algorithm starts with the initial lower threshold (the x-intercept) and runs through a search of all possible upper threshold values. Each line represents a different initial expected value for the background, and together these lines map all the possible combinations of lower and upper thresholds for 3D CMF. The y-axis includes the PPV value for each combination of lower and upper thresholds.

From Fig. 4(a), the maximum PPV is 40% for the 3D CMF segmentation of the static silicone valve. In Fig. 4(b), the maximum PPV value is 31%, calculated over 8 averaged dynamic silicone volumes. Finally, in Fig. 4(c), the 3D CMF segmentation of the dynamic bovine model yields a maximum averaged PPV value of 49%. Given that this is an advanced and commonly used technique for mitral valve segmentation, these results highlight the limitations of an intensity-only approach [12].

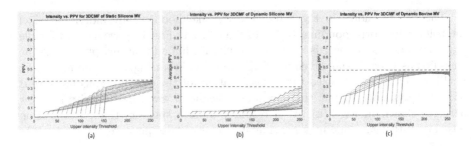

**Fig. 4.** Plots of positive predictive value (PPV) for a range of lower and upper intensity thresholds for segmentation using 3DCMF of (a) static silicone valve model (b) dynamic silicone valve model and (c) dynamic bovine valve model. The coloured lines indicate different lower threshold values for a range of upper thresholds, tracked along the x-axis (Color figure online)

### 3.3    Center Line Based Analysis

Following the description in Sect. 2.2, we performed a center line analysis on the static silicone image pairs. As an example, the overlayed 2D sections are shown in Fig. 5, where Fig. 5(a) demonstrates the overlayed centerlines and Fig. 5(b) demonstrating the overlaid segmentations. The CT segmentation and center line is shown as green and the TEE segmentation and center line is shown as pink. As demonstrated by Fig. 5(b), there are large distances between the edges of the TEE and CT segmentations, while from Fig. 5(a), the distance between CT and TEE center lines is significantly smaller. Quantitatively, from 3D studies, the surface Root Mean Squared Distance (RMSD) between surfaces of the TEE and CT segmentations is 1.80 mm whereas the RMSD between the center lines extracted from the TEE and CT segmented volumes is 0.81 mm. Thus, these preliminary results support our hypothesis that the center line of the TEE segmentation is a better representation of the valve morphology than the solely intensity-based methods of leaflet edge delineation.

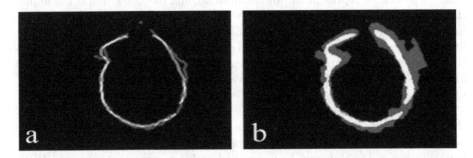

**Fig. 5.** Overlaid segmentations of static silicone valve TEE (pink) and CT (green) images, and complete agreement between them (white) (a) with center line extraction from 3DCMF segmentation and (b) with 3DCMF segmentation only (Color figure online)

# 4  Conclusions

Comprehensive visual and quantitative analysis of dynamic mitral valve morphology is necessary for identifying and correcting regurgitation. In this study, we evaluated the accuracy of TEE mitral valve imaging through construction of quasi-static silicone and bovine valve phantoms used to compare CT ground truth with simultaneously-acquired TEE images. The results provide evidence that reliance on intensity alone for segmentation cannot fully capture the valve structure. The curves in Fig. 3 indicate that individual pixel intensities in TEE do not have a clear correspondence to underlying tissue. Comparison across the three valve models (static silicone, dynamic silicone and dynamic bovine) demonstrates that the segmentation of the dynamic bovine TEE with 3DCMF is more accurate than both static and dynamic silicone. Further studies should be done incorporating patient imaging data for a more comprehensive analysis of TEE for mitral valve imaging. The general trend present across all graphs in Fig. 3 is that higher intensity values have a higher probability of identifying valve tissue. Although, with a maximum 50% likelihood of valve classification at any intensity, these results are compelling and motivate the use of new segmentation techniques that do not rely solely on intensities. Rather, valve segmentation should also incorporate some level of shape information about the valve structure. We demonstrate this by comparing an intensity-based segmentation with a center line analysis. These demonstrate the efficacy of augmenting an intensity-based segmentation with information about the shape and structure of the valve.

A limitation of this work is that the TEE images analyzed in this study were acquired from an older generation system. As the technology continues to improve, a similar analysis should be conducted on future systems. Further, we constructed our ground truth with the assumption that any motion artifacts due to valve motion in dynamic CT would be negligible. In reality, the resulting images had streaking artifacts and duplicate structures, with higher prevalence of artefact at higher simulated heart rates. To accomodate for and avoid these motion artifacts in our ground truth images, we selected 4DCT data from the 30bpm valve simulation. In addition, the analysis was restricted to 3 pairs of images due to limited acquisition time and TEE image quality. Further studies will evaluate a larger range of corresponding CT and TEE images of valve models at different cardiac phases.

This work has the potential to both inform current techniques for TEE imaging and segmentation as well as motivate the development of new methods for identifying valve structure in TEE from known correspondence to ground truth CT images. In future, being able to pair up ground truth CT images with actual TEE images may form the basis for improved reconstructions from TEE data. With CT data as a labelled training set, deep learning techniques can further extend this work to map between CT and TEE, and incorporate deep learning to predict the TEE segmentation from pre-existing CT segmentations.

**Acknowledgments.** The authors would like to acknowledge the follow individuals for their technical expertise: Aaron So and Jennifer Hadway for acquisition and reconstruction using the Revolution CT, Joy Dunmore-Buyze for preparation of the iodine solution and Olivia Ginty for dissection and suturing of the bovine valve.

# References

1. d'Arcy, J., Prendergast, B., Chambers, J., et al.: Valvular heart disease: the next cardiac epidemic. Heart **97**(2), 91–93 (2011)
2. Apostolidou, E., Maslow, A., Poppas, A.: Primary mitral valve regurgitation: update and review. Glob. Cardiol. Sci. Pract. **2017**, 3 (2017)
3. Desai, M., Grigioni, F., Di Eusanio, M., et al.: Outcomes in degenerative mitral regurgitation: current state-of-the-art and future directions. Prog. Cardiovasc. Dis. **60**(3), 370–385 (2017)
4. Flameng, W., Herijgers, P., Bogaerts, K.: Recurrence of mitral valve regurgitation after mitral valve repair in degenerative valve disease. Circulation **107**, 1609–1613 (2003)
5. Garbi, M., Monaghan, M.J.: Quantitative mitral valve anatomy and pathology. Echo Res. Pract. **2**(3), 63–72 (2015)
6. Eibel, S., Turton, E., Mukherjee, C., et al.: Feasibility of measurements of valve dimensions in en-face-3D transesophageal echocardiography. Int. J. Cardiovasc. Imaging **33**(10), 1503–1511 (2017)
7. Mashari, A., Montealegre-Gallegos, M., Knio, Z., et al.: Making three-dimensional printing with echocardiographic data. Echo Res. Pract. **3**(4), 57–64 (2016)
8. Noble, J., Boukerroui, D.: Ultrasound image segmentation: a survery. IEEE Trans. Med. Imaging **25**(8), 987–1010 (2006)
9. Chew, P., Bounford, K., Plein, S., et al.: Multimodality imaging for the quantitative assessment of mitral valve regurgitation. Quant. Imaging Med. Surg. **8**(3), 342–359 (2018)
10. Dumore-Buyze, J., Tate, E., Xiang, F., et al.: Three-dimensional imaging of the mouse heart and vasculature using micro-CT and whole-body perfusion of iodine or phosophotungstic acid. Contrast Media Mol. Imaging **9**(5), 383–390 (2014)
11. Yuan, J., Bae, E., Tai, X., et al.: A study on continuous max-flow and min-cut approaches. In: IEEE Computer Society Conference on Computer Vision and Pattern Recognition, pp. 2217–2224 (2010)
12. Schnieder, R., Perrin, D., Vasilyev, N., et al.: Mitral annulus segmentation from 3D ultrasound using graph cuts. IEEE Trans. Med. Imaging **29**(9), 1676–1687 (2010)

# Stochastic Model-Based Left Ventricle Segmentation in 3D Echocardiography Using Fractional Brownian Motion

Omar S. Al-Kadi[1,2(✉)], Allen Lu[3], Albert J. Sinusas[2], and James S. Duncan[2,3]

[1] King Abdullah II School for Information Technology,
University of Jordan, Amman, Jordan
[2] Radiology and Biomedical Imaging, Yale University, New Haven, CT, USA
omar.al-kadi@yale.edu
[3] Biomedical Engineering, Yale University, New Haven, CT, USA

**Abstract.** A novel approach for fully-automated segmentation of the left ventricle (LV) endocardial and epicardial contours is presented. This is mainly based on the natural physical characteristics of the LV shape structure. Both sides of the LV boundaries exhibit natural elliptical curvatures by having details on various scales, i.e. exhibiting *fractal-like* characteristics. The fractional Brownian motion (fBm), which is a non-stationary stochastic process, integrates well with the stochastic nature of ultrasound echoes. It has the advantage of representing a wide range of non-stationary signals and can quantify statistical local self-similarity throughout the time-sequence ultrasound images. The locally characterized boundaries of the fBm segmented LV were further iteratively refined using global information by means of second-order moments. The method is benchmarked using synthetic 3D echocardiography time-sequence ultrasound images for normal and different ischemic cardiomyopathy, and results compared with state-of-the-art LV segmentation. Furthermore, preliminary results on real data from canine cases is presented.

## 1 Introduction

Accurate and reliable segmentation of the left ventricle (LV) is important for cardiac function analysis. In clinical practice, the segmentation task involves the delineation of LV endocardium and epicardium contours. This process, however, is tedious and time consuming. An automatic and robust approach for segmenting cardiac ultrasound time-sequences would highly facilitate the routine clinical work [1].

Several key challenges in the automated segmentation of the LV in cardiac ultrasound datasets exist. Namely, *speckle intensity heterogeneity*: as in LV blood pool due to blood flow or the dynamic motion of the heart; *obscureness*: close proximity of the papillary muscles tend to show speckle intensities similar to that of the myocardium; *spatial complexity of anatomy*: the separating border between right and left ventricle, and the low contrast between the myocardium and lung air may pose additional challenges for LV segmentation.

© Springer Nature Switzerland AG 2019
M. Pop et al. (Eds.): STACOM 2018, LNCS 11395, pp. 77–84, 2019.
https://doi.org/10.1007/978-3-030-12029-0_9

To address these challenges, it is advantageous to have an efficient segmentation algorithm that is objective and reproducible to accelerate and facilitate the process of diagnosis. Regarding ultrasound B-mode segmentation efforts, the endocardium and epicardium boundaries are delineated using a variety of strategies. In particular, statistical models which encode high-level knowledge, as the parametric or geometric deformable models, can handle topological changes robustly [2]. However, the model needs to be initialized sufficiently close to the object to converge and is sometimes prone to local minima. Taking advantage of unlabelled data for detecting data-driven features has drawn increased attention in prior probabilistic maps [3] and deep learning techniques [4]. However, limited training data is a common obstacle in the latter techniques, and regularization of the training data with a large amount of human-annotated data or anatomical models from large data sets is not always available. An intuitive approach would be to incorporate the spatiotemporal domain for improving structure and inter-dependencies of the output.

Dealing with the LV segmentation problem from a spatiotemporal perspective can give further information on the shape boundaries. Due to the nature of the speckle pattern, it is hard to draw conclusions about the boarder of the LV from still frames. Thus, cardiologist usually examine videos of the deformation of the LV wall during the echocardiographic examination. It is logical to assume the speckle pattern structure is better localized when the spatiotemporal coherence is considered. In this regard, Huang et al. exploits the spatiotemporal coherence of individual data for cardiac contour estimation [5]. Others embarked on introducing temporal consistency in the tracked speckle pattern using optical flow [7] or gradient vector flow approach [6]. Nevertheless, the spatiotemporal structures of the speckle patterns are usually stochastic in nature. Neglecting the inherent heterogeneity might not characterize the structure in space and time efficiently. The fractional Brownian motion (fBm), which is a non-stationary stochastic process, integrates well with the stochastic nature of ultrasound echoes [8]. It has the advantage of representing a wide range of non-stationary signals and can quantify statistical self-similarity in time-sequence ultrasound images.

To address the aforementioned challenges, we present a novel physically motivated stochastic model for improved epicardium and endocardium boundary segmentation. Motion roughness, reflected in the speckle pattern, is characterized by fBm for local boundary delineation – which is theoretically invariant to intensity transformations [9]. Using second-order moments for LV shape global information complements the local characterization of the fBm process.

## 2   Methods

### 2.1   Fractional Brownian Motion

Brownian motion $B(t)$ is a *Markovian* process, whose conditional transition density function is time-homogeneous. Based on the Lagrangian representation, the generalization of a standard $B(t)$ is a Fractional Brownian motion $B_H(t)$, which is a continuous Gaussian self-similar process in $\mathbb{R}$ with stationary increments [10]. $B_H(t)$ can be modeled via stochastic integral equation, given by

$$B_H(t) - B_H(0) = \frac{1}{\Gamma(H + \frac{1}{2})} \left\{ \int_{-\infty}^{0} (t - s)^{H-1/2} - (-s)^{H-1/2} dB(s) \right.$$

$$\left. + \int_{0}^{t} (t - s)^{H-1/2} dB(s) \right\}, \tag{1}$$

where $\Gamma(x)$ and $B(t)$ are the gamma function and standard Brownian motion, respectively, and $H \in (0,1)$ is called the Hurst index which describes the scaling behaviour of the process and the roughness of the resultant motion or trajectory [11]. For our case, $H$ characterizes the deformation of the left ventricle wall motion, with lower values leading to a heterogeneous motion and vice versa. From (1), the standard $B(t)$ is recovered when $H = \frac{1}{2}$. But in contrast, fBm has dependent increments when $H \neq \frac{1}{2}$. By allowing $H$ to differ from $\frac{1}{2}$, a fBm process is achieved, where for $H > \frac{1}{2}$ increments are positive correlated, and for $H < \frac{1}{2}$ increments are negatively correlated.

A normalized fractional Brownian motion $\{B_H(t), t \in \mathbb{R}^N\}$ with $0 < H < 1$ is uniquely characterized by the following properties: (a) $B_H(t)$ has stationary increments; (b) $B_H(0) = 0$, and $\mathbb{E}[B_H(t)] = 0$ for $t \geq 0$; and (c) $\mathbb{E}[B_H^2(t)] = t^{2H}$ for $t \geq 0$; then it follows that the covariance function is given by

$$\rho(s,t) = \mathbb{E}(B_H(s)B_H(t)) = \frac{1}{2}\left[|t|^{2H} + |s|^{2H} - |t - s|^{2H}\right], \quad \forall \ s, t \in \mathbb{R}^N, \tag{2}$$

for $0 < s \leq t$, where $\mathbb{E}$ denotes the expectation operator with respect to probability space, and $|t|$ is the Euclidean norm of $t \in \mathbb{R}^N$.

## 2.2   Fractal Dimension Estimation

There are several ways to estimate the fractal dimension (FD) of a stochastic process modeled by a fBm [9]. All of them are based on the formula

$$\mathbb{E}\left[(\tilde{B}_H(n + t) - \tilde{B}_H(n))^2\right] = \sigma^2(t)^{2H}, \tag{3}$$

for the variogram of fBm, which follows from (2), and $\sigma^2$ represents the variance of the increments. The fBm and power-law variogram fits were used to estimate $H$ as a measure of self-similarity. Then the discrete-time fBm process $\tilde{B}_H(n)$ after taking the logarithm of (3) yields the linear equation

$$\log \mathbb{E}\left[(\tilde{B}_H(n + s) - \tilde{B}_H(n))^2\right] = H \log |s| + \log c. \tag{4}$$

where $c$ is proportional to the standard deviation $\sigma$.

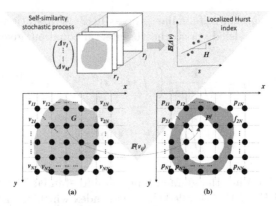

**Fig. 1.** Multiscale parametric mapping based on fractional Brownian motion $(B_H)$ for voxel-based segmentation. (a) Object $G$ having a spatial support $v_{1,1} \times v_{N,N}$ in source image $f_{x,y}$, and (b) constructed fractal parametric map $\mathcal{F}$ at different scales $(r)$ representing localized Hurst indices $(H)$.

The linearly-related $H$ can be calculated from the slope of the average absolute difference plotted as a function of the increments $n$ with step-size $k$ on a log-log plot. The FD of an $m$-dimensional fBm is related to the Hurst index by $FD = m - H$, where $m$ is the Euclidean dimension, or the number of independent variables, and $H$ quantifies the self-affinity of the process. Thereby, closer $H$ is to one, the lower the FD, and the smoother the process becomes and vice versa. Finally the parametric fractal dimension volume $(\mathcal{F})$ is generated voxel-by-voxel, represented as:

$$
\mathcal{F}^{(i,r)} =
\begin{pmatrix}
m_1 \\
m_2 \\
\vdots \\
\vdots \\
m_M
\end{pmatrix}
-
\begin{pmatrix}
h^i_{11L} & h^i_{12L} & \cdots\cdots & h^i_{1NL} \\
h^i_{21L} & h^i_{22L} & \cdots\cdots & h^i_{2NL} \\
\vdots & \vdots & \ddots & \vdots \\
\vdots & \vdots & & \ddots & \vdots \\
h^i_{M1L} & h^i_{M2L} & \cdots\cdots & h^i_{MNL}
\end{pmatrix}
\tag{5}
$$

where $M$, $N$ and $L$ are the volume elements defined in 3D space for each ultrasound time-sequence $i$, and $r = 1, \ldots, j$ is the resolution limits of $\mathcal{F}^{(i,r)}$ which represents the mean absolute intensity difference to center voxels defined as local range dependence (i.e. how far the resolution of $r$ can be deeply probed). Figure 1 illustrates the fractional Brownian motion (fBm) segmentation process.

**Classification:** Patches of $6.3 \times 8.1$ mm$^2$ – representing myocardium vs blood pool – were selected randomly from a single ultrasound volume transformed to $\mathcal{F}$, and used for training data. Such that, the training set, $S_T$, is composed of a set of $N$ patches for which the feature vector $[V-(5)]$, and the classification result ($C_1$ or $C_2$: myocardium or blood pool) are known $S_T = \{(V^{(n)}, C_k^{(n)} | n = 1, \ldots, N; k \in \{1, 2\}\}$. The Hurst index is a useful statistical method for inferring the properties of a time series without making assumptions about stationarity. It is most useful when used in conjunction with other techniques. Thus, features representing the mean FD, lacunarity – which defines the sparsity of the fractal pattern – and first and higher order statistics (skewness, kurtosis and variance) were derived from $\mathcal{F}$ to form a vector in a 5-D feature space $V = (v_1, v_2, \ldots, v_5)$. Then in the classification procedure, one of the classes $C_1$ (myocardium) or $C_2$ (blood pool) would be assigned to each candidate voxel when its representation is known. A Bayesian classifier was used for classification, although SVM or random forests would have also served the purpose.

## 2.3   Contour Structuring Moments

Image moments are defined as weighted averages of voxel intensities. For the time-sequence 3D ultrasound images, at time $t$ and depth $z$, the raw $(p, q)$-moment $m_{p,q}$ for an fBm segmented object $G$ (LV for our case) is given by:

$$m_{pq} = \sum_{x=1}^{N} \sum_{y=1}^{N} x^p y^q f(x, y), \tag{6}$$

where $N$ is the size of $G$ and $f(x, y)$ are the gray levels of individual voxels.

The first-order moments $m_{1,0}$ and $m_{0,1}$, when normalized by $m_{0,0}$ give the coordinates of the binary object – $x_c$ and $y_c$ of the endocardium or epicardium. Accordingly, second-order moments describe the "distribution of mass" with respect to the coordinate axes and define the orientation $\theta$ of $G$. In order to extracting the parameters of the equivalent ellipse from the second-order moments $m_{2,0}$, $m_{1,1}$, and $m_{0,2}$, the central moments can be defined as

$$\mu_{pq} = \sum_{p,q \in G}^{N} (x - x_c)^p (y - y_c)^q, \tag{7}$$

such that

$$\mu_{2,0} = \frac{m_{2,0}}{m_{0,0}} - x_c^2, \quad \mu_{1,1} = 2\left(\frac{m_{1,1}}{m_{0,0}} - x_c y_c\right), \quad \mu_{0,2} = \frac{m_{0,2}}{m_{0,0}} - y_c^2, \tag{8}$$

where $x_c$ and $y_c$ are the coordinates of the centroid $c$ of $G$. These moments are invariants to translation. Then the eigen vectors of the covariance matrix of $G$ correspond to the major $l$ and minor $w$ axes lengths with a certain angle $\theta$ for the equivalent ellipse. In this work, $c$ was first identified on the fBm segmented LV base image, and used afterwards for alignment – through the rest of image sequence – towards the apex.

## 3   Experimental Results

**Synthetic and Animal Data.** Validation was done on 5 different 3D ultrasound time-sequences from the KU Leuven synthetic data set [12], representing 1 normal and 4 ischemic. Each image had $224 \times 176 \times 208$ voxels of size $0.7 \times 0.9 \times 0.6$ mm$^3$. For each sequence, ground truth motion trajectories were provided at 2250 mesh points. The endocardium and epicardium meshes were then converted into segmentation masks and used for benchmarking. Also, 3D ultrasound sequence images were acquired from 2 acute canine studies (open chested) following a severe occlusion of the left anterior descending coronary artery. A Philips iE33 ultrasound system with an X7-2 probe was used for image acquisition. Images typically had $400 \times 140 \times 120$ voxels of $0.25 \times 0.85 \times 0.85$ mm$^3$ with an average of 23 temporal frames. All experiments were conducted in compliance with the Institutional Animal Care and Use Committee policies.

**Evaluating Model Goodness-of-Fit:** The residual sum of squares (RSS) was used to measure the discrepancy in the estimated $H$ indices of the parametric volume maps ($\mathcal{F}$). A smaller value indicates the model has a smaller random error component. Figure 2 shows where errors most likely occur when locally computing $H$, and hence the estimation of $\mathcal{F}$. The size of the fBm localized range dependence in (5) was adjusted in the range between $0.7 \times 0.9$ and $9.1 \times 11.7$ mm$^2$. Higher RSS was encountered in longer range dependence as the introduced error may get accumulated due to echo artifacts. Also, Fig. 2 shows how $H$ values vary at different fBm local dependence ranges. Selecting the best fBm dependence range can avoid unnecessary computational time and give better accuracy.

**Quantitative Evaluation:** The proposed fBm segmentation method was benchmarked using Huang et al.'s method [5], which is based on employing spatiotemporal coherence under a dynamic appearance model (C-DAM).

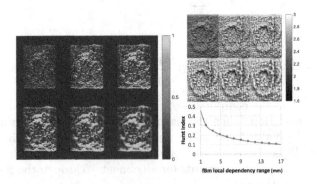

**Fig. 2.** fBm parameter estimation discrepancy by [left] residual sum of squares (RSS) for [right-top] corresponding fractal dimension parametric images; [right-bottom] change of RSS error for different fBm range dependence.

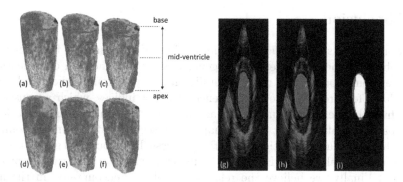

**Fig. 3.** [Left] 3D segmentation of left ventricle for time-sequence ultrasound images illustrating (a)–(c) ground truth, fBm, C-DAM for normal condition, and (d)–(f) ground truth, fBm, C-DAM for ischemic-Ladprox; [Right] canine endocardium segmentation of LV mid-slice for (g)–(i) manual, fBm-based, and comparing segmentation accuracy, respectively.

**Table 1.** Endocardium ($E_n$) and Epicardium ($E_p$) segmentation quality using Dice coefficient for normal and ischemic cardiomyopathy

| Method | Left ventricle condition | | | | |
|---|---|---|---|---|---|
| | Normal | LADprox | LADdist | LCX | RCA |
| Our fBm ($E_n$) | $0.87 \pm 0.02$ | $0.85 \pm 0.02$ | $0.84 \pm 0.04$ | $0.87 \pm 0.02$ | $0.88 \pm 0.02$ |
| C-DAM ($E_n$) | $0.84 \pm 0.07$ | $0.86 \pm 0.03$ | $0.84 \pm 0.05$ | $0.82 \pm 0.04$ | $0.85 \pm 0.05$ |
| Our fBm ($E_p$) | $0.92 \pm 0.02$ | $0.90 \pm 0.02$ | $0.88 \pm 0.04$ | $0.92 \pm 0.02$ | $0.93 \pm 0.02$ |
| C-DAM ($E_p$) | $0.91 \pm 0.03$ | $0.88 \pm 0.02$ | $0.93 \pm 0.02$ | $0.89 \pm 0.04$ | $0.88 \pm 0.02$ |

The evaluation of the segmentation quality was performed using Dice coefficient, and computed for both end-diastolic and -systolic and averaged over all cases, see Table 1.

One of the key properties of fBm is that it can exhibit persistence ($H > \frac{1}{2}$) or anti-persistence ($H < \frac{1}{2}$). Persistence is the property that the LV wall motion tends to be smooth, e.g. near to normal. Anti-persistence is the property that the relative stochastic process is very noisy, and hence LV wall motion trajectories tend to be heterogeneous. The latter case is an example of cardiac ischemia, such that displacements over one temporal or spatial interval are cancelled out by displacements over another time interval. From Table 1, the fBm segmentation method showed improved performance for 4 out 5 of both tested LV endo- and epicardium deformation conditions. Moreover, to visually demonstrate practical relevance, Fig. 3 shows the 3D segmentation of the LV and the mid-slice for the synthetic and canine data, respectively. Improved delineation of the epi- and endocardium boundaries can be seen, especially for the LV base slice.

## 4 Conclusion

This work presents a novel method for improving LV segmentation by addressing the problem of speckle pattern heterogeneity, where (a) the fBm classification-based segmentation method relies naturally on the spatiotemporal dependencies of the local features; (b) the 3D sequence of Hurst indices, which was used to derive fractal dimension volume maps, are invariant to intensity transformations; (c) global information about the LV shape using second-order moments complements the local characterization of the fBm process. Both local and global boundary information about the LV shape boundaries was captured with improved precision. Finally, we believe the relative complexity encountered in the animal study can be better dealt with if the segmentations were based on the radio-frequency envelope detected echos, which may better reflect the intrinsic properties of the myocardium tissue.

## References

1. Frangi, A., Niessen, W., Viergever, M.: Three-dimensional modeling for functional analysis of cardiac images: a review. IEEE Trans. Med. Imaging **20**(1), 2–25 (2001)
2. Carneiro, G., Nascimento, J.C.: Combining multiple dynamic models and deep learning architectures for tracking the left ventricle endocardium in ultrasound data. IEEE Trans. Pattern Anal. Mach. Intell. **35**(11), 2592–2607 (2013)
3. Hansson, M., et al.: Segmentation of B-mode cardiac ultrasound data by Bayesian probability maps. Med. Image Anal. **18**(7), 1184–1199 (2014)
4. Carneiro, G., Nascimento, J.C., Freitas, A.: The segmentation of the left ventricle of the heart from ultrasound data using deep learning architectures and derivative-based search methods. IEEE Trans. Image Process. **21**(3), 968–982 (2012)
5. Huang, X.J., et al.: Contour tracking in echocardiographic sequences via sparse representation and dictionary learning. Med. Image Anal. **18**(2), 253–271 (2014)
6. Zagrodsky, V., Walimbe, V., et al.: Registration-assisted segmentation of real-time 3-D echocardiographic data using deformable models. IEEE Trans. Med. Imaging **24**(9), 1089–1099 (2005)
7. Barbosa, D., Friboulet, D., D'hooge, J., Bernard, O.: Fast tracking of the left ventricle using global anatomical affine optical flow and local recursive block matching. In: Proceedings of MICCAI CETUS Challenge (2014)
8. Al-Kadi, O.S., Chung, D.Y.F., et al.: Quantification of ultrasonic texture intra-heterogeneity via volumetric stochastic modeling for tissue characterization. Med. Image Anal. **21**(1), 59–71 (2015)
9. Al-Kadi, O.S.: Fractals for biomedical texture analysis. In: Biomedical Texture Analysis: Fundamentals, Tools and Challenges, pp. 131–161. Academic Press, London (2017)
10. Mandelbrot, B.: Fractals and Chaos: The Mandelbrot Set and Beyond. Springer, New York (2004). https://doi.org/10.1007/978-1-4757-4017-2
11. O'Malley, D., Cushman, J.: Two-scale renormalization-group classification of diffusive processes. Phys. Rev. E **86**(1), 011126–7 (2012)
12. Alessandrini, M., Heyde, B., et al.: Detailed evaluation of five 3D speckle tracking tlgorithms using synthetic echocardiographic recordings. IEEE Trans. Med. Imaging **35**(8), 1915–1926 (2016)

# Context Aware 3D Fully Convolutional Networks for Coronary Artery Segmentation

Yongjie Duan[1,2,3], Jianjiang Feng[1,2,3(✉)], Jiwen Lu[1,2,3], and Jie Zhou[1,2,3]

[1] Department of Automation, Tsinghua University, Beijing, China
jfeng@tsinghua.edu.cn
[2] State Key Lab of Intelligent Technologies and Systems,
Tsinghua University, Beijing, China
[3] Beijing National Research Center for Information Science and Technology,
Beijing, China

**Abstract.** Cardiovascular disease caused by coronary artery disease (CAD) is one of the most common causes of death worldwide. Coronary artery segmentation has attracted increasing attention since it is useful for better visualization and diagnosis. Conventional lumen segmentation methods basically describe vessels by a rough tubular model, thus presenting inferiority on abnormal vascular structures and failing to distinguish exact coronary arteries from vessel-like structures. In this paper, we propose a context aware 3D fully convolutional network (FCN) for vessel enhancement and segmentation in coronary computed tomography angiography (CTA) volumes. Combining the superior capacity of CNN in extracting discriminative features and satisfactory suppression of vessel-like structures by spatial prior knowledge embedded, the proposed approach significantly outperforms conventional Hessian vesselness based approach on a dataset of 50 coronary CTA volumes.

**Keywords:** Spatial prior knowledge · Deep learning ·
Lumen segmentation · Coronary computed tomography angiography

## 1 Introduction

Over the past decades, coronary artery disease (CAD), which is caused by vessel calcification or atherosclerosis, is among the most common causes of human deaths in the world [1]. Coronary computed tomography angiography (CTA) has been widely utilized to diagnose CAD, such as plaque evaluation and stenosis assessment, thanks to the advantage of noninvasive nature and high resolution image acquisition. Most plaque evaluation and stenosis assessment, as well as estimation of fractional flow reserve (FFR), which is used to measure the influence of stenosis impeding oxygen delivery to the heart muscle, rely on a semiautomatic or automatic precise coronary artery segmentation. An inaccurate artery tree extraction will result in improper stenosis assessment while manual correction of wrong artery trees is very time-consuming.

© Springer Nature Switzerland AG 2019
M. Pop et al. (Eds.): STACOM 2018, LNCS 11395, pp. 85–93, 2019.
https://doi.org/10.1007/978-3-030-12029-0_10

Accurate vascular segmentation in medical image is a widely researched topic, yet remaining a challenging task because of the presence of calcifications, image artifacts, insufficient contrast, and large anatomical variations among patients. A complete vessel tree is usually considered as a combination of multi-scale and multi-orientation tubular structures, in view of the appearance the vessels present in CTA. This property is directly applied for vessel segmentation [2], centerline extracting, and lumen diameter estimation. However, such approach and similar ones fail to detect abnormal structures, namely bifurcations and lesions (e.g., calcifications, atherosclerosis, aneurysms, stents, and stenoses), due to the inferiority of using a rough tubular model [3]. To improve the segmentation performance, machine learning was applied to capture more powerful and discriminative features [4,5]. Nonetheless, since the segmentation is regarded as a voxel-wise classification problem, these approaches may generate a lot of leaks (false positives) or holes (false negatives) in the final extraction result. Therefore, fully convolutional network (FCN), U-Net and its 3D extension [6,7] have earned increasing attention in medical image segmentation and represented better performance because of the superiority of integrating high-level constraints within the upsampling step. However, to the best of our knowledge, these networks have not yet considered corresponding holistic anatomical information, such as spatial distribution of coronary artery tree. In other words, the priori anatomical knowledge is rarely incorporated to guide the segmentation procedure.

Inspired by the conventional vessel extraction algorithms and deep learning framework, we propose an improved segmentation approach that adopts a 3D fully convolutional network and spatial prior knowledge of coronary artery tree, to predict the probability map of coronary arteries from the whole coronary CTA images. We call it context aware because unlike original 3D U-Net, our network utilizes the location information of input patches. Briefly, the main contributions of our approach are summarized as: (1) a 3D UNet-like network is customized to achieve coronary artery tree segmentation in the whole coronary CTA, and the architecture of our network is tailored to identify small objects in 3D patches by introducing a shortcut from low-level to high-level feature maps; (2) the spatial prior knowledge of coronary artery tree, estimated from training images, is incorporated to guide the segmentation within each local patch, thus reducing the complexity of model learned and increasing the performance at the same time. We evaluate our algorithm on 50 coronary CTA volumes by a five-fold cross validation. The experimental results demonstrate that our framework is robust for coronary artery segmentation and outperforms conventional Hessian vesselness based approach.

## 2    Methods

In this paper, we aim to extract complete coronary artery tree from coronary CTA volume. A statistical model is obtained first to estimate the spatial distribution of coronary arteries, which contributes to the reduction of false positives and false negatives. Then our proposed segmentation network utilizes this prior knowledge as an additional input channel and produce voxel-wise probability map, which is thresholded to obtain the final segmentation result. No morphological post-processing is used to clean up small disconnected components. The schematic illustration of our framework is shown in Fig. 1.

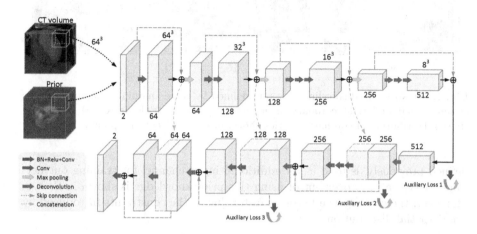

**Fig. 1.** Network architecture of our proposed framework.

## 2.1 Spatial Distribution Prior

Normally, the coronary arteries start from the coronary ostia, then bifurcating into two main branches along the heart surface. Obviously, this anatomical knowledge can be introduced as an auxiliary constraint to help to reject most wrong extractions scattered out of the heart or inside. In order to acquire the spatial distribution model, all the annotated labels in training dataset, are aligned to the same coordinate system defined by three anatomical landmarks, namely, the ostium of left coronary artery, the ostium of right coronary artery, and the left ventricle apex. These landmarks are extracted using a boosting-based detection algorithm [8]. As shown in Fig. 2, the origin is defined as the middle point between the ostia of left and right coronary artery. The $z$ axis is defined as the direction pointing from the origin to the left ventricle apex. The $y$ axis is defined as the vector perpendicular to the $z$ axis and lies inside the plane determined by three landmarks. Then the $x$ axis is defined as the cross product of previous two axes. After the alignment, a spatial probability map of coronary arteries is obtained by some non-parametric kernel density estimation, such as Parzen window. Therefore, given a test volume, the corresponding coordinate system attached is estimated following detecting the three cardiac landmarks. Then the statistical priori distribution, transformed based on the coordinate system

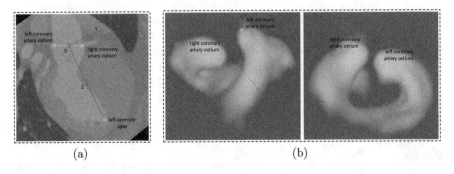

(a)                                        (b)

**Fig. 2.** (a) Normalized coordinate system defined by three anatomical landmarks: the ostium of left and right coronary artery and the left ventricle apex. (b) Generated spatial distribution of coronary arteries shown in two different viewpoints.

defined previously, is used as an additional input channel for training our network. A typical improvement of using this prior knowledge is shown in Sect. 3. Moreover, the convergence becomes faster due to the constraints introduced by priori spatial distribution.

## 2.2   Network Architecture

Illustrated as Fig. 1, our proposed context aware network is customized from the original 3D U-Net [6]. Similar to its standard version, our main network is composed of 3D convolution, max pooling, up sampling (deconvolution), and shortcut connections from layers in down-sampling path to the ones in up-sampling path with equal resolution. However, considering the growth of model depth and number of trainable parameters in 3D kernels, training the network mentioned can be extremely difficult owing to the vanishing of forward and backward propagating signals. Therefore, residual skip connection is added between the input and output of each layer block to preserve flowing signals in our network. Each layer is composed of batch normalization, ReLU activation, and convolution in order, according to its superiority demonstrated in previous study [9]. Moreover, max pooling rather than convolution with stride 2 is utilized to reduce the resolution of feature maps, thus leading signals flowing directly from feature maps with high-resolution to the ones with low-resolution. In addition, the kernel size adopted in our network is set to $3^3$ to keep the number of trainable parameters at a low level while preventing performance declining significantly.

Apparently, class imbalance, namely the background in cardiac CT volume vastly outnumbering the coronary arteries to be extracted, results in serious misclassifications. To deal with the problem and mitigate easy example dominant, the focal loss, proposed in [10], is modified as follows in our framework.

$$\mathcal{L}_{\alpha Focal}(\mathcal{X}; \mathcal{W}) = -\frac{1}{|\mathcal{X}|} \sum_{y} \sum_{x} \alpha(1 - p_t)^{\gamma} \log(p_t) \tag{1}$$

$$p_t = yp + (1 - y)(1 - p), \quad \alpha = 1 - \frac{|\mathcal{X}_y|}{|\mathcal{X}|}$$

Given an input patch $\mathcal{X}$, $|\mathcal{X}|$ is the size of patch $\mathcal{X}$, and $|\mathcal{X}_y|$ is the size of class $y$ within patch $\mathcal{X}$ similarly. $\mathcal{W}$ are the parameters to be trained within the network proposed. $y$ and $p$ denote the ground truth label and corresponding probability prediction of sample $x$ after softmax, respectively. Besides, we use a trade-off parameter $w$ to balance the importance of positive/negative examples, and $\gamma$, which is set as 2 in this paper, is introduced to mitigate easy example dominant and focus on hard examples consequently. In addition, we also employ a two-stage training strategy to train such an imbalance problem. We extract patches whose centers have a 0.8 probability of being on foreground initially, and decrease it to 0.5 after certain epochs.

In deep network, early layers are always under-tuned because of gradient vanishing problem. The residual structure mentioned above alleviates this problem by adding skip paths. On the other hand, enhancing the gradient flow for shallow layers with deep supervision is also demonstrated to be effective in segmentation task. Similar to [11], we incorporate three side-paths auxiliary loss to shorten the backpropagation path of gradient flow signals. So as to obtain the final formulation of loss function in our proposed network.

$$\mathcal{L}(\mathcal{X}; \mathcal{W}) = \mathcal{L}_{\alpha Focal}(\mathcal{X}; \mathcal{W}) + \sum_{s=1,2,3} \beta_s \mathcal{L}^s_{\alpha Focal}(\mathcal{X}; w^s) \tag{2}$$

where $\beta_s$ is the weight of different side-paths $w^s$ and set as 0.3, 0.6, 0.9 from coarse to fine.

## 3 Experiments and Results

Our proposed approach is evaluated on a total of 50 coronary CTA volumes, which are collected from different scanners, by a five-fold cross validation. The images are randomly divided into 5 groups, and one of them is selected for testing then the rest for training in turn. In term of ground truth, some public datasets, such as Rotterdam Coronary Artery Algorithm Evaluation Framework, only provide annotated contours on cross-sections of some main branches. It is not feasible to use these annotations for training or testing our networks. In this paper, therefore, the complete coronary artery trees in 50 CTA volumes are annotated by radiologists from cooperative hospital. And the spatial prior calculated uses only training cases in each of the five folds of cross validation. Considering the variations of image spacing, we resampled these images to the same voxel size of $0.4^3$ mm$^3$. Moreover, data augmentation is applied on 30 percent training samples, including $\pm 25$ degrees rotation around $z$ axis and $\pm 10$ pixels translation along orthogonal axes.

### 3.1    Implementation Details

We implemented our 3D network using a NVIDIA GeForce 1080Ti in *Tensor-flow*. Considering the limitation of available GPU memory and superiority of training on mini-batch, cropped $64 \times 64 \times 64$ sub-volumes are randomly selected as the input to train our network with a batch size of 6. We update the weights of network using an Adam optimizer with initial learning rate of 0.001 and momentum of 0.5. To avoid overfitting, a L2 regularization is used in our network with the value of $5e^{-4}$. Finally, the whole network is trained from scratch, and the weights are initialized from a normal distribution $\mathcal{N}(0, 0.01)$. Within the test stage, we adopted sliding windows with an overlapping ratio of 0.4 to generate input patches from the whole volume, and consequently overlap-tiling strategy is utilized to reconstruct the whole predicted probability map.

### 3.2    Results

Three metrics are used to evaluate the segmentation performance of our proposed approach, including dice similarity coefficient (DSC), precision, and recall. Quantitative comparison of our proposed network with/without spatial priori distribution (**Ours with/no prior**) and Hessian vesselness based method (**Frangi**) [2][1], is shown in Fig. 3 and Table 1.

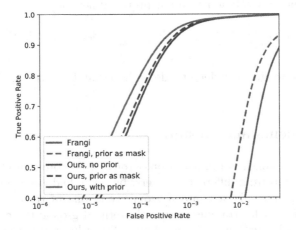

**Fig. 3.** The quantitative performance comparison of our proposed algorithm and multi-scale Hessian vesselness based method [2]. The curves of Frangi filtering are consistent with its performance reported in [4].

---

[1] Public implementation by Kroon, D.J. https://ww2.mathworks.cn/matlabcentral/fileexchange/24409-hessian-based-frangi-vesselness-filter.

**Table 1.** Quantitative evaluation for coronary segmentation

| Metrics | Frangi | Frangi + mask | Ours no prior | Ours + mask | Ours with prior |
|---|---|---|---|---|---|
| DSC[%] | 2.5 ± 0.9 | 9.5 ± 3.6 | 74.7 ± 5.6 | 75.8 ± 4.7 | **79.5 ± 3.6** |
| Precision[%] | 1.3 ± 0.5 | 5.2 ± 3.2 | 70.2 ± 8.5 | 72.0 ± 7.2 | **78.5 ± 6.0** |
| Recall[%] | 74.1 ± 10.9 | 74.0 ± 10.9 | 80.9 ± 7.6 | 80.9 ± 7.6 | **81.3 ± 3.6** |

To explore the effectiveness of spatial prior knowledge usage as well as the way of incorporating prior knowledge, our generated priori spatial distribution is thresholded as a binary mask to reduce negative voxels (**Frangi + mask, Ours + mask**), thereby achieving a fair comparison with our proposed context aware method. It is observed that our proposed network gets an obvious improvement, compared with Hessian vesselness based method. Concentrating on the usage of spatial prior knowledge, we notice that utilizing the priori spatial distribution by just using it as a binary mask is inferior to incorporating it as an additional input channel.

In Fig. 4, the probability maps predicted by different approaches are thresholded to visually compare the performance. Similarly, we could observe that satisfactory performance is achieved by our proposed approach, in which the false extractions are significantly less than the ones in Hessian vesselness based method with binary priori mask. Besides, some disconnected segments caused by artifacts or insufficient contrast could also be reconnected due to the prior knowledge incorporated. It is reasonable to assume that the priori spatial distribution, estimated from finite samples, may not be consistent with the coronary arteries distribution within a new image exactly. Therefore, compared with using the prior knowledge as a binary mask, taking it as an additional input channel achieves better performance, especially on those segments referred by red arrows in Fig. 4.

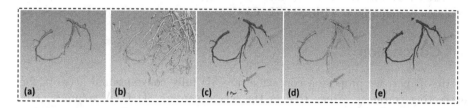

**Fig. 4.** Qualitative comparison of different algorithms on one example. No morphological post-processing is used to clean up small disconnected components. (a) annotated label, (b) filtered result of Frangi [2] using mask, (c) result of our network without prior knowledge, (d) filtered result of (c) using mask, (e) result of proposed method. (Color figure online)

For a better view of results, contours of arteries segmented by different approaches on some 2D planes, including common, calcified, and bifurcated, are shown in Fig. 5. Compared to Hessian vesselness based method, our proposed

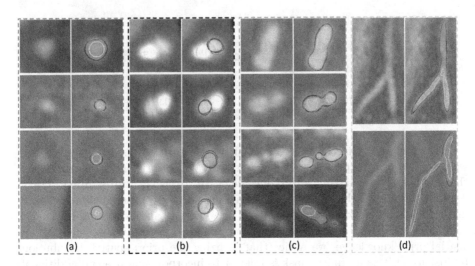

**Fig. 5.** Comparison of segmented lumen on 2D planes, including annotations (green), Hessian vesselness (yellow), our segmentation with (red) and without (blue) prior. (a) common, (b) calcifications, (c)(d) bifurcations from different projections. (Color figure online)

method is highly consistent with manual annotations, especially on those abnormal vascular structures. Note that the DSC and Precision in Table 1 are not that high. It is because the radiologists are somewhat conservative, thereby omitting many thin vessels in annotations, while the algorithms extract the complete coronary tree (including thin vessels).

Average segmentation time of our approach for the complete coronary Artery tree is about 159 seconds per volume on a computer with a NVIDIA GeForce 1080Ti GPU. The Hessian vesselness method consumes about 226 seconds per volume on a computer with 2.1 GHz CPU.

## 4 Conclusion

In this paper, we propose a context aware 3D fully convolutional network for extracting coronary artery tree in the whole cardiac 3D CTA volumes. The proposed approach integrates the strength of deep networks in extracting effective features and spatial prior knowledge constraint in guiding segmentation procedure, thus reducing vast majority negative voxels. Our evaluation on 50 CTA volumes shows obvious suppression of vessel-like structures and accurate segmentation of coronary arteries. For further improvement, the spatial prior distribution could be estimated by a more accurate heart alignment algorithm. Besides, Zheng et al. have proved that it is effective to use extracted heart surface as a constraint of coronary artery segmentation [4]. We intend to introduce the heart surface as another anatomical prior knowledge to achieve a further improvement.

**Acknowledgement.** This work is supported by the National Natural Science Foundation of China under Grant 61622207.

# References

1. Roger, V.L., et al.: Heart disease and stroke statistics-2012 update: a report from the american heart association. Circulation **125**(1), e2–e220 (2012)
2. Frangi, A.F., Niessen, W.J., Vincken, K.L., Viergever, M.A.: Multiscale vessel enhancement filtering. In: Wells, W.M., Colchester, A., Delp, S. (eds.) MICCAI 1998. LNCS, vol. 1496, pp. 130–137. Springer, Heidelberg (1998). https://doi.org/10.1007/BFb0056195
3. Lesage, D., Angelini, E.D., Bloch, I., Funka-Lea, G.: A review of 3D vessel lumen segmentation techniques: models, features and extraction schemes. Med. Image Anal. **13**(6), 819–845 (2009)
4. Zheng, Y., Loziczonek, M., Georgescu, B., Zhou, S.K., Vega-Higuera, F., Comaniciu, D.: Machine learning based vesselness measurement for coronary artery segmentation in cardiac CT volumes. In: Proceedings of the SPIE 7962, Medical Imaging 2011: Image Processing. p. 79621K (2011)
5. Chen, F., Li, Y., Tian, T., Cao, F., Liang, J.: Automatic coronary artery lumen segmentation in computed tomography angiography using paired multi-scale 3D CNN. In: Proceedings of the SPIE 10578, Medical Imaging 2018: Biomedical Applications in Molecular, Structural, and Functional Imaging, p. 105782R (2018)
6. Çiçek, Ö., Abdulkadir, A., Lienkamp, S.S., Brox, T., Ronneberger, O.: 3D U-Net: learning dense volumetric segmentation from sparse annotation. MICCA **I**, 424–432 (2016)
7. Milletari, F., Navab, N., Ahmadi, S.A.: V-Net: Fully convolutional neural networks for volumetric medical image segmentation. In: Fourth International Conference on 3D Vision (3DV), pp. 565–571. IEEE (2016)
8. Zhou, S.K.: Discriminative anatomy detection: classification vs regression. Pattern Recognit. Lett. **43**, 25–38 (2014)
9. He, K., Zhang, X., Ren, S., Sun, J.: Deep residual learning for image recognition. In: CVPR, pp. 770–778 (2016)
10. Lin, T.Y., Goyal, P., Girshick, R., He, K., Dollar, P.: Focal loss for dense object detection. In: ICCV, pp. 2980–2988 (2017)
11. Yang, X., Bian, C., Yu, L., Ni, D., Heng, P.A.: Hybrid loss guided convolutional networks for whole heart parsing. In: International Workshop on Statistical Atlases and Computational Models of the Heart, pp. 215–223 (2017)

# Learning Associations Between Clinical Information and Motion-Based Descriptors Using a Large Scale MR-derived Cardiac Motion Atlas

Esther Puyol-Antón[1]([✉]), Bram Ruijsink[1,2], Hélène Langet[3],
Mathieu De Craene[3], Paolo Piro[4], Julia A. Schnabel[1], and Andrew P. King[1]

[1] School of Biomedical Engineering and Imaging Sciences,
King's College London, London, UK
esther.puyol_anton@kcl.ac.uk
[2] St Thomas' Hospital NHS Foundation Trust, London, UK
[3] Philips Research, Medisys, Paris, France
[4] Sanofi R&D, Paris, France

**Abstract.** The availability of large scale databases containing imaging and non-imaging data, such as the UK Biobank, represents an opportunity to improve our understanding of healthy and diseased bodily function. Cardiac motion atlases provide a space of reference in which the motion fields of a cohort of subjects can be directly compared. In this work, a cardiac motion atlas is built from cine MR data from the UK Biobank ($\approx 6000$ subjects). Two automated quality control strategies are proposed to reject subjects with insufficient image quality. Based on the atlas, three dimensionality reduction algorithms are evaluated to learn data-driven cardiac motion descriptors, and statistical methods used to study the association between these descriptors and non-imaging data. Results show a positive correlation between the atlas motion descriptors and body fat percentage, basal metabolic rate, hypertension, smoking status and alcohol intake frequency. The proposed method outperforms the ability to identify changes in cardiac function due to these known cardiovascular risk factors compared to ejection fraction, the most commonly used descriptor of cardiac function. In conclusion, this work represents a framework for further investigation of the factors influencing cardiac health.

**Keywords:** Cardiac motion atlas · Non-imaging data ·
Dimensionality reduction · Multivariate statistics

## 1 Introduction

Although much is known about the factors influencing healthy and diseased cardiac development, the recent availability of large scale databases such as the UK Biobank represents an excellent opportunity to extend this knowledge. The wide

M. Pop et al. (Eds.): STACOM 2018, LNCS 11395, pp. 94–102, 2019.
https://doi.org/10.1007/978-3-030-12029-0_11

range of clinical information contained in such databases can be used to learn new associations with indicators of cardiac function. In this paper we propose a framework for performing such an investigation.

Left-ventricular (LV) function has been traditionally assessed using global indicators such as ejection fraction and stroke volume. The main drawback of such indicators is that they do not provide localised information about ventricular function, and so their diagnostic power is limited. Motion parameters such as displacements allow for a more spatially and temporally localised assessment of LV function. Statistical cardiac motion atlases have been proposed as a way of making direct comparisons between motion parameters from multiple subjects [1,10]. We use the UK Biobank database to build a cardiac motion atlas from cine MR images using around 6000 subjects with no known cardiovascular disease. To ensure that the estimated motion parameters are accurate, two automated quality control strategies are proposed to reject subjects with insufficient image quality. We use motion descriptors extracted from the atlas to identify relationships between cardiac function and other clinical information.

Previously, several papers have tried to analyse the influence of clinical information such as smoking, alcohol and body weight on cardiovascular disease [5–7]. However, these studies were based on a limited number of factors and used relatively crude indicators of cardiac health, such as categorical variables indicating disease state. In contrast, we exploit the large scale atlas to make use of more localised motion parameters and also examine a wider range of clinical information.

**Contributions:** Our scientific contributions in this paper are threefold. First, we present, for the first time, a large scale cardiac motion atlas built from the UK Biobank database. Second, we propose an automated quality control approach to ensure the robustness of the extracted motion descriptors. Third, we use the atlas to learn associations between descriptors of cardiac function and non-imaging clinical information. This automatic framework represents an ideal tool to further explore factors influencing cardiac health.

## 2    Materials

The UK Biobank data set contains multiple imaging and non-imaging information from more than half a million 40–69 year-olds. From this data set, only participants with MR imaging data were considered for analysis. Furthermore, participants with non-Caucasian ethnicity, known cardiovascular disease, respiratory disease, diabetes, hyperlipidaemia, haematological disease, renal disease, rheumatological disease, malignancy, symptoms of chest pain or dyspnoea were excluded.

MR imaging was carried out on a 1.5T scanner (Siemens Healthcare, Erlangen, Germany) [11]. Short-axis (SA) stacks covering the full heart, and two orthogonal long-axis (LA) planes (2-chamber (2Ch) and 4-chamber (4Ch) views) were available for each subject (TR/TE = 2.6/1.10ms, flip angle = 80°).

In-plane resolution of the SA stack and LA images was 1.8mm, with slice thickness of 8mm and 6mm for SA and LA respectively. 50 frames were acquired per cardiac cycle (temporal resolution $\approx$14–24 ms/frame). In addition, the following non-imaging parameters were used for the statistical analysis: age, smoking status (smoker/non-smoker), body mass index (BMI), body fat percentage (BFP), basal metabolic rate (BMR), hypertension and alcohol intake frequency (occasionally/regularly). Demographics of the analysed cohort are presented in Table 1.

**Table 1.** Baseline characteristics for the healthy cohort from the UK Biobank data set. All continuous values are reported as mean (SD), while categorical data are reported as number (percentage).

|  | Total |
| --- | --- |
| Number of participants | 6002 |
| Male gender (n(%)) | 3046 (50.8%) |
| Age(years) | 61 (8) |
| Body mass index (BMI) $(kg/m^2)$ | 26.4 (4.2) |
| Body fat percentage (BFP) (%) | 28.2 (7.7) |
| Basal metabolic rate (BMR) $(KJ)$ | 6650.1 (1324.9)) |
| Smokers (n(%)) | 1011 (38.2%) |
| Regular alcohol intake (n(%)) | 1407 (48.9%) |
| Hypertension (n(%)) | 2903 (48.4%) |
| Ejection fraction (EF)(%) | 59.2 (6.3%) |

## 3   Methods

In the following Sections we describe the estimation and analysis of the atlas-based motion descriptors. Details of the atlas formation procedure are reported in Sect. 3.1. Section 3.2 details the quality control (QC) strategies proposed. Section 3.3 describes the different dimensionality reduction algorithms used to characterise cardiac function. Section 3.4 presents the statistical methods used to identify relationships between the motion descriptors and non-motion clinical information.

### 3.1   Motion Atlas Formation

This section outlines the procedure used to build the motion atlas, which is based on the work published in [1] and [10].

**LV Geometry Definition.** A fully-convolutional network (FCN) with a 17 convolutional layer VGG-like architecture was used for automatic segmentation of the LV myocardium and blood-pool at end-diastole (ED) for SA and LA slices

[2,14]. Each convolutional layer of the network is followed by batch normalisation and ReLU, except the last one, which is followed by the softmax function. In the case of the SA stack, each slice is segmented independently, i.e. in 2D. From the segmentations, a bounding box was generated and used to crop the image to only include the desired field of view, improving pipeline speed and reducing errors in motion tracking. The MR SA and LA segmentations were used to correct for breath-hold induced motion artefacts using the iterative registration algorithm proposed in [14]. The motion-corrected LA/SA segmentations were fused to form a 3D smooth myocardial mask. An open-source statistical shape model (SSM) was optimised to fit to this LV binary mask following an initial rigid registration using anatomical landmarks [1]. The use of landmarks ensured that the LV mesh was aligned to the same anatomical features for each patient, and the endocardial and epicardial surfaces of the mesh were smoothly fitted to the myocardial mask, providing surface meshes with point correspondence for each of the subjects and a 17 AHA regions segmentation.

**Motion Tracking.** A 3D GPU-based B-spline free-form deformation (FFD) was used [13] to estimate LV motion between consecutive frames. Subsequently, the inter-frame transformations were estimated and then composed using a $3D + t$ B-spline to estimate a full cycle $3D + t$ transformation.

**Spatial Normalisation.** Similar to [10], we transform each mesh to an unbiased atlas coordinate system using a combination of Procrustes alignment and Thin Plate Spline transformation. We denote the transformation for each subject $n$ from its subject-specific coordinate system to the atlas by $\phi_n$.

**Medial Surface Generation.** A medial surface mesh with regularly sampled vertices ($\approx 1000$) was generated from the SSM epicardial and endocardial meshes using ray-casting and homogeneous downsampling followed by cell subdivision.

**Motion Reorientation.** The displacements ($\boldsymbol{u}_n$) for each subject $n$ were reoriented into the atlas coordinate system under a small deformation assumption using a push-forward action: $\boldsymbol{u}_n^{atlas} = \boldsymbol{J}(\phi_n(\boldsymbol{r}_n))\boldsymbol{u}_n$ [10], where $\boldsymbol{J}(\cdot)$ refers to the Jacobian matrix, and $\boldsymbol{r}_n$ are the positions of the vertices in the ED frame.

**Transform to Local Coordinate System.** For each subject $n$ the displacements in the atlas coordinate system $\boldsymbol{u}_n^{atlas}$ were projected onto a local cylindrical coordinate system $\boldsymbol{x}_n^{atlas}$, providing radial, longitudinal, and circumferential information. The long axis of the LV ED atlas medial mesh was used as the longitudinal direction.

## 3.2   Quality Control

To ensure that the atlas was formed only from subjects with high quality cine MR images and robust motion descriptors, two QC methods were implemented to automatically reject subjects with insufficient quality or incorrect segmentations and/or motion tracking. These two methods are:

**QC1.** SA images with insufficient image quality were excluded using two deep learning based approaches. First the 3D convolutional neural network (CNN) described in [8] was used to automatically identify images with motion-related artefacts such as mistriggering, arrhythmia and breathing artefacts. Second, subjects with incorrectly planned 4Ch LA images were automatically excluded using the CNN described in [9]. This pipeline identified images containing the left ventricular outflow tract, which is a common feature of poorly planned 4Ch images. A total of 806 subjects were rejected using the QC1 step.

**QC2.** Our second QC measure aimed to detect segmentation and/or motion tracking errors. LV volume curves over the cardiac cycle were computed based on the LV geometry and motion fields. Based on these curves, an automatic pipeline was used to identify curves with unrealistic properties. The LV curve was divided into diastole and systole based on the ED and end-systolic (ED) points, where ED corresponds to the point with maximum volume, and ES corresponds to the point with minimum volume. These two curves were treated independently. In the systole part of the curve, the downslope should be smooth and continuous and therefore the first derivative should only contain one peak. Cases with more than one (automatically identified using the first and the second derivative of the LV curve) peak were therefore excluded. In the diastole part the curves were divided into rapid inflow, diastasis and atrial systole phases. In the rapid inflow phase the upslope should be smooth and continuous and the first derivative should only contain one peak. Cases with more than one peak were therefore excluded. In the diastasis phase the curve should be flat and with no variation, so the first and second derivatives were used to exclude any cases with changes of volume higher than 10%. Finally, in the atrial systole phase the upslope should be smooth and continuous and the first derivative should only contain one peak. Cases with more than one peak were therefore excluded. The QC2 step excluded a total of 64 cases.

### 3.3   Dimensionality Reduction

In this work, we propose to learn descriptors of cardiac function from the motion atlas using dimensionality reduction techniques. More specifically, three techniques were compared: Principal Component Analysis (PCA) [3], Locally Linear Embedding (LLE) [12], and stacked denoising autoencoders (SDA) [15]. To form the inputs to these three algorithms, the local displacements per AHA region were averaged and these were concatenated into a column vector such that for subject $n$, $x^{atlas} \in \mathbb{R}^L$, where $L = 3AT$, $T$ is the number of cardiac phases and $A$ is the number of AHA regions (17) in the atlas medial surface mesh. The column vectors for each subject were combined to produce a matrix $X = [x_1^{atlas}, ..., x_N^{atlas}]$, where $N$ is the number of subjects. The matrix $X$ is the input for all of the proposed dimensionality reduction techniques, and the output of all of them is a matrix $D \in \mathbb{R}^{(d \times N)}$, where $d$ is the size of the low dimensional space.

## 3.4  Statistical Analysis

Multivariate statistics were used to find relationships between the low-dimensional motion descriptors and non-imaging clinical information. For categorical variables (i.e. smoking status, hypertension and alcohol intake) a multivariate Hotelling's $T^2$ test (assuming unequal variance) [4] was used to test for differences between groups, and the $p$-value was recorded. For continuous variables (age, BMI, BFP, BMR) a multivariate regression was used to compute the $R$ statistic that measures the degree of relationship between the data. The $R$ statistic and associated $p$-values were recorded. All the recorded $p$-values were adjusted using Bonferroni's correction for multiple tests.

Furthermore, we analysed the impact of using a cardiac motion atlas compared to standard clinical global measures such as ejection fraction (EF), the current most used clinical global measure of cardiac function. To this end, the EF for each subject was computed from the tracked meshes. For categorical variables, similar to the previous case, a paired $t$-test (assuming unequal variance) was used to test for differences between groups, and the $p$-value was recorded. For continuous variables a linear regression was used to compute the $R$ statistic and the associated $p$-value. As before, all the recorded $p$-values were adjusted using Bonferroni's correction for multiple tests.

# 4  Experiments and Results

For all of the proposed dimensionality reduction algorithms the number of dimensions $d$ retained was optimised. In the case of PCA, $d$ was selected to retain 99% of the variance; in the case of LLE a grid search algorithm was used to vary $d$ between 2 and 256 and the number of neighbours $s$ was fixed to 10. Then, $d$ and $s$ were selected to minimise the reconstruction error $\epsilon$. Finally, SDA was applied with 5 layers with size $[L, 2000, 1000, 500, d]$, a learning rate of 0.001, a corruption level of 50%, and number of epochs 500. The value of $d$ was optimised using a grid search between 2 and 256, and $d$ was selected to minimise the reconstruction error $\epsilon$. For the three dimensionality reduction algorithms, the optimised $d$ were: $d_{PCA} = 932$; $d_{LLE} = 32$ and $\epsilon_{LLE} = 0.37\,\mathrm{mm}$; $d_{SDA} = 50$ and $\epsilon_{SDE} = 0.15\,\mathrm{mm}$.

Table 2 shows the computed $p$-values with Bonferroni's correction and $R$ statistics for the different clinical data for the different dimensionality reduction algorithms, and also using only EF as a motion descriptor. Bold numbers show $p$-values less than 0.05 (i.e. that showed statistical significance at 95% confidence). The results show that for all of the dimensionality reduction methods there was no statistical significance for age and BMI, while BFP, BMR, smoking status, hypertension and alcohol consumption are shown to be associated with cardiac function. Furthermore, the R statistics for BFP and BMR for all of the non-linear dimensionality reduction algorithms are around 60%, which indicates a clear linear relationship with the data. Using only EF we can see that there is no statistical significance with any of the clinical information, and the $R$ values are lower compared to the use of the cardiac motion atlas. Non-linear dimensionality reduction algorithms, i.e. LLE and SDA show lower $p$-values and higher $R$ statistics compared to linear PCA for all of the clinical information analysed. For BFP, BMR, smoking status, hypertension and alcohol consumption, LLE shows a better linear fitting, with slightly higher $R$ statistics compared to SDA. However, the difference between the two methods is not significant.

**Table 2.** Association of clinical information with estimated motion descriptors from the atlas represented by computed $p$-values and $R$ statistics using multivariate statistics. The $p$-values and $R$ statistics for EF were computed using paired t-tests and linear regression respectively. Bold numbers show $p$-values after Bonferroni's correction less than 0.05, which are considered to be statistically significant at 95% confidence.

|  | $R$ statistics (%) | | | | $p$-value | | | |
|---|---|---|---|---|---|---|---|---|
|  | PCA | LLE | SDA | EF | PCA | LLE | SDA | EF |
| Age | 48.45 | 49.03 | 56.11 | 39.67 | 0.31 | 0.29 | 0.28 | 0.38 |
| BMI | 45.19 | 57.47 | 59.83 | 25.87 | 0.20 | 0.16 | 0.18 | 0.31 |
| BFP | 48.22 | 64.20 | 64.63 | 23.98 | **0.05** | **0.03** | **0.05** | 0.68 |
| BMR | 55.67 | 64.60 | 55.84 | 21.96 | **0.05** | **0.01** | **0.04** | 0.21 |
| Smoking |  |  |  |  | **0.05** | **0.02** | **0.04** | 0.56 |
| Alcohol |  |  |  |  | **0.05** | **0.03** | **0.02** | 0.62 |
| Hypertension |  |  |  |  | **0.03** | **0.01** | **0.03** | 0.15 |

## 5  Discussion

In this work, we have demonstrated the derivation of motion atlas-based descriptors of cardiac function and investigated their association with a range of clinical information. This is the first time that a cardiac motion atlas has been formed from a large number of subjects from the UK Biobank, and this has enabled us to produce a framework for investigating factors influencing cardiac health. We show that our method was able to identify factors known to impact cardiac health (directly or indirectly), such as hypertension, smoking, alcohol intake and body composition. Contrary to previous smaller studies, age was not related to large changes in cardiac function. We were unable to assess whether the apparent lack of association was influenced by a selection bias due to anticipated better survival of healthy versus relatively unhealthy individuals, the relatively small spread in ages (45–70) compared to other studies, the clustering of most participants in the middle age range (55–65), or was a consequence of the dimensionality reduction algorithms. We will investigate these possibilities in future work.

In addition, we demonstrated that the use of a cardiac motion atlas has added value over using only global indicators such as EF. The proposed method represents an important contribution to furthering our understanding of the influences on cardiac function. We used the selected categories as an illustration of the capabilities of the atlas method. However, the univariate analysis that we performed does not capture the complex interrelation of health factors, age and cardiac function. Future work will focus on a deeper analysis of the impact of these parameters and will also include further clinical characteristics. In this work, displacements were used as a representation of cardiac motion. However, other representations such as velocity and strain can characterise motion in different ways. Our future work also includes extension of the motion atlas to incorporate strain parameters. In conclusion, this work represents an important contribution to furthering our understanding of the influences on cardiac function and opens up the possibility of learning the relationships between motion descriptors and further non-imaging information.

**Acknowledgements.** This work is funded by the Kings College London & Imperial College London EPSRC Centre for Doctoral Training in Medical Imaging (EP/L015226/1) and supported by the Wellcome EPSRC Centre for Medical Engineering at Kings College London (WT 203148/Z/16/Z). This research has been conducted using the UK Biobank Resource under Application Number 17806.

# References

1. Bai, W., Shi, W., et al.: A bi-ventricular cardiac atlas built from 1000+ high resolution MR images of healthy subjects and an analysis of shape and motion. Med. Image Anal. **26**(1), 133–145 (2015)
2. Bai, W., et al.: Semi-supervised learning for network-based cardiac MR image segmentation. In: Descoteaux, M., Maier-Hein, L., Franz, A., Jannin, P., Collins, D.L., Duchesne, S. (eds.) MICCAI 2017. LNCS, vol. 10434, pp. 253–260. Springer, Cham (2017). https://doi.org/10.1007/978-3-319-66185-8_29
3. Hotelling, H.: Analysis of a complex of statistical variables into principal components. J. Educ. Psychol. **24**(6), 417 (1933)
4. Hotelling, H.: The generalization of student's ratio. In: Kotz, S., Johnson, N.L. (eds.) Breakthroughs in Statistics. SSS, pp. 54–65. Springer, New York (1992). https://doi.org/10.1007/978-1-4612-0919-5_4
5. Jitnarin, N., Kosulwat, V., Rojroongwasinkul, N., et al.: The relationship between smoking, body weight, body mass index, and dietary intake among Thai adults: results of the National Thai Food Consumption Survey. Asia Pac. J. Public Health **26**(5), 481–493 (2014)
6. Mukamal, K., Chiuve, S., Rimm, E.: Alcohol consumption and risk for coronary heart disease in men with healthy lifestyles. Arch. Intern. Med. **166**(19), 2145–2150 (2006)
7. Nadruz, W., Claggett, B., Gonçalves, A., et al.: Smoking and cardiac structure and function in the elderlyclinical perspective: the ARIC study (atherosclerosis risk in communities). Circ. Cardiovasc. Imaging **9**(9), e004950 (2016)

8. Oksuz, I., et al.: Deep learning using K-space based data augmentation for automated cardiac MR motion artefact detection. In: Frangi, A.F., Schnabel, J.A., Davatzikos, C., Alberola-López, C., Fichtinger, G. (eds.) MICCAI 2018. LNCS, vol. 11070, pp. 250–258. Springer, Cham (2018). https://doi.org/10.1007/978-3-030-00928-1_29

9. Oksuz, I., Ruijsink, B., Puyol-Antón, E., et al.: Automatic left ventricular outflow tract classification for accurate cardiac MR planning, pp. 462–465 (2018)

10. Peressutti, D., Sinclair, M., et al.: A framework for combining a motion atlas with non-motion information to learn clinically useful biomarkers: application to cardiac resynchronisation therapy response prediction. Med. Image Anal. **35**, 669–684 (2017)

11. Petersen, S., Matthews, P., Francis, J., et al.: UK biobank's cardiovascular magnetic resonance protocol. J. Cardiovasc. Magn. Reson. **18**(1), 8 (2016)

12. Roweis, S., Saul, L.: Nonlinear dimensionality reduction by locally linear embedding. Science **290**(5500), 2323–2326 (2000)

13. Rueckert, D., Sonoda, L., et al.: Nonrigid registration using free-form deformations: application to breast MR images. IEEE Trans. Med. Imag. **18**(8), 712–721 (1999)

14. Sinclair, M., Bai, W., Puyol-Antón, E., Oktay, O., Rueckert, D., King, A.P.: Fully automated segmentation-based respiratory motion correction of multiplanar cardiac magnetic resonance images for large-scale datasets. In: Descoteaux, M., Maier-Hein, L., Franz, A., Jannin, P., Collins, D.L., Duchesne, S. (eds.) MICCAI 2017. LNCS, vol. 10434, pp. 332–340. Springer, Cham (2017). https://doi.org/10.1007/978-3-319-66185-8_38

15. Vincent, P., Larochelle, H., Lajoie, I., Bengio, Y., Manzagol, P.A.: Stacked denoising autoencoders: learning useful representations in a deep network with a local denoising criterion. J. Mach. Learn. Res. **11**, 3371–3408 (2010)

# Computational Modelling of Electro-Mechanical Coupling in the Atria and Its Changes During Atrial Fibrillation

Sofia Monaci$^{(\boxtimes)}$, David Nordsletten, and Oleg Aslanidi

School of Biomedical Engineering and Imaging Sciences,
King's College London, London, UK
sofia.monaci@kcl.ac.uk

**Abstract.** Atrial fibrillation (AF) is a very common, multifaceted disease that affects atrial structure as well as electrophysiological and biomechanical function. However, the mechanistic links between structural and functional factors underlying AF are poorly understood. To explore these mechanisms, a 3D atrial electro-mechanical (EM) model was developed that includes 3D atrial geometry based on the Visible Female dataset, rule-based fibre orientation, CRN human atrial electrophysiology model and activation-based mechanical contraction model. Electrical activation in the 3D atria was simulated under control condition and two AF scenarios: sinus rhythm (SR), functional re-entry in the right atrium (RA) and structural re-entry around fibrotic patches in the left atrium (LA). Fibrosis distributions were obtained from patient LGE MRI data. In both AF scenarios, re-entrant behaviours led to substantial reductions in the displacement at peak contraction compared to SR. Specifically, high-frequency re-entry led to a decrease in maximal displacement from 6.8 to 6.1 mm in the posterior RA, and a larger decrease from 7.8 to 4.5 mm in the LA in the presence of fibrotic patches. The simulated displacement values agreed with available clinical data. In conclusion, the novel model of EM coupling in the 3D human atria provided new insights into the mechanistic links between atrial electrics and mechanics during normal activation and re-entry sustaining AF. Re-entry in the RA and LA resulted in weaker contractions compared to SR, with additional effect of fibrosis on the atrial wall stiffness further reducing the contraction.

**Keywords:** Atrial fibrillation · Electro-mechanical coupling · Sinus rhythm · Re-entry · Fibrosis · Mechanical contraction

## 1 Introduction

Atrial Fibrillation (AF), one of the most common forms of arrhythmia [1], is characterized by a significant increase of the atrial rate (up to 400–600 beats per minute) and electrical desynchronization, which causes the atria to quiver instead of contracting. In a healthy heart, the electrical activity of the atria is driven by action potentials (APs) generated in the Sino-atrial node (SAN), which then spread as waves through all chambers of the heart. This provides basis for electrical coupling and synchronization

M. Pop et al. (Eds.): STACOM 2018, LNCS 11395, pp. 103–113, 2019.
https://doi.org/10.1007/978-3-030-12029-0_12

of the myocardium and allows various regions of the heart, from the atria to the ventricles, to inflate and contract as a unit. However, during AF, the desynchronization of electrical activity causes asynchronous contraction of cardiac fibers and a reduction in the atrial mechanical work, affecting blood flow and cardiac output. In the long term, this can predispose to stroke, heart failure and sudden death, increasing the rates of mortality and morbidity associated with AF [2, 3].

The complexity of AF and the multiple challenges in understanding and treating this self-sustained and treatment-resistant condition warrant the use of multi-disciplinary and multi-physics approaches to elucidate its complex mechanisms. Over the years, computational modelling has led to important insights into the mechanisms of AF, primarily focusing on the electrophysiological (EP) aspects. Thus, atrial models facilitated the understanding of the main drivers of AF, high-frequency re-entrant electrical waves called rotors [4], and the mechanisms of their initiation and sustenance. Despite of the importance of the excitation-contraction coupling for an efficient pumping and mixing of the blood, very few studies have considered electro-mechanical processes during AF. Understanding how the presence of rotors and underlying structural and functional factors affect atrial mechanics is crucial for dissecting the mechanistic links between the EP and mechanical atrial functions, and can provide biomarkers for the quantification of AF state. Specific factors that can influence atrial EP function, as well as atrial stiffness and contractions, include major effects of fibrotic tissue on the rotor behaviour – which have been characterized recently [5, 6].

This study aims to couple electrical and mechanical processes using the 3D atrial models to understand how electrical drivers for AF influence contraction of atrial myocardium and how fibrotic patches can facilitate not only electrical but also mechanical dysfunction.

## 2   Methods

### 2.1   3D Atrial Datasets

The EP simulations were performed using the 3D atrial geometry reconstructed from the Visible Female Dataset (Fig. 1a). The resolution of the latter images is $0.3 \times 0.3 \times 0.3$ mm$^3$. In addition, detailed orientation of atrial fibers was included into the 3D models, as illustrated in Fig. 1b. Such orientation has been previously introduced using a rule-based approach [7]. Fibrosis distributions were derived from late gadolinium enhanced magnetic resonance images (LGE MRI); such a distribution can be seen in Fig. 1c. Fibrosis was segmented based on the image intensity ratio (IIR), computed by dividing individual voxel intensities by the mean blood pool intensity. Transition from diffuse to dense fibrosis was represented by labelling segmented fibrotic tissue from 1 to 5 according to LGE MRI intensity, where level 5 corresponded to dense fibrosis (IIR > 1.32) and level 1–4 corresponded to variable degree of diffuse fibrosis (1.08 < IIR < 1.32 split into four equal intervals). Fibrotic patches were then mapped into the 3D atrial geometry [5] to evaluate their role in the genesis of AF.

Mechanical simulations were performed on a tetrahedral 3D mesh with 394598 elements, derived from the 3D atrial geometry (Fig. 1d). Such a mesh was generated by using iso2mesh toolbox in Matlab [8]. The average edge length was 3.8 mm. All the relevant data, including fiber orientation and fibrosis (translating into EP and mechanical tissue properties, see below), as well as simulated activation times, were mapped onto the mesh.

## 2.2   Modelling Atrial Electrics and Mechanics

EM coupling simulations were performed by solving the two problems independently, EP and then mechanics, using the models described below. This involved simulating AP propagation in the 3D atrial geometry first, so that the atrial activation sequences could then be mapped on the respective 3D mesh and used to simulate the resulting mechanics.

Atrial AP propagation was modelled using the standard reaction-diffusion equation:

$$\frac{\partial V}{\partial t} = \nabla \cdot (D\nabla V) + \frac{I_{tot}}{C_m} \tag{1}$$

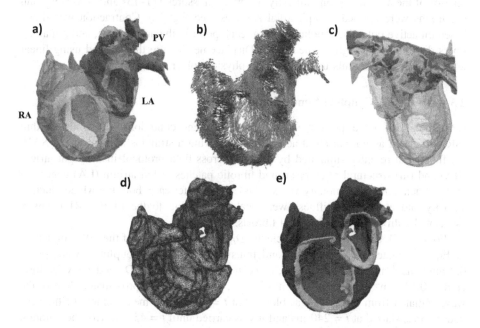

**Fig. 1.** 3D atrial datasets used for EP and mechanical simulations. (a) 3D atrial geometry; right atrium (RA), left atrium (LA) and pulmonary veins (PVs) are shown; (b) atrial fibre orientation; (c) distribution of fibrosis in the 3D atria. Level of fibrosis in the LA gradually changes from high (dark red) to low (dark blue); the rest is healthy tissue (grey); (d) 3D atrial mesh used for mechanics simulations; (e) mesh surface boundaries, derived from the segmentation shown in (a). Red: Endocardium, Blue: Epicardium, Yellow: Atrio-Ventricular Rings, Light Blue: Superior and Inferior Vena Cava, Green: PVs. (Color figure online)

Here $V$ represents the cell membrane potential, $I_{tot}$ is the total membrane ionic current and $C_m$ is the membrane capacitance. Courtemanche-Ramirez-Nattel (CRN) model was used to describe all the main currents characterizing atrial myocytes [9]. The CRN model was solved numerically using forward Euler method on the 3D human atrial geometry with zero-flux boundary conditions. The spatial and temporal steps were 0.3 mm and 0.005 ms, respectively.

Atrial mechanics was modelled using a reduced, nearly-incompressible version of Holzapfel-Ogden (HO) formulation [10] developed previously to describe the Cauchy stress tensor in the ventricular myocardium:

$$\boldsymbol{\sigma} = ae^{b(I_1-3)}\mathbf{B} + KJ(J-1)\mathbf{I} + 2a_f(I_{4f}-1)e^{b_f(I_{4f}-1)^2}(\boldsymbol{f}\otimes\boldsymbol{f}) + T_a(t)(\boldsymbol{f}\otimes\boldsymbol{f}) \qquad (2)$$

Here $I_1$ and $I_{4f}$ (first and fourth invariant of the Right Cauchy Green strain $\mathbf{C}$) describe the non-muscular matrix surrounding the fibres and the fibre direction respectively; $f$ represents the fibre field and $T_a(t)$ the active component, dependent on activation times and fiber orientation.

The quasi-static conservation of momentum ($\nabla \cdot \boldsymbol{\sigma} = 0$) was solved throughout the cardiac cycle. Zero Dirichlet conditions were set on the valve rings and inlet veins feeding both atrial chambers. Pressure boundary conditions were imposed on the interior of the atria based on clinically observed pressures [11–13] and zero Neumann conditions were applied on epicardial surfaces (see Fig. 1e). Contraction was timed based on active potential depolarization at each point in the atrial tissue, and ran along fibers distributed throughout the model. Displacements were then solved using linear tetrahedral finite elements using the multi-physics solver, CHeart [14].

## 2.3    Atrial EM Coupling Simulations

EP simulations were performed for three different conditions: (i) sinus rhythm (SR) simulated as normal atrial activations following a stimulus applied in the SAN, (ii) functional re-entry simulated by applying cross-field protocol in the right atrium (RA) and (iii) structural re-entry around fibrotic patches in left atrium (LA) generated by fast-pacing at the pulmonary veins (PVs). In the latter case, both the AP conduction velocity and atrial wall stiffness were scaled according to the LGE MRI intensity associated with the accumulation of fibrosis.

For case (i), two diffusion coefficients, $D_L$ and $D_T$ (elements of the diffusion tensor in Eq. (1), related to longitudinal and transverse cell-to-cell coupling), were set to 0.3 mm$^2$ ms$^{-1}$ and 0.03 mm$^2$ ms$^{-1}$, respectively. For case (ii), $D_L$ and $D_T$ were lowered to 0.12 mm$^2$ ms$^{-1}$ and 0.012 mm$^2$ ms$^{-1}$, respectively. Approximately half of the wave initiated from the SAN was blocked at $t = 50$ ms and measurement of the activation times started at $t = 240$ ms and it was carried until $t = 450$ ms (for the duration of the whole period of re-entry). Activation times were calculated according to the steepest rate of AP change (upstroke) in the cell.

For case (iii), re-entry was facilitated by rescaling the diffusion coefficients according to the level of fibrosis [5]: level 0 corresponded to healthy tissue and $D_L = 0.3$ mm$^2$ ms$^{-1}$; inside the fibrotic patch, the value of $D_L$ for levels 1 to 5 was

$0.25$ mm$^2$ ms$^{-1}$, $0.20$ mm$^2$ ms$^{-1}$, $0.15$ mm$^2$ ms$^{-1}$, $0.10$ mm$^2$ ms$^{-1}$ and $0.05$ mm$^2$ ms$^{-1}$, respectively. We did not used the value $D_L = 0$ for level 5, because there is no experimental evidence suggesting that dense fibrotic regions are completely non-conductive. The initial S1 stimulus was applied to the left posterior PV, and it was followed by a second S2 at $t = 147$ ms to initiate a rotor around the fibrotic patches.

As for the mechanical simulations, the main parameters were set to be $a = 650$ Pa, $b = 5$, $a_f = 3260$ Pa and $b_f = 5$ for all conditions by adjusting computational values used for mechanical simulations of the ventricular tissue [15]; for case (iii), $T_a(t)$ was scaled according to the LGE MRI intensity associated with the fibrotic patches (levels 1–5, see above). The reference value of $T_a(t)$ was calculated from Laplace's Law, $T_a(t) = \frac{PR}{2h}$, which approximates the geometry to a thin-walled sphere, simplifying calculation of the tangential stress acting of the walls of such a geometry. In this study, values of the atrial radius $R$ and myocardial thickness $h$ were taken from Pironet et al. [11]: $R = 30$ mm and $h = 2$ mm, respectively.

CHeart solver [14] was used to perform all mechanical simulations. These were run on a local cluster (32 core with AMD Opteron 6128 processors, 125 Gb RAM) using 20 cores. The EP model was developed in C++ and ran on a local High Performance Computing cluster (SGI UV, 640 cores with Intel Xeon 8837 processors, 5 Tb RAM) using 128 cores. The duration of EP simulations was about 2 hours per 1 second of activity, whereas mechanical contractions were simulated in about 4 h per atrial contraction cycle.

## 3   Results

3D atrial model was simulated for three cases: (i) normal atrial activation in SR (Fig. 2a), (ii) functional re-entry in the RA (Fig. 2b) and (iii) structural re-entry in the LA in the presence of fibrotic patches (Fig. 2c). Activations in the three scenarios showed different patterns in SR and AF, with excitation waves in cases (ii) and (iii) forming rotors that persisted for several periods. The activation sequences corresponding to one period of the rotor can be seen in Fig. 2b and c, respectively, with the RA activated before the LA in the former, and the LA activated before the RA in the latter. The rotation period was around 210 ms for case (ii) and around 180 ms for case (iii). In the control condition (i), the total atrial activation time was 130 ms, with the latest activation in the RA occurring after 92 ms and the latest activation in the LA after 120 ms, as summarised in Table 1. These results are in agreement with the experimental observations by Lemery et al. [16].

Mechanical contractions in the three cases considered were also substantially different. Figure 3 shows the differences in atrial displacements between control condition (i) and case (ii) at t = 250 ms, and Fig. 4 shows the respective difference in the displacements between control condition (i) and case (iii) at t = 300 ms. The maximum relative displacements are mainly located in the RA for case (ii) and in LA for case (iii): the respective values are −5.5 mm and −5.9 mm, both showing substantially reduced contractions. Such maxima occur at around t = 250 ms in case (ii) and around t = 300 ms in case (iii). Figure 5 shows changes of the absolute displacement over the period of contraction for each of cases (i), (ii) and (iii), as well as regional differences in the magnitudes of the displacement between the LA and RA.

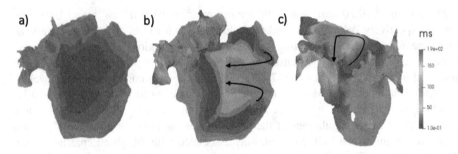

**Fig. 2.** Activation sequences in the 3D atrial model simulated in SR and different AF scenarios. (a) control case (i); (b) and (c) re-entry in the RA and LA in case (ii) and (iii), respectively. Atrial activation times are shown using the rainbow palette, and the black arrows indicate the trajectories of the rotors.

**Table 1.** Activation times in the RA and LA for control condition and periods of re-entry during AF. The results for the control condition are compared here with the experimental results obtained by Lemery et al., whereas re-entry has been validated against clinical studies by Haïssaguerre et al. [17].

|  | Simulated activation time | Experiment (Lemery et al. [8]) |
|---|---|---|
| Latest RA (case (i)) | 95 ms | 97 ± 17 ms |
| Latest LA (case (i)) | 120 ms | 116 ± 18 ms |
|  | Simulated rotor period | Experiment (Haïssaguerre et al. [17]) |
| Structural re-entry (case (ii)) | 210 ms | 230 ms |
| Functional re-entry (case (iii)) | 180 ms | 175 ± 30 ms |

The major differences in contraction during re-entry for case (ii) and (iii) were seen in the RA near SAN and in the LA near fibrotic patches, respectively; these were the locations where the rotors were initiated and sustained their rotation (Figs. 3 and 4).

The displacements at peak contraction, calculated with respect to the baseline values during the maximum expansion, are shown in Table 2 and compared with experimental values obtained by Quintana et al. [18] in healthy subjects. The values of displacements in the RA agreed with the experimental values. In the LA, the simulated results were significantly higher than experimental values [9] for case (i) and (ii). However, another study [19] reported substantially higher displacement values in the LA during contraction.

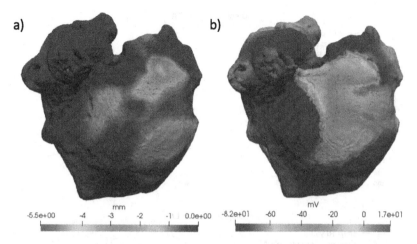

**Fig. 3.** (a) Difference in displacement between control condition (i) and case (ii) at t = 250 ms. The most affected area was the RA, where the rotor was initiated and perpetuated (maximum reduction in contraction up to −5.5 mm). However, contraction of parts of LA was also affected substantially. (b) shows the 3D voltage distribution at the beginning of one period of the rotor.

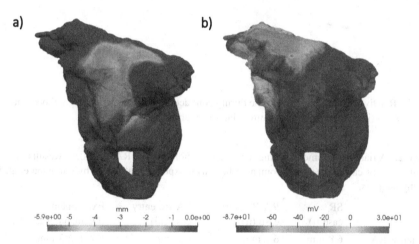

**Fig. 4.** (a) Difference in displacement between control condition (i) and case (iii) at t = 300 ms (maximum reduction in contraction up to −5.9 mm); major differences can be seen to be around the fibrotic patches. (b) shows the 3D voltage distribution half way through the period of the rotor.

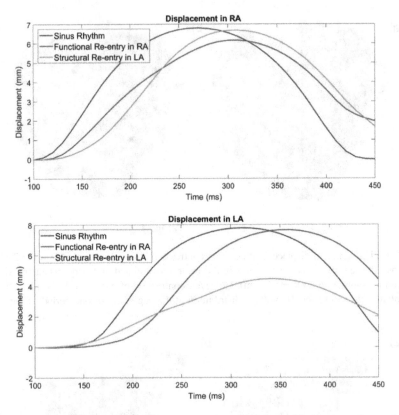

**Fig. 5.** Relative displacement over time during contraction in the posterior RA and posterior LA for cases (i)–(iii). Reference: maximum value of inflation.

**Table 2.** Atrial displacement during peak contraction in the RA and LA. Results of EM simulations for cases (i)–(iii) are compared here with experimental values by Quintana et al. for healthy subjects.

|              | SR<br>case (i) | RA Re-entry<br>case (ii) | LA Re-entry<br>case (iii) | Experiment<br>(Quintana et al. [18]) |
|--------------|----------------|--------------------------|---------------------------|--------------------------------------|
| Posterior RA | 6.8 mm         | 6.1 mm                   | 6.7 mm                    | 6.7 ± 2.3 mm                         |
| Posterior LA | 7.8 mm         | 7.6 mm                   | 4.5 mm                    | 4.0 ± 0.9 mm                         |

## 4   Conclusion

In this study, a novel coupled EM model of both atria was developed and applied to investigate changes in mechanical contraction from normal to abnormal electrical activity. The modelling enabled (1) simulation of mechanical contraction in the 3D atria following normal SR, with both electrophysiological and mechanistic results validated by experimental values; (2) simulation of EM coupling in the 3D atria during AF in two different scenarios (functional rotor in the RA and structural re-entry in the LA due to fibrotic tissue) and (3) identification of the differences in maximal atrial displacements between AF scenarios with respect to the normal activation: 6.1 mm (functional re-entry) against 6.8 mm (SR) in the RA near the SAN, and 4.5 mm (structural re-entry) against 7.8 mm (SR) in the LA near fibrotic patches. Hence, modelling showed that the weakest contraction occurred in the LA in presence of fibrosis.

Validation of EP results was carried by comparing activation sequences derived from the simulations with the respective experimental values obtained previously [10, 11].

Validation and interpretation of mechanical results were more challenging due to the smaller number of existing studies. Hence, our analysis focused on changes in atrial displacement during contraction in the posterior LA and RA, validating the control simulation against results of an experimental study by Quintana et al. [18], and then focusing on the difference between normal and abnormal contractions.

Novelty of this EM model lies in the introduction of fibrotic distribution and analysis of its impact on the EM coupling. A previous atrial EM model by Adeniran et al. [20] focused on the analysis of AF-induced electrical remodelling leading to impaired atrial mechanics without, however, emphasising the structural remodelling caused by fibrosis. Our study not only accounts for such structural remodelling effects, but also suggests that greatest decrease in peak displacement can be used as a bio-marker to identify atrial regions where functional and structural re-entry can be located. In the future, regional peak displacements could be further analysed to produce simple binary indices for the stratification of AF patients.

Future EM models could improve on some limitations of this work by considering (1) additional effects of AF-induced remodelling [21]; (2) AF scenarios based on re-entry around the PVs [22]; (3) patient-specific MR imaging data [23] to enable clinical translation of the results and (4) other mechanical properties of the myocardium together with atrial displacements (such as strain and volume profiles). Future research could also include the mechano-electric feedback, integrate calcium dynamics to the excitation-contraction coupling model and incorporate patient-specific mechanical properties and hemodynamic effects, ultimately creating truly multi-physics models of the human atria.

**Acknowledgements.** This work was supported by the Engineering and Physical Sciences Research Council (EPSRC) [EP/L015226/1], the British Heart Foundation [PG/15/8/31138] and the Wellcome/EPSRC Centre for Medical Engineering [WT 203148/Z/16/Z].

# References

1. Nattel, S.: New ideas about atrial fibrillation 50 years on. Nature **415**, 219–226 (2002)
2. Camm, A.J., Kirchhof, P., Lip, G.Y.H., Schotten, U., Savelieva, I., et al.: Guidelines for the management of atrial fibrillation. Eur. Heart J. **31**, 2369–2429 (2010)
3. Ball, J., Carrington, M.J., McMurray, J.J.V., Stewart, S.: Atrial fibrillation: profile and burden of an evolving epidemic in the 21st century. Int. J. Cardiol. **167**, 1807–1824 (2013)
4. Pandit, S.V., Jalife, J.: Rotors and the dynamics of cardiac fibrillation. Circ. Res. **112**, 849–862 (2013)
5. Morgan, R., Colman, M.A., Chubb, H., Seemann, G., Aslanidi, O.V.: Slow conduction in the border zones of patchy fibrosis stabilizes the drivers for atrial fibrillation: insights from multi-scale human atrial modeling. Front. Physiol. **7**, 474 (2016)
6. Zahid, S., et al.: Patient-derived models link re-entrant driver localization in atrial fibrillation to fibrosis spatial pattern. Cardiovasc. Res. **110**, 443–454 (2016)
7. Krueger, M.W., et al.: Modeling atrial fiber orientation in patient-specific geometries: a semi-automatic rule-based approach. In: Metaxas, D.N., Axel, L. (eds.) FIMH 2011. LNCS, vol. 6666, pp. 223–232. Springer, Heidelberg (2011). https://doi.org/10.1007/978-3-642-21028-0_28
8. Fang, Q., Boas, D.A.: Tetrahedral mesh generation from volumetric binary and gray-scale images. In: Proceedings Sixth IEEE International Conference Symposium Biomedical Imaging: From Nano to Macro, pp. 1142–1145 (2009)
9. Courtemanche, M., Ramirez, R.J., Nattel, S.: Ionic mechanisms underlying human atrial action potential properties: insights from a mathematical model Ionic mechanisms underlying human atrial action potential properties: insights from a mathematical model. Am. Physiol. Soc. **275**, H301–H321 (1998)
10. Holzapfel, G.A., Ogden, R.W.: Constitutive modelling of passive myocardium. Philos. Trans. R. **367**, 3445–3475 (2009)
11. Pironet, A., et al.: Simulation of left atrial function using a multi-scale model of the cardiovascular system. PLoS One **8**(6), e65146 (2013)
12. Leischik, R., et al.: Echocardiographic evaluation of left atrial mechanics: function, history, novel techniques, advantages, and pitfalls. Biomed. Res. Int. **2015**, 1–12 (2015)
13. Cambier, B.A., et al.: Influence of the breathing mode on the time course and amplitude of the cyclic inter-atrial pressure reversal in postoperative coronary bypass surgery patients. Eur. Heart J. **14**, 920–924 (1993)
14. Lee, J., et al.: Multiphysics computational modeling in CHeart. SIAM J. Sci. Comput. **38**, 150–178 (2016)
15. Issa, O., et al.: Left atrial size and heart failure hospitalization in patients with diastolic dysfunction and preserved ejection fraction. J. Cardiovasc. Echogr. **27**, 1 (2017)
16. Lemery, R., et al.: Normal atrial activation and voltage during sinus rhythm in the human heart: an endocardial and epicardial mapping study in patients with a history of atrial fibrillation. J. Cardiovasc. Electrophysiol. **18**, 402–408 (2007)
17. Haïssaguerre, M., et al.: Spontaneous initiation of atrial fibrillation by ectopic beats originating in the pulmonary veins. N. Engl. J. Med. **339**, 659–666 (1998)
18. Quintana, M., et al.: Assessment of atrial regional and global electromechanical function by tissue velocity echocardiography: a feasibility study on healthy individuals. Cardiovasc. Ultrasound. **3**, 1–12 (2005)

19. Patel, A.R., et al.: Cardiac cycle-dependent left atrial dynamics: implications for catheter ablation of atrial fibrillation. Hear. Rhythm. **5**, 787–793 (2008)
20. Adeniran, I., Maciver, D.H., Garratt, C.J., Ye, J., Hancox, J.C., Zhang, H.: Effects of persistent atrial fibrillation- induced electrical remodeling on atrial electro-mechanics - insights from a 3D model of the human atria. PLoS One **10** (2015)
21. Colman, M.A., et al.: Pro-arrhythmogenic effects of atrial fibrillation-induced electrical remodelling: insights from the three-dimensional virtual human atria. J. Physiol. **591**, 4249–4272 (2013)
22. Aslanidi, O.V., et al.: Heterogeneous and anisotropic integrative model of pulmonary veins: computational study of arrhythmogenic substrate for atrial fibrillation. Interface Focus. **3**, 20120069 (2013)
23. Varela, M., et al.: Novel MRI technique enables non-invasive measurement of atrial wall thickness. IEEE Trans. Med. Imaging. **36**, 1607–1614 (2017)

# High Throughput Computation of Reference Ranges of Biventricular Cardiac Function on the UK Biobank Population Cohort

Rahman Attar[1], Marco Pereañez[1], Ali Gooya[1], Xènia Albà[2], Le Zhang[1],
Stefan K. Piechnik[3], Stefan Neubauer[3], Steffen E. Petersen[4],
and Alejandro F. Frangi[1(✉)]

[1] Center for Computational Imaging and Simulation Technologies in Biomedicine,
Department of Electronic and Electrical Engineering,
The University of Sheffield, Sheffield, UK
a.frangi@leeds.ac.uk
[2] Center for Computational Imaging and Simulation Technologies in Biomedicine,
Universitat Pompeu Fabra, Barcelona, Spain
[3] Oxford Center for Clinical Magnetic Resonance Research (OCMR),
Division of Cardiovascular Medicine, University of Oxford,
John Radcliffe Hospital, Oxford, UK
[4] Cardiovascular Medicine at the William Harvey Research Institute,
Queen Mary University of London and Barts Heart Center,
Barts Health NHS Trust, London, UK

**Abstract.** The exploitation of large-scale population data has the potential to improve healthcare by discovering and understanding patterns and trends within this data. To enable high throughput analysis of cardiac imaging data automatically, a pipeline should comprise quality monitoring of the input images, segmentation of the cardiac structures, assessment of the segmentation quality, and parsing of cardiac functional indexes. We present a fully automatic, high throughput image parsing workflow for the analysis of cardiac MR images, and test its performance on the UK Biobank (UKB) cardiac dataset. The proposed pipeline is capable of performing end-to-end image processing including: data organisation, image quality assessment, shape model initialisation, segmentation, segmentation quality assessment, and functional parameter computation; all without any user interaction. To the best of our knowledge, this is the first paper tackling the fully automatic 3D analysis of the UKB population study, providing reference ranges for all key cardiovascular functional indexes, from both left and right ventricles of the heart. We tested our workflow on a reference cohort of 800 healthy subjects for which manual delineations, and reference functional indexes exist. Our results show statistically significant agreement between the manually obtained reference indexes, and those automatically computed using our framework.

© Springer Nature Switzerland AG 2019
M. Pop et al. (Eds.): STACOM 2018, LNCS 11395, pp. 114–121, 2019.
https://doi.org/10.1007/978-3-030-12029-0_13

# 1    Introduction

Cardiovascular diseases (CVDs) are recognised as the number one cause of death worldwide [1]. Diagnosis of cardiovascular disease is often made at late symptomatic stages, which leads to late interventions and decreased efficacy of medical care. Thus, mechanisms for early and reliable quantification of cardiac function is of utmost importance.

Analysis and interpretation of cardiac structural and functional indexes in large-scale population image data can reveal patterns and trends across population groups, and allow insights into risk factors before CVDs develop. UKB is one of the world's largest population-based prospective studies, established to investigate the determinants of disease.

In terms of population sample size, experimental setup, and quality control, the most reliable reference ranges for cardiovascular structure and function in adult caucasians aged 45–74 found in the literature are those reported in [2]. In [2], cardiovascular magnetic resonance (CMR) scans were manually delineated and analysed using cvi42 post-processing software (Version 5.1.1, Circle Cardiovascular Imaging Inc., Calgary, Canada). These reference values are used in this paper to validate the proposed workflow.

In this paper, we present a fully automatic 3D image parsing workflow with quality control modules to analyse CMR images in the UKB and corroborate their validity compared to their manual counterpart. The proposed workflow is capable of segmenting the cardiac ventricles and generating clinical reference ranges that are statistically comparable to those obtained by human observers. The main contribution of this paper is in its clinical impact, resulting from the analysis of left ventricle (LV) and right ventricle (RV) of the heart, as well as the extraction of key cardiac functional indexes from large CMR datasets.

# 2    Methods

Figure 1 shows the architecture of the proposed workflow addressing the issue of large-scale analysis of CMR images. It consists of eight main modules to analyse every single subject of the database. To create a modular workflow and enable processing multiple subjects in parallel, a workflow manager software package is required. This provides an infrastructure for the set-up, performance and monitoring of a defined sequence of tasks, regardless their programming language. In our implementation, the Nipype package [3] has been used. It allows us to combine a heterogeneous set of software packages within a single and highly efficient workflow, processing several subjects in parallel using cloud computing platforms provided by Amazon (high performance processors and S3 storage services).

## 2.1    Data Organisation (DO)

The Data Organisation (DO) module was developed to hierarchically organise image series from raw DICOM data. It is important to organise the data files

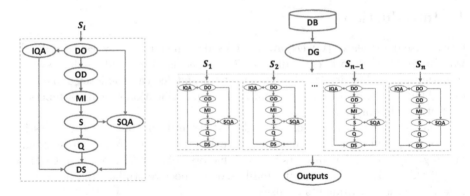

**Fig. 1.** The proposed fully automatic image parsing workflow for the analysis of cardiac ventricles in parallel. Left: The workflow includes the following modules: DO: Data Organisation, IQA: Image Quality Assessment, OD: Organ Detection, MI: Model Initialisation, S: Segmentation, SQA: Segmentation Quality Assessment, Q: Quantification, DS: Data Sink. Right: The quantitative functional analysis of a large database in parallel mode. DB: Database, DG: Data Grabber, n: number of subjects, and $S_i$: $i^{th}$ subject of the dataset.

to minimise redundancy and inconsistency. As a result, the organised data provides improved searchability and identification of contents. Clear, descriptive, and unique file names have been used to reflect the contents of the file, uniquely identify the data, and enable precise accessibility and data retrieval. Each subject's DICOM data are organised according to cardiac cycle phase, and into short axis (SAX) and long axis (LAX) views.

## 2.2   Image Quality Assessment (IQA)

Low image quality can not be fully avoided, particularly in large-scale imaging studies. To ensure that the quality of collected data is optimal for statistical analysis, having an IQA module is of paramount. This allows the automatic detection of abnormal images, whose analysis would otherwise impair the aggregated statistics over the cohort. Since the lack of basal and/or apical slices is the most common problem affecting image quality in CMR images, and has a major impact on the accuracy of quantitative parameters of cardiac function, our IQA module is designed to detect missing apical and basal slices of the CMR input. Thus, every top and bottom short axis view of input image volumes is analysed using two convolutional neural networks, each particularly trained for detection of missing slices in the basal and/or apical positions. The details of the architecture used can be found in [4].

## 2.3   Organ Detection (OD)

To segment the image, we use a Sparse Active Shape Models framework (SPASM) [5], which requires model initialisation. We achieve this automatically by extending the method proposed in [6] for LV initialisation to biventricular initialisation. In [6], the location of the LV is determined by a rough estimation of the intersection of slices from different views (SAX and LAX). Then, a Random Forest regressor trained with two complementary feature descriptors (i.e. the Histogram of Oriented Gradients and Gabor Filters) is used to predict the final landmark positions. This method is LV specific and therefore we have extended it to take into account image features corresponding to the RV, and obtain optimal initialisations for biventricular segmentation.

### 2.4 Model Initialization (MI)

The landmarks obtained in Sect. 2.3 are used (1) to suitably place the initial shape inside the image volume (translation), (2) to scale the initial shape along the main axis of the heart (scaling); and (3) to define the initial orientation of the heart based on the relative position of the mitral valve (rotation). These initial pose parameters are estimated by registering the obtained landmarks to their corresponding points on the mean model shape. As we segment all timepoints in the CINE sequence, we initialise the first image timepoint with the model mean, however, subsequent cardiac phases are initialised with the resulting segmentation from the previous timepoint.

### 2.5 Segmentation (S)

Cardiac LV and RV segmentation is performed with a modified 3D-SPASM segmentation method [5]. The main components of the 3D-SPASM are a Point Distribution Model (PDM), an Intensity Appearance Model (IAM), and a Model Matching Algorithm (MMA).

In this work, the PDM is a surface mesh representing the endocardial and epicardial surfaces for the LV and the endocardial surface for the RV. The PDM is built during training by applying Principal Component Analysis to a set of aligned shapes and maintaining eigenvectors corresponding to a predefined percentage of shape variability. The learned shape variability can be modeled as $\hat{x} = \bar{x} + \Phi b$ where $\hat{x}$ is a shape model instance, $\bar{x}$ is the mean shape, $\Phi$ is an eigenvector matrix and $b$ is a vector of scaling values for each principal component. By modifying $b$, we can generate shapes from the shape distribution.

The IAM is trained by learning the graylevel intensity distribution along perpendiculars to boundary points on the cardiac shape. An appearance mean and covariance matrix $\{\bar{g}, \Sigma_{gg}\}$ is computed for each landmark by sampling the intensity around each point over the image training set.

The last element of the segmentation process is the MMA, whose role is iterating between finding the optimal location of boundary points by distance minimisation between sampled image profiles and the IAM, and projection of these points onto the valid shape space defined by the PDM.

## 2.6    Segmentation Quality Assessment (SQA)

Due to varying image quality, image artefacts, or extreme anatomical variations found in large-scale studies it is essential to have a self-verification capabilities to automatically detect incorrect results, either to reprocess those images, or disregard them. This becomes even more important when automated segmentation methods are applied to large-scale datasets, and the segmentation results are to be used for further statistical population analysis [7]. In our pipeline we incorporate the SQA proposed in [6]. The SQA uses Random Forest classifiers trained on intensity features associated to blood pool and myocardium, and is able to detect successful segmentations.

## 2.7    Quantification (Q)

After successful SQA, we compute a thorough set of functional parameters based on blood-pool and myocardial volumes. To reproduce the reference ranges reported in [2], our quantification module performs volume computations using the Simpson's rule. The principle underlying this method is that total volume can be approximated by the summation of stacks of elliptical disks.

## 3    Experiments and Results

We use the same dataset exploited in [2], and evaluate the performance of the proposed automatic workflow in two ways: (1) applying common metrics for segmentation accuracy assessment i.e., Dice Similarity Coefficient (DSC), Mean Contour Distance (MCD) and Hausdorff distance (HD), against ground truth values obtained through manual delineation by clinicians. (2) comparing cardiac biventricular function indexes derived from manual and automatic segmentations such as ventricular end diastolic/systolic volumes and myocardial mass. Additionally, quantitative evaluation of human performance i.e., the inter-observer variability, is measured among the manual segmentations of different clinical experts. A set of 50 subjects was randomly selected and each subject was analysed by three expert observers (O1, O2, O3) independently. We compare the result of segmentation on the same set of subjects to show how close the performance of the automatic segmentation is to human performance and also the performance of the proposed workflow on a large dataset.

Image volumes at end diastolic and end systolic timepoints of 250 random subjects (500 images in total) were used for training the PDM and IAM. The test dataset contains 800 subjects (not included in training) used for evaluation of the proposed automatic approach. The input images and output segmentation contours were automatically quality controlled to ensure that image volumes included both basal and apical slices, and to verify the automatic segmentation results. After IQA, all 800 images were classified as having full coverage. After SQA, 21 segmentations were deemed suboptimal. Since the aim of the results presented in Sect. 3.1 is the evaluation of segmentation accuracy, all 800 segmentation results (including 21 outliers) were included in the statistics. In contrast, those results presented in Sect. 3.2 are based on 779 good quality segmentations, i.e. excluding those deemed suboptimal by SQA.

### 3.1 Segmentation Accuracy

Table 1 reports mean and standard deviation for DSC, MCD, and HD comparing between automatic and manual segmentations performed on test sets of 50 and 800 subjects never seen before by the PDM and IAM. The set of 50 subjects is the same set used for the evaluation of inter-observer variability. The set of 800 subjects is the same set used to generate reference ranges in [2].

The reported DSC values show excellent agreement ($\geq 0.87$) between manual delineations and automatic segmentations. MCD errors are smaller than the in-plane pixel spacing range of 1.8 mm to 2.3 mm found in the UKB. Although HD is larger than the in-plane pixel spacing, it is still within an acceptable range when compared with the distance range seen between different human observers. Table 1 shows that the segmentation accuracy of our method is within error ranges observed between different human raters. This indicates that our workflow performs with human-like reliability, and can fully automatically segment large scale datasets where manual inputs are infeasible.

**Table 1.** Segmentation accuracy expressed in terms of DSC, MCD and HD comparing the automatic (Auto.), manual (Man.), and observers (O1-O3) segmentations. $LV_{endo}$: LV endocardium. $LV_{myo}$: LV myocardium, $RV_{endo}$: RV endocardium. Values indicate mean $\pm$ standard deviation.

(a) DSC

|  | O1 vs O2 (n = 50) | O2 vs O3 (n = 50) | O3 vs O1 (n = 50) | Auto. vs Man. (n = 50) | Auto. vs Man. (n = 800) |
|---|---|---|---|---|---|
| $LV_{endo}$ | $0.94 \pm 0.04$ | $0.92 \pm 0.04$ | $0.93 \pm 0.04$ | $\mathbf{0.93 \pm 0.03}$ | $\mathbf{0.93 \pm 0.04}$ |
| $LV_{myo}$ | $0.88 \pm 0.02$ | $0.87 \pm 0.03$ | $0.88 \pm 0.02$ | $\mathbf{0.88 \pm 0.03}$ | $\mathbf{0.87 \pm 0.03}$ |
| $RV_{endo}$ | $0.87 \pm 0.06$ | $0.88 \pm 0.05$ | $0.89 \pm 0.05$ | $\mathbf{0.87 \pm 0.06}$ | $\mathbf{0.89 \pm 0.05}$ |

(b) MCD (mm)

|  | O1 vs O2 (n = 50) | O2 vs O3 (n = 50) | O3 vs O1 (n = 50) | Auto. vs Man. (n = 50) | Auto. vs Man. (n = 800) |
|---|---|---|---|---|---|
| $LV_{endo}$ | $1.00 \pm 0.25$ | $1.30 \pm 0.37$ | $1.21 \pm 0.48$ | $\mathbf{1.28 \pm 0.39}$ | $\mathbf{1.17 \pm 0.32}$ |
| $LV_{myo}$ | $1.16 \pm 0.34$ | $1.19 \pm 0.25$ | $1.21 \pm 0.36$ | $\mathbf{1.20 \pm 0.34}$ | $\mathbf{1.16 \pm 0.40}$ |
| $RV_{endo}$ | $2.00 \pm 0.79$ | $1.78 \pm 0.45$ | $1.87 \pm 0.74$ | $\mathbf{1.79 \pm 0.80}$ | $\mathbf{1.81 \pm 0.67}$ |

(c) HD (mm)

|  | O1 vs O2 (n = 50) | O2 vs O3 (n = 50) | O3 vs O1 (n = 50) | Auto. vs Man. (n = 50) | Auto. vs Man. (n = 800) |
|---|---|---|---|---|---|
| $LV_{endo}$ | $2.84 \pm 0.70$ | $3.31 \pm 0.90$ | $3.25 \pm 0.96$ | $\mathbf{3.21 \pm 0.97}$ | $\mathbf{3.21 \pm 0.99}$ |
| $LV_{myo}$ | $3.70 \pm 1.16$ | $3.82 \pm 1.07$ | $3.76 \pm 1.21$ | $\mathbf{3.91 \pm 1.20}$ | $\mathbf{3.92 \pm 1.30}$ |
| $RV_{endo}$ | $7.56 \pm 5.51$ | $7.35 \pm 2.19$ | $7.14 \pm 2.20$ | $\mathbf{7.41 \pm 4.11}$ | $\mathbf{7.31 \pm 3.32}$ |

## 3.2    Estimation of Cardiac Function Indexes

We evaluate the accuracy of cardiac function indexes derived from automatic segmentation versus gold standard reference ranges derived from manual segmentation. We calculate the LV end-diastolic volume (LVEDV) and end-systolic volume (LVESV), LV Stroke Volume (LVSV), LV Ejection-Fraction (LVEF), LV myocardial mass (LVM), RV end-diastolic volume (RVEDV) and end-systolic volume (RVESV), RV Stroke Volume (RVSV) and RV Ejection-Fraction (RVEF) from automated segmentation and compare them to measurements from manual segmentation.

Table 2 shows excellent agreement between the mean and standard deviation of ventricular parameters of a healthy population obtained through both automatic and manual segmentations. Furthermore, we performed two-sample Kolmogorov-Smirnov (K-S) tests to show that ventricular parameters obtained through manual and automatic approaches are drawn from the same population, under the null hypothesis that the manual and automatic methods are from the same continuous distribution in terms of clinical indexes. K-S test on different indexes does not reject the null hypothesis of being from same distribution at the 5% significance level.

**Table 2.** Cardiac function indexes derived from manual (Man.) vs automatic (Auto.) segmentation on 779 subjects. Values indicate mean ± standard deviation.

|  | LVEDV (ml) | LVESV (ml) | LVSV (ml) | LVEF (%) | LVM (g) | RVEDV (ml) | RVESV (ml) | RVSV (ml) | RVEF (%) |
|---|---|---|---|---|---|---|---|---|---|
| Man. | 144 ± 34 | 59 ± 18 | 85 ± 20 | 60 ± 6 | 86 ± 24 | 154 ± 40 | 69 ± 24 | 85 ± 20 | 56 ± 6 |
| Auto. | 146 ± 31 | 60 ± 18 | 86 ± 18 | 60 ± 7 | 87 ± 23 | 154 ± 40 | 71 ± 26 | 83 ± 21 | 54 ± 7 |

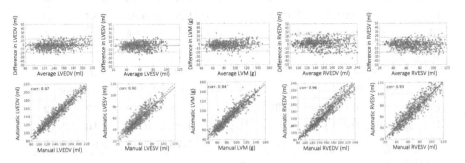

**Fig. 2.** Repeatability of various cardiac functional indexes: manual vs automatic analysis on the test dataset. The first row shows **Bland-Altman** plots. The solid line denotes the mean difference (bias) and the two dashed lines denote ±1.96 standard deviations from the mean. The second row shows **correlation** plots. The dashed and solid line denote the identity and linear regression lines, respectively.

Figure 2 shows Bland-Altman (top) and correlation (bottom) plots of ventricular parameters computed using the proposed automatic method and the manual reference on the test dataset. The Bland-Altman plots show good limits of agreement and also the mean difference line nearly at zero, which suggests that the clinical indexes obtained through the automatic approach have little bias. The correlation plots and their correlation coefficient (corr) indicate a strong relationship between the manual and automatic approaches.

## 4 Conclusion

In this paper, we propose a fully automatic workflow capable of performing high throughput end-to-end 3D cardiac image analysis. We tested our workflow on a reference cohort of 800 healthy subjects for which manual delineations, and reference functional indexes exist. Our results show statistically significant agreement between the manually obtained reference indexes, and those computed automatically using the proposed workflow. As future work, we plan to analyse all available UKB datasets including both healthy and pathological subjects and report the regional and global cardiac function indexes.

**Acknowledgements.** R. Attar was funded by the Faculty of Engineering Doctoral Academy Scholarship, University of Sheffield. This work has been partially supported by the MedIAN Network (EP/N026993/1) funded by the Engineering and Physical Sciences Research Council (EPSRC), and the European Commission through FP7 contract VPH-DARE@IT (FP7-ICT-2011-9-601055) and H2020 Program contract InSilc (H2020-SC1-2017-CNECT-2-777119). The UKB CMR dataset has been provided under UK Biobank Application 2964.

## References

1. Roth, G.A., et al.: Global, regional, and national burden of cardiovascular diseases for 10 causes, 1990 to 2015. J. Am. Coll. Cardiol. **70**(1), 1–25 (2017)
2. Petersen, S.E., et al.: Reference ranges for cardiac structure and function using cardiovascular magnetic resonance (CMR) in caucasians from the UK Biobank population cohort. J. Cardiovasc. Mag. Reson. **19**(1), 18 (2017)
3. Gorgolewski, K., et al.: Nipype: a flexible, lightweight and extensible neuroimaging data processing framework in python. Frontiers Neuroinformatics **5**, 13 (2011)
4. Zhang, L., et al.: Automated quality assessment of cardiac MR images using convolutional neural networks. In: Tsaftaris, S.A., Gooya, A., Frangi, A.F., Prince, J.L. (eds.) SASHIMI 2016. LNCS, vol. 9968, pp. 138–145. Springer, Cham (2016). https://doi.org/10.1007/978-3-319-46630-9_14
5. Van Assen, H.C., et al.: SPASM: a 3D-ASM for segmentation of sparse and arbitrarily oriented cardiac MRI data. Med. Image Anal. **10**(2), 286–303 (2006)
6. Albà, X., Lekadir, K., Pereañez, M., Medrano-Gracia, P., Young, A.A., Frangi, A.F.: Automatic initialization and quality control of large-scale cardiac MRI segmentations. Med. Image Anal. **43**, 129–141 (2018)
7. Valindria, V.V., et al.: Reverse classification accuracy: predicting segmentation performance in the absence of ground truth. IEEE Trans. Med. Imaging (2017)

# Lumen Segmentation of Aortic Dissection with Cascaded Convolutional Network

Ziyan Li[1,2,3], Jianjiang Feng[1,2,3(✉)], Zishun Feng[1], Yunqiang An[4], Yang Gao[4], Bin Lu[4], and Jie Zhou[1,2,3]

[1] Department of Automation, Tsinghua University, Beijing, China
jfeng@tsinghua.edu.cn
[2] State Key of Intelligent Technologies and Systems,
Tsinghua University, Beijing, China
[3] Beijing National Research Center for Information
Science and Technology, Beijing, China
[4] Department of Radiology, Fu Wai Hospital, Beijing, China

**Abstract.** For the diagnosis and treatment of aortic dissection, where blood flows in between the layers of the aortic wall, the segmentation of true and false lumens is necessary. This is a challenging task because the intimal flap separating true and false lumens is thin, discontinuous and has a complex shape. In this paper we formulate lumen segmentation of aortic dissection as the extraction of aortic adventitia (the contour of aorta) and intima (the contour of true lumen). To this end, we propose a cascaded convolutional network for contour extraction on 2-D cross-section images, and then construct a 3-D adventitia and intima shape model. We performed a five-fold cross-validation on 45 aortic dissection CT volumes. The proposed method demonstrated a good performance for both aorta and true lumen segmentation, and the mean Dice similarity coefficient was 0.989 for aorta and 0.925 for true lumen.

**Keywords:** Aortic dissection · Lumen segmentation ·
Contour extraction · Cascaded convolutional network

## 1 Introduction

Aortic dissection (AD) is a vascular disease with a high mortality rate. Without timely treatment, 1%–2% of the patients with acute type A dissection die in an hour, and nearly half of them die within a week [1]. AD occurs when the aortic intima gets ruptured and detached from the aortic wall, separating aortic lumen into true lumen and false lumen. At present, CT angiography is the standard reference for AD preoperative imaging. The choice of the stent-graft type and size requires accurate quantitative measurements [2]. Therefore, an accurate segmentation of AD CT volumes is important for better visualization and automatic measurement.

However, AD segmentation is a quite challenging task because the intimal flap is a thin membrane structure with an irregular shape, and the intensity

© Springer Nature Switzerland AG 2019
M. Pop et al. (Eds.): STACOM 2018, LNCS 11395, pp. 122–130, 2019.
https://doi.org/10.1007/978-3-030-12029-0_14

distribution inside the lumens varies due to the blood flow variation and the inhomogeneous contrast agent.

Existing AD segmentation approaches can be categorized into two types based on problem formulation: intimal flap segmentation and lumen segmentation (see Fig. 1). Kovács et al. [3] extracted intimal flaps with a sheetness measure, followed by an extrapolation algorithm to connect the intimal flap with the aortic wall. Krissian et al. [4] presented a comprehensive segmentation of AD by integrating multiple methods (such as level-set algorithm and multiscale vesselness extraction based on gradient and Hessian-matrix), which required manual centerline selection and parameter tuning. Duan et al. [5] extracted intimal flaps by combining Hessian-matrix with spatial continuity between CT slices. However, since Hessian-matrix based methods have little effect on enhancing indistinct intimal flap or blob-like connection point of intimal flap and aortic wall, the intimal flaps obtained from these studies [4,5] were often not continuous or did not completely separate true lumen from false lumen as illustrated in Fig. 1(b), which require further post-processing.

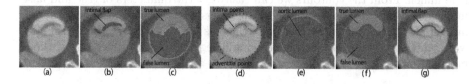

**Fig. 1.** Three different formulations of AD segmentation: (a) original slice, (b) intimal flap segmentation, (c) true and false lumen segmentation, and (d) our proposed adventitia and intima contour extraction. Segmentation results of (b) and (c) require complex post-processing to be converted into each other, while contours in (d) can be easily converted into other representations: (e) aortic lumen, (f) true and false lumens, and (g) intimal flap.

On the other hand, it is essential to distinguish the true lumen from the false lumen, since the stent-graft should be placed in the true lumen in endovascular treatments [2]. Lee et al. [6] focused on the true-false lumen boundary segmentation by wavelet analysis and generative-discriminative model, without distinguishing the two lumens. Kovács et al. [7] applied an identification method to distinguish between true and false lumens based on a single feature (lumen area) after the segmentation task. Fetnaci et al. [8] segmented true and false lumens separately with a transformed fast marching algorithm. It required a manual seed point to identify the two lumens. These AD segmentation methods, which are based on conventional algorithms, are not robust enough when dealing with blurry intimal flap or noise.

Different from existing approaches, we formulate AD segmentation as the extraction of adventitia and intima contours in order to facilitate quantitative measurement of AD and enable automatic true lumen recognition on crosssections. In addition, from the adventitia and intima contours we can easily

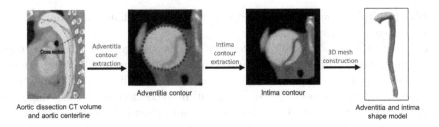

**Fig. 2.** Flowchart of the proposed method

obtain other representations of AD (see Fig. 1). The flowchart of the proposed method is illustrated in Fig. 2.

Inspired by shape regression based on convolutional neural network (CNN)[9], we propose a three-level cascaded convolutional network for a two-step contour extraction. The proposed approach achieved a good performance with the mean Dice similarity coefficient (DSC) reaching 0.989 for aortic lumen and 0.925 for true lumen. From the contours we can easily get the diameters of true and false lumens on different cross-sections, which brings convenience to the quantitative measurement. Another contribution is that, as far as we know, this is the first work where true lumen is distinguished and segmented simultaneously, without manual assistance.

## 2   Methodology

Our proposed method contains three parts (see Fig. 2): First, we intercept cross-sections along the aortic centerlines from CT volumes using the Visualization Toolkit (VTK) (https://www.vtk.org), where the centerlines are pre-extracted (see Sect. 3). Then we focus on the two-step contour extraction of adventitia and intima. At last the adventitia and intima shape model is constructed in three-dimension space with VTK.

### 2.1   Contour Model of Adventitia and Intima

True lumen segmentation is meaningful for the treatment of AD, and our main concern is to segment the true lumen by extracting the intima contour. While actually extracting the adventitia contour is much easier, and the intima contour is highly related to the adventitia contour, therefore we build the adventitia and intima model as a whole. In the proposed contour model, the adventitia and intima on each cross-section are represented by a fixed number $M$ ($M = 38$) of contour points respectively, as illustrated in Fig. 3. Three experienced radiologists helped annotate the cross-section images.

As the adventitia contour always approximates to a circle, we number the adventitia contour points from a fixed starting position counterclockwise at an equal angle interval, so that each point has a certain location on the circle.

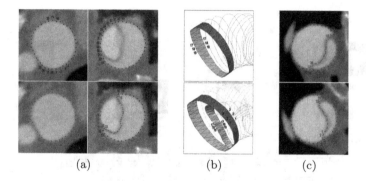

(a)                    (b)                    (c)

**Fig. 3.** Blue points represent adventitia contour. Red points represent intima contour. (a) First column: a cross-section without dissection, second column: a cross-section with dissection. (b) Connecting method of contour points on adjacent cross-sections (upper: only adventitia, lower: adventitia and intima). (c) Rotation transform. (Color figure online)

For aortic intima, if there is no dissection, we index the contour points in the same way as adventitia. Otherwise the contour points of the normal part of intima keep unchanged and the others get detached along with the detached intima (since AD forms because part of the intima gets detached from the aortic wall, as introduced in Sect. 1). In this way we can connect two adjacent cross-sections with a triangle-strip by linking the corresponding points and finally construct a regular adventitia and intima shape model (see Fig. 3(b)).

## 2.2    Contour Extraction

Comparing to aortic lumen segmentation, true lumen segmentation is more difficult due to its irregular shape and varying intensity on cross-sections. Some previous works [4,7] segmented the aorta as ROI before intimal flap segmentation. Inspired by it, we extract the adventitia contour first to guide the intima contour extraction by a scale normalization. For the $i^{th}$ image $I_i$ in the dataset, a scale factor $\lambda_i$ is calculated with its predicted adventitia contour points $p_{ij}$ ($p_{ij} \in \mathbb{R}^{2 \times 1}$) by:

$$\lambda_i = \frac{r * \mathrm{M}}{\sum_{j=1}^{\mathrm{M}} \|p_{ij} - c_i\|_2} , \tag{1}$$

where the image center $c_i$ is the intersection point of the centerline and $I_i$. We set standard radius $r$ as one-quarter of the image length for a proper foreground size. Then the intima contour point estimation is performed with scaled images.

In the training phase adventitia contour points are initialized by the mean contour of the training set. For intima, we randomly choose a ground-truth from the training set and rescale it as the initial contour. The combinations of different initial contours and training images improve the generalization ability of intima network. In the testing phase both adventitia and intima contour points are initialized by their mean contours of the training set.

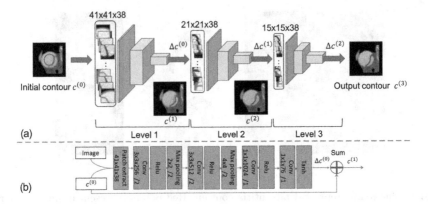

**Fig. 4.** (a) Proposed cascaded convolutional network contains three similar levels. (b) Structure of level 1. (e.g. "Conv $3 \times 3 \times 256/2$" means convolutional with $3 \times 3$ kernels, 256 output features and a stride of 2).

To train a robust model, we augment training data on-the-fly by mirror transform, rotation transform and the combination of them. The rotation transform shown in Fig. 3(c) is custom-designed for the contour model. We rotate an image and its contour points by an angle of $\theta = t * \theta_0$, where $t$ is randomly chosen from $\{1, 2, ..., \text{M}\}$ and $\theta_0 = 2\pi/\text{M}$, and then adjust the indexes of contour points in accordance with the contour model in Sect. 2.1. More specifically, the indexes of adventitia points correspond to fixed locations on the circle and will not rotate with the adventitia points, and the indexes of intima points are adjusted according to the indexes of adventitia points as introduced in Sect. 2.1.

**Network Architecture.** The proposed adventitia network and intima network have the same architecture, as shown in Fig. 4. For both adventitia and intima contour points, their positions are associated with the positions of their neighbors. To utilize the coordinate dependences for jointly contour point regression, the first layer takes local patches of contour points as a multi-channel input.

The lumen area and *beak sign*[1] are main visual features to distinguish true and false lumens on cross-sections [10]. To extract these non-local features for rough localization and balance the computation expense, we set the patch size as $41 \times 41$ pixels in level 1, which is $21 \times 21$ in level 2 and $15 \times 15$ in level 3. Three similar cascaded levels with a shrink patch size enable a coarse-to-fine regression of contour points. There are four convolutional layers in level 1 (see Fig. 4(b)), and the last layer with $1 \times 1$ kernels is followed by *tanh* layer to predict the deviation $\Delta c^{(0)}$ of the input contour points $c^{(0)}$ of this level, where

---

[1] The beak sign is an acute angle between the intimal flap and the aortic wall. It only exists in the false lumen.

$c^{(0)} = [x_1, y_1, x_2, y_2, ..., x_M, y_M]' \in \mathbb{R}^{2M \times 1}$ denotes the coordinates of the contour points. For an image, we define the loss function as:

$$\mathcal{L}\left(c^*, c^{(1)}, c^{(2)}, c^{(3)}\right) = \sum_{i=1}^{3} \left\| c^{(i)} - c^* \right\|_2^2 , \qquad (2)$$

where $c^{(i)}$ denotes the contour predicted by level $i$, and $c^*$ is the ground-truth. In this way the deviations of contour points of three levels all contribute to the parameter update of our model by back-propagation algorithm.

## 3   Experiments and Results

We collected 45 contrast-enhanced AD CT volumes acquired with a Philips iCT256 scanner at 100 kVp, from 45 different patients with type A dissection (29 patients) or type B dissection (16 patients). The size of volumes is $512 \times 512 \times Z$ (723–1023) and the voxel spacing is $X \times X \times 0.8$ mm$^3$ ($X$ from 0.5507 to 0.7774). To the best of our knowledge, this is the largest evaluation dataset of AD segmentation. The evaluation framework of our lumen segmentation algorithm is similar to [11]. We labeled binary volumes for the dataset with ITK-SNAP (http://www.itksnap.org) and then extracted aortic centerlines with VMTK (http://www.vmtk.org). There were 4963 cross-section images extracted from the dataset. The pixel spacing of all images was resized to 0.4 mm and the image size is $144 \times 144$ pixels. In the experiments, we did not consider the starting part of the ascending aorta due to the motion artifact of CT volumes.

Our method was evaluated by a five-fold cross-validation on all 45 cases. The dataset was partitioned into five groups (9 cases per group). We trained the adventitia network first and then the intima network. Network parameters were initialized randomly and learned by SGD with weight decay of 0.0005 and momentum of 0.9. For adventitia network, the learning rate was 0.001 for 8K iterations and then 0.0001 for 2K iterations with the batch size of 256. For intima network, after the same training process as adventitia network, we fine-tuned it with the batch size of 128 and learning rate from $10^{-4}$ to $10^{-6}$ (decaying to 1/10 every 4K iterations). The proposed network was built with MatConvNet (https://github.com/vlfeat/matconvnet). The training time was 21 h for adventitia network and 50 h for intima network, using GTX1080 Ti GPU.

It is hard to make a fair comparison with previous AD segmentation algorithms for the following reasons:

1. The outputs of algorithms are different. The outputs of [4,5] are intimal flaps.
2. Certain study [8] did not report quantitative result.
3. Study [6] did not consider distinguishing true and false lumens.
4. The datasets used in all these studies are not publicly available.
5. No public implementation is available and a faithful reimplementation is hard.

Hence, we compared our proposed method to the well-known U-net [12], which has achieved high performance in many medical image segmentation tasks and is publicly available. The U-net was trained to output pixel-wise segmentation results for cross-section images, and for fair comparison, our contour representation is also converted into pixel-wise segmentation. Table 1 reported the segmentation performance of the two methods.

**Table 1.** Evaluation and comparison of lumen segmentation. Mean and relative standard deviation of DSC, sensitivity and specificity are reported.

|  | Method | DSC | Sensitivity | Specificity |
|---|---|---|---|---|
| Aortic lumen | U-net | $0.960 \pm 0.026$ | $0.958 \pm 0.031$ | $0.993 \pm 0.008$ |
|  | Proposed | $\mathbf{0.989 \pm 0.006}$ | $\mathbf{0.990 \pm 0.009}$ | $\mathbf{0.997 \pm 0.003}$ |
| True lumen | U-net | $0.317 \pm 0.808$ | $0.356 \pm 0.789$ | $0.933 \pm 0.061$ |
|  | Proposed | $\mathbf{0.925 \pm 0.106}$ | $\mathbf{0.921 \pm 0.114}$ | $\mathbf{0.995 \pm 0.008}$ |

Comparing to manual labeled ground-truth, our method obtained a mean DSC of 0.989 for aortic lumen segmentation and 0.925 for true lumen segmentation. Comparing to the U-net method, the mean DSC of our method was about 3% higher for aortic lumen segmentation. As for true lumen segmentation, we observed that the U-net method was unable to distinguish true lumen from false lumen on cross-sections for pixel-wise segmentation, and therefore resulted in a low DSC, while our method reported a mean DSC of 0.925 with no post-processing.

Figure 5 shows the extracted contours of some cross-section images. In the cases of bifurcation point and blurry intimal flap, our method can also extract accurate contours. The 3-D adventitia and intima shape models of four cases are presented in Fig. 6. We can see that most parts of AD obtained an accurate segmentation. Only a few positions had poor results, which is mainly because the true and false lumens on cross-sections of these positions had nearly the same shape, area and intensity.

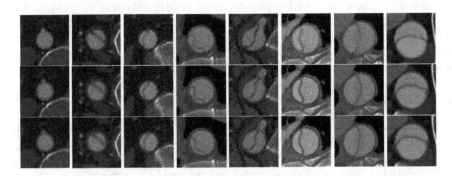

**Fig. 5.** Three rows show the original cross-sections, ground-truth contours and contours extracted by our method, respectively.

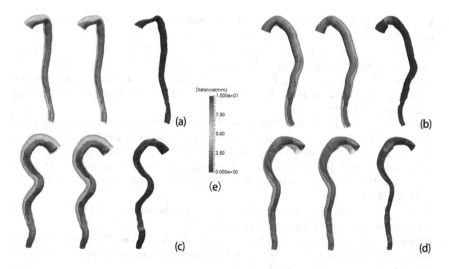

**Fig. 6.** In (a) (b) (c) (d), the adventitia and intima shape models are built by ground-truth (left) and extracted contours (middle), where the inner yellow shape model represents intima and the outer red shape model represents adventitia. The colors of intima shape models (right) represent the distances of intima points between ground-truth and extracted results, with (e) a distance lookup table. (Color figure online)

## 4   Conclusion

In this paper, we transform region segmentation of AD into contour extraction by constructing an adventitia and intima shape model, which enables automatic true lumen recognition on cross-sections. Then we present a two-step contour extraction with the proposed cascaded convolutional network. The shape model is built with a fixed number of contour points on each cross-section by a regular connecting method. It suggests their promising applications in quantitative analysis of AD, for instance, measuring diameters of key positions and the volumes of the true and false lumens during follow-ups. Although the closed contour model is beneficial for quantification, AD has locations where the real contour is not closed and may even have complex shape, especially around intima tears. For future work we intend to include these special cases in our model.

**Acknowledgement.** This work is supported by the National Natural Science Foundation of China under Grant 61622207.

## References

1. Nienaber, C.A., Clough, R.E.: Management of acute aortic dissection. Lancet **385**(9970), 800–811 (2015)
2. Chiesa, R., Melissano, G., Zangrillo, A., Coselli, J.S.: Thoraco-Abdominal Aorta: Surgical and Anesthetic Management, vol. 783. Springer, Milan (2011)

3. Kovács, T., Cattin, P., Alkadhi, H., Wildermuth, S., Székely, G.: Automatic segmentation of the aortic dissection membrane from 3D CTA images. In: Yang, G.-Z., Jiang, T.Z., Shen, D., Gu, L., Yang, J. (eds.) MIAR 2006. LNCS, vol. 4091, pp. 317–324. Springer, Heidelberg (2006). https://doi.org/10.1007/11812715_40

4. Krissian, K., Carreira, J.M., Esclarin, J., Maynar, M.: Semi-automatic segmentation and detection of aorta dissection wall in MDCT angiography. Med. Image Anal. **18**(1), 83–102 (2014)

5. Duan, X., et al.: Visual three-dimensional reconstruction of aortic dissection based on medical CT images. Int. J. Digit. Multimed. Broadcast. **2017**(14), 1–8 (2017)

6. Lee, N., Laine, A.F.: True-false lumen segmentation of aortic dissection using multiscale wavelet analysis and generative-discriminative model matching. In: Proceedings of SPIE - The International Society for Optical Engineering, vol. 6915, p. 69152V (2008)

7. Kovács, T.: Automatic segmentation of the vessel lumen from 3D CTA images of aortic dissection. Ph.D. thesis, ETH Zurich (2010)

8. Fetnaci, N.: 3D segmentation of the true and false lumens on CT aortic dissection images. In: Proceedings of SPIE - The International Society for Optical Engineering, vol. 8650, no. 4, p. 86500M (2013)

9. Zhang, J., Shan, S., Kan, M., Chen, X.: Coarse-to-fine auto-encoder networks (CFAN) for real-time face alignment. In: Fleet, D., Pajdla, T., Schiele, B., Tuytelaars, T. (eds.) ECCV 2014. LNCS, vol. 8690, pp. 1–16. Springer, Cham (2014). https://doi.org/10.1007/978-3-319-10605-2_1

10. LePage, M.A., Quint, L.E., Sonnad, S.S., Deeb, G.M., Williams, D.M.: Aortic dissection: CT features that distinguish true lumen from false lumen. Am. J. Roentgenol. **177**(1), 207–211 (2001)

11. Kirişli, H., Schaap, M., Metz, C., et al.: Standardized evaluation framework for evaluating coronary artery stenosis detection, stenosis quantification and lumen segmentation algorithms in computed tomography angiography. Med. Image Anal. **17**(8), 859–876 (2013)

12. Ronneberger, O., Fischer, P., Brox, T.: U-Net: convolutional networks for biomedical image segmentation. In: Navab, N., Hornegger, J., Wells, W.M., Frangi, A.F. (eds.) MICCAI 2015. LNCS, vol. 9351, pp. 234–241. Springer, Cham (2015). https://doi.org/10.1007/978-3-319-24574-4_28

# A Vessel-Focused 3D Convolutional Network for Automatic Segmentation and Classification of Coronary Artery Plaques in Cardiac CTA

Jiang Liu[1], Cheng Jin[1,2,3], Jianjiang Feng[1,2,3(✉)], Yubo Du[1], Jiwen Lu[1,2,3], and Jie Zhou[1,2,3]

[1] Department of Automation, Tsinghua University, Beijing, China
jfeng@tsinghua.edu.cn
[2] State Key Lab of Intelligent Technologies and Systems,
Tsinghua University, Beijing, China
[3] Beijing National Research Center for Information Science and Technology,
Beijing, China

**Abstract.** The segmentation and classification of atherosclerotic plaque (AP) are of great importance in the diagnosis and treatment of coronary artery disease. Although the constitution of AP can be assessed through a contrast-enhanced coronary computed tomography angiography (CCTA), the interpretation of CCTA scans is time-consuming and tedious for radiologists. Automation of AP segmentation is highly desired for clinical applications and further researches. However, it is difficult due to the extreme unbalance of voxels, similar appearance between some plaques and background tissues, and artefacts. In this paper, we propose a vessel-focused 3D convolutional network for automatic segmentation of AP including three subtypes: calcified plaques (CAP), non-calcified plaques (NCAP) and mixed calcified plaques (MCAP). We first extract the coronary arteries from the CT volumes; then we reform the artery segments into straightened volumes; finally, a 3D vessel-focused convolutional neural network is employed for plaque segmentation. The proposed method is trained and tested on a dataset of multi-phase CCTA volumes of 25 patients. We further investigate the effect of artery straightening through a comparison experiment, in which the network is trained on original CT volumes. Results show that by artery extraction and straightening, the training time is reduced by 40% and the segmentation performance of non-calcified plaques and mixed calcified plaques gains significantly. The proposed method achieves dice scores of 0.83, 0.73 and 0.68 for CAP, NCAP and MCAP respectively on the test set, which shows potential value for clinical application.

**Keywords:** Coronary artery plaques · Plaque segmentation · CCTA · 3D CNN

J. Liu and C. Jin—These two authors contribute equally to this study.

© Springer Nature Switzerland AG 2019
M. Pop et al. (Eds.): STACOM 2018, LNCS 11395, pp. 131–141, 2019.
https://doi.org/10.1007/978-3-030-12029-0_15

# 1  Introduction

Coronary artery disease (CAD) is one of the biggest causes of mortality in the world. It is usually caused by atherosclerosis, of which the plaques are divided into three subtypes: calcified (CAP), non-calcified (NCAP) and mixed calcified (MCAP). The plaque composition is an important indicator for the diagnosis and treatment of CAD [1].

Contrast-enhanced coronary CT angiography (CCTA) allows assessment of AP composition, while it is demanding to interpret CCTA scans due to the large volume of data and the numerous segments of coronary arteries. Therefore, automatic segmentation and classification of AP is highly desirable. It can not only facilliate the interpretation of CCTA scans, but also provides a quantitative measurement of AP. However, AP segmentation on voxel level is difficult due to extreme unbalance of voxels (much more background voxels than plaque voxels), similar appearance between some plaques and background tissues, and artefacts.

Many studies have focused on CAP segmentation in 3D CT volumes. CAP is characterized by bright appearance in CT images, which can be easily discerned in both non-contrast-enhanced and contrast-enhanced cardiac CT. Wolterink *et al.* [2] summarized CAP segmentation methods. In non-contrast cardiac CT scans, CAP can be detected by a threshold of 130 HU [3] and subsequent classifiers, while in CCTA the detection threshold may vary depending on protocols, scanners and contrast agents, and thus special techniques are needed to determine the threshold. CNN-based methods have emerged recently, which typically consist of two network: the first performs a rough segmentation to restrict the area of interest and the second performs a meticulous segmentation [4,5].

Although fewer studies have covered the segmentation of NCAP and MCAP, the detection of them is important because they are more prone to rupture and result in acute coronary syndromes such as stroke and sudden death. In contrast to CAP, NCAP and MCAP show similar intensity with surrounding tissues, which makes segmentation of NCAP and MCAP more challenging (Fig. 1). [6] adopts a two-fold methodology for NCAP segmentation, in which the first step extracts the coronary arteries and NCAP is detected based on the extracted arteries. To our knowledge, there is no existing method that segments all three subtypes of AP at the same time.

(a) CAP          (b) NCAP          (c) MCAP

**Fig. 1.** Atherosclerosis plaques in CT images. CAP is characterized by bright appearance while NCAP shows similar intensity with surrounding tissues. MCAP is a mixture of bright and gray area.

Recently, fully convolutional neural networks (FCNs) have demonstrated state-of-the-art performance on many challenging image segmentation tasks. 3D U-Net [7] is one example that is especially suitable for medical image analysis.

In this paper, we propose a robust method based on 3D FCNs for automatic coronary artery plaques segmentation including all three types. To begin with, a bounding box encasing the coronary arteries are automatically generated for the purpose of reducing computation. Then we extract the coronary arteries and reform artery segments into straightened volumes as inputs of the network using multi-planar reformation (MPR) technique. Finally, a vessel-focused 3D convolutional network with attention layers [8] is trained to segment subtypes of AP. We further investigate the effect of artery straightening through a comparison experiment in which the network is trained on original data.

## 2    Method

The main challenges for the segmentation of artery plaques include:

- class imbalance (a lot more background voxels than plaque voxels)
- high variability of the plaque appearance
- high similarity between non-calcified plaques and background

To address the first two problems, we first extract and straighten vessel segments along artery centerlines as inputs, which restricts the volume-of-interest that alleviates the class imbalance and simplifies the analysis of lumen curvature variation and surrounding tissues. We also use a multi-class Dice loss function that increases the cost of segmentation mistakes on the plaques.

To better distinguish plaques from background tissues, we design a U-Net [9] like encoder-decoder network with residual blocks that preserve the signals from shallow layers, deep supervision that encourages multi-scale segmentation and attention layers [8] that helps to locate the pathologies.

Figure 2 shows the workflow of the proposed method. The network architecture is described in detail in Sect. 2.3.

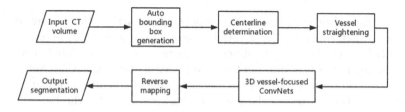

**Fig. 2.** Flowchart of the proposed method.

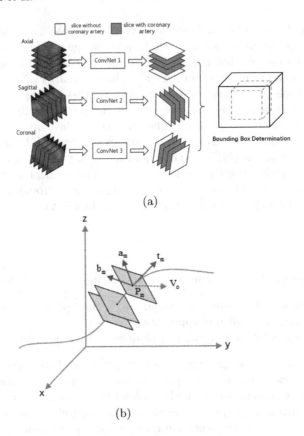

(a)

(b)

**Fig. 3.** (a) Auto bounding box generation. Three convolutional neural networks (ConvNets) are trained on axial, sagittal and coronal planes respectively and determine the boundary of coronary artery on each axis. (b) Illustration of the 3D MPR technique. Cross-sectional planes are extracted along the centerline and stacked into a rectangular volume.

## 2.1   Auto Bounding Box Detection

To restrict the volume-of-interest for computation reduction, three ConvNets are employed to determine the bounding box around the heart for the comparison method (Fig. 3(a)). Slices along different directions are given a label (1 for slice with coronary). We use 100 additional volumes to train the ConvNets. Each ConvNet is a pretrained AlexNet [10], while we added an additional neuron with sigmoid activation on top of the original AlexNet output for the binary classification and fine-tuned the networks on three orthogonal planes (axial, sagittal, coronal). We chose the utmost predicted foreground slices as the boundary of the bounding box to make sure all the coronary arteries were enclosed.

## 2.2 Coronary Artery Extraction and Straightening

The coronary artery trees are extracted by MSCAR-DBT [11] which only requires two manually identified seed points. Vessels with diameter less than 2 mm are left out. The centerlines is then extracted based on the coronary artery extraction results.

We reform the artery segments into straightened volumes using 3D multiplanar reformation technique, which is illustrated in Fig. 3(b). We first subdivide the centerline into n segments with a equal length of 0.5 mm, and obtain $n+1$ evenly spaced control points $P_m$ $(1 \leq m \leq n)$ along the centerline. Then we extract cross-sectional planes with a size of $32 \times 32$ pixels at each control points and stack them into a rectangular volume. We define the reference vector $\vec{r_0}$ as a unit vector parallel to the y-axis of $\sum_P$. Let $\vec{t_m}$ to be the normalized tangent to the centerline at $P_m$. The bases of the cross-sectional plane at $P_m$ are calculated by:

$$\vec{a_m} = \vec{r_0} \times \vec{t_m}, \quad \vec{b_m} = -\vec{t_m} \times \vec{a_m}. \tag{1}$$

Equation 2 defines the mapping of point $P(i, j, k)$ in the straightened space $\sum_S$ to the original physical coordinate $\sum_P$.

$$\forall i, j, k \in \mathbb{Z} : -32 \leq i, j < 32, 0 < k \leq n, P(x, y, z) = P_k + \frac{1}{2} i \times \vec{a_k} + \frac{1}{2} j \times \vec{b_k}. \tag{2}$$

The reformed voxel values are obtained by bicubic interpolation. We cut the volumes that are longer than 128 voxels into several overlapping segments. The segmentation results for overlapped areas are obtained by averaging the output of the overlapping segments.

## 2.3 Network Architecture

The network is an encoder-decoder network shown in Fig. 4. The left part shows the encoding path. In each stage, one to three convolutional layers extract abstract context information. Each layer is with kernels of $3 \times 3 \times 3$, symmetric padding, instance normalization and PRelu non linearity. We formulate each stage as a residual block, that is the input of each stage is added to the output of the last layer of the stage. Then convolutions with stride two are applied to halve the resolution and double the number of feature channels. The right part of the network is the decoding path. Each stage has a similar structure with the left but consists of a concatenation with the corresponding feature maps from the attention layers of the contracting path. De-convolution is applied to increase feature map resolution and halve the number of feature channels.

The idea of attention layer is to use attention gates to implicitly learn to suppress irrelevant regions in the input while highlighting salient features useful for plaque segmentations, which is described detailedly in [8].

We employ deep supervision in the expansive path by combining output of different stages via element-wise summation to form the final network output, which forces the network to produce an accurate segmentation in an early stage.

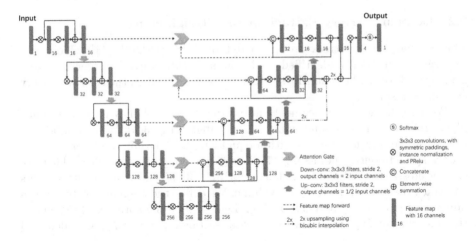

**Fig. 4.** The proposed network architecture. The networks is an encoder-decoder style network, with attention gates to select the most useful features from encoder path. Deep supervision is achieved by combining output of different stages via element-wise summation.

To address the extreme class unbalance in our data, we use a multiclass dice loss function [12] instead of the conventional categorical crossentropy loss:

$$L_{dc} = -\frac{2}{|K|} \sum_{k \in K} \frac{\sum_i u_i^k v_i^k}{\sum_i u_i^k + \sum_i v_i^k} \tag{3}$$

where $u$ is the output of the network and $v$ is a one hot encoding of the ground truth. For both $u$ and $v$, $i$ is the number of pixels and $k \in K$ being the classes.

## 3    Experiment Results

### 3.1    Dataset

We collected ECG-gated 4D-CTA scans by Philip 256-iCT from 25 patients, among which 13 were diagnosed with AP. The 4D-CT data sets are constructed in 20 phases: 5%, 10%,..., 100%. The size of each slice is $512 \times 512$ pixels with an isotropic resolution of 0.414 mm. The number of slices in each volume ranged from 213 to 358 with thickness of 0.335 mm. We pick 4 phases (25%, 45%, 55% and 75%) from each patient and obtain a dataset consist of 100 scans, which allows the network to learn the variance introduced by cardiac motion and enhances the generalization ability. We select 80 scans as training set and the rest as test set.

The APs were annotated and classified by 5 trained radiologists, each scan is only annotated by one radiologist and then examined by a second one. The annotations serve as ground truth.

**Table 1.** Experiment results

| Method | Dice score | | | Sensitivity | | | PPV | | |
|---|---|---|---|---|---|---|---|---|---|
| | CAP | NCAP | MCAP | CAP | NCAP | MCAP | CAP | NCAP | MCAP |
| Proposed method | 0.83 | **0.73** | **0.68** | 0.85 | **0.76** | **0.72** | 0.82 | **0.69** | **0.62** |
| Comparison method | **0.87** | 0.61 | 0.58 | **0.89** | 0.63 | 0.68 | **0.86** | 0.58 | 0.52 |

## 3.2 Comparison Experiment

To investigate the effect of artery straightening, we train a second network with the same architecture on the original data. Figure 5 shows the workflow of the method for comparison experiment.

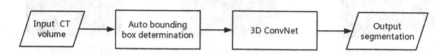

**Fig. 5.** Flowchart of the method for comparison experiment.

## 3.3 Training Procedure

We implemented the proposed network in Keras with TensorFlow backend. For the proposed method, the input size is $64 \times 64 \times 128$; for the comparison method, the input size is $128 \times 128 \times 128$. The initial learning rate was 0.0001, and we reduced the learning rate by 50% if the validation loss did not improve after 10 epochs. The weights were updated by stochastic gradient descent with Adam optimizer. We ran more than 200 epochs on four NVIDIA 1080 GPUs.

We use extensive data augmentation techniques including rotation, scaling, flipping, and smooth dense deformations on both data and ground truth labels.

## 3.4 Results

We remapped the output of the 3D network to the original space with nearest interpolation for visualization. Figures 6 and 7 show some examples. We evaluate the performance of the proposed method by true positive rate (sensitivity), positive predictive value (PPV) and dice score. The dice score of two sets A and B is evaluated as $2|A \cap B|/(|A| + |B|)$.

The proposed method achieved dice scores of 0.83, 0.73 and 0.68 for CAP, NCAP and MCAP respectively on the test set. Table 1 compares the performance of the proposed method and comparison experiment. It is shown that both methods perform well on the segmentation of CAP while proposed achieves significant

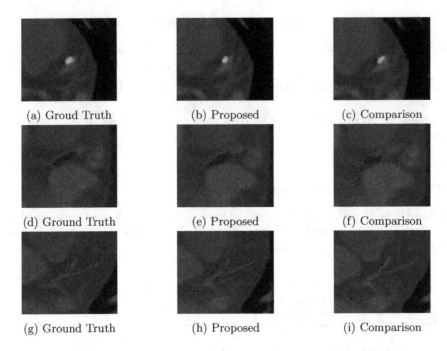

(a) Groud Truth    (b) Proposed    (c) Comparison

(d) Ground Truth    (e) Proposed    (f) Comparison

(g) Ground Truth    (h) Proposed    (i) Comparison

**Fig. 6.** Segmentation Result. (a)–(c): CAP; (d)–(f): NCAP; (g)–(i): MCAP. CAP is easily segmented by both methods. However, NCAP shows similar intensity with surrounding tissues, and comparison method fails to distinguish a majority of the plaque from background in (f), while proposed method successfully segments the whole structure in (e). MCAP is a mixture of bright and gray area while comparison method only captures the bright part in (i).

gains in the segmentation of NCAP and MCAP. In addition, the training time for proposed method is around 40% less than the comparison method because of the smaller input size, which also demonstrates the benefits of artery straightening. It is also observed during training that the comparison method is more

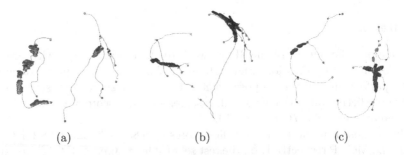

(a)    (b)    (c)

**Fig. 7.** Examples of segmentation results in 3D visualization (gray: CAP, red: MCAP). (Color figure online)

prone to overfit due to the small dataset while the proposed method benefits from the MPR transformation that makes the data more uniform and reduces the gap between training and testing data.

Existing methods only segment one type of the plaques and mostly reports sensitivity. For calcified plaques(CAP), [13] reports a sensitivity of 91.24%. For non-calcified plaques(NCAP), [14] reports a sensitivity of 92.5%. The result of proposed method is comparable to existing results while generalizes to three subtypes.

## 4   Discussion and Conclusion

Studies has shown that multi-planar reconstructions are useful in the evaluation of CT scans of coronary arteries for human radiologists [15]. The comparison experiment shows that MPR also contributes to a better segmentation performance for convolutional neural networks.

The detection and segmentation of artery plaques is very challenging due to class imbalance, high similarity between non-calcified plaques and high variability of the plaque appearance. The proposed method achieved dice scores of 0.83, 0.73 and 0.68 for CAP, NCAP and MCAP respectively on the test set. However, there are several drawbacks of this study. First, the current dataset only consists of 100 scans from 25 patients, which may not be representative considering the large variability of plaques. Second, the label for each scan is annotated by only one radiologist, and thus the network learns the segmentation regarding inter-observer error, which may hinder the network performance. Third, the location priors of the plaques are not utilized. In addition, although the MPR helps to boost the segmentation performance, it relies on the centerline of coronary artery which can be difficult to extract with the presence of plaques; and the coronary artery extraction method we are using now is semi-automatic and can be replaced with recent fully automatic state-of-the-art coronary artery segmentation algorithms.

To summarize, in this study we propose a pipeline for automatic plaque segmentation, which shows potential value in clinical use. Results show that by artery extraction and straightening, the training time is reduced by 40% and the segmentation performance of NCAP and MCAP gains significantly. The proposed method remains to be improved and tested on larger datasets.

**Acknowledgement.** This work is supported by the National Natural Science Foundation of China under Grant 61622207.

# References

1. Pasterkamp, G., Falk, E., Woutman, H., Borst, C.: Techniques characterizing the coronary atherosclerotic plaque: influence on clinical decision making? J. Am. Coll. Cardiol. **36**(1), 13–21 (2000)
2. Wolterink, J.M., Leiner, T., de Vos, B.D., van Hamersvelt, R.W., Viergever, M.A., Išgum, I.: Automatic coronary artery calcium scoring in cardiac CT angiography using paired convolutional neural networks. Med. Image Anal. **34**, 123–136 (2016)
3. Agatston, A.S., Janowitz, W.R., Hildner, F.J., Zusmer, N.R., Viamonte, M., Detrano, R.: Quantification of coronary artery calcium using ultrafast computed tomography. J. Am. Coll. Cardiol. **15**(4), 827–832 (1990)
4. Wolterink, J.M., Leiner, T., Viergever, M.A., Išgum, I.: Automatic coronary calcium scoring in cardiac CT angiography using convolutional neural networks. In: Navab, N., Hornegger, J., Wells, W.M., Frangi, A.F. (eds.) MICCAI 2015. LNCS, vol. 9349, pp. 589–596. Springer, Cham (2015). https://doi.org/10.1007/978-3-319-24553-9_72
5. Lessmann, N., et al.: Automatic calcium scoring in low-dose chest CT using deep neural networks with dilated convolutions. IEEE Trans. Med. Imaging **37**, 615 (2017)
6. Renard, F., Yang, Y.: Coronary artery extraction and analysis for detection of soft plaques in MDCT images. In: 2008 15th IEEE International Conference on Image Processing, pp. 2248–2251. IEEE (2008)
7. Çiçek, Ö., Abdulkadir, A., Lienkamp, S.S., Brox, T., Ronneberger, O.: 3D U-Net: learning dense volumetric segmentation from sparse annotation. In: Ourselin, S., Joskowicz, L., Sabuncu, M.R., Unal, G., Wells, W. (eds.) MICCAI 2016. LNCS, vol. 9901, pp. 424–432. Springer, Cham (2016). https://doi.org/10.1007/978-3-319-46723-8_49
8. Oktay, O., et al.: Attention U-Net: learning where to look for the pancreas. arXiv preprint arXiv:1804.03999 (2018)
9. Ronneberger, O., Fischer, P., Brox, T.: U-Net: convolutional networks for biomedical image segmentation. In: Navab, N., Hornegger, J., Wells, W.M., Frangi, A.F. (eds.) MICCAI 2015. LNCS, vol. 9351, pp. 234–241. Springer, Cham (2015). https://doi.org/10.1007/978-3-319-24574-4_28
10. Krizhevsky, A., Sutskever, I., Hinton, G.E.: Imagenet classification with deep convolutional neural networks. In: Advances in Neural Information Processing Systems, pp. 1097–1105 (2012)
11. Zhou, C., et al.: Pulmonary vessel segmentation utilizing curved planar reformation and optimal path finding (crop) in computed tomographic pulmonary angiography (CTPA) for CAD applications. In: Medical Imaging 2012: Computer-Aided Diagnosis, vol. 8315, p. 83150N. International Society for Optics and Photonics (2012)
12. Isensee, F., Kickingereder, P., Wick, W., Bendszus, M., Maier-Hein, K.H.: Brain tumor segmentation and radiomics survival prediction: contribution to the BRATS 2017 challenge. In: Crimi, A., Bakas, S., Kuijf, H., Menze, B., Reyes, M. (eds.) BrainLes 2017. LNCS, vol. 10670, pp. 287–297. Springer, Cham (2018). https://doi.org/10.1007/978-3-319-75238-9_25

13. Santini, G., et al.: An automatic deep learning approach for coronary artery calcium segmentation. In: Eskola, H., Visnen, O., Viik, J., Hyttinen, J. (eds.) EMBEC & NBC 2017. EMBEC 2017, NBC 2017. IFMBE Proceedings, vol. 65, pp. 374–377. Springer, Singapore (2018). https://doi.org/10.1007/978-981-10-5122-7_94
14. Wei, J., et al.: Computerized detection of non-calcified plaques in coronary CT angiography: evaluation of topological soft gradient prescreening method and luminal analysis. In: Medical Physics, vol. 41, no. 8Part1 (2014)
15. AChenbaCh, S., Moshage, W., Ropers, D., BaChmann, K.: Curved multiplanar reconstructions for the evaluation of contrast-enhanced electron beam CT of the coronary arteries. Am. J. Roentgenol. **170**(4), 895–899 (1998)

# Semi-automated Image Segmentation of the Midsystolic Left Ventricular Mitral Valve Complex in Ischemic Mitral Regurgitation

Ahmed H. Aly[1,5(✉)], Abdullah H. Aly[4], Mahmoud Elrakhawy[6],
Kirlos Haroun[1], Luis Prieto-Riascos[7], Robert C. Gorman Jr.[5],
Natalie Yushkevich[2], Yoshiaki Saito[3], Joseph H. Gorman III[3],
Robert C. Gorman[3], Paul A. Yushkevich[2], and Alison M. Pouch[2]

[1] Perelman School of Medicine, University of Pennsylvania, Philadelphia, PA, USA
ahmedaly@pennmedicine.upenn.edu
[2] Department of Radiology, Perelman School of Medicine,
University of Pennsylvania, Philadelphia, PA, USA
[3] Department of Surgery, Perelman School of Medicine,
University of Pennsylvania, Philadelphia, PA, USA
[4] Department of Computer Science, University of Virginia, Charlottesville, VA, USA
[5] Department of Bioengineering, University of Pennsylvania, Philadelphia, PA, USA
[6] Lewis Katz School of Medicine, Temple University, Philadelphia, PA, USA
[7] Department of Anesthesiology, Vagelos College of Physicians and Surgeons,
Columbia University, New York, NY, USA

**Abstract.** Ischemic mitral regurgitation (IMR) is primarily a left ventricular disease in which the mitral valve is dysfunctional due to ventricular remodeling after myocardial infarction. Current automated methods have focused on analyzing the mitral valve and left ventricle independently. While these methods have allowed for valuable insights into mechanisms of IMR, they do not fully integrate pathological features of the left ventricle and mitral valve. Thus, there is an unmet need to develop an automated segmentation algorithm for the left ventricular mitral valve complex, in order to allow for a more comprehensive study of this disease. The objective of this study is to generate and evaluate segmentations of the left ventricular mitral valve complex in pre-operative 3D transesophageal echocardiography using multi-atlas label fusion. These patient-specific segmentations could enable future statistical shape analysis for clinical outcome prediction and surgical risk stratification. In this study, we demonstrate a preliminary segmentation pipeline that achieves an average Dice coefficient of 0.78 ± 0.06.

**Keywords:** Multi-atlas label fusion · Ischemic mitral regurgitation · 3D echocardiography

© Springer Nature Switzerland AG 2019
M. Pop et al. (Eds.): STACOM 2018, LNCS 11395, pp. 142–151, 2019.
https://doi.org/10.1007/978-3-030-12029-0_16

# 1    Introduction

Ischemic mitral regurgitation (IMR) is a disease in which the mitral valve is dysfunctional due to left ventricular remodeling after a myocardial infarction. The two most common surgical treatments of severe IMR are mitral valve repair with annuloplasty and mitral valve replacement. The trade-off between these treatments presents a surgical dilemma. Mitral valve repair avoids risks related to mechanical and prosthetic valve replacement and, in the largest and first randomized control trial, has been shown to produce the best outcomes when patients do not experience disease recurrence post-repair [1,2]. However, the same trial demonstrated that approximately 60% of patients undergoing mitral valve repair experience disease recurrence within 24 months [1,2]. Thus, there is an urgent need to understand and predict post-repair IMR recurrence in the immediate pre-operative setting. The problem of post-surgical recurrence can be significantly reduced if repair is only applied to patients in which durable repair is likely.

Recent studies have attempted to solve this problem by extracting and quantifying mitral valve features, such as leaflet tethering angles, from pre-operative 3D transesophageal echocardiography (TEE) for prediction of IMR in the early post-operative period [4]. While these features show great promise in predicting post-surgical disease recurrence, the most predictive features, thus far, have been ventricular features, such as the presence of a basal aneurysm [3,9]. This is expected as the etiology of IMR is primarily ventricular, and mitral valve dysfunction is secondary to geometric ventricular distortions. To our knowledge, the study of ventricular wall and shape parameters has either been primarily qualitative or has involved laborious manual measurements [3]. As a next step, it is imperative to study the integrated left ventricular mitral valve (LVMV) complex in a consistent and reproducible manner rather than to study the anatomical structures in isolation. It is equally important to automate the segmentation of TEE images in order for LVMV analysis to integrate into the operating room work flow. Previous studies have regionally segmented the left ventricle in isolation with a variety of methods, such as deep learning [13] and active surfaces [14], from different modalities, such as 3D computed tomography (CT) [6] and magnetic resonance imaging (MRI) [7,12], with excellent results. The majority of this research has not been focused on the IMR patient population or pre-operative TEE imaging and has been targeted towards making rapid measurements of left ventricular volume [14]. While these efforts represent the state-of-the-art, our goal is to evaluate patient-specific features of IMR with a comprehensive analysis of the LVMV in the setting of surgical decision making.

The objective of this study is to automate and evaluate the accuracy of LVMV segmentation of 3D TEE images that are acquired pre-operatively in the operating room. Multi-atlas label fusion has proven successful for mitral valve segmentation even with a limited amount of data (i.e., 20 atlases) [5]. Given its previous success and that the LV and MV are relatively consistent in appearance in 3D TEE images acquired from adult patients, we hypothesize that the multi-atlas label fusion approach is an appropriate method for tackling this problem.

## 2    Methods and Materials

### 2.1    Materials

**Data.** A total of 14 pre-operative real-time 3D TEE images were acquired at the Hospital of University of Pennsylvania. Of these, 8 images were acquired from patients with severe IMR and 6 were acquired from patients with normal MV functionality using a transesophageal matrix array transducer and the iE33 platform (Philips Healthcare). In this work, "normal" cases refer to patients who underwent cardiac surgery unrelated to the mitral valve and who have no morphological or functional abnormalities of the MV. In this study, image segmentation focused on the mid-systolic frame of the cardiac cycle, which has been shown to be valuable for predicting post-repair disease recurrence [4]. As a reference for evaluation of multi-atlas segmentation, manual segmentations were generated, and then finalized and reviewed by an expert cardiac surgeon (co-author YS). The manual and multi-atlas segmentations (see Fig. 1) include three labels: the anterior (red) and posterior (green) leaflets of the mitral valve and the left ventricular wall with the exclusion of trabeculae and papillary muscles (blue). In order to initialize atlas-based registration, 5 landmarks were identified in the mid-systolic frame. As seen in Fig. 1, these 5 landmarks consist of the anterior aortic peak (AAOP) of the annulus, anterior commissure (ACM), mid-posterior annulus (MPA), posterior commissure (PCM) and the midpoint of the coaptation zone (COAP). In this study, the commissures are defined as landmarks on the mitral annulus that are points of transition between the anterior and posterior leaflets. We approximate the commissures to be on the annulus since these points are most consistently identifiable in mid-systolic TEE images.

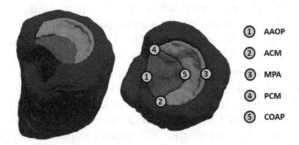

**Fig. 1.** (left) The left ventricle mitral valve manual segmentation with three components: anterior leaflet (red), posterior leaflet (green) and left ventricle (blue). (right) These 5 mitral valve anatomical landmarks are used for the affine transform initialization. (Color figure online)

### 2.2    Multi-atlas Label Fusion

Given the limited data set and the consistency in LVMV appearance in 3D TEE, we can utilize multi-atlas label fusion (MALF) [8]. The MALF algorithm is composed of two steps: (1) registration of a collection of atlases to a target image to generate candidate segmentations and (2) joint label fusion of the registered atlas segmentations form the consensus multi-atlas segmentation. First, we register the atlases to a target image using a combination of affine and diffeomorphic transformations. The affine transformation is generated from user-defined landmarks, and the diffeomorphic transformation is calculated using the Advanced Normalization Tools (ANTs) diffeomorphic deformation framework [10]. We apply joint label fusion to combine the resulting candidate segmentations to form the consensus label map. Joint label fusion is particularly advantageous because it is a locally varying weighted voting scheme that optimizes voting weights based on similarities between the atlases and target image. These optimal voting weights account for similarities in the atlases themselves, which helps to reduce bias associated with redundancy in the atlas set. Figure 2 summarizes the multi-atlas pipeline.

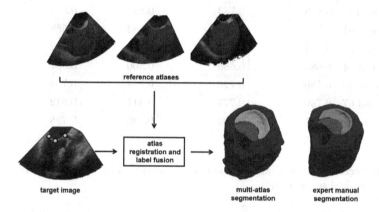

**Fig. 2.** Multi-atlas label fusion pipeline shown. The first step uses the manual input of 5 landmarks on the target image followed by atlas registration to generate the candidate segmentations. The second step applies joint label fusion to output the multi-atlas segmentation.

## 3   Results

A leave-one-out evaluation was performed in order to assess the accuracy of the resulting segmentations. For each of the 14 semi-automated image segmentations performed, the images from the other 13 subjects were used as atlases for multi-atlas segmentation, and the multi-atlas segmentation result was evaluated against the manual segmentation from the same subject. This process was performed for each subject. In order to evaluate reproducibility of the semi-automated method, we repeated this process for a set of target images that were initialized by a second observer.

The Dice coefficients with reference to the manual segmentations for all three individual labels and for the combined labels were used as the metric for comparison. From Tables 1 and 2, we see that while the Dice coefficients for each

**Table 1.** Accuracy of multi-atlas segmentation initialized by Observer 1 with respect to expert manual segmentation

| ID | Dice coefficients | | | Overall Dice |
|---|---|---|---|---|
| malf seg | Anterior leaflet | Posterior leaflet | Left ventricle | LVMV |
| imr 01 | 0.736 | 0.693 | 0.765 | 0.767 |
| imr 02 | 0.746 | 0.626 | 0.764 | 0.767 |
| imr 05 | 0.772 | 0.724 | 0.827 | 0.828 |
| imr 06 | 0.785 | 0.777 | 0.824 | 0.825 |
| imr 08 | 0.851 | 0.805 | 0.781 | 0.782 |
| imr 09 | 0.839 | 0.813 | 0.759 | 0.761 |
| imr 11 | 0.821 | 0.820 | 0.839 | 0.840 |
| imr 15 | 0.843 | 0.783 | 0.810 | 0.812 |
| nrml 01 | 0.798 | 0.770 | 0.584 | 0.588 |
| nrml 03 | 0.774 | 0.714 | 0.790 | 0.792 |
| nrml 04 | 0.683 | 0.666 | 0.800 | 0.801 |
| nrml 06 | 0.760 | 0.726 | 0.752 | 0.754 |
| nrml 07 | 0.790 | 0.777 | 0.811 | 0.812 |
| nrml 08 | 0.777 | 0.794 | 0.786 | 0.788 |

**Table 2.** Accuracy of multi-atlas segmentation initialized by Observer 2 with respect to expert manual segmentation

| ID | Dice coefficients | | | Overall Dice |
|---|---|---|---|---|
| malf seg | Anterior leaflet | Posterior leaflet | Left ventricle | LVMV |
| imr 01 | 0.740 | 0.698 | 0.779 | 0.780 |
| imr 02 | 0.762 | 0.599 | 0.765 | 0.768 |
| imr 05 | 0.771 | 0.730 | 0.826 | 0.826 |
| imr 06 | 0.762 | 0.776 | 0.841 | 0.841 |
| imr 08 | 0.846 | 0.811 | 0.754 | 0.756 |
| imr 09 | 0.847 | 0.830 | 0.762 | 0.764 |
| imr 11 | 0.776 | 0.775 | 0.844 | 0.845 |
| imr 15 | 0.780 | 0.752 | 0.769 | 0.772 |
| nrml 01 | 0.764 | 0.743 | 0.703 | 0.705 |
| nrml 03 | 0.771 | 0.711 | 0.776 | 0.778 |
| nrml 04 | 0.674 | 0.663 | 0.800 | 0.800 |
| nrml 06 | 0.756 | 0.731 | 0.724 | 0.727 |
| nrml 07 | 0.756 | 0.738 | 0.800 | 0.800 |
| nrml 08 | 0.791 | 0.802 | 0.789 | 0.791 |

**Table 3.** Inter-observer Dice coefficients generated from the overlap between multi-atlas segmentations initialized by Observers 1 and 2

| ID | Dice coefficients | | | Overall Dice |
|---|---|---|---|---|
| malf seg | Anterior leaflet | Posterior leaflet | Left ventricle | LVMV |
| imr 01 | 0.927 | 0.922 | 0.905 | 0.906 |
| imr 02 | 0.890 | 0.869 | 0.880 | 0.881 |
| imr 05 | 0.924 | 0.917 | 0.905 | 0.906 |
| imr 06 | 0.892 | 0.889 | 0.897 | 0.898 |
| imr 08 | 0.885 | 0.877 | 0.846 | 0.847 |
| imr 09 | 0.916 | 0.904 | 0.907 | 0.907 |
| imr 11 | 0.872 | 0.880 | 0.892 | 0.893 |
| imr 15 | 0.856 | 0.864 | 0.841 | 0.843 |
| nrml 01 | 0.893 | 0.869 | 0.674 | 0.679 |
| nrml 03 | 0.923 | 0.915 | 0.888 | 0.889 |
| nrml 04 | 0.920 | 0.918 | 0.901 | 0.901 |
| nrml 06 | 0.912 | 0.905 | 0.879 | 0.880 |
| nrml 07 | 0.875 | 0.871 | 0.885 | 0.885 |
| nrml 08 | 0.936 | 0.933 | 0.930 | 0.931 |

segmented label may vary, the performance of MALF is consistent. The average Dice coefficients for the anterior and posterior leaflets and LV across both experiments are $0.78 \pm 0.04$, $0.74 \pm 0.06$, $0.78 \pm 0.05$, respectively. From Table 3, we see that when examining the inter-observer Dice coefficients (i.e., the overlap between the multi-atlas segmentations initialized by Observers 1 and 2) demonstrate high reproducibility in semi-automated segmentation. The average inter-observer Dice coefficients for the anterior and posterior leaflets and LV are $0.90 \pm 0.02$, $0.90 \pm 0.02$, $0.87 \pm 0.06$, respectively.

Figure 4 plots the Dice coefficients with reference to manual segmentations for the multi-atlas and registered atlas segmentations. Figure 3 shows the "Dice box", a qualitative summary of the overall Dice coefficients, which allows for evaluation of the registration of each atlas segmentation onto its target image. The presented Dice coefficients are calculated in reference to the expert manual segmentations of the target subject. Of note, the demonstrated Dice coefficients of the MALF segmentation is consistently greater when examining each row of the "Dice box". This trend is quantitatively confirmed in Fig. 4.

## 4    Discussion

The ability to automatically generate accurate representations of the LVMV complex from 3D TEE images is a critical step towards predicting post-surgical outcomes of treatment for severe IMR in the operation room. Previous work

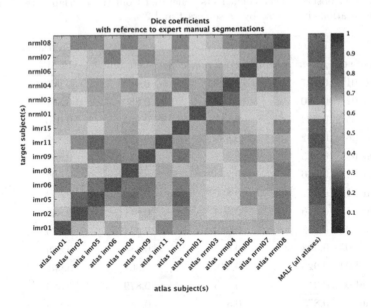

**Fig. 3.** Demonstrates a qualitative summary of registration performance and mean Dice coefficients across both experiments for each of the candidate segmentations, in order to highlight outliers.

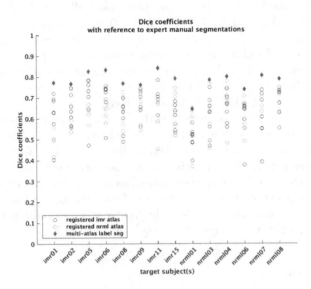

**Fig. 4.** Demonstrates a quantitative summary of the registration performance and mean Dice coefficients across both experiments for each of the candidate segmentations.

has shown that geometric features such as left ventricular basal aneurysm had the highest predictive value (AUC = 0.94 [9] and AUC = 0.83 [3]) when combined with mitral valve-derived features. The automation of the segmentation is a significant contribution to accelerating intraoperative decision making and allowing for a more quantitative analysis of the LVMV complex in severe IMR for improving patient risk stratification.

Towards this goal, this work demonstrates promising results in semi-automated LVMV segmentation with average Dice scores of 0.78 ± 0.06 (all labels) and 0.79 ± 0.03 (excluding nrml01, an outlier) relative to manual segmentation. The method also has high reproducibility with average inter-observer Dice scores for the semi-automated combined LVMV segmentation of 0.87 ± 0.06 and 0.89 ± 0.02 (excluding nrml01). Atlas-based segmentation accuracy is related to the effectiveness of the registration of each atlas to the target image. The advantage of utilizing label fusion for combining the candidate segmentations of all atlases and correcting biases in the atlas set is clearly demonstrated in Fig. 4. Here, the MALF segmentations have the highest Dice coefficient values compared to the individual candidate segmentations. Overall the quality of the semi-automated segmentations was both quantitatively and qualitatively high (see Fig. 3) with the exception of the segmentation of nrml01, which was the outlier of the group. For the nrml01 dataset, the average Dice coefficient of individual atlas registrations was 0.56 ± 0.15 and the Dice coefficient of the multi-atlas label fusion was 0.65. This was due to poor image quality caused by signal attenuation that led to a visibly occluded left ventricular apex region, as seen in Fig. 5. This problem could potentially be addressed in further studies by adding a template-based shape modeling component to the pipeline as discussed in [5], which has been shown to correct for missing components in some mitral valve segmentations.

The results of this study, which were produced with a limited dataset of 13 reference atlases, represents significant progress towards generating statistical atlases for further clinical analyses. The next step is to extract measurements and analyze the LVMV structure in terms of morphology. To capture the variations of the joint LVMV complex we could potentially utilize continuous medial representation (cm-rep), which is an excellent candidate for geometric modeling that has allowed for previous success in modeling the mitral valve accurately and for reliable feature extraction and statistical analysis [15, 16]. Finally, the current algorithm is a CPU implementation and is not optimized in terms of speed, having a runtime of approximately 3 h. The runtime for the LVMV segmentation is much larger than that for the segmentation of the valve alone since the field of view is larger. This can be improved in the future by using a GPU implementation for the registration steps, which has been shown to decrease the runtime by a factor of 55 [11]. If successful, the combination of accurate segmentation and modeling will allow for more sophisticated morphometric measurements and, ultimately, a more informed decision in the immediate surgical setting.

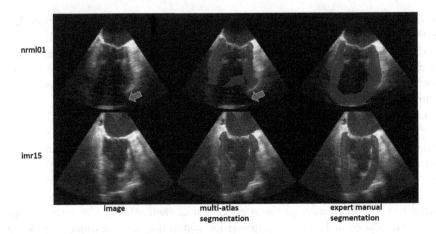

nrml01

imr15

image            multi-atlas            expert manual
                 segmentation           segmentation

**Fig. 5.** nrml01 (top) image exhibits signal dropout (red arrow) leading to the poor registration of candidate segmentations that affect the multi-atlas segmentation. We can clearly see this in comparison to imr15 (bottom), which is a representative segmentation in the leave-one-out study. (Color figure online)

In conclusion, multi-atlas segmentation with joint label fusion is a promising tool for segmentation of the LVMV from 3D TEE pre-operative images. Semi-automated segmentation was demonstrated both in patients with severe IMR and in patients with normal MV function using TEE images acquired in the pre-operative setting. In the future, we will fully automate the segmentation and employ geometric modeling for comprehensive morphological assessment of the LVMV complex.

**Acknowledgements.** This research was funded by the National Institutes of Health (R01 EB017255).

# References

1. Acker, M.A., et al.: Mitral-valve repair versus replacement for severe ischemic mitral regurgitation. N. Engl. J. Med. **370**(1), 23–32 (2014)
2. Goldstein, D., et al.: Two-year outcomes of surgical treatment of severe ischemic mitral regurgitation. N. Engl. J. Med. **374**(4), 344–353 (2016)
3. Kron, I., et al.: Predicting recurrent mitral regurgitation after mitral valve repair for severe ischemic mitral regurgitation. J. Thorac. Cardiovasc. Surg. **149**, 752 (2015)
4. Bouma, W., et al.: Preoperative three-dimensional valve analysis predicts recurrent ischemic mitral regurgitation after mitral annuloplasty. Ann. Thorac. Surg. **101**, 567 (2016)
5. Pouch, A.M., et al.: Fully automatic segmentation of the mitral leaflets in 3D transesophageal echocardiographic images using multi-atlas joint label fusion and deformable medial modeling. Med. Image Anal. **18**(1), 118–129 (2014)

6. Zheng, Y., et al.: Four-chamber heart modeling and automatic segmentation for 3D cardiac CT volumes using marginal space learning and steerable features. IEEE Trans. Med. Imaging **36**(11), 2287–2296 (2008)

7. Shahzad, R., et al.: Fully-automatic left ventricular segmentation from long-axis cardiac cine MR scans. Med. Image Anal. **39**, 44 (2017)

8. Wang, H., et al.: Multi-atlas segmentation with joint label fusion. IEEE Trans. Pattern Anal. Mach. Intell. **7**, 27 (2013)

9. Wijdh-den Hamer, I.J., et al.: The value of preoperative 3-dimensional over 2-dimensional valve analysis in predicting recurrent ischemic mitral regurgitation after mitral annuloplasty. J. Thorac. Cardiovasc. Surg. **152**, 847 (2016)

10. Avants, B.B., et al.: Symmetric diffeomorphic image registration with cross-correlation: evaluating automated labeling of elderly and neurodegenerative brain. Med. Image Anal. **12**, 26–41 (2008)

11. Mousazadeh, H., Marami, B., Sirouspour, S., Patriciu, A.: GPU implementation of a deformable 3D image registration algorithm. In: 2011 Annual International Conference of the IEEE Engineering in Medicine and Biology Society (2011)

12. Wake, N., et al.: Whole heart self-navigated 3D radial MRI for the creation of virtual 3D models in congenital heart disease. J. Cardiovasc. Magn. Reson. **18**(Suppl 1), P185 (2016)

13. Zheng, Q., et al.: 3D consistent & robust segmentation of cardiac images by deep learning with spatial propagation. IEEE Trans. Med. Imaging **37**(9), (2018). https://ieeexplore.ieee.org/document/8327905

14. Pedrosa, J., et al.: Fast and fully automatic left ventricular segmentation and tracking in echocardiography using shape-based B-spline explicit active surfaces. IEEE Trans. Med. Imaging **36**(11), 2287–2296 (2017)

15. Pouch, A.M., et al.: Statistical assessment of normal mitral annular geometry using automated three-dimensional echocardiographic analysis. Ann. Thorac. Surg. **97**, 71 (2014)

16. Pouch, A.M., et al.: Development of a semi-automated method for mitral valve modeling with medial axis representation using 3D ultrasound. Med. Phys. **39**, 933 (2012)

# Atrial Scar Segmentation via Potential Learning in the Graph-Cut Framework

Lei Li[1,2], Guang Yang[3], Fuping Wu[2], Tom Wong[3], Raad Mohiaddin[3],
David Firmin[3], Jenny Keegan[3], Lingchao Xu[4],
and Xiahai Zhuang[2(✉)]

[1] School of Biomedical Engineering,
Shanghai Jiao Tong University, Shanghai, China
[2] School of Data Science, Fudan University, Shanghai, China
zxh@fudan.edu.cn
[3] National Heart and Lung Institute, Imperial College London, London, UK
[4] School of NAOCE, Shanghai Jiao Tong University, Shanghai, China

**Abstract.** Late Gadolinium Enhancement Magnetic Resonance Imaging (LGE MRI) emerges as a routine scan for patients with atrial fibrillation (AF). However, due to the low image quality automating the quantification and analysis of the atrial scars is challenging. In this study, we proposed a fully automated method based on the graph-cut framework, where the potential of the graph is learned on a surface mesh of the left atrium (LA), using an equidistant projection and a deep neural network (DNN). For validation, we employed 100 datasets with manual delineation. The results showed that the performance of the proposed method was improved and converged with respect to the increased size of training patches, which provide important features of the structural and texture information learned by the DNN. The segmentation could be further improved when the contribution from the t-link and n-link is balanced, thanks to the interrelationship learned by the DNN for the graph-cut algorithm. Compared with the existing methods which mostly acquired an initialization from manual delineation of the LA or LA wall, our method is fully automated and has demonstrated great potentials in tackling this task. The accuracy of quantifying the LA scars using the proposed method was 0.822, and the Dice score was 0.566. The results are promising and the method can be useful in diagnosis and prognosis of AF.

## 1 Introduction

Atrial fibrillation (AF) is the most common arrhythmia of clinical significance, occurring in up to 2% of the population and rising fast with advancing age. It is associated with structural, contractile and electrical remodeling, and also related to increased morbidity and mortality. Catheter ablation (CA) using the pulmonary vein (PV) isolation technique has emerged as the most common methods for the AF patients who are not responding to pharmacological treatments.

This work was supported by Science and Technology Commission of Shanghai Municipality (17JC1401600).

M. Pop et al. (Eds.): STACOM 2018, LNCS 11395, pp. 152–160, 2019.
https://doi.org/10.1007/978-3-030-12029-0_17

Generally, the clinical reference standard technique for the assessment of atrial scars is the electro-anatomical mapping (EAM) [1], performed during an electro-physiological (EP) study prior to CA. However, due to its invasiveness, the use of ionizing radiation and suboptimal accuracy, a non-invasive Late Gadolinium Enhancement Magnetic Resonance Imaging (LGE MRI) is a promising alternative and potentially a more accurate imaging technique. LGE MRI allows the detection of native fibrosis and ablation induced scaring based on the mechanism of slow washout kinetics of the gadolinium in the scar area.

Before the delineation of atrial scars, an accurate segmentation of the atrium wall is required for the exclusion of outer structures with the similar intensities as the atrial scars and the inclusion of all the scar areas. However, automating this segmentation remains challenging due to the possible poor quality of the LGE MRI images, the thin thickness of the left atrium (LA) wall ($\sim 1$–$3$ mm) and the various LA shape. Most of the methods for scar segmentation used the manual segmentation of the LA or LA wall to provide an accurate initialization for scar delineation [2, 3]. There is a benchmark paper which evaluated different methods for the LA segmentation [4]. Vein et al. [5] proposed an algorithm named ShapeCut, which combined a shape-based system and the graph cut approach to make a Bayesian surface estimation for the LA wall segmentation. Currently, the most widespread method for scar segmentation is still based on thresholding. However, the choice of an appropriate threshold relies on subjective opinions from a domain expert, eventually limiting the external validation and reproducibility of this method. A reliable and reproducible method to detect and segment atrial scars on the LA wall remains an open question.

In this paper, we present a novel fully automatic framework for the segmentation and quantification of atrial scars. We tackled the quantitative analysis of atrial scars by developing a graph-based segmentation with the learned t-/n-link potential in the flatten LA surface. It starts by the segmentation of the LA endocardium from whole heart segmentation (WHS) by combining LGE MRI and the anatomical 3D MRI, referred to as Ana-MRI based on b-SSFP sequence. Then the segmented LA wall is projected onto the surface for a graph cut. Finally, the two weights (n-link and t-link potentials) of the graph are designed and explicitly learned from a Deep Neural Work (DNN).

## 2   Method

Figure 1 presents the overall workflow of the proposed method, which consists of three major components: (1) 2D graph-cut with well-defined n/t link weights was performed to automatically segment the atrial scars from the LA wall, (2) an equidistant projection of the 3D LA endocardium to a 2D map and (3) explicit learning of the potentials for edge weights of the graph.

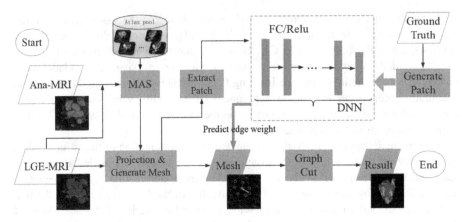

**Fig. 1.** Flowchart of the proposed segmentation framework.

## 2.1 Initialization of Atrial Endocardium

Multi-atlas segmentation (MAS) of whole heart was used to generate the initial segmentation of LA endocardium, which includes the multi-atlas propagation based on image registration and label fusion [6]. WHS offers adequate information of the LA boundary and there's a multi-modality WHS (MM-WHS) challenge which was held in conjunction with STACOM and MICCAI 2017 [7]. As the LGE MRI images could have relatively poor image quality, we propose to propagate the segmentation result of Ana-MRI to the LGE MRI using a primitive registration [8]. After the atlas propagation, we adopt the local weighted label fusion to cover the different texture patterns of LGE MRI and b-SSFP images. At the same time, a multi-scale multi-modality patch-based label fusion algorithm was used [9], because the intensity distribution of LA blood pool is almost the same as that of the blood pool from other chambers.

Having finished the WHS for LA and PV delineation (mean Dice = 0.89), we then generate an initial LA image labeled with blood pool and mitral valve. The classical marching cubes algorithm was then used to obtain a mesh surface of LA wall which excludes the mitral valve.

## 2.2 Graph Formulation for Scar Segmentation via Equidistant Projection

Classification and quantification of scars can be formulated as a 2D graph-cut problem with well-defined n/t link weights. In our study, the graph is formed by a set of vertices and edges defined in a surface, which is obtained from the 3D LA endocardium by an equidistant projection (details in Sect. 2.3). It should be noted preservation of distance is required in the definition of n-link weights in the proposed graph cut framework. Associated with each vertex is an intensity profile that is consisted of a patch defined along the normal direction of the target surface. The cost function of the graph cut problem includes both region and boundary terms of the segments, which can be explicitly learned (details in Sect. 2.4).

Let $G = \{S, \mathcal{N}\}$ denote a graph, where $S = \{p_i\}$ indicates the set of graph nodes, $\mathcal{N} = \{\langle p_i, p_j \rangle\}$ is the set of edges. Let $l_{p_i} \in \{0, 1\}$ be the label assigned to $p_i$ and $l = \{l_{p_i}, p_i \in S\}$ be the label vector that defines a segmentation. Then the segmentation energy can be defined as:

$$E(l|\theta^0, \theta^1) = E_D(l|\theta^{l_{p_i}}) + \lambda E_B(l)$$
$$= \sum_{p_i \in S} W_{p_i}^{t-link}(l_{p_i}) + \lambda \sum_{(p_i, p_j) \in \mathcal{N}} W_{\{p_i, p_j\}}^{n-link} \delta(l_{p_i}, l_{p_j}) \tag{1}$$

where $\delta(l_{p_i}, l_{p_j})$ is the Kronecker delta function, namely the function is 1 if the two variables are equal, and 0 otherwise.

In conventional graph-based segmentation [10], the appearance models $(\theta^0, \theta^1)$ are considered as a prior by manual defining some seed points (i.e., graph-cut) or giving a bounding box for interactive segmentation (i.e., grab-cut). Different from these interactive methods, we propose to directly learns the t- and n-link weights using a DNN.

### 2.3   Projection of the Atrial Myocardium

Localization and quantification of the atrial scars can be made using the EAM system, which only focuses on the surface area of the LA. Inspired by this, we proposed a method to allow simultaneous representation of multiple parameters on a surface based on the template of an average LA mesh that is similar as Williams et al. [11]. In this scheme, the atrial scar can be classified on the flattened surface projected from the 3D LA geometry, thereafter both the errors due to LA wall thickness and the misregistration of the WHS can be mitigated. In this formulation, each vertex (or node) on the surface should include an intensity profile that represents the texture information of corresponding location in the LGE MRI. This idea is incorporated into the learning of the graph representing the surface.

### 2.4   Explicit Learning of Graph Potentials

The atrial scars are classified using the graph-cut algorithm on the projected surface graph, whose t-/n-link potentials are learned using a DNN. Figure 2 provides the pipeline of the training and testing phases.

For the training samples, which are nodes on the graph with intensity profiles, we propose to extract a patch at the corresponding location in the LGE MRI by back projection. The local orientation of the patches is defined on the local coordinate of the pixel, such that the x- and y- axis of the patch are aligned with the tangent plane of the surface in the 3D geometry of the LA myocardium. To further mitigate the effect of misalignment of the LA surface, which is computed from the WHS, an internal and external extension along the normal direction of the surface is considered.

After the construction of patch library, the labeling of them is based on the ground truth of scar and the WHS. For training of the t-link potentials, patches with labels are considered. For training of the n-link, a sample is defined with a patch pair (associated with two connected nodes), the geodesic distance between the two nodes, and their

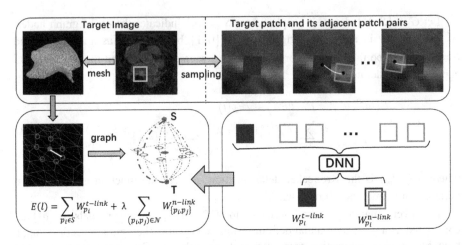

**Fig. 2.** Pipeline of our proposed method at the training and testing phases

similarity. Here, the similarity is computed from their labels or probability of being a label, with higher similarity value if they have the same label or higher probability of being the same label, and lower similarity otherwise.

To learn the t-link potentials, we use a fully connected neural network. Feeding the patch around a node into the framework, one can obtain the probability of the node belonging to scar or background separately. Similarly, we designed another neural network to calculate the n-link of two connected nodes, with the corresponding two patches and distance as the input. To account the distance, the training procedure begins by projecting each patch into a multi-feature space and then combines with the distance feature as a whole to obtain the final weight.

## 3 Experiment

### 3.1 Data Acquisition

We acquired 100 datasets for experiments. The patients were with longstanding persistent AF on a 1.5T Siemens Magnetom Avanto scanner (Siemens Medical Systems, Erlangen, Germany). Transverse navigator-gated 3D LGE-CMRI was performed using an inversion prepared segmented gradient echo sequence (TE/TR 2.2 ms/5.2 ms) 15 min after gadolinium administration. The inversion time was set to null the signal from normal myocardium. Detailed scanning parameters are 30–34 slices at $(1.4–1.5) \times (1.4–1.5) \times 4$ mm$^3$, reconstructed to 60–68 slices at $(0.7–0.75) \times (0.7–0.75) \times 2$ mm$^3$, field-of-view $380 \times 380$ mm$^2$. For each patient, prior to contrast agent administration, coronal navigator-gated 3D b-SSFP (TE/TR 1 ms/2.3 ms) data were also acquired with the following parameters: 72–80 slices at $(1.6–1.8) \times (1.6–1.8) \times 3.2$ mm$^3$, reconstructed to 144–160 slices at $(0.8–0.9) \times (0.8–0.9) \times 1.6$ mm$^3$, field-of-view $380 \times 380$ mm$^2$. Both LGE-CMRI and b-SSFP data were acquired during free-breathing using a crossed-pairs navigator positioned over the dome of the right hemidiaphragm with navigator acceptance window size of 5 mm and CLAWS respiratory motion control.

## 3.2 Gold Standard and Evaluation

The 100 LGE MRI images were all manually segmented by experienced cardiologists specialized in cardiac MRI to label the enhanced atrial scar regions. The manual delineation is regarded as gold standard for experiments. The 100 datasets were randomly divided into two equal-size sets, one for training and the other for test. For the evaluation, we computed the accuracy, sensitivity, specificity and Dice scores between the classification sets, respectively from ground truth and the automatic segmentation.

We first investigated the influence of different patch sizes. Then, the performance of the proposed method with different value of the balancing parameter $\lambda$ was studied. Finally, the optimal parameters were used for the comparison study. It should be noted that in the comparison study, the proposed method is equivalent to a classification method which is purely based on the DNN for prediction of scars when $\lambda = 0$.

## 4 Result and Discussion

Figure 3(a) presents the mean Dice scores of the proposed method using different sizes of the patches. One can see from the figure that the Dice coefficient first increases dramatically, then tends to converge after the patch size becoming more than $9 \times 9 \times 13$, and the best Dice score was obtained when the method used the largest size of patches. This is reasonable, as the larger size of the patches is used, the richer intensity profile is included for feature training and detection. When the patch size is larger than it is necessary, the DNN will simply discard the useless information in the training and detection, thanks to the well-designed architecture of the DNN.

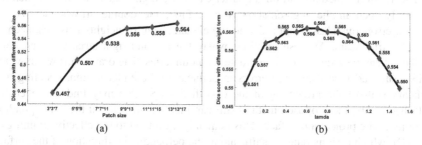

**Fig. 3.** Dice scores of the proposed method with different parameterizations: (a) performance against different patch sizes; (b) performance against the balancing parameter $\lambda$ to weight the t-link term and n-link term in the graph-cut framework.

The result of the parameter study on $\lambda$ is presented in Fig. 3(b). One can see that the best performance in terms of Dice score was obtained when $\lambda$ is set to 0.6, meaning inter-node relation is important and should be considered for scar classification, and the weight between the t-link and n-link terms should be balanced, in order to achieve the optimal results.

For comparisons, we evaluated two conventional algorithms, i.e. the Ostu intensity threshold [12] and the Gaussian mixture model (GMM) [13] for intensity classification. Both methods required the initialization of the LA wall segmentation from the WHS of the corresponding Ana-MRI image in each dataset. The quantitative results of the two methods are presented in Table 1, compared with the proposed learning graph-cut algorithm, referred to as LearnGC in the table. By compared the performance of pre- and post-ablation scans separately, we conclude that the quantification of the pre-ablation cases is more challenging. It could be due to the fact that the fibrosis appears more diffuse and has greater overlap with normal wall. In addition, we implemented the threshold (UTA) and graph cuts (KCL), both benchmarked in the Challenge [3], but initialized by our automatic LA segmentation. Their Dice scores, UTA: 0.360 and KCL: 0.485, were both significantly lower than ours (0.566) ($p < 0.0001$).

**Table 1.** Summary of the quantitative evaluation methods.

| Method | Accuracy | Sensitivity | Specificity | Dice |
|---|---|---|---|---|
| Ostu | $0.496 \pm 0.235$ | $0.853 \pm 0.225$ | $0.290 \pm 0.188$ | $0.360 \pm 0.164$ |
| GMM | $0.716 \pm 0.096$ | $0.961 \pm 0.049$ | $0.370 \pm 0.160$ | $0.467 \pm 0.155$ |
| LearnGC | $0.822 \pm 0.065$ | $0.932 \pm 0.048$ | $0.515 \pm 0.167$ | $0.566 \pm 0.140$ |
| LearnGC (post) | $0.807 \pm 0.070$ | $0.921 \pm 0.046$ | $0.552 \pm 0.149$ | $0.652 \pm 0.084$ |
| LearnGC (pre) | $0.839 \pm 0.054$ | $0.987 \pm 0.049$ | $0.360 \pm 0.135$ | $0.465 \pm 0.125$ |

Figure 4 visualizes three examples from the test set, i.e. the first quarter, median and third quarter cases in terms of Dice scores using the proposed method. One can see that the method could provide promising results for localization and quantification of atrial scars. In the median and third quarter cases, we highlighted the errors which are commonly seen in the segmentation results. First, the double arrows identify the potential erroneous classification due to the enhancement of neighbor tissues, such as the right atrial scars. At these locations where the left and right atrial walls are joint together, it can be arduous to differentiate the boundaries. The orange arrows show the WHS result had relatively large errors in delineating the endocardium, resulting in a boundary shift of the projecting of the classified scars. One may find that even the WHS had errors, the proposed method still can identify the scars at the corresponding locations of the projecting surface. This is mainly attributed to the effective training of the DNN, which assigns random shifts along the perpendicular direction of the surface when extracting the training patches. This capability contributes to less demanding of the WHS accuracy for providing initialization and reconstruction of the LA surface mesh.

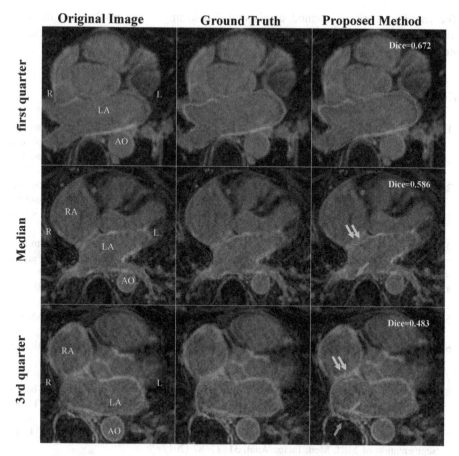

**Fig. 4.** 2D visualization of the final atrial scar segmentation results of the first quarter (a), median (b) and third quarter cases (c) in the test dataset by the proposed method. Double yellow arrows in (b) and (c) show the enhancement due to the effect of the scar from right atrium (RA). And single orange arrows in (b) and (c) show the errors due to the misregistration of WHS. (Color figure online)

## 5   Conclusion

We proposed a fully automated method for segmentation and quantification of atrial wall scars. Two major technical contributions are introduced. One is the graph-cut framework combined with the idea of projecting the 3D LA wall onto a surface graph for graph-based segmentation. The other is the adoption of the DNN, which extracts the features from local intensity profiles of LGE MRI for learning the relations or similarity of node pairs in the constructed graph. In addition, we achieved the automatic segmentation of LA, which can provide an accurate enough initialization for scar segmentation. The proposed method performs better when the size of extracted patches increases, but the performance converges when the size is larger than a certain value.

The learning graph-cut method demonstrated evidently better results than the two conventional methods (p < 0.0001). The Dice scores for quantifying LA and LA scars are respectively 0.891 and 0.566.

# References

1. Calkins, H., Kuck, K., Cappatoet, R., et al.: 2012 HRS/EHRA/ECAS expert consensus statement on catheter and surgical ablation of atrial fibrillation: Recommendations for patient selection, procedural techniques, patient management and follow-up, definitions, endpoints, and research trial design. Heart Rhythm **4**(6), 816–861 (2012)
2. Perry, D., Morris, A., Burgon, N., McGann, C., MacLeod, R., Cates, J.: Automatic classification of scar tissue in late gadolinium enhancement cardiac MRI for the assessment of left-atrial wall injury after radiofrequency ablation. In: Proceedings of SPIE, vol. 8315, pp. 8351D–8351D-9 (2012)
3. Karim, R., et al.: Evaluation of current algorithms for segmentation of scar tissue from late Gadolinium enhancement magnetic resonance of the left atrium: an open-access grand challenge. J. Cardiovasc. Magn. Reson. **15**(1), 105–121 (2013)
4. Tobon-Gomez, C., et al.: Benchmark for algorithms segmenting the left atrium from 3D CT and MRI datasets. IEEE Trans. Med. Imaging **34**(7), 1460–1473 (2015)
5. Veni, G., Elhabian, S., Whitaker, R.: ShapeCut: Bayesian surface estimation using shape-driven graph. Med. Image Anal. **40**, 11–29 (2017)
6. Zhuang, X.: Challenges and methodologies of fully automatic whole heart segmentation: a review. J. Healthc. Eng. **4**(3), 371–407 (2013)
7. http://www.sdspeople.fudan.edu.cn/zhuangxiahai/0/mmwhs/
8. Zhuang, X., Rhode, K., Razavi, R., Hawkes, D.J., Ourselin, S.: A registration-based propagation framework for automatic whole heart segmentation of cardiac MRI. IEEE Trans. Med. Imaging **29**(9), 1612–1625 (2010)
9. Zhuang, X., Shen, J.: Multi-scale patch and multi-modality atlases for whole heart segmentation of MRI. Med. Image Anal. **31**, 77–87 (2017)
10. Boykov, Y.Y., Jolly, M.P.: Interactive graph cuts for optimal boundary & region segmentation of objects in N-D images. In: IEEE International Conference on Computer Vision, pp. 105–112 (2001)
11. Williams, S.E., Tobon-Gomez, C., et al.: Standardized unfold mapping: a technique to permit left atrial regional data display and analysis. J. Interv. Card. Electrophysiol. **50**(1), 125–131 (2017)
12. Otsu, N.: A threshold selection method from gray-level histograms. IEEE Trans. Syst. Man Cybern. **SMC-9**(1), 62–66 (1979)
13. Leemput, K.V., Maes, F., Vandermeulen, D., Suetens, P.: Automated model-based tissue classification of MR images of the brain. IEEE Trans. Med. Imaging **18**(10), 897–908 (1999)

# 4D Cardiac Motion Modeling Using Pair-Wise Mesh Registration

Siyeop Yoon[1,2] ⓘ, Stephen Baek[3] ⓘ, and Deukhee Lee[1,2(✉)] ⓘ

[1] Korea University of Science and Technology,
Hwarang-ro 14gil 5, Seoul 02792, South Korea
[2] KIST, Hwarang-ro 14gil 5, Seoul 02792, South Korea
{H14515,dkylee}@kist.re.kr
[3] University of Iowa, Iowa City, IA 52242, USA
stephen-baek@uiowa.edu

**Abstract.** In this paper, we present a novel method for the real-time cardiac motion compensation. Our method generates interpolated cardiac motion using segmented mesh models from preoperative 3D+T computed tomography angiography (CTA). We propose a pair-wise mesh registration technique for building correspondence and interpolating the control points over a cardiac cycle. The key contribution of this work is a rapid creation of a deformation field through a concise mathematical formulation while maintaining desired properties. These are $C^2$ continuity, invertibility, incompressibility of cardiac structure and capability to handling large deformation. And we evaluated the proposed method using different conditions, such as deformation resolution, temporal sampling rates, and template model selection.

**Keywords:** Cardiac · Spatiotemporal · Registration · Deformation · 4D motion modeling

## 1 Introduction

Cardiovascular diseases (CVDs) are a leading cause of death in developed countries [3]. Not only CVDs is one of the most mortality but also it is reported as a major cost burden on health-care system in North America [14]. Health-care cost of CVDs is continuously growing as developed countries become an aging society [14]. In practical treatment of CVDs, cardiovascular imaging systems and techniques became a gold standard for diagnosis and functional analysis [4,12]. Analyzing a cardiac motion over cycle, it helps to reduce morbidity and mortality induced from cardiovascular disease [11].

A number of studies for cardiac motion have been conducted using spatiotemporal images, such that ECG-gated CT angiography [8] and tagged MRI [7].

This work was supported by Industrial Strategic Technology Development Programs (10052980, 10077502) funded By the Korea Ministry of Trade, Industry and Energy (MI, Korea).

M. Pop et al. (Eds.): STACOM 2018, LNCS 11395, pp. 161–170, 2019.
https://doi.org/10.1007/978-3-030-12029-0_18

In order to estimate a cardiac motions, registration method is crucial to building a correspondence between anatomic land-marks and to computing a deformation fields between different cardiac phases [6, 9, 13]. Although many of authors have studied motion modeling of using registration methods, these are still suffered from the heavy computation of displacement field and lack of correspondence between scenes. Also, a large amount of data is required for statistical population-based approach to cover a variability of shapes.

In this study, we propose a novel 4D cardiac motion modeling method using cage deformation scheme. For efficient computation, we formulate a representation of an object using control points and constant weight values. Thus deformation is only governed by the position of control points. In order to estimate a cardiac motion, shape interpolation method is addressed with pair-wise mesh registration. Proposed motion modeling method resolves lack of correspondence and heavy calculation of displacement field, therefore, it is suitable for real-time use.

## 2   Related Work

We categorize the studies of cardiac motion analysis methods into their (1) motion modeling method and (2) applications.

**Motion Modeling Methods.** In cardiac motion modeling, the one of the desired property is spatiotemporal continuity. In the aspect of temporal continuity, group-wise registration method defines a cost function accumulatively or globally. This approaches register multiple objects simultaneously, therefore it improves consistency and continuity between scenes [9, 13, 15]. In spatially, the spatial transformation model governs the way of regional displacement between scenes. Basis functions and parametric models use a basis function or parametric geometries, such as splines, curves, and surfaces, to represent the regional movement of the anatomical object, smoothly. Also, data-driven methods use a plenty of data to model a deformation of the anatomical object [1]. The variability of shape can be analyzed by a statistical method.

**Applications.** The main application of cardiac motion modeling is image guided intervention. Recently, Yabe et al. report the motion compensation and real-time visualization of coronary arteries significantly reduce the contrast volume and fluoroscopic time while percutaneous coronary intervention [17]. In ablation therapy, Wilson et al. studied an electrophysiological (EP) mapping method that is essential for arrhythmia treatment [16], they introduced the patient-specific 4D cardiac mapping method to improve EP mapping under the dynamic situation. In the perspective of diagnosis and monitoring, Gilbert et al. proposed a rapid quantification method for biventricular function using standard MRI [5].

**Creating Motion Model for Complex Structure.** Tiny and complex structures of the heart, such as coronary arteries, are very challenging to model dynamically. Due to the motion artifacts and topological/geometrical variations, Baka et al. shifted the problem of motion modeling from coronary arteries to the attached muscle [1]. Representation of lumen changes is limited because they focus on the centerline of the coronary artery.

In this context, we propose a novel 4D motion modeling method for either convex/non-convex structures using pair-wise registration and shape interpolation. Our contribution is to simplify a shape deformation, therefore it is possible to real-time usage.

## 3   Method

Proposed 4D shape modeling method mainly consists of pair-wise registration and shape blending method. In order to build the correspondence between sampled cardiac phases from 4D CT, pair-wise registration is conducted between segmented meshes from 4D CT. After pair-wise registration, control points governing the shape deformation are interpolated for the whole cardiac phase (Fig. 1). The result of interpolation creates a 4D shape model.

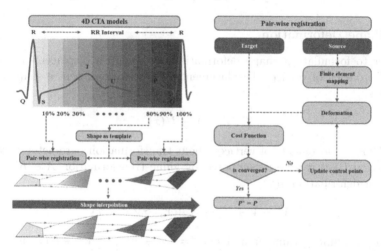

**Fig. 1.** Generating a 4D motion model for the cardiac structure periodically changing according to the motion of the heart (left). Transforming the source object into target object by optimizing the cost function (right).

### 3.1   Shape Representation for Cardiac Structures

The shape of the cardiac structure is obtained by segmentation from volumetric images, such as CT, 3D ultrasound, and MRI. In this paper, we use a mesh representation for shape in order to model a specific cardiac structure, such as a coronary artery, atrium, or ventricle. A shape $V$ is defined as a mesh that consists of vertices, edges and faces. Therefore, $V = \{v_1, \ldots, v_{I_V}\}$ and $E = V \times V$, where $v_k \subset R^d$, $I_V$ is number of vertices, and $d$ is a embedding dimension of an object. Let us consider a sequence of objects from spatiotemporal image, such as 4D CT, then a sequence is $V_s = \{V_1, \ldots, V_n\}$ where $V_i$ is a shape.

## 3.2   Landmark Base Registration

The purpose of registration is to find the best alignment from source to target, source is referred as moving mesh by transformation and target is destination. Now, let mesh $V_s \in V$ be a source, and mesh $V_t \in V$ be a target, respectively. The goodness of alignment is measured by cost function. Cost function $C$ is defined as

$$C := \sum d(T(x) \circ V_s), V_t) \tag{1}$$

where $d$ is a disparity measure between meshes $V_s$ and $V_t$. For simplicity, we use the Euclidean distance. $T$ is a transformation model, and $x$ is a transformation parameter. The best transformation parameter $x^*$ is optimized by minimizing cost function,

$$x^* = \arg \min_x C = \arg \min_x \sum d(T(x) \circ V_s, V_t) \tag{2}$$

## 3.3   Shape Deformation

In order to formulate a shape deformation, we assign a displacement vector, $u_i \in R^d$, to each of vertices. Displacements of vertices represent a shape deformation as

$$T : V \rightarrow V + U(x) \tag{3}$$

where $U(x)$ is movements of vertices about transformation parameter $x$. Shape deformation $U(x)$ can be decomposed by affine transformation $A \in R^{d \times d}$ and non-linear deformation $u(x)$,

$$T : V \rightarrow V + U(x) = A(x_a) \circ V + u(x) \tag{4}$$

where $x_a$ is affine parameter and $u(x) = (V - A(x_a) \circ V) + U(x)$ is a local deformation after global affine transformation.

## 3.4   Finite Element Mapping for Deformation Modeling

In order to represent the object $V$ using control points, let domain $\Omega$ be $\Omega = [min(V), max(V)]$, where $\Omega$ defines a bounding box of object $V$. $\Omega$ is partitioned by $m^d$ number of evenly spaced grid and $(m+1)^d$ control points.

A cage $P$ consist of control points $V_P = \{p_i\}_{i \in I_P} \subset R^d$, where $I_P$ is number of control points and $p_i$ is a control point. If a point $v \in R^d$ is inside cage $P$, such that $v \subset \Omega_P = [min(P), max(P)]$, then given vertex $v \in R^d$ can be represented by control points as below,

$$v = F(v; P) = \sum_{k \in I_P} \varphi_k(v) p_k \tag{5}$$

where $F(\cdot, \cdot)$ is a finite element mapping and $\varphi_k(\cdot)$ is a weighting function with respect to given vertex $v$. For example, a 3D weighting functions are $\varphi_1(x, y, z) = \frac{(1-x)(1-y)(1-z)}{8}$, ..., and $\varphi_8(x, y, z) = \frac{(1-x)(1+y)(1+z)}{8}$. After weighting functions are determined, the movement of vertex $v$ only depends on control points,

$$v' = F(v; P') = \sum_{k \in I_P} \varphi_k(v) p'_k \tag{6}$$

where $P'$ represents the changed cage $P$, such that $V_{P'} = \{p'_k\}_{k \in I_P} \subset R^d$. The moved vertex $v'$ in (6) can be rewritten using (4) as below,

$$v' = A(x_a) \circ v + u(x) = \sum_{k \in I_P} \varphi_k(v)(A(x_a) \circ p_k + u_k(x)) \tag{7}$$

## 3.5   Optimization

As we mentioned in Sect. 3.2, registration problem is rewritten as optimization of cost function. We define dissimilarity measurement, as below

$$d(V_s, V_t, x) = \sum_{i \in I_{V_s}} \min_{j \in I_{V_t}} |T(x) \circ v_i - v_j|^2 \tag{8}$$

where $v_i \in V_s$ and $v_j \in V_t$.

The regularization supports that the deformation became a diffeomorphism, therefore deformation is invertible and smooth. We are inspired from [2]. The hyper-elasticity regularization is easily extended to proposed deformation setting. The hyper-elasticity regularization is defined as below

$$S^{hyper}(x) = \int \alpha_1 \varphi_{vol}(x) + \alpha_2 \varphi_{sur}(x) + \alpha_3 \varphi_{len}(x) d\Omega \tag{9}$$

where $\alpha_i$ are balancing parameter and $\varphi_{vol}$, $\varphi_{sur}$, $\varphi_{len}$ are cost functions that penalizing a change of volume, area, length. Control points of cage $P$ act like corner points of hexahedron of [2]. By introducing the hyper-elasticity regularization, cost function has an infinity value for non-diffeomorphic transformation.

The cost function is combination of dissimilarity measurement and regularization as below,

$$C(V_s, V_t, x) = d(V_s, V_t, x) + S^{hyper}(x) \tag{10}$$

We decouple optimization process into two steps, (1) finding global affine transform $(x_a^*)$ and (2) finding non-linear transformation $(x^*)$ with regularization. In step (1), we neglect regularization energy. Updating control points is simply iterative closest point (ICP). In step (2), we use Levenberg-Marquardt (LM) method to optimize a non-linear least square problem. In order to use gradient decent optimization, the derivative of dissimilarity measurement is given as below,

$$\frac{\partial}{\partial x} d(V_s, V_t, x) = 2 \sum_{i \in I_{V_s}} \sum_{k \in I_P} \varphi_k(v_i)(v_i^T - v_j^T) \tag{11}$$

where $v_i \in V_s$ and $v_j \in V_t$.

### 3.6   Interpolation of Shapes

Let a vector $P_k = \{p_{k1}, p_{k2}, \ldots, p_{k((m+1)^3-1)}, p_{k(m+1)^3}\}$ be set of control points at $k$-th object, where $p_i \in R^d$, $d$ is embedding dimension. By registering a template shape to the $k$-th sequence, the control points are updated $P_{(k)} = A_k P_k + U_k$, where $A_k$ is an affine transformation and displacement of control points is $U_k = \{\Delta p_{k1}, \Delta p_{k2}, \ldots, \Delta p_{k((m+1)^3-1)}, \Delta p_{k(m+1)^3}\}$.

From sets of control points, $S(t) = \{s_1(t), \ldots, s_{(m+1)^3}(t)\}$ is interpolated, where $s_i(t)$ is a spline function of corresponding control points. In our implementation, closed cubic spline is used. Therefore, spline vector $S(t)$ has $C^2$ continuity with respect to $t$. A vector function $S(t)$ is a interpolated control point of time, which maps $S : R \rightarrow R^{(m+1)^3 \times d}$. Then shape is function of time as below

$$v(t) = F(v; S(t)) = \sum_{i \in I_V} \phi_i(v) s_i(t) \tag{12}$$

## 4   Evaluation and Discussion

In order to evaluate the performance of 4D modeling, we used anonymous patient data. A patient data was evenly sampled from 4D CT within R-R peak (5% sampling interval), therefore, 20 volumes ($M_{0\%}$, $M_{5\%}$, ..., $M_{95\%}$) were available. The coronary artery models were manually segmented from CT volumes using ITK-Snap [18]. Through the evaluation, we observed (1) the effect of cage partitioning, (2) the effect of the phase sampling, and (3) the effect of the template model. We define the discrepancy of modeling by an average Euclidean distance of bijective closest point pairs as below

$$d = \frac{1}{I_{Bi}} \sum_{i \in I_{V_s}} \min_{j \in I_{V_t}} |T(x) \circ v_i - v_j|^2 \tag{13}$$

where $I_{Bi}$ is the number of bijective pairs.

### 4.1   Effect of Cage Partitioning

In this section, we observed the effect of cage partitioning. Cage partitioning governs the degree of freedom for deformation. The spatial partitioning is defined by the number of cage control points. For example, let a region $\Omega_S$ be the bounding box of a source model $M_{source}$, then a cage partitioning produces the cages that $\Omega_S = \Omega_{\{0,0,0\}} \cup \Omega_{\{1,0,0\}} \cup \cdots \cup \Omega_{\{i,j,k\}}$, where $i, j, k$ are the number grid partitioning in 3D. We compared different partitioning resolutions $[i, j, k]$ as 4 cases that are $[2, 2, 2], [4, 4, 4], [8, 8, 8]$ and $[12, 12, 12]$. We measured the discrepancy between the registered mesh model $M'_{i\%}$ and manually segmented model $M_{i\%}$. As Fig. 2, the resolution of cage partitioning affect the accuracy of registration, the higher resolution provide the lower error. Time consumption for creating 4D model was 15 min for $[2, 2, 2], [4, 4, 4]$, 20 min for $[8, 8, 8]$, and 30 min $[12, 12, 12]$ using Intel i7-7700K CPU.

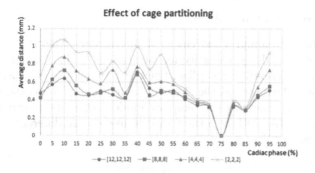

**Fig. 2.** Comparison of different partitioning resolutions $[i, j, k]$ as 4 cases that are $[2, 2, 2], [4, 4, 4], [8, 8, 8]$ and $[12, 12, 12]$.

## 4.2  Effect of the Phase Sampling

In order to observe the effect of sampling, the proposed 4D model was compared to manually segmented model under different sampling conditions. Our data contain 20 volumes per each patient, we were able to select some volumes to create the 4D model. For example, if a sampling step is 10%, then $M_{0\%}$, $M_{10\%}$, ..., $M_{90\%}$ are selected. We selected the 75% phase as the template model. The missed models are generated from 4D model. For example, if a sampling step is 10%, then missed models are $M'_{5\%}$, $M'_{15\%}$, ..., $M'_{95\%}$). We calculated the discrepancy between the interpolated model $M'_{i\%}$ and manually segmented model $M_{i\%}$. As Fig. 3, the temporal sampling largely influences the accuracy of 4D modeling, the higher temporal sampling rate provides the lower error.

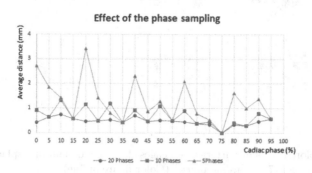

**Fig. 3.** The effect of sampling was observed by comparing manually segmented models and interpolated models under different conditions (5 samples $\{15, 35, 55, 75, 95\}$, 10 samples $\{5, 15, 25, 35, 45, 55, 65, 75, 85, 95\}$ and 20 samples). We use $[8, 8, 8]$ partitioning and 75% phase as template

## 4.3   Effect of Template Model

Because of the very complex movement of the cardiac structure, quality of the cardiac image depends on the selection of phase [10]. This variation of image quality affects the visibility of cardiac structures, therefore, segmentation results are non-consist. We observed the select of template model section, we selected the 35% and 75% phases as the template models, that are known as the least movement phase [10]. Although the template models are different, trend of discrepancy is similar (Figs. 4 and 5).

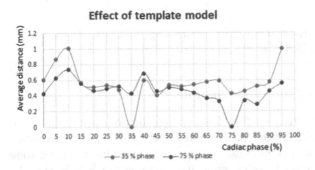

**Fig. 4.** The discrepancy between the registered mesh models and manually segmented mesh models under different template model selection (35% and 75%). We use [8, 8, 8] partitioning.

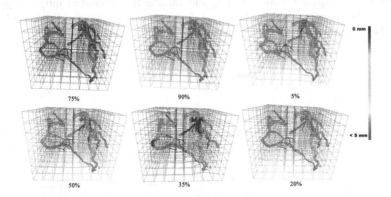

**Fig. 5.** The colored coronary arteries and cages show the amount of displacement from the template model (75% cardiac phase) (Color figure online)

## 5   Conclusion

In this study, we present a pair-wise registration and shape interpolation method for 4D motion modeling. In order to formulate the shape deformation concisely, we use a finite element mapping. A shape is represented by cages that consist of control points and their weighting constants. Therefore, the deformation only depends on the change of control point. Also, the deformation is decomposed by global affine transform and local non-linear deformation to cover large deformation.

However, local non-linear deformation can't ensure to be diffeomorphic deformation. By using hyperelastic regularization of [2], the non-diffeomorphic deformation is penalized. Thus, our deformation model is able to cover large deformation while preserving diffeomorphism. One of desired property of cardiac modeling is the incompressibility, this is provided by regularization term and decomposition of motion. The correspondence of control points is given by pair-wise registration between different objects, then the corresponding control points are interpolated by spline function over the time domain. The spline interpolation of control points is smooth and invertible, because of the inherited property from spline function.

Despite we use a regularization term for the method, this regularization term doesn't ensure that the volume preservation while interpolating shapes. The motion of the right coronary artery is very dynamic within the cardiac interval of 0% to 20%, therefore, it is hard to accurately segment. These non-consist segmentation results produce the large error at that interval.

Combination of pair-wise registration and control points interpolation reconstructs the cardiac model at any time point. The proposed method enables to simplify the shape deformation by introducing finite element mapping. Although [12, 12, 12] partitioning spend 30 min to create 4D motion modeling, it is able to real-time rendering (60 fps) during the cardiac intervention. Because shape deformation is simply represented by control points that change over time.

# References

1. Baka, N., et al.: Statistical coronary motion models for 2D+t/3D registration of X-ray coronary angiography and CTA. Med. Image Anal. **17**(6), 698–709 (2013)
2. Burger, M., Modersitzki, J., Ruthotto, L.: A hyperelastic regularization energy for image registration. SIAM J. Sci. Comput. **35**(1), B132–B148 (2013)
3. Clark, H.: NCDs: a challenge to sustainable human development. Lancet **381**(9866), 510–511 (2013)
4. Giesler, T., et al.: Noninvasive visualization of coronary arteries using contrast-enhanced multidetector CT: influence of heart rate on image quality and stenosis detection. Am. J. Roentgenol. **179**(4), 911–916 (2002)
5. Gilbert, K., Pontre, B., Occleshaw, C., Cowan, B., Suinesiaputra, A., Young, A.: 4D modelling for rapid assessment of biventricular function in congenital heart disease. Int. J. Cardiovasc. Imaging **34**(3), 407–417 (2018)
6. Guyader, J.M., Bernardin, L., Douglas, N.H., Poot, D.H., Niessen, W.J., Klein, S.: Influence of image registration on apparent diffusion coefficient images computed from free-breathing diffusion MR images ofthe abdomen. J. Magn. Reson. Imaging **42**(2), 315–330 (2015)
7. Huang, J., Abendschein, D., Davila-Roman, V.G., Amini, A.A.: Spatio-temporal tracking of myocardial deformations with a 4-D B-spline model from tagged MRI. IEEE Trans. Med. Imaging **18**(10), 957–972 (1999)

8. Metz, C.T., et al.: Patient specific 4D coronary models from ECG-gated CTA data for intra-operative dynamic alignment of CTA with X-ray images. In: Yang, G.-Z., Hawkes, D., Rueckert, D., Noble, A., Taylor, C. (eds.) MICCAI 2009. LNCS, vol. 5761, pp. 369–376. Springer, Heidelberg (2009). https://doi.org/10.1007/978-3-642-04268-3_46

9. Metz, C., Klein, S., Schaap, M., van Walsum, T., Niessen, W.J.: Nonrigid registration of dynamic medical imaging data using nD+ t B-splines and a groupwise optimization approach. Med. Image Anal. 15(2), 238–249 (2011)

10. Ohnesorge, B.M., Flohr, T.G., Becker, C.R., Knez, A., Reiser, M.F.: Multi-slice and Dual-Source CT in Cardiac Imaging: Principles-Protocols-Indications-Outlook. Springer, Heidelberg (2006). https://doi.org/10.1007/978-3-540-49546-8

11. Otto, C.M.: Textbook of Clinical Echocardiography E-Book. Elsevier Health Sciences, St. Louis (2013)

12. Peng, P., Lekadir, K., Gooya, A., Shao, L., Petersen, S.E., Frangi, A.F.: A review of heart chamber segmentation for structural and functional analysis using cardiac magnetic resonance imaging. Magn. Reson. Mater. Phys. Biol. Med. 29(2), 155–195 (2016)

13. Rohé, M.M., Sermesant, M., Pennec, X.: Low-dimensional representation of cardiac motion using baryncetric subspaces: a new group-wise paradigm for estimation, analysis, and reconstruction. Med. Image Anal. 45, 1–12 (2017)

14. Tarride, J.E., et al.: A review of the cost of cardiovascular disease. Can. J. Cardiol. 25(6), e195–e202 (2009)

15. Wachinger, C., Navab, N.: Simultaneous registration of multiple images: similarity metrics and efficient optimization. IEEE Trans. Pattern Anal. Mach. Intell. 35(5), 1221–1233 (2013)

16. Wilson, K., Guiraudon, G., Jones, D.L., Peters, T.M.: Mapping of cardiac electrophysiology onto a dynamic patient-specific heart model. IEEE Trans. Med. imaging 28(12), 1870–1880 (2009)

17. Yabe, T., et al.: The impact of percutaneous coronary intervention using the novel dynamic coronary roadmap system. J. Am. Coll. Cardiol. 71(11), A1103 (2018)

18. Yushkevich, P.A., et al.: User-guided 3D active contour segmentation of anatomical structures: significantly improved efficiency and reliability. Neuroimage 31(3), 1116–1128 (2006)

# ISACHI: Integrated Segmentation and Alignment Correction for Heart Images

Benjamin Villard$^{(\boxtimes)}$, Ernesto Zacur, and Vicente Grau

Institute of Biomedical Engineering, University of Oxford,
Old Road Campus Research Building, Oxford, UK
benjamin.villard@eng.ox.ac.uk

**Abstract.** We address the problem of cardiovascular shape representation from misaligned Cardiovascular Magnetic Resonance (CMR) images. An accurate 3D representation of the heart geometry allows for robust metrics to be calculated for multiple applications, from shape analysis in populations to precise description and quantification of individual anatomies including pathology. Clinical CMR relies on the acquisition of heart images at different breath holds potentially resulting in a misaligned stack of slices. Traditional methods for 3D reconstruction of the heart geometry typically rely on alignment, segmentation and reconstruction independently. We propose a novel method that integrates simultaneous alignment and segmentation refinements to realign slices producing a spatially consistent arrangement of the slices together with their segmentations fitted to the image data.

**Keywords:** Computational geometry · Slice misalignment · CMR

## 1 Introduction

Current research in Cardiovascular Magnetic Resonance (CMR) images is moving towards three dimensional (3D) applications, with 3D geometrical models used for various purposes, such as diagnosis, surgical planning, patient risk stratification, or device innovation [1,2]. To quantitatively assess cardiac functions, clinicians typically segment CMR slices in 2D, which are then used for the calculation of relevant 3D parameters such as volumes, ejection fraction, or myocardial mass. 3D reconstructions also allow for more sophisticated tools such as shape analysis, multi-modality fusion, or mechanical and electrophysiological simulations [3,4]. In order to accurately preserve the underlying anatomy, information obtained from magnetic resonance imaging (MRI) acquisitions needs to be spatially consistent in 3D. However, spatial inconsistencies are common due to various artefacts including patient motion inside the scanner. In CMR, the acquisition of slices at different breath holds and the natural cardiac motion introduces additional artifacts, which may introduce distortion in the generated 3D models and the subsequent quantification. Different methods can be used

© Springer Nature Switzerland AG 2019
M. Pop et al. (Eds.): STACOM 2018, LNCS 11395, pp. 171–180, 2019.
https://doi.org/10.1007/978-3-030-12029-0_19

to limit those artefacts: at acquisition via the use of respiratory gating [5], for example, or more clinically relevant, through image processing methods after acquisition.

Current methods for generating personalised CMR mesh models typically involve the sequential steps of misalignment correction, segmentation and interpolation. Typically, misalignments between slices are first corrected for, after which the segmented contours, obtained using manual or (semi-) automated methods, are used to create the mesh. However the sparsity of the data make registration and interpolation methods sensitive to local minima requiring strong prior assumptions. With contouring protocols extracting the object of interest in 2D, these contours suffer from spatially inconsistency [6], which is further extended by segmentation variability. Our aim is to develop an integrated method that combines the alignment of slices and a fine correction of the contours, for which an initial segmentation is obtained using any state of the art algorithm, in order to arrange the segmented data consistently in a 3D environment, ultimately, leading to accurate geometrical meshes representing the heart anatomy.

There exists prior work that simultaneously correct for spatial misalignment and quantify morphological and dynamic information [7]. However, [7] rely on prior shape models which can bias the alignment. Marchesseau et al. [8], use all of the multi-frame segmentations from cardiac data to generate an average mesh which they deform using estimated transformation obtained via registration, to quantify regional volume change. To our knowledge, only one prior work has attempted to achieve the integration of segmentation, spatial correction and 3D reconstruction from CMR images [9,10]. Paiement et al. demonstrate a technique that integrates these steps, making use of level sets to automatically segment and construct a 3D surface of the left ventricle on which they perform the alignment correction. By constraining the level sets to grow by using the combined information found in the intersecting slices, they allow the integration of the alignment and level set's segmentation. Validation is performed on registration results but not, to our knowledge, on the segmentation.

We propose a novel methodology to correct for spatially inconsistent contours, allowing for a subsequent precise *patient-specific* surface mesh reconstruction, whilst aligning the image stacks. The method makes use of long axis (LAX) and short axis (SAX) views to resolve the alignment. Our approach is image driven, and does not necessitate any expert contouring, although an initial segmentation is required. By combining information from both the alignment as well as the segmentation steps in an integrated manner, we are able to obtain an accurate shape that is spatially consistent with the information from the images. By using an initial mesh, generated by using automatically segmented contours via a convolutional neural network (CNN), we compute image derived forces that rigidly transform the image poses whilst deforming the segmentations towards image edges. The resulting segmentation combines both global and local information from the stack of short axis images as well as from the long axis view, aiming at maximizing consistency at the intersection of slices.

## 2    Proposed Framework

Our proposed method, ISACHI (Integrated Segmentation and Alignment Correction for Heart Images), aims to correct for segmentation errors in order to obtain 3D spatially consistent segmentations, allowing us to produce an accurate, and spatially consistent surface mesh. This is done by relying on an image-derived metric. Many different metrics would work in this context, as long as they are able to guide the segmentation towards the likely location of contours in each of the intersecting images, with our proposed framework incorporating 3D consistency. The image-derived metric is introduced in the shape of an *edgemap*, with areas with high *edgness* reflecting borders and edges. In our experiments we demonstrate the use of a deep learning approach for this initial segmentation, without affecting the generality of the framework.

### 2.1    Automatic Segmentation

Our method requires an initial segmentation to create an initial 3D surface mesh "$\mathcal{M}$" using the method described in [11,12]. The segmentation can be obtained in many ways; algorithmically via modern convolutional neural networks (CNN), for example, or via manual contouring. In order to render our method fully automatic, we used a U-Net CNN based on [13] as they have recently been shown to provide state-of-the-art segmentation results [14] and can work well on limited training data. While other potentially more accurate methods such as ResNet [15], our aim in this paper is to demonstrate the accuracy of the method and its robustness to the initial segmentation.

Since most current CNN segmentation techniques, including the one we use here, rely on 2D segmentation, the aforementioned 3D inconsistencies remain. While 3D CNN segmentation algorithms have been recently proposed [16–18], these require access to a large well aligned ground-truth-labelled volumetric cardiac dataset, as well as require high computational and memory requirements [16]. Standard clinical CMR does not provide 3D consistent data, and datasets should be considered as an independent collection of 2D slices.

### 2.2    Methodology

We compute the line intersections between all cross-sectional planes and obtain the points where the lines intersect with the generated mesh. Let $l_{ij}$ be the line intersection between two planes containing contours $C_i$ and $C_j$, with $i, j \in [1...N]$, $N$ being the total number of contours. Let $S_i$ be the plane containing contour $C_i$, with $\{S\}$ is the set of all planes containing respective contours. We denote $l_{ij} = S_i \bigcap S_j$; with $i < j$, or $l_{ij} \equiv l_{ji}$. Let $\{P^n\}$ be a set of sampled points where the lines $l_{ij}$, intersect with the mesh $\mathcal{M}$, i.e. $\{P^n\} = \{\mathcal{M} \bigcap l_{ij}, i < j\}$ with $n \in [1...T]$ and T being equal to the maximum amount of intersection points. At these points *patches* are created. Let $D^n$ be the patch lying on the plane tangent to $\mathcal{M}$ with center $P^n$. Thus, $D^n = (\top_{P^n}(\mathcal{M})) \bigcap (B_{P^n} R)$, where $B$ is a sphere of radius $R$ and $\top_{P^n}(\mathcal{M})$ the tangent to $\mathcal{M}$ at point $P^n$. Let $X_i$ be a segment of

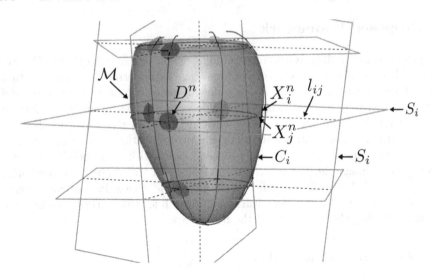

**Fig. 1.** Illustration of ISACHI definitions.

length $2R$ and lying on $S_i$. Therefore, $X_i = D^n \bigcap S_i$ only if $D^n$ has been built from $S_i$. It should be noted that a patch contains 2 segments, belonging to each of the respective intersecting planes, and that there are as many patches as there are intersections between the mesh and line intersections $l_{ij}$. This is illustrated in Fig. 1.

In order to obtain relevant patches, or slices during our optimization process, we define two functions: $g(n)$ and $p(i)$ that return slice indices when given a patch number, and all the patches built from a specific slice $S_i^n$, respectively. Let $g(n) = \{i, j\}$ and $p(i) = \{n : i \in g(n)\}$. We now define our energy function to be:

$$J_i = \sum_{n \in p(i)} \sum_{k \in g(n)} \xi(X_k^n, \Psi_k) \tag{1}$$

$$\xi(s, \rho) = \frac{\int_{x \in s} \Psi_x}{2R} \tag{2}$$

where $\Psi$ is the edge map of an image, $X_k^n$ is the $n - th$ segment belonging to slice $k$, and $\xi$ represents the energy function between a segment and an edge map. The method to compute $\Psi$ is left to the user, and is not detrimental to our method as long as it describes features of interests including edges and borders, which reflect structural information, such as the epicardium and endocardium. The method can work with fundamental techniques such as the magnitude of the gradient, although newer techniques can be used such as the one proposed by [19].

One of our aims is to align slices in order to obtain spatial consistency between the slices and inherently the contours. We assume that slices taken from a subject are triggered at the same cardiac phase and thus there is a global optimal alignment between the slices which can be modelled by an individual rigid transformation for each slice. We thus consider 3D rigid body transformations with 6 degrees of freedom, represented by: $\theta = \{t_x, t_y, t_z, \alpha, \beta, \gamma\}$. Our optimization can then be defined as: $\underset{\theta_i \in SE(3)}{\operatorname{argmin}} J_i(\theta_i)$, where SE(3) stands for *special Euclidean* and is the set of rigid transformations of dimension 3 and

$$J_i(\theta_i) = \sum_{n \in p(i)} \sum_{k \in g(n)} \xi(s_k^n(\theta_{g(n)}^1, \theta_{g(n)}^2), \theta_k \star \Psi_k) \tag{3}$$

where $\star$ denotes the application of $\theta$ on $\Psi$. Once we have iterated through all the slices until convergence, we consider the slices to be aligned.

We define a global energy

$$\tilde{J} = \sum_{i=1}^{T} J_i \tag{4}$$

where $T$ is the maximum amount of patches, and inherently the number of intersection points. The Energy $J_i$ represents a measure of how well each of the segments agrees with the image boundary information. It relies on all the segments lying on the slice, as well as the segments lying on the respective intersecting slice. By using the edgeness profiles, calculated from $\Psi$, the slices can be realigned by taking the information evaluated at the patches.

As the initial contours might not be a correct representation of the segmentation, we proceed to transform individual patches $\{D^n\}$ spatially. As the patches lie on the mesh, which lies close to the non intersecting contours, we consider them to be a good initial approximation of their spatially consistent location. To refine the segmentation independently from slice misalignments, we only allow the patches to move along the intersection lines. In this way, we search for the best spatial location of a patch taking into account the information from its respective intersecting slices. Each patch is optimized iteratively following the same method previously mentioned, but taking into account both intersecting segments. Once the position of the patch has been optimized, the deformation is extrapolated to the contours via approximated thin-plate splines (TPS) [20]. This is to ensure that the patches lie on their optimum calculated positions whilst approximating the mesh in a *smooth* manner

$$\xi_{TPS} = reg(u) + \lambda \sum \|u(m) - F\|^2. \tag{5}$$

where $u$ represents the deformation space, $m$ the moving points ($\{C\}$), $F$ the target points ($D^n(A^*)$), and the operator $\|\cdot\|$ the Frobenius norm. The parameter $\lambda$ is a smoothing parameter controlling the trade-off between the smoothness of the deformation and the data reliability. Our choice of $\lambda$ was chosen empirically using a data driven approach. Once deformed, we proceed to recompute a mesh $\mathcal{M}$ to obtain a representation in accordance with the new spatial alignment.

## 3    Experiments and Results

To properly assess our methodology, synthetic and real data were used. As our algorithm encompasses both alignment and segmentation refinement, synthetic data allowed us to evaluate each component of our method, as we do not possess a *true* spatially consistent clinical ground truth. ISACHI was also applied to a pathological case study, which we present the result for qualitative evaluation. An average MRI heart data obtained from [21][1] was used for our synthetic data. Biventricular average shapes and modes of variation at end-systole and end-diastole are available with their respective surface mesh models. We treated these meshes along with their respective image templates as our ground truth. Using these surface meshes, contours were synthesized by slicing the mesh to form a short axis stack (4 slices), as well as at two (vertical and horizontal) long axis. Random translations and rotations, sampled from uniform distributions with ranges $[-2:2]°$ for each angle and $[-5:5]$ mm for each spatial coordinate were applied to the planes containing the images. Furthermore, non-rigid deformations were applied to the contours to simulate non spatial consistency. For this purpose, eleven uniformly sampled points were extracted from each contour and randomly displaced following a uniform distribution of width 2 mm. The new, deformed, contour points were then calculated using a thin-plate spline interpolation. Using this approach, we generated 10 new cases.

In order to validate the first aspect of our methodology, *i.e.* the ability to correct for slice misalignment, we computed the cross correlation between the line intensities for all slice intersections (as per [6]). This was used as a surrogate measure, assuming that the information at the slice intersection from two slices should be identical. To evaluate the second aspect of our methodology, the accuracy of the segmentation, we applied the iterative closest point (ICP) [22] method between our output mesh and the ground truth mesh, to align them. This allowed to bring the two meshes back to the same spatial location allowing quantitative comparison, in case slices have shifted during optimization. We then compute the average point-to-mesh distance between the two meshes.

The ISACHI algorithm was also applied on real data. A study case having abnormal physiology and severe misalignment was used, and can be seen in Fig. 3. In this case, the slices were initially segmented using our CNN.

### 3.1    Results

Figure 2 shows a representative result of ISACHI after correcting the misalignments and refining the initial segmentation of the end-diastolic dataset. The slice misalignment for this case ranged from 3–5 mm. Figure 2 shows the mean (red squares), and standard deviations for the normalized cross correlation (NCC) of the intersection line intensities between slices, calculated using the method described in [6], as a function of ISACHI iterations. At the left of the vertical red line is the mean and standard deviation for the NCC values for

---

[1] http://wp.doc.ic.ac.uk/wbai/data/.

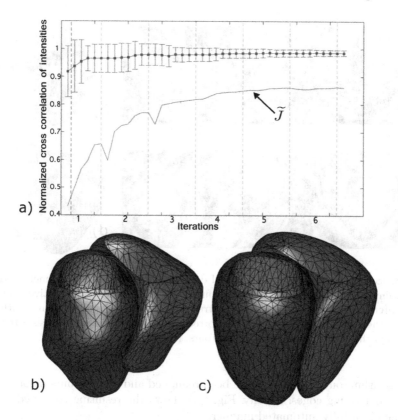

**Fig. 2.** Alignment correction validation for a representative case. (a) Mean and standard deviation values for the NCC of the intersection line intensities, at every step of ISACHI. The left of the vertical red line represents the synthetically misaligned dataset. The bottom line represents the energy function $\widetilde{J}$. (b) Synthetic mesh: the introduced misalignments are clearly visible. (c) Mesh after ISACHI. (Color figure online)

the misaligned slices. The right of the red line shows the subsequent evaluations for each slice update. The grey vertical lines represent the iterations. In this case, it took 6 iterations to converge. The bottom blue line represents our energy function $\widetilde{J}$. Figure 2(b) shows the mesh generated from the misaligned slices (left) and the result of ISACHI (right). A mesh-to-mesh comparison was performed between the resulting mesh and the ground truth template mesh, between the epi-, endo-, and right ventricle meshes. For the 10 synthetic cases, the mean errors between the misaligned and ground truth meshes were: 2.03 mm, 1.94 mm, 2.20 mm, respectively. After ISACHI, the respective errors were: 1.12 mm, 0.83 mm, 0.83 mm. Figure 3 represents the application of ISACHI on a real case study having abnormal physiology. Figure 3(a) shows the initial CNN segmentation of the slices. The coloured dots in each image correspond to the intersection with a contour on an intersecting slice. In this way, the misalignment between slices can clearly be visualized. Figure 3(c) shows the output of

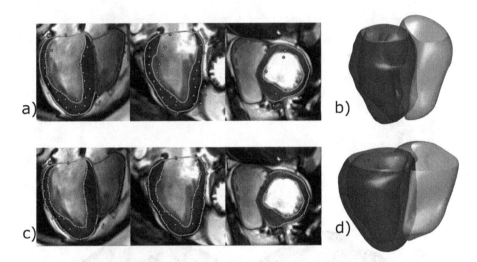

**Fig. 3.** ISACHI application on abnormal data. (a) The initial CNN segmented data, containing severe misalignment. The coloured dots represent the respective contours from intersecting slices. (b) Mesh reconstructed from the original contours, with the slices at their original locations. (c) The corrected contours and aligned slices. (d) Mesh built from the aligned and deformed contours. (Color figure online)

the algorithm, once the slices have been realigned and the contours deformed to match underlying edges. Finally, Fig. 3(d) shows the resulting corrected mesh, obtained in a fully automated manner.

We presented a novel methodology to perform simultaneous alignment and segmentation correction for cardiac images. Our methodology combines both information from image data as well as consistency in the geometry. The premise of our method lies in that for slices to be aligned, the image features in a slice must match respective features in cross-sectional slices. Furthermore, structural information must also lie on those features of interest. Based on initial slice segmentations, which can be obtained using any method, a surface mesh is generated and used to combine both structural and feature information allowing our optimization to incorporate both global and local data. We show that the method is able to realign the images, as well as recover a realistic mesh. A potential limitation of our method is its reliance on first order patches; it is possible that the use of higher order patches (e.g. quadratic) would improve results.

# References

1. Vukicevic, M., Mosadegh, B., Min, J.K., Little, S.H.: Cardiac 3D printing and its future directions. JACC Cardiovasc. Imaging **10**(2), 171–184 (2017)
2. Lopez-Perez, A., Sebastian, R., Ferrero, J.M.: Three-dimensional cardiac computational modelling: methods, features and applications. BioMed. Eng. OnLine **14**(1), 35 (2015)

3. Carminati, M.C., Maffessanti, F., Caiani, E.G.: Nearly automated motion artifacts correction between multi breath-hold short-axis and long-axis cine CMR images. Comput. Biol. Med. **46**, 42–50 (2014)
4. Peng, P., Lekadir, K., Gooya, A., Shao, L., Petersen, S., Frangi, A.: A review of heart chamber segmentation for structural and functional analysis using cardiac magnetic resonance imaging. Comput. Aided Surg. **29**, 155–195 (2016)
5. Ehman, R.L., McNamara, M.T., Pallack, M., Hricak, H., Higgins, C.: Magnetic resonance imaging with respiratory gating: techniques and advantages. Am. J. Roentgenol. **143**(6), 1175–1182 (1984)
6. Villard, B., Zacur, E., Dall'Armellina, E., Grau, V.: Correction of slice misalignment in multi-breath-hold cardiac MRI scans. In: Mansi, T., McLeod, K., Pop, M., Rhode, K., Sermesant, M., Young, A. (eds.) STACOM 2016. LNCS, vol. 10124, pp. 30–38. Springer, Cham (2017). https://doi.org/10.1007/978-3-319-52718-5_4
7. Georgescu, B., et al.: Model based automated 4D analysis for real-time free-breathing cardiac MRI. Proc. Int. Soc. Magn. Reson. Med. (ISMRM) **21**, 4498 (2013)
8. Marchesseau, S., Duchateau, N., Delingette, H.: Segmentation and registration coupling from short-axis cine MRI: application to infarct diagnosis. In: Mansi, T., McLeod, K., Pop, M., Rhode, K., Sermesant, M., Young, A. (eds.) STACOM 2016. LNCS, vol. 10124, pp. 48–56. Springer, Cham (2017). https://doi.org/10.1007/978-3-319-52718-5_6
9. Paiement, A., Mirmehdi, M., Xie, X., Hamilton, M.C.K.: Integrated segmentation and interpolation of sparse data. IEEE Trans. Image Process. **23**(1), 110–125 (2014)
10. Paiement, A., Mirmehdi, M., Xie, X., Hamilton, M.C.K.: Registration and modeling from spaced and misaligned image volumes. IEEE Trans. Image Process. **25**(9), 4379–4393 (2016)
11. Villard, B., Carapella, V., Ariga, R., Grau, V., Zacur, E.: Cardiac mesh reconstruction from sparse, heterogeneous contours. In: Valdés Hernández, M., González-Castro, V. (eds.) MIUA 2017. CCIS, vol. 723, pp. 169–181. Springer, Cham (2017). https://doi.org/10.1007/978-3-319-60964-5_15
12. Villard, B., Grau, V., Zacur, E.: Surface mesh reconstruction from cardiac MRI contours. J. Imaging **4**(1), 16 (2018)
13. Vigneault, D.M., Xie, W., Bluemke, D.A., Noble, J.A.: Feature tracking cardiac magnetic resonance via deep learning and spline optimization. In: Pop, M., Wright, G.A. (eds.) FIMH 2017. LNCS, vol. 10263, pp. 183–194. Springer, Cham (2017). https://doi.org/10.1007/978-3-319-59448-4_18
14. Ronneberger, O., Fischer, P., Brox, T.: U-Net: convolutional networks for biomedical image segmentation. In: Navab, N., Hornegger, J., Wells, W.M., Frangi, A.F. (eds.) MICCAI 2015. LNCS, vol. 9351, pp. 234–241. Springer, Cham (2015). https://doi.org/10.1007/978-3-319-24574-4_28
15. He, K., Zhang, X., Ren, S., Sun, J.: Deep residual learning for image recognition. CVPR (2016)
16. Dolz, J., Desrosiers, C., Ayed, I.B.: 3D fully convolutional networks for subcortical segmentation in MRI: a large-scale study. CoRR abs/1612.03925 (2016)
17. Patravali, J., Jain, S., Chilamkurthy, S.: 2D-3D fully convolutional neural networks for cardiac MR segmentation. CoRR abs/1707.09813 (2017)
18. Zotti, C., Luo, Z., Lalande, A., Humbert, O., Jodoin, P.: Novel deep convolution neural network applied to MRI cardiac segmentation. CoRR (2017)
19. Xie, S., Tu, Z.: Holistically-nested edge detection. In: Proceedings of the 2015 IEEE International Conference on Computer Vision (ICCV), pp. 1395–1403 (2015)

20. Rohr, K., Stiehl, H., Sprengel, R., Buzug, T., Weese, J., Kuhn, M.: Landmark-based elastic registration using approximating thin-plate splines. IEEE Trans. Med. Imaging **20**(6), 526–534 (2001)
21. Bai, W., et al.: A bi-ventricular cardiac atlas built from 1000+ high resolution MR images of healthy subjects and an analysis of shape and motion. Med. Image Anal. **26**(1), 133–145 (2015)
22. Besl, P., McKay, N.: A method for registration of 3-D shapes. IEEE Trans. Pattern Anal. Mach. Intell. **14**(2), 239–256 (1992)

# 3D LV Probabilistic Segmentation in Cardiac MRI Using Generative Adversarial Network

Dong Yang[1]([✉]), Bo Liu[1], Leon Axel[2], and Dimitris Metaxas[1]

[1] Department of Computer Science, Rutgers University, Piscataway, NJ 08854, USA
don.yang.mech@gmail.com
[2] Department of Radiology, New York University, New York, NY 10016, USA

**Abstract.** Cardiac magnetic resonance imaging (MRI) is one of the most useful techniques to understand and measure cardiac functions. Given the MR image data, segmentation of the left ventricle (LV) myocardium is the most common task to be addressed for recovering and studying LV wall motion. However, most of segmentation methods heavily rely on the imaging appearance for extracting the myocardial contours (epi- and endo-cardium). These methods cannot guarantee a consistent volume of the heart wall during cardiac cycle in reconstructed 3D LV wall models, which is contradictory to the assumption of approximately constant myocardial tissue. In the paper, we propose a probability-based segmentation method to estimate the probabilities of boundary pixels in the trabeculated region near the solid wall belonging to blood or muscle. It helps avoid artifactually moving the endocardium boundary inward during systole, as commonly happens with simple threshold-based segmentation methods. Our method takes s stack of 2D cine MRI slices as input, and produces the 3D probabilistic segmentation of the heart wall using a generative adversarial network (GAN). Based on numerical experiments, our proposed method outperformed the baseline method in terms of evaluation metrics on a synthetic dataset. Moreover, we achieved very good quality reconstructed results with on a real 2D cine MRI dataset (there is no truly independent 3D ground truth). The proposed approach helps to achieve better understanding cardiovascular motion. Moreover, it is the first attempt to use probabilistic segmentation of LV myocardium for 3D heart wall reconstruction from 2D cardiac cine MRI data, to the best of our knowledge.

## 1 Introduction

Segmentation of the endocardial heart walls in MR images is a key part of calculating cardiac global function measures, such as stroke volume or ejection fraction. By standard convention, we regard the transitional zone of trabeculae and papillary muscles between the solid part of the wall and the clear blood-filled cavity as part of the cavity, and seek to define the endocardial contour of the solid portion of the heart wall, to use for calculation. While this is usually

© Springer Nature Switzerland AG 2019
M. Pop et al. (Eds.): STACOM 2018, LNCS 11395, pp. 181–190, 2019.
https://doi.org/10.1007/978-3-030-12029-0_20

relatively easy at end-diastole, when these structures are well defined. It is often more difficult near end-systole, when the blood is squeezed out from between the trabeculae, making the intensity of the transitional zone similar to that of the solid wall. This frequently necessitates time-consuming manual correction of any contours automatically generated from simple intensity threshold-based segmentation methods. As an alternative, we propose a "fuzzy" approach for segmentation of the endocardium in 3D space, in order to achieve more robust segmentation of both the solid heart wall and the transitional zone, compared with conventional methods. The pixels in the transitional zone commonly have mixed contributions from both muscle and blood; we can consider them as characterized by voxel-wise percentages in calculating the segmentation results. This can provide better understanding of cardiac wall motion, comparing with the conventional 2D cine MRI analysis, as it allows for a more realistic modeling of the transition region near the solid wall. It can also be used for both improved calculation of global function measures, and the dynamic characterization of the transitional zone itself.

Starting at end-diastole (when the ventricular cavity is most open), we can generally produce a good segmentation of the solid portion of the wall from the rest of the cavity with the use of simple conventional thresholding for the muscle intensity, combined with a smoothness constraint to suppress the derived muscle boundary from sticking to smaller structures within the cavity. This can be augmented with the use of machine learning or deep learning approaches, trained on expertly segmented images. We can then define the transition zone as the region between the solid boundary and the clear cavity. The trabeculae (which can be smaller than or on the order of the size of the voxels) and blood are mixed inside the transition zone, which can be seen as a blurry appearance in 2D cine MRI. It is a more challenging task to maintain a consistent definition of the boundary of the solid wall, when the heart moves into the later systolic and early diastolic phases of the cardiac cycle. Because the blood will be ejected from the spaces between the muscle structures in the transition zone as we progress into systole, this has the effect of causing these structures to appear to merge with each other. Thus, tracking the apparent boundary of the solid portion of the wall in the images is likely to over-estimate the degree of inward motion (squeeze) of the solid wall boundary, even if using some sort of deformable contour for the boundary. Therefore, a robust probabilistic segmentation approach is necessary for the estimation of the extent of the transitional zone, to separate it more reliably from the solid portion of the wall. Such a segmentation approach would assign probability values to each pixel according to how likely they are to contain myocardium. Thus, we add a partial label (probability) to the pixels during the segmentation process, such that the total amount of muscle in the transition zone should remain about the same during the cardiac cycle (neglecting through-plane motion effects). The total amount of muscle in the transition zone would be given by summing the area of the potential pixels in the zone, weighted by their probability of being muscle rather than blood. Then, as the heart contracts, the initial inner solid wall contour derived from simple intensity thresholding for a given cardiac phase would be moved outward (expansion) until about the right amount of muscle is included in the transition zone, in order to match the initial condition. That can provide a more reliable way to segment the solid portion of the wall, using this probabilistic segmentation scheme as part of the process.

Assuming that the "conservation of transition zone muscle" approach provides a reasonable way to approximately correct the simple initial segmentation of the apparent contour of the solid wall for the changing appearance of the transition zone, we can then use the statistics of the transition zone to estimate the corresponding changing blood content of the transition zone. This could potentially be a new sort of regional "ejection fraction" rate calculated, which would provide a novel way to characterize cardiovascular functions. While it would likely correlate overall with the regional wall motion, it provides a somewhat different way to assess function (independent of an eternal reference frame). Codella et al. proposed a fuzzy segmentation approach purely based on image intensity for the global blood content of the ventricle [2]. It did not provide any regional information about blood content of the transitional zone. Also, their approach could be problematic because it was only applied on the 2D cine MRI data (neglecting through-plane motion effect), not in 3D space.

It is desirable to use the probabilistic segmentation approach to provide 3D results for the transitional zone at each cardiac phase, because the through-plane motion of the tapered LV wall introduces systematic artifacts into the simple 2D segmentation. In order to recover the full 3D information from multiple separately acquired 2D cine slices, it is necessary to understand the motion in space. Recently, generative adversarial networks (GAN) have been proposed to model the appearance distribution of the image domain for many applications. Thus, fully-connected or fully-convolutional neural networks have been utilized successfully for image generation from 1D noise vectors [6,7]. The idea of GAN can be further extended for image understanding, completion, reconstruction, segmentation [4], and other similar applications.

In this paper, we propose a novel 3D probabilistic cardiac segmentation approach given only 2D cardiac cine MRI acquisitions, as shown in Fig. 1. In the proposed approach, 2D epi- and endo-cardial contours are first extracted to provide the 2D in-plane probabilistic segmentation, and to compensate spatial offset artifacts caused by inconsistent respiration. Each pixel is assigned a probability value characterizing how likely it is to belong to the myocardium. Then, we adopt a generative adversarial network (GAN) to transform multiple 2D probabilistic segmentation maps in space into a fully 3D probabilistic segmentation. With the results of the 3D segmentation over the cardiac cycle, we would have a better understanding of the cardiac wall motion. Our work is the first attempt to reconstruct such a 3D probabilistic segmentation from 2D cine MRI, to the best of our knowledge.

The motivation of using GAN to reconstruct 3D probabilistic segmentation in our method is as follows. GAN is capable of describing the distribution of contextual appearances, and it has achieved state-of-the-art results on several image-related tasks, for instance, image super-resolution, image denoising, MRI reconstruction, etc. We treat the probabilistic segmentation as one kind of

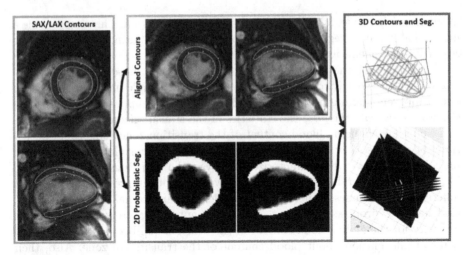

**Fig. 1.** 2D probabilistic segmentation and respiration offset artifact removal, using in-plane contours for both short-axis (SAX) and long-axis (LAX) cine MRI that may initially not be well aligned. The output is the aligned contours and the aligned 2D probabilistic segmentation in space.

contextual appearance, and our goal is to process the partial signals for full-information reconstruction based on the appearance population of training samples, which is equivalent to the aforementioned tasks. Although GAN is not the only way to process such tasks, it is a more efficient way than many others, in terms of accuracy and speed.

## 2    Probabilistic Segmentation on 2D Cine MRI and Respiration Compensation

Initially, we adopt a LV segmentation framework using 2D/3D U-Net and multi-component deformable model enforcing the spatial and temporal smoothness on short-axis cardiac cine MRI [8]. The output from the framework is the epi- and endo-cardium contours during cardiac cycle. Similarly, we apply the framework to the long-axis cine MRI to extract the LV wall with the single-component deformable model. Once the LV contours are finalized from the segmentation framework and verified by doctors, they are adopted for generating 2D probabilistic segmentation and compensating artifacts caused by respiration.

Based on the epi- and endo-cardial contours, we adopted a novel approach to estimate the presence of boundary pixels close to trabeculae, papillary muscles, and solid wall, and to estimate the corresponding classification probability [8]. Initially, to compute the probability of belonging to blood or myocardium for the boundary pixels in the LV cavity, a 2D U-Net [1] is trained with pure LV cavity blood and pure myocardium regions (determined by intensity). Next, the features from the second-to-last layer of the U-Net are extracted from each pixel accordingly, and a logistic regression classifier is further trained using those features. The clear myocardium pixels are labeled as 1, and the clear blood pixels are labeled as 0. Finally, the trained classifier assigns the probability values to the pixels inside the transitional zones with fuzzy appearance between myocardium and cavity shown in Fig. 1. Normally, in the transitional zones, with mixed blood and muscle, we cannot reliably determine the regional proportion of blood or muscle from their appearance alone. Using the proposed approach, we are able to track the mixed spaces and estimate the regional percentage of blood/muscle using the fuzzy classifier. By adjusting the threshold of the classifier within the range [0,1] at different cardiac phases (especially near end-systole), the sensitivity to the relatively small amount of blood near boundary pixels is increased. The adjustment of the threshold can be made based on the criterion of maintaining a consistent amount of muscle at each cardiac phase.

The epi- and endo-contours are further utilized to remove imaging artifacts caused by inconsistent respiratory state between image acquisitions [3,5]. The artifacts are seen as apparent mis-alignment between different cine MRI slices in space, which increases the difficulty of recovering the full 3D heart motion. Different cine MRI image slices are typically acquired at different suspended respiration phases, with associated different spatial offsets, even though they are all synchronized by ECG according to cardiac phases; Conventional breath-holding MRI cannot completely diminish this effect, even with cooperative subjects. We assume that the mis-alignment primarily causes an in-plane translation for 2D MR images. Thus, we iteratively minimize the overall distance between the intersectional points of contours recovered from slices perpendicular to a slice and contours within the slices, by serially translating all contours within the planes of their slices. After several iterations, the intersecting contour points would approximately meet the in-plane contours. When we have thus derived estimates of the translation vectors of the MRI slices that best align the contours, we then apply them to shift the aforementioned 2D probabilistic segmentations for better spatial consistency between slices.

## 3   3D Label Propagation Using Generative Adversarial Network

In order to propagate the label information (probabilistic segmentation) of 2D slices to the whole 3D volume, we propose a generative adversarial network (GAN) [6] based model for the propagation. The proposed GAN based model is shown in Fig. 2. The generator ($G$) is fed with the 3D volume containing multiple 2D slices of probabilistic segmentation, denoted by $x$. The in-plane label is probability of belonging to myocardium region of each pixel, obtained from Sect. 2, the voxels outside slices are valued to be 0 by default. With the 3D U-Net, we predict the label for the whole 3D volume. The discriminator ($D$) is fed by ground truth of 3D volume label as real sample, and the output of the $G$ as fake sample. It aims to discriminate the ground truth from the prediction results. The learning

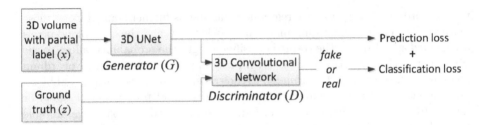

**Fig. 2.** The proposed GAN model for label propagation.

objective is designed to be the weighted combination of (1) the prediction error of $G$, evaluated by the MSE error with ground truth; (2) the minimax loss of the GAN model. Those two parts are weighted by $\lambda$, which is

$$
\min_G \max_D V(G, D) = \mathbb{E}_{x \sim p_{data}(x)}[\log(D(x))] + \mathbb{E}_{z \sim p_z(z)}[\log(1 - D(G(z)))]
$$
$$
+ \lambda L_{MSE}(Y, G(z)) \tag{1}
$$

where $\mathbb{E}$ is the empirical estimate of expected value of the probability; $Y$ denotes the ground truth of the 3D volume (probabilistic segmentation). The model parameters in $G$ and $D$ are learned by iteratively maximizing $V(G, D)$ with respect to $D$ and minimizing $V(G, D)$ with respect to $G$. After the model learning, the 3D U-Net in $G$ is used for label propagation. The probabilistic estimation of myocardium region is obtained from label propagation result.

For the 3D U-Net in generator $G$, we adopt the structure defined in [10], with the same layer number and convolutional filter number in each layer. The 3D CNN in discriminator $D$ is designed to be a 7-layer convolutional neural network including five convolutional layers, single fully connected layer, and a sigmoid layer. For both networks, batch normalization is used between each two neighboring layers and leaky ReLU are used as the activation functions.

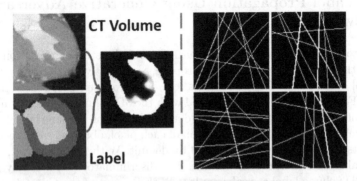

**Fig. 3.** Left: synthetic data of 3D probabilistic segmentation using CT volume and its label; right: four samples of pattern masks used for synthetic data (probabilistic segmentation of CT volumes).

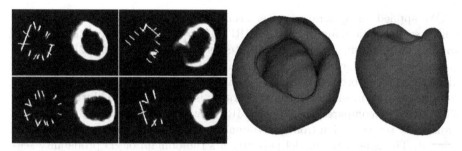

**Fig. 4.** Left: the cross sections of 2D probabilistic segmentation maps in space before and after 3D reconstruction; right: 3D myocardium model from the 3D probabilistic segmentation with threshold 0.5.

## 4    Experiments

There is no ground truth for 3D probabilistic segmentation for 2D MRI acquisitions. Because the 2D MRI is sparsely sampled around the LV area. So we adopt a synthetic dataset from CT to train our neural networks. The CT dataset contains both appearance volumes and the corresponding labels, including both blood pool and myocardium [9]. All volumes and labels are re-sampled to 1.0 mm isotropically for simplicity. We follow the similar strategy in Sect. 2 to generate ground truth of 3D probabilistic segmentation, and use a 3D U-Net instead for feature extraction. Then, we randomly sample 10 planes (with arbitrary angles and locations) crossing the volume as a pattern, and make them the binary masks with the same size shown in Fig. 3. The input to the neural network is the element-wise multiplication of 3D probabilistic segmentation and a pattern, and the output is the probabilistic segmentation itself. The problem formulation is similar to the task of image completion. During training, we have 15 segmentation maps and 1000 different patterns, which gives us 15000 training samples. We collect another 5 segmentation maps for testing, and in total we have 5000 testing samples. The training details follows the setting in [7]. $\lambda$ is set as 1000 in our experiments.

In Table 1, we can see that our proposed method using GAN performs much better than the baseline 3D U-Net in terms of peak signal-to-noise ratio (PSNR) and mean squared error (MSE). If we provide more planes of probabilistic segmentation (e.g. 15 planes) as input to the neural network, and the output would be closer to the ground truth. It fits the intuition that more information introduces better local and global constraints. After the model is finalized, we deploy it to the real 2D cine MRI dataset. The multiple 2D probabilistic maps in space as a volume are directly treated as input of the generator. The output is full 3D probabilistic segmentation shown in Fig. 4. We can achieve myocardium segmentation simply by placing a threshold to the 3D probabilistic segmentation.

We applied our generative adversarial network in the domain of 3D probabilistic segmentation, instead of CT/MRI volume domains, because the synthetic training data is difficult to generate in terms of appearance. Both 2D cine MRI and high-dose CT have excellent image quality, from which the details of trabeculae can be well-observed. Directly acquired 3D cardiac MRI commonly suffers from respiratory motion, and the appearance is generally more blurry and noisy with artifacts, compared to 2D cine MRI. However, we are able to obtain 3D probability segmentation from CT volumes, and can learn the generative model using it. The generative model perfectly fits the domain of 2D probability segmentation from 2D cine MRI in space. Therefore, it would be ideal to study the 3D cardiac motion using 2D cine MRI in this way (normally 3D CT acquisitions have low temporal resolution).

Figure 5 shows the calculated myocardial volumes at different cardiac phases, using two different methods, from 2D MRI sequences of a patient with cardiac dyssynchrony. The orange curve is from the volumes of shells built using 3D deformable models recovered with simple thresholding [5], and the blue curve is computed by summing up myocardium probabilities over all voxels. We can observe that our 3D probabilistic segmentation clearly has larger volumes comparing to the deformable shell, because our probabilistic model is able to capture the fractional volume components of trabeculae and papillary muscles in the transition zone inside LV cavity. Moreover, the volume quantity of the probabilistic segmentation is more stable temporally, which meets the assumption that myocardium volume should be close to a constant value during cardiac cycle. The appearance of our 3D probabilistic segmentation is qualitatively verified by clinicians, and it has a potential for improved basic understanding and clinical study of cardiac function (e.g., improving the estimation of ejection fraction and related global measures).

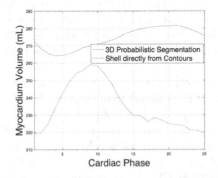

**Fig. 5.** LV myocardium volumes at different cardiac phases, for conventional and probabilistic segmentations. (Color figure online)

**Table 1.** Evaluation on test dataset.

| PSNR (dB) | | | |
|---|---|---|---|
| | 5 Planes | 10 Planes | 15 Planes |
| U-Net | 13.57 | 14.13 | 14.61 |
| GAN | **15.01** | **15.22** | **15.34** |
| MSE | | | |
| | 5 Planes | 10 Planes | 15 Planes |
| U-Net | 0.0442 | 0.0388 | 0.0348 |
| GAN | **0.0331** | **0.0316** | **0.0301** |

# 5    Conclusion

To sum up, we have proposed a probability-based segmentation approach to estimate the fractional blood/muscle classification of boundary pixels near the trabeculae, between the solid wall and the clear blood, in full 3D space, from 2D cardiac MRI acquisitions. Our results are quantitatively validated on a synthetic dataset, and also visually validated on a real MRI dataset. This is the first attempt to reconstruct such a 3D probabilistic segmentation from 2D cardiac cine MRI, to the best of our knowledge. The proposed approach has a good potentials for providing improved cardiac motion understanding and clinical applications.

# References

1. Ronneberger, O., Fischer, P., Brox, T.: U-Net: convolutional networks for biomedical image segmentation. In: Navab, N., Hornegger, J., Wells, W.M., Frangi, A.F. (eds.) MICCAI 2015. LNCS, vol. 9351, pp. 234–241. Springer, Cham (2015). https://doi.org/10.1007/978-3-319-24574-4_28
2. Codella, N.C., et al.: Improved left ventricular mass quantification with partial voxel interpolation: In-Vivo and Necropsy Validation of a Novel Cardiac MRI Segmentation Algorithm. Circulation: Cardiovascular Imaging, CIRCIMAGING-111, vol. 31(4), pp. 845–853 (2011)
3. Sinclair, M., Bai, W., Puyol-Antón, E., Oktay, O., Rueckert, D., King, A.P.: Fully automated segmentation-based respiratory motion correction of multiplanar cardiac magnetic resonance images for large-scale datasets. In: Descoteaux, M., Maier-Hein, L., Franz, A., Jannin, P., Collins, D.L., Duchesne, S. (eds.) MICCAI 2017. LNCS, vol. 10434, pp. 332–340. Springer, Cham (2017). https://doi.org/10.1007/978-3-319-66185-8_38
4. Yang, D., et al.: Automatic liver segmentation using an adversarial image-to-image network. In: Descoteaux, M., Maier-Hein, L., Franz, A., Jannin, P., Collins, D.L., Duchesne, S. (eds.) MICCAI 2017. LNCS, vol. 10435, pp. 507–515. Springer, Cham (2017). https://doi.org/10.1007/978-3-319-66179-7_58
5. Yang, D., Wu, P., Tan, C., Pohl, K.M., Axel, L., Metaxas, D.: 3D motion modeling and reconstruction of left ventricle wall in cardiac MRI. In: Pop, M., Wright, G.A. (eds.) FIMH 2017. LNCS, vol. 10263, pp. 481–492. Springer, Cham (2017). https://doi.org/10.1007/978-3-319-59448-4_46
6. Goodfellow, I., et al.: Generative adversarial nets. In Advances in Neural Information Processing Systems, pp. 2672–2680 (2014)
7. Radford, A., Metz, L., Chintala, S.: Unsupervised representation learning with deep convolutional generative adversarial networks. arXiv preprint arXiv:1511.06434 (2015)
8. Yang, D., Huang, Q., Axel, L., Metaxas, D.: Multi-component deformable models coupled with 2D-3D U-Net for automated probabilistic segmentation of cardiac walls and blood. In: 2018 IEEE 15th International Symposium on Biomedical Imaging (ISBI 2018), pp. 479–483. IEEE, April 2018

9. Zhuang, X., Shen, J.: Multi-scale patch and multi-modality atlases for whole heart segmentation of MRI. Med. Image Anal. **31**, 77–87 (2016)
10. Çiçek, Ö., Abdulkadir, A., Lienkamp, S.S., Brox, T., Ronneberger, O.: 3D U-Net: learning dense volumetric segmentation from sparse annotation. In: Ourselin, S., Joskowicz, L., Sabuncu, M.R., Unal, G., Wells, W. (eds.) MICCAI 2016. LNCS, vol. 9901, pp. 424–432. Springer, Cham (2016). https://doi.org/10.1007/978-3-319-46723-8_49

# A Two-Stage U-Net Model for 3D Multi-class Segmentation on Full-Resolution Cardiac Data

Chengjia Wang[1,2]([✉]), Tom MacGillivray[2], Gillian Macnaught[1,2], Guang Yang[3], and David Newby[1,2]

[1] BHF Centre for Cardiovascular Science, University of Edinburgh, Edinburgh, UK
chengjia.wang@ed.ac.uk
[2] Edinburgh Imaging Facility QMRI, University of Edinburgh, Edinburgh, UK
[3] National Heart and Lung Institute, Imperial College London, London, UK

**Abstract.** Deep convolutional neural networks (CNNs) have achieved state-of-the-art performances for multi-class segmentation of medical images. However, a common problem when dealing with large, high resolution 3D data is that the volumes input into the deep CNNs has to be either cropped or downsampled due to limited memory capacity of computing devices. These operations can lead to loss of resolution and class imbalance in the input data batches, thus downgrade the performances of segmentation algorithms. Inspired by the architecture of image super-resolution CNN (SRCNN), we propose a two-stage modified U-Net framework that simultaneously learns to detect a ROI within the full volume and to classify voxels without losing the original resolution. Experiments on a variety of multi-modal 3D cardiac images have demonstrated that this framework shows better segmentation performances than state-of-the-art Deep CNNs with trained with the same similarity metrics.

**Keywords:** Image segmentation · Convolutional neural networks · High resolution · Cardiac CT/MR

## 1 Introduction

Segmenting the whole heart structures from CT and MRI data is a necessary step for pre-precedural planing of cardiovascular diseases. Although it is the most reliable approach, manual segmentation is very labor-intensive and subject to user variability [1]. High anatomical and signal intensity variations make automatic whole heart segmentation a challenging task. Previous methods that separately segment specific anatomic structure [2] are often based on active deformation models. Others perform multi-class segmentation where atlas-based models play an important role. Active deformation models can suffer from limited ability to

This work is funded by BHF Centre of Cardiovascular Science and MICCAI 2017 Multi-Modality Whole Heart Segmentation (MM-WHS) challeng.

M. Pop et al. (Eds.): STACOM 2018, LNCS 11395, pp. 191–199, 2019.
https://doi.org/10.1007/978-3-030-12029-0_21

decouple pose variation [3], and the main disadvantage of atlas-based methods is requiring complex procedures to construct the atlas or non-rigid registration [4]. Recently, deep convolutional neural networks (DCNNs), such as U-Net [5], have been vastly used for cardiac segmentation and achieved start-of-the-art results. But due to limited computational power, the data input into U-Net are always down-sampled to a uniform resolution. The purpose of this study is to develop a DCNN which can perform multi-class segmentation on full-resolution volumetric CT and MR data with no post-prediction resampling or subvolume-fusion operations. This is necessary due to the loss of information introduced by interpolation and extra complexity of post-processing.

The original U-Net is entirely an 2D architecture. So are some DCNN-based full heart segmentation methods [6,7]. To process volumetric data, one solution is to take three perpendicular 2D slices as input and fuse the multi-view information abstraction for 3D segmentation [8]. Or one can directly replace 2D operations in the original U-Net with their 3D counterparts [9]. This 3D U-Net has been applied to multi-class whole-heart segmentation on MR and CT data [10]. However, due to limited memory of GPUs, these methods have to either down-sampled the input volumes, which leads to loss of resolution in the final results, or predict subvolumes of the data make extra efforts to merge the overlapped subvolume predictions. A strategy to reduce the memory load of 3D U-Nets has been used in [11], where a region of interests (ROI) was first extracted by a localization network then input to a segmentation U-Net. However, this method still requires to work on intensively downsampled data. Methods that preserve the original data resolution have to use a relatively shallow U-Net architecture. Inspired by the network structure of [11], we propose a two-stage DCNN framework which consists of two concatenated U-Net-like networks. A new multi-stage learning pipeline was adopted to the training process. This framework also segment 3D full-resolution MR and CT data within a ROI that is dynamically extracted, but the original resolution of the image is kept. Our method outperformed the well-trained 3D U-Net in our experiments.

## 2   Method

**Architecture.** As shown in Fig. 1 the model consists of two concatenated modified U-Nets. Each basic U-Net block has two convolutional layers, followed by a nonlinear activation and a pooling layer. Both the encoding and the decoding paths of the two U-Nets have 4 basic U-Net convolution blocks. The final output of each network is produced by a softmax classification layer. The first network (*Net1* in Fig. 1) aim to produce a coarse segmentation on down-sampled full 3D volume. We use dilated $5 \times 5 \times 5$ convolutional kernel with zero-padding which preserves shapes of feature maps. In the $n$th block of the encoding path, the dilation rate of the convolutional kernel is $2n$. This pattern is reversed in the expansive path. Each convolutional layer is followed by a rectified linear unit (ReLU), and a dropout layer with a 0.2 dropout rate is attached to each U-Net block. In the test phase, a dynamic-tile layer is introduced between *Net1* and

*Net2* to crop out a ROI from both the input and output volume of *Net1*. This layer is removed when performing end-to-end training. The output of *Net1* is resampled to the original resolution before input into *Net2*. *Net2* is a 3D-2D U-Net segmenting the central slice of a subvolume. The structure of *Net2* is inspired by the deep Super-Resolution Convolutional Neural Network (SRCNN) [12] with skip connections [13]. The input of this network is a two-channel 4D volume consist of the output of *Net1* and the original data. The convolutional kernel size in the encoding path is $3 \times 3 \times 3$, and $5 \times 5 \times 5$ in the decoding path. The size of the 3D pooling kernels in the contracting path of *Net2* is $2 \times 2 \times 1$ so that the number of axial slices is preserved. A 3D-2D slice-wise convolution block with $1 \times 1 \times (K-1)$ convolutional kernels is introduced before the decoding path, where $K$ is the number of neighboring slices used to label one single axial slice. No zero-padding is used to ensure that every $K$ input slices will generate only one axial feature map. Furthermore, $K$ should always be an odd number to prevent generating labels for interpolated slices. The layers in the decoding path of *Net2* perform 2D convolutions and pooling.

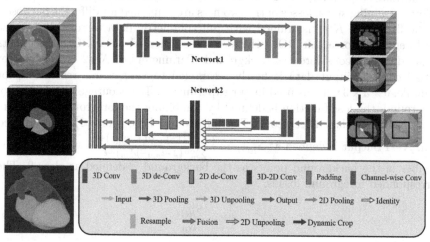

**Fig. 1.** The concatenated U-Net architecture proposed in this work. The nonlinear activation and pooling layer within each U-Net block are not shown for demonstration purpose.

**Training Pipeline and Losses.** The duo-U-Net framework can be trained either separately or end-to-end. We combined both approaches into a four-step training procedure. At the beginning, *Net1* is pre-trained for initial localization of the object. Then the whole framework is trained with different combinations of *Net1* and *Net2* loss functions for quick convergence. Details of training batches and loss functions used in each step are shown in Table 1. A commonly used similarity metric for single-class segmentation is soft Dice score. Let $p_{n,c}^i$ denote the probability that a voxel belongs to class $c, c \in \{0, \cdots, C\}$ (background is defined by $c = 0$), given by the softmax layer of *Neti*, and $t_{n,c} \in \{0, 1\}$ represent the ground truth one-hot label. The soft Dice score can be defined as

**Table 1.** Purposes and loss functions of each step in the training process

| Step | Input | Purpose | Loss |
|------|-------|---------|------|
| 1 | full volumetric data | foreground localization | $\mathcal{L}^1_{ROI}$ |
| 2 | partial volumetric data | coarse multi-class segmentation | $\mathcal{L}^1_{ROI} + \mathcal{L}^1$ |
| 3 | partial volumetric data | coarse+fine segmentation | $\mathcal{L}^1 + \mathcal{L}^2$ |
| 4 | stack of full axial slices | fine multi-class segmentation | $\mathcal{L}^2$ |

$\mathcal{S}^i_c = \frac{2\sum_n^{N_c} t_{n,c} p^i_{n,c} + \epsilon}{\sum_n^{N_c}(t_{n,c} + p^i_{n,c}) + \epsilon}$, where $N_c$ is number of voxels labeled as class $c$ and $\epsilon$ is a smooth factor. To perform multi-class segmentation, we just define our loss function using weighted Dice scores weighted by voxel counts for simplicity:

$$\mathcal{L}^i = 1 - \sum_c^C \frac{\mathcal{S}^i_c}{N_c}. \tag{1}$$

But nothing stops using a more sophisticated loss functions as shown in [14]. In different training steps, losses of the two nets are combined for different purposes. In the first step, *Net1* is trained with full volumetric data to roughly localize the foreground, or a soft ROI, which is the segmented object. Other contents in the data are considered as background. Parameters of *Net2* is frozen after initialized. The input data is firstly resampled to very coarse resolution, for example, $3 \times 3 \times 3 \, \text{mm}^3$ as used in our experiments. To encourage localization of foreground, the loss function is defined by combining the foreground Dice score with the multi class Dice score. The foreground Dice score $\mathcal{L}^1_{ROI}$ computed from *Net1* output is defined as $\mathcal{S}^1_{ROI} = \frac{2\sum_n^{N_0}(1-t_{n,0})(1-p^1_{n,0}) + \epsilon}{\sum_n^{N_0}(2-t_{n,0}-p^1_{n,0}) + \epsilon}$, where $N_0$ is the number of the background points as the background is defined as class 0. The corresponded foreground loss is:

$$\mathcal{L}^1_{ROI} = 1 - \frac{\mathcal{S}^1_{ROI}}{N_0}. \tag{2}$$

We use reversed label to calculate foreground score rather than the Dice score of the background class ($c = 0$) to reduce the imbalance introduced by large background. *Net1* is trained to minimize the loss $\mathcal{L}^1_{ROI}$ for locating the foreground of the object. After pre-training with $\mathcal{L}^1_{ROI}$, $\mathcal{L}^1 + \mathcal{L}^1_{ROI}$ is used as the loss for coarse multi-class segmentation in the second step, where $\mathcal{L}^1$ is the *Net1* loss defined by Eq. 1. In this step, *Net1* is trained using subvolumes of the data. Dimensions of the data are varied in different training batches as an augmentation strategy. In the third step, the whole framework (both *Net1* and *Net2*) is trained end-to-end with the loss $\mathcal{L}^1 + \mathcal{L}^2$ to evolve both coarse 3D segmentation and the fine-level axial 2D segmentation. In the final step, inputs of the framework are subvolumes, each consists of $K$ axial slices. The output of *Net2* is the segmentation of the $\frac{K+1}{2}$th slice of a input subvolume. The parameters of *Net1* are frozen, and *Net2* is finetuned using the loss $\mathcal{L}^2$.

**Implementation Details.** Because the framework is mostly trained with sub-volumes of the 3D data except the first step, we use a hierarchical sampling strategy similar with [15]. Each batches are generated from a small number of data. As the ground-truth labels are imbalanced, we sample the label of the central voxel in the subvolume from a uniform distribution. Once the label of the central voxel is fixed, the subvolume is generated by randomly picking its centre from all voxels labeled as the selected class. In this way, the probabilities that the central voxel belongs to any of the classes should be $\frac{1}{C+1}$. For optimization, we use Adam optimizer with initial learning rate 0.0001. We use only 1 full volume in each batch in the first step. In step 2 and 3, the size of each subvolume is randomly selected from $\{64, 128, 256\}$. In the final training step, we set $K = 9$, which means $Net2$ take a subvolume that contains 9 axial slices and predict the labels for the 5th slice of this subvolume. Tensorflow code was adopted on Microsoft Azure virtual machine with one NVidia Tesla K40 GPU. Data augmentation includes random rotation, translation, sheering, scaling, flipping and elastic deformations.

## 3 Experiments and Data

The MICCAI 2017 Multi-Modality Whole Heart Segmentation (MM-WHS) challenge recently benchmarks existing whole heart segmentation algorithms. For training, the challenge provides 40 volumes (20 cardiac CT and 20 cardiac MR). The data were acquired with different scanners, which leads to varying voxel sizes, resolutions and imaging qualities. An extra 80 testing images are available from the challenge for a final test. In this dataset, anatomical structures which are manually delineated include, the left ventricle blood cavity (LV), the myocardium of the left ventricle (Myo), the right ventricle blood cavity (RV), the left atrium blood cavity (LA), the right atrium blood cavity (RA), the ascending aorta (AA) and the pulmonary artery (PA).

One may argue that the proposed framework can be trained directly using the final training step. To demonstrate effects of the proposed training procedure, we trained our model by omitting one of the first three steps, and visually assessed the segmentation results which can be found in the next section. The first three steps do not necessarily need to converge as long as the final step converges. In this step-wise experiment, all results were obtained with 200 epochs in each step, and each epoch includes 16 iterations of backpropogation. The whole training process contains 12800 iterations in total.

Besides qualitatively evaluating the visualized segmentation, for each modality, we use 15 volumes for 3-fold cross-validation training, and 5 volumes for validation. To compare our framework with state-of-the-art U-Net-based models, we trained two 3D U-Nets for each modality which predict on data resampled to resolution of $2 \times 2 \times 2\,\mathrm{mm}^3$. Then the output volumes are resampled to the original resolution using 2nd order BSpline interpolation. Intensities of all images are rescaled to $[-1, 1]$ with no further preprocessing. Three metrics were used to assess segmentation quality for each class: binary Dice score (Dice), binary Jaccard index (Jaccard).

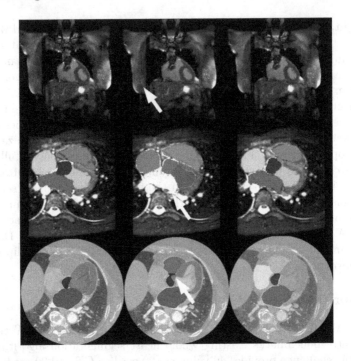

**Fig. 2.** Visualization of segmentation results overlapped with the input images: Ground-truth segmentations are shown on the left; the middle column shows the results obtained by omitting the first, second and third training step (from top to bottom); on the right is the results obtained with the proposed training process. White arrows point to misclassifications caused by omitting a training step.

## 4    Results

Examples of visualized segmentation results are shown in Fig. 2. Omitting the foreground localization step in the first training step may lead to misclassification of the background voxels, as shown in the top row of Fig. 2. The middle row shows that without the coarse segmentation (second step) the model failed to label left atrium, and produced inhomogeneous segmentation for aorta when skipping the joint training of $Net1$ and $Net2$ (step three).

Tables 2 and 3 show the binary Dice and Jaccard scores for all assessed structures obtained by $Net1$ and $Net2$ from the proposed framework, compared to individually trained U-Nets. $Net2$ produced highest Dice and Jaccard scores for all segmented structures in CT data, $Net1$ gave better results than the individual U-Net trained on the same downsampled data. As the MR data have relatively lower resolution, the volume size changed less after resampling. $Net2$ still produced better segmentation accuracy except for RV and AA. $Net1$ gave better segmentations for 4 out of all 7 classes.

The quantitative results of $Net2$ obtained using the test dataset of MM-WHS competition are shown in Table 4. The proposed framework achieved obviously

**Table 2.** Comparison of CT segmentation results obtained by 3D U-Net, and our proposed *Net*1 and *Net*2.

|  | Metrics | LV | Myo | RV | LA | RA | AA | PA |
|---|---|---|---|---|---|---|---|---|
| N3D U-Net | Dice | 0.6451 | 0.8301 | 0.7873 | 0.7768 | 0.6784 | 0.8306 | 0.7123 |
|  | Jaccard | 0.4889 | 0.7126 | 0.6572 | 0.6397 | 0.5217 | 0.7143 | 0.5560 |
| Net1 | Dice | 0.6774 | 0.8107 | 0.8136 | 0.8118 | 0.7997 | 0.8889 | 0.8086 |
|  | Jaccard | 0.5399 | 0.6979 | 0.6977 | 0.6908 | 0.6717 | 0.8030 | 0.6802 |
| Net1+Net2 | Dice | **0.8374** | **0.8588** | **0.8600** | **0.8613** | **0.8620** | **0.9176** | **0.8846** |
|  | Jaccard | **0.7823** | **0.8210** | **0.8233** | **0.8256** | **0.8268** | **0.8534** | **0.7959** |

**Table 3.** Comparison of MR segmentation results obtained by 3D U-Net, and our proposed *Net*1 and *Net*2.

|  | Metrics | LV | Myo | RV | LA | RA | AA | PA |
|---|---|---|---|---|---|---|---|---|
| N3D U-Net | Dice | 0.8296 | 0.9141 | **0.9173** | 0.8946 | 0.8792 | **0.9202** | 0.8847 |
|  | Jaccard | 0.7106 | 0.8419 | **0.8479** | 0.8107 | 0.7894 | **0.8526** | 0.7938 |
| Net1 | Dice | 0.8811 | 0.9367 | 0.9131 | 0.9334 | 0.8572 | 0.8750 | 0.9204 |
|  | Jaccard | 0.7877 | 0.8813 | 0.8430 | 0.8757 | 0.7694 | 0.7833 | 0.8528 |
| Net1+Net2 | Dice | **0.8813** | **0.9377** | 0.9125 | **0.9338** | **0.9220** | 0.8758 | **0.9210** |
|  | Jaccard | **0.7879** | **0.8829** | 0.8422 | **0.8764** | **0.8568** | 0.7847 | **0.8539** |

**Table 4.** Quantitative results obtained using MM-WHS test data, evaluation metrics include: average of Dice score (Dice), average of Jaccard score (Jaccard), and Average surface distance (ASD).

| Modality | Metrics | LV | Myo | RV | LA | RA | AA | PA | WH |
|---|---|---|---|---|---|---|---|---|---|
| CT | Dice | 0.7995 | 0.7293 | 0.7857 | 0.9044 | 0.7936 | 0.8735 | 0.6482 | 0.8060 |
|  | Jaccard | 0.6999 | 0.6091 | 0.6841 | 0.8285 | 0.6906 | 0.8113 | 0.5169 | 0.6970 |
|  | ASD | 4.4067 | 5.4854 | 4.8816 | 1.3978 | 4.1707 | 3.7898 | 6.0041 | 4.1971 |
| MR | Dice | 0.8632 | 0.7443 | 0.8485 | 0.8524 | 0.8396 | 0.8236 | 0.7876 | 0.8323 |
|  | Jaccard | 0.7693 | 0.6049 | 0.7469 | 0.7483 | 0.7404 | 0.7095 | 0.6657 | 0.7201 |
|  | ASD | 1.9916 | 2.3106 | 1.8925 | 1.7081 | 2.7566 | 4.2610 | 2.9296 | 2.4718 |

higher accuracy for LA and AA. This is a sign of premature termination of training, although the framework still obtained much better results than the baseline algorithm of the competition. For MR data, the average Dice score of *Net*2 is 0.8323, which is comparable to the winner of the competition. However, our model was only trained 12800 iterations which is only 1/5 of the winner model. Notice that the purpose of this study is to generate the model gave better performance than state-of-the-art U-Net when segmenting high resolution data. This has been shown in the experiment described above.

198    C. Wang et al.

## 5  Conclusion and Discussion

In this paper, we described a two-stage U-Net-like framework for multi-class segmentation. Unlike other U-Net based 3D data segmentation DCNN, the proposed method can directly make prediction for data with original resolution due to its SRCNN-inspired architecture. A novel 4-step training procedure were applied to the framework. Validated using data from MM-WHS2017 competition, it produced more accurate multi-class segmentation results than state-of-the-art U-Net. With much less training iterations and without any further post-processing, our method achieved segmentation accuracies comparable to the winner of MM-WHS2017 competition.

## References

1. Pace, D.F., Dalca, A.V., Geva, T., Powell, A.J., Moghari, M.H., Golland, P.: Interactive whole-heart segmentation in congenital heart disease. In: Navab, N., Hornegger, J., Wells, W.M., Frangi, A.F. (eds.) MICCAI 2015. LNCS, vol. 9351, pp. 80–88. Springer, Cham (2015). https://doi.org/10.1007/978-3-319-24574-4_10
2. Arrieta, C., Uribe, S., Sing-Long, C., Hurtado, D., Andia, M., Irarrazaval, P., Tejos, C.: Simultaneous left and right ventricle segmentation using topology preserving level sets. Biomed. Sig. Process. Control 33, 88–95 (2017)
3. Gonzalez-Mora, J., De la Torre, F., Murthi, R., Guil, N., Zapata, E.L.: Bilinear active appearance models. In: IEEE 11th International Conference on Computer Vision, ICCV 2007, pp. 1–8. IEEE (2007)
4. Marsland, S., Twining, C.J., Taylor, C.J.: Groupwise non-rigid registration using polyharmonic clamped-plate splines. In: Ellis, R.E., Peters, T.M. (eds.) MICCAI 2003. LNCS, vol. 2879, pp. 771–779. Springer, Heidelberg (2003). https://doi.org/10.1007/978-3-540-39903-2_94
5. Ronneberger, O., Fischer, P., Brox, T.: U-Net: convolutional networks for biomedical image segmentation. In: Navab, N., Hornegger, J., Wells, W.M., Frangi, A.F. (eds.) MICCAI 2015. LNCS, vol. 9351, pp. 234–241. Springer, Cham (2015). https://doi.org/10.1007/978-3-319-24574-4_28
6. Wolterink, J.M., Leiner, T., Viergever, M.A., Išgum, I.: Dilated convolutional neural networks for cardiovascular MR segmentation in congenital heart disease. In: Zuluaga, M.A., Bhatia, K., Kainz, B., Moghari, M.H., Pace, D.F. (eds.) RAMBO/HVSMR -2016. LNCS, vol. 10129, pp. 95–102. Springer, Cham (2017). https://doi.org/10.1007/978-3-319-52280-7_9
7. Moeskops, P., et al.: Deep learning for multi-task medical image segmentation in multiple modalities. In: Ourselin, S., Joskowicz, L., Sabuncu, M.R., Unal, G., Wells, W. (eds.) MICCAI 2016. LNCS, vol. 9901, pp. 478–486. Springer, Cham (2016). https://doi.org/10.1007/978-3-319-46723-8_55
8. Mortazi, A., Burt, J., Bagci, U.: Multi-planar deep segmentation networks for cardiac substructures from MRI and CT. arXiv preprint arXiv:1708.00983 (2017)
9. Roth, H.R., et al.: Hierarchical 3D fully convolutional networks for multi-organ segmentation. arXiv preprint arXiv:1704.06382 (2017)
10. Yu, L.: Automatic 3D cardiovascular MR segmentation with densely-connected volumetric convnets. In: Descoteaux, M., Maier-Hein, L., Franz, A., Jannin, P., Collins, D.L., Duchesne, S. (eds.) MICCAI 2017. LNCS, vol. 10434, pp. 287–295. Springer, Cham (2017). https://doi.org/10.1007/978-3-319-66185-8_33

11. Payer, C., Štern, D., Bischof, H., Urschler, M.: Multi-label whole heart segmentation using CNNs and anatomical label configurations. In: Pop, M., et al. (eds.) STACOM 2017. LNCS, vol. 10663, pp. 190–198. Springer, Cham (2018). https://doi.org/10.1007/978-3-319-75541-0_20
12. Dong, C., Loy, C.C., He, K., Tang, X.: Image super-resolution using deep convolutional networks. IEEE Trans. Pattern Anal. Mach. Intell. **38**(2), 295–307 (2016)
13. Kim, J., Kwon Lee, J., Mu Lee, K.: Deeply-recursive convolutional network for image super-resolution. In: Proceedings of the IEEE Conference on Computer Vision and Pattern Recognition, pp. 1637–1645 (2016)
14. Berger, L., Hyde, E., Cardoso, J., Ourselin, S.: An adaptive sampling scheme to efficiently train fully convolutional networks for semantic segmentation. arXiv preprint arXiv:1709.02764 (2017)
15. Girshick, R.: Fast R-CNN. arXiv preprint arXiv:1504.08083 (2015)

# Centreline-Based Shape Descriptors of the Left Atrial Appendage in Relation with Thrombus Formation

Ibai Genua[1], Andy L. Olivares[1], Etelvino Silva[2], Jordi Mill[1],
Alvaro Fernandez[1], Ainhoa Aguado[1], Marta Nuñez-Garcia[1],
Tom de Potter[2], Xavier Freixa[3], and Oscar Camara[1(✉)]

[1] DTIC (Department of Information and Communication Technologies),
Universitat Pompeu Fabra, Barcelona, Spain
oscar.camara@upf.edu
[2] Arrhythmia Unit, Department of Cardiology, Cardiovascular Center,
Aalst, Belgium
[3] Department of Cardiology, Hospital Clinic de Barcelona,
Universitat de Barcelona, Barcelona, Spain

**Abstract.** The majority of thrombi in non-valvular atrial fibrillation (AF) patients are formed in the left atrial appendage (LAA). The shape of the LAA is highly variable and complex morphologies seem to favour the development of thrombus since they may induce blood stasis. Nevertheless, the relation between LAA shape and risk of thrombus has not been rigorously studied due to the lack of appropriate imaging data and robust tools to characterize LAA morphology. The main goal of this work was to automatically extract simple LAA morphological descriptors and study their relation with the presence of thrombus. LAA shape characterization was based on the computation of its centreline combining a heat transfer-derived distance map and a marching algorithm, once it LAA was segmented from 3D medical images. From the LAA centreline, several morphological descriptors were derived such as its length, tortuosity and major bending angles, among others. Other LAA parameters such as surface area, volume and ostium shape parameters were also obtained. A total of 71 LAA geometries from AF patients were analysed; 33 of them with a history of a thromboembolic event. We performed a statistical analysis to identify morphological descriptors showing differences between patients with and without thrombus history. Ostium size and centreline length were significatively different in both cohorts, with larger average values in thromboembolic cases, which could be related to slower blood flow velocities within the LAA. Additionally, some of the obtained centreline-based LAA parameters could be used for a better planning of LAA occluder implantations.

**Keywords:** Left atrial appendage · Centreline-based shape descriptors · Thrombus formation

© Springer Nature Switzerland AG 2019
M. Pop et al. (Eds.): STACOM 2018, LNCS 11395, pp. 200–208, 2019.
https://doi.org/10.1007/978-3-030-12029-0_22

# 1 Introduction

The left atrial appendage (LAA) is a sac-shaped sub-structure of the left atrium, which physiological role in the cardiovascular system is not fully known yet. In patients affected by atrial fibrillation (AF), the left atria (LA) becomes rigid over the years and loses the ability to contract properly. In those cases, blood flow might have low velocities in the LAA, which increases the risk of thrombus generation [1]. In fact, around 90% of thrombus in non-valvular atrial fibrillation patients takes place in the LAA [2]. The contractility of the LAA has a strong effect in the thrombus formation in patients suffering non-valvular AF, despite having low CHADS2 scores [3].

Another interesting characteristic of LAA is its high morphological variability. Ernst et al. [4] published a pioneering LAA morphology analysis on necropsy synthetic resin casts, proving LAA shape heterogeneity among individuals in parameters such as LAA volume, ostium diameters (i.e. diameters of the orifice interface between the LA and LAA) and number of secondary/protrusion lobes. Veinot et al. [5] later measured LAA lengths and widths on 500 autopsy hearts, finding ostium increases with age. More recent studies (e.g. [6]), based on medical imaging techniques, also provide LAA morphology analysis, including a classification based on qualitative parameters manually obtained. These subjective descriptors are prone to high inter-observer variability, specially, if derived from different medical imaging modalities. For instance, Lopez-Minguez et al. [7] found that only in 21% of the studied cases, there was an agreement on optimal sizing of the LAA between Computed Tomography (CT), transesophageal echocardiography and angiography images.

Further studies performing Computer Fluid Dynamic (CFD) analyses, pretended to identify potential relationships between the different shapes and the risk of thrombotic events [8–10], using different haemodynamic descriptors like velocities, wall shear stress and a shape classification proposed by Di Biase et al. 2012 [6]. Nevertheless, the shape classification used for these studies was based on subjective criteria proposed by Di Biase et al. [6], were the variability of LAA are not represented and the anatomic knowledge of clinicians are very subjective when making the classification Thrombus formation is a complex phenomenon were not only the shape of the anatomy and the stimuli associated to the blood flow are involving, but also the blood coagulable properties.

The centreline of an 3D object can be defined as a real or imaginary line that is equidistant from the surface or sides of a geometry, providing a simple 1D description of the geometry. It has been typically used to characterize vessel-like structures [11], including aneurysms [12], but also in more complex organs such as the liver [13]. Moreover, for complex geometries, a mean curvature flow technique was used to create the skeleton of different shapes [6]. The main goal of this work was to compute LAA centrelines to establish a common reference system and derive parameters such as the length and bending angles, which will be a first description of the geometry. Other objective and robust LAA shape parameters were also included such as the volume, the area and the ostium size. The whole set of LAA shape descriptors were tested on clinical data from 71 AF patients, with and without a history of a thromboembolic event, to evaluate their relation with thrombus formation.

## 2   Materials and Methods

The computation of the LAA centrelines was performed by combining a heat transfer-derived distance map and a marching algorithm, once it was segmented from 3D medical images. Consequently, some descriptors were extracted from the centreline, which were complemented by other LAA morphology parameters. Finally, a statistical analysis was performed to identify which shape descriptors were different in AF patients with and without thrombus history.

### 2.1   Data Acquisition

A total of 71 LAA geometries were obtained from 3D Rotational Angiography (3DRA) images, after applying semi-automatic thresholding and region-growing algorithms available at the scanner console at OLV Hospital in Aalst, Belgium. All patients considered in this study had AF and underwent ablation. The data was anonymized and a written consent was obtained from all patients. The patient cohort consisted of 33 patients with a history of thromboembolic event and 38 without, by the time the study was performed.

### 2.2   Computation of the Centreline

The centreline was defined as the line that goes from the centre of the ostium to the furthest point of the LAA. In order to obtain it, several steps were required: (1) a *marching algorithm;* (2) application of a *heat transfer equation* to aid with the direction of the parameter (see Fig. 1).

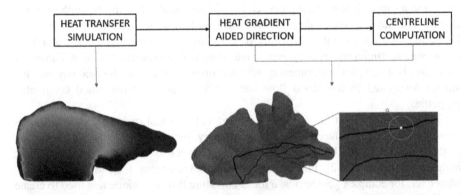

**Fig. 1.** Computational pipeline for centreline (black line) estimation. Left: Heat transfer simulation with red and blue colors representing high and low temperatures, respectively. The black line (centre, right) represents the obtained LAA centreline, while the red line follows the heat gradient. Right: zoom of a region showing the analysis performed at each centreline point to find the next point following heat gradient direction out of the different possibilities (blue arrows). (Color figure online)

*Marching Algorithm:* This technique consists on iteratively adding consecutive points that define a line. The criterion used to choose centreline points was to consider a point close to the previous one (distance smaller than 0.05 mm) such that it is located as furthest as possible to any point in the surface of the geometry. However, the high irregularities (i.e. secondary lobes) in the surfaces of different LAA posed scenarios where the centreline went through narrowing pathways. To solve this problem and avoid centreline loops or abrupt changes, a certain direction was forced along the centreline. The chosen direction was set up as the straight line from the furthest point within the LAA (i.e. so-called apex; starting point) to the ostium centre (ending point).

*Heat Transfer Simulation:* The starting point of the centreline, i.e. the furthest anatomical point of the LAA going through the cavity, was automatically obtained after applying a heat transfer propagation simulation in the LAA geometry. High temperatures were set in the ostium and low temperatures in the LAA surface. A steady state heat transfer simulation was then performed. The coldest non-null point of the resulting map was the one corresponding to the furthest anatomical point in the LAA. Subsequently, the direction of the centreline was forced to follow the heat gradient direction in each point, from the coldest point to the hottest.

**Shape Descriptors.** Several shape descriptors (see Table 1) were obtained from the centreline to globally characterize the LAA geometry, including length, tortuosity and bending. Bi-directional shortest distances between the centreline and the LAA surface were also computed, which generated secondary lines as branches of the initial centreline.

**Table 1.** Shape descriptors obtained from the centreline. $n$ refers to the number of points of the centreline; $m$ is the number of points of the surface of the LAA; $u$ is the director vector previous to a point of the centreline; and $v$ is the director vector posterior to a point in the centreline. $i$ takes integer values from 1 to $n$ and $j$ from 1 to $m$.

| Shape descriptors | Description | Equation |
|---|---|---|
| Length | Total length of the centreline | $L = \sum_{i=1}^{n-1} d(p_i, p_{i+1})$ |
| Tortuosity | Tortuosity of the centreline, total length divided by the linear distance between extreme points | $T = \frac{d(p_1, p_n)}{L}$ |
| Bending | Local bending of the centreline | $b_i = \cos^{-1}\left(\frac{u \cdot v}{|u||v|}\right)$ |
| Distance centreline-surface | Distance from every point of the centreline to the closest point in the surface | $D_i = \min(d_j(p_i, p_j))$ |
| Distance surface-centreline | Distance from every point of the surface to the closest point of the centreline | $D_j = \min(d_i(p_i, p_j))$ |

Additionally, the centreline was also used to define several transversal planes (i.e. perpendicular planes to the centreline direction) along the LAA cavity to have local shape descriptors. These transversal planes provide useful information to the clinician for planning left atrial appendage occlusion (LAAO) implantations such as optimal depth and orientation to position the device in the cavity. We have computed the maximum and minimum diameters (D1 and D2, respectively) of several transversal planes along the LAA centreline to characterize their shapes, as can be seen in Fig. 2. For a visual and comparable representation of local LAA shape descriptors, a triangular common reference space was created based on the computed centrelines (see Fig. 4). The left part of the triangle (its base) corresponds to diameters close to the ostium while the right part represents the most apical (i.e. furthest) point of the LAA. Therefore, the x dimension shows the length of the centreline for each LAA, which was normalized (from 0 to 1) so that different LAA could be compared. The y dimension (width of triangle) does not plot any particular parameter but we assigned larger widths to the triangle base to represent that ostium diameters are usually among the largests and are progressively reduced until reaching the apical LAA point. A normalized colormap was used to show in the triangular LAA reference space any local shape descriptor value such as maximum and minimum diameters of each transversal plane.

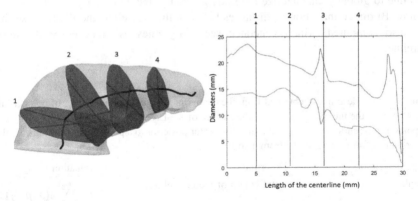

**Fig. 2.** Transversal planes (red slices) in a left atrial appendage (LAA) geometry estimated from the centreline (black line). Blue lines in the transversal planes depict the maximum and minimum local diameters. The graph (right) shows the evolution of local diameters (maximum and minimum in blue and red, respectively) along the centreline length, showing a constant reduction until the ending point of the LAA. (Color figure online)

### 2.3 Statistical Analysis

In order to assess the significance of some of the parameters obtained from the centreline in our database, a statistical analysis by means of a Student t-test (normal variables) or Mann-Whitney U (non-normal) was performed. Normality of the variables was assessed using a Shapiro-Wilks test. In addition, a Levene's test was used to analyze the homoscedasticity of the data. P values < 0.05 were considered statistically significant and all the statistical tests were 2-sided. The software used to carry out the analysis was SPSS (IBM SPSS Statistics for Windows, Version 25.0. Armonk, NY: IBM Corp).

## 3 Results

Firstly, the centreline algorithm was verified on simple synthetic geometries (cylinder, torus), which accurately reproduced the expected equidistant lines from the surface. Consequently, the centreline algorithm was applied to each of the 71 LAA geometries of our clinical database. Some examples of the obtained centrelines on different LAA morphologies are illustrated in Fig. 3 (black lines).

**Fig. 3.** Some examples of centrelines computed on different left atrial appendages, showing their high inter-individual shape variability.

The global shape parameters that showed significant differences between the two groups were the centreline length and the ostium diameters. Specifically, the thromboembolic group showed larger mean values than the non-thromboembolic one for centreline length ($37.21 \pm 8.20$ vs. $34.47 \pm 7.06$; p = 0.023), the maximum ostium diameter ($29.71 \pm 4.56$ vs. $25.76 \pm 4.40$; p < 0.001) and the minimum ostium diameter ($18.96 \pm 3.20$ vs. $17.45 \pm 3.66$; p = 0.008). However, significant differences were not found in the centreline tortuosity parameter ($0.72 \pm 0.09$ vs. $0.76 \pm 0.10$; p = 0.387), Fig. 4 illustrates a boxplot representation of these results.

We also computed the local shape descriptors (e.g. bending angles, diameters of transversal planes) in all LAA geometries. Figure 5 shows the different types of local parameters and visualizations to characterize LAA shapes.

## 4 Discussion and Conclusions

A total of 71 LAA geometries were geometrically characterized by obtaining their main centreline (with its corresponding parameters) and transversal cuts. A statistical analysis was performed between LAA geometries corresponding to patients with a history of thromboembolic events and those who did not. According to our study, the only LAA shape parameters showing statistically significant differences in patients with thrombus were the centreline length and ostium diameters, larger comparing to patients without thrombus history. These results make sense from the mechanistic point of view and are in agreement with the literature: larger ostium and LAA lengths would likely be associated to lower blood flow velocities, then to more favourable conditions to develop thrombus.

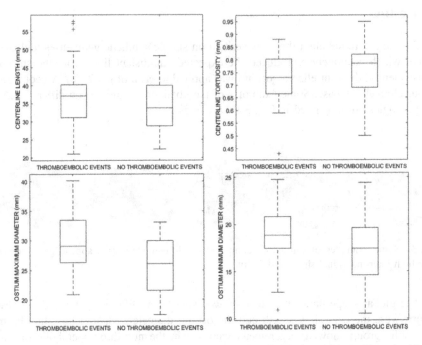

**Fig. 4.** Boxplots showing the difference of the shape descriptors for patients with and without thrombus history. Top: Centreline length (left) and tortuosity (right). Bottom: maximum (left) and minimum (right) ostium diameters.

Bending angles and transversal planes obtained from the centreline can be useful to support decisions for LAAO implantations, including the optimal size of the device. The bending angle curves easily identify where the main lobe orientation is changing (see Fig. 5), which could help to determine where to implant the device (i.e. define the so-called landing zone) and to preview the need of complicated maneuvers depending on the LAA geometry (e.g. not enough space for the LAAO device). Additionally, transversal planes can provide information on how irregular is the LAA shape and to define the maximum depth that a LAAO device with a given size could reach in the LAA.

The main characteristics of the LAA have been calculated automatically. Unlike Di Biase et al.'s method, this approach is objective since was based on the calculation of the centerline allowing define parameters such as: central lobe, secondary lobes, dimensions of diameters along the entire LAA cavity, tortuosity and bending of the LAA.

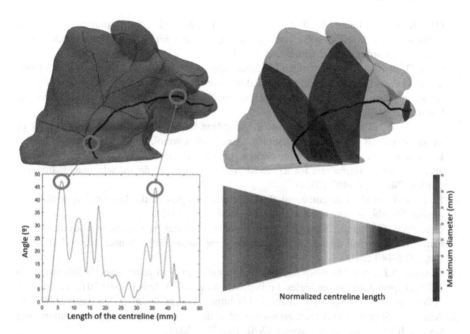

**Fig. 5.** The full geometrical characterization performed for a single LAA geometry. The upper left geometry shows an LAA geometry with the centreline (black) and secondary lines (blue) linking secondary lobes with the ostium. The graph (bottom, left) represents the bending angles of the centreline. The upper right geometry represents the transversal cuts with the corresponding main diameters. The colormap (bottom, right) represents the diameters of the transversal cuts along the centreline (ostium in the left). (Color figure online)

In summary, a complete characterization of LAA morphology is required to better understand the mechanisms under thrombus formation in atrial fibrillation patients and to support clinical decisions such as for planning of LAAO device implantation. Future work will be focused on analyzing these LAA morphological descriptors together with blood flow haemodynamics in the whole database and to develop clinical-friendly visualization tools of the resulting descriptors.

**Acknowledgments.** This work was partially supported by COMPILAAO Retos I+D project (DPI2015-71640-R), funded by the Spanish Government of Economy and Competitiveness.

# References

1. Goldman, M.E., et al.: Pathophysiologic correlates of thromboembolism in nonvalvular atrial fibrillation: I. reduced flow velocity in the left atrial appendage. J. Am. Soc. Echocardiogr. **12**(12), 1080–1087 (1999)
2. Johnson, W.D., Ganjoo, A.K., Stone, C.D., Srivyas, R.C., Howard, M.: The left atrial appendage: our most lethal human attachment! Surgical implications. Eur. J. Cardio-thoracic Surg. **17**(6), 718–722 (2000)

3. Ono, K., et al.: Motion of left atrial appendage as a determinant of thrombus formation in patients with a low CHADS2 score receiving warfarin for persistent nonvalvular atrial fibrillation. Cardiovascular Ultrasound 10, 50 (2012)
4. Ernst, G., et al.: Morphology of the left atrial appendage. Anat. Rec. **242**(4), 553–561 (1995)
5. Veinot, J.P., et al.: Anatomy of the normal left atrial appendage: a quantitative study of age-related changes in 500 autopsy hearts: implications for echocardiographic examination. Circulation **96**(9), 3112 (1997)
6. Di Biase, L., et al.: Does the left atrial appendage morphology correlate with the risk of stroke in patients with atrial fibrillation? J. Am. Coll. Cardiol. **60**(6), 531–538 (2012)
7. López-Mínguez, J.R., et al.: Comparison of imaging techniques to assess appendage anatomy and measurements for left atrial appendage closure device selection. J. Invasive Cardiol. **26**(9), 462–467 (2014)
8. Bosi, G.M., et al.: Computational fluid dynamic analysis of the left atrial appendage to predict thrombosis risk. Front. Cardiovasc. Med. **5**, 34 (2018)
9. Guadalupe, G.I., et al.: Sensitivity analysis of geometrical parameters to study haemodynamics and thrombus formation in the left atrial appendage. Int. J. Numer. Method. Biomed. Eng., 122–140 (2018)
10. Olivares, A.L., et al.: In silico analysis of haemodynamics in patient-specific left atria with different appendage morphologies. In: Pop, M., Wright, G.A. (eds.) FIMH 2017. LNCS, vol. 10263, pp. 412–420. Springer, Cham (2017). https://doi.org/10.1007/978-3-319-59448-4_39
11. Antiga, L., Steinman, D.A.: Robust and objective decomposition and mapping of bifurcating vessels. IEEE Trans. Med. Imaging **23**(6), 704–713 (2004)
12. Hernandez, S.V.: Anatomic Registration based on Medial Axis Parametrizations. Universitat Autònoma de Barcelona (2015)
13. O'Flynn, P.M., O'Sullivan, G., Pandit, A.S.: Methods for three-dimensional geometric characterization of the arterial vasculature. Ann. Biomed. Eng. **35**(8), 1368–1381 (2007)
14. Alli, O., Holmes, D.R.: Left atrial appendage occlusion for stroke prevention. Curr. Probl. Cardiol. **37**(10), 405–441 (2012)
15. Fender, E.A., Kiani, J.G., Holmes, D.R.: Left atrial appendage closure for stroke prevention in atrial fibrillation. Curr. Atheroscler. Rep. **18**(11), 65 (2016)
16. Tagliasacchi, A., Alhashim, I., Olson, M., Zhang, H.: Mean curvature skeletons. In: Eurographics Symposium on Geometry Processing, vol. 31, no. 5, pp. 1735–1744 (2012)
17. Beigel, R., Wunderlich, N.C., Ho, S.Y., Arsanjani, R., Siegel, R.J.: The left atrial appendage: anatomy, function, and noninvasive evaluation. JACC Cardiovasc. Imaging **7**(12), 1251–1265 (2014)

# 3D Atrial Segmentation Challenge

# Automatic 3D Atrial Segmentation from GE-MRIs Using Volumetric Fully Convolutional Networks

Qing Xia[1(✉)], Yuxin Yao[2], Zhiqiang Hu[3], and Aimin Hao[1]

[1] State Key Lab of Virtual Reality Technology and Systems,
Beihang University, Beijing, China
{xiaqing,ham}@buaa.edu.cn
[2] School of Automation Science and Electrical Engineering,
Beihang University, Beijing, China
yyxdawang78@buaa.edu.cn
[3] School of Electronics Engineering and Computer Science,
Peking University, Beijing, China
huzq@pku.edu.cn

**Abstract.** In this paper, we propose an approach for automatic 3D atrial segmentation from Gadolinium-enhanced MRIs based on volumetric fully convolutional networks. The entire framework consists of two networks, the first network is to roughly locate the atrial center based on a low-resolution down-sampled version of the input and cut out a fixed size area that covers the atrial cavity, leaving out other pixels irrelevant to reduce memory consumption, and the second network is to precisely segment atrial cavity from the cropped sub-regions obtained from last step. Both two networks are trained end-to-end from scratch using 2018 Atrial Segmentation Challenge (http://atriaseg2018.cardiacatlas.org/) dataset which contains 100 GE-MRIs for training, and our method achieves satisfactory segmentation accuracy, up to 0.932 in Dice Similarity Coefficient score evaluated on the 54 testing samples, which ranks 1st among all participants.

**Keywords:** Automatic atrial segmentation ·
Fully convolutional networks · Gadolinium-enhanced-MRI

## 1 Introduction

Atrial fibrillation (AF) is one of the most common type of cardiac arrhythmia, which greatly affects human health throughout the world [11]. But it is still challenging to develop successful treatment because of the gaps in understanding the mechanisms of AF [4]. Magnetic resonance images (MRI) can produce pictures of different structures within the heart, and gadolinium contrast agencies are usually used to improve the clarity of these images. These Gadolinium-enhanced MRIs (GE-MRIs) are widely used to study the extent of fibrosis across

© Springer Nature Switzerland AG 2019
M. Pop et al. (Eds.): STACOM 2018, LNCS 11395, pp. 211–220, 2019.
https://doi.org/10.1007/978-3-030-12029-0_23

the atria [9] and recent studies on human atria imaged with GE-MRIs have suggested the atrial structure may hold the key to understanding and reversing AF [4,15]. However, due to the low contrast between the atrial tissue and surrounding background, it is very challenging to directly segment the atrial chambers from GE-MRIs. Most of the existing atrial structural segmentation methods are based on hand-crafted shape descriptors or deformable models on non-enhanced MRIs [14], which can not be directly applied on GE-MRIs because of the low contrast. As for GE-MRIs, current atrial segmentation approaches are still labor-intensive, error/bias-prone, which are obviously not suitable for practical and clinical medical use.

In the past decade, deep learning techniques, in particular Convolutional Neural Networks (CNNs), have achieved great progress in various computer vision tasks, and rapidly become a methodology of choice for analyzing medical images [7]. Ciresan et al. [3] firstly introduced CNNs to medical image segmentation by predicting a pixel's label based on the raw pixel values in a square window centered it. But this method is quite slow because the network must run separately for every pixel within every single image and there is a lot of redundancy due to overlapping windows actually. Later on, Ronneberger et al. proposed U-Net [13], which consists of a contracting path to capture context and a symmetric expanding path that enables precise localization and can be trained end-to-end from very few images built upon the famous Fully Convolutional Network (FCN) [8]. Then, Çiçek et al. [2] replaced the convolution operations in 2D U-Net with 3D counterparts and proposed 3D U-Net for volumetric segmentation. Furthermore, Milletari et al. [10] proposed V-Net, wherein they introduce a novel loss function based on Dice coefficient and learn a residual function inspired by [6] which ensures convergence in less training time and achieves good segmentation accuracy.

In this paper, we develop an automatic 3D atrial segmentation framework using volumetric fully convolutional networks for 2018 Atrial Segmentation Challenge. The overall pipeline of our method is shown in Fig. 1, it consists of two main stages: (1) in the first stage, we use a segmentation based localization strategy to estimate a fixed size target region that covers the whole atria, and leave out pixels outside this region to cut down memory consumption; (2) in the second stage, we train a fine segmentation network based on the cropped target region obtained in the first stage, and transform the predicted masks in target region to the original size volume. The segmentation networks in these two stages are both adapted from V-Net, which can be trained end-to-end and used to segment the atrial cavity fully-automatically.

## 2   Method

### 2.1   Dataset and Preprocessing

Our framework is trained and tested using 2018 Atrial Segmentation Challenge dataset, which contains 100 3D GE-MRIs for training (with both images and masks) and 54 ones for testing (only with images). Each 3D MRI was acquired

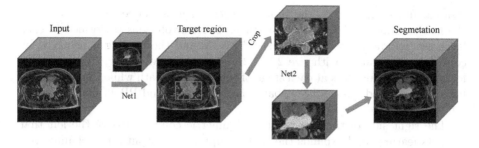

**Fig. 1.** The overall pipeline of our automatic 3D atrial segmentation framework.

using a clinical whole-body MRI scanner and contained raw MRI scan and the corresponding ground truth labels for the left atrial (LA) cavity. The original resolution of these data is $0.625 * 0.625 * 1.25 \, \text{mm}^3$, some of them are with $576 * 576 * 88$ voxels and the others are with $640 * 640 * 88$ voxels, and it is very hard to apply neural networks to directly segment from such high-resolution volumes on a normal personal computer due to memory restriction. Actually, the LA cavity, even the whole heart, takes only a very small fraction of the entire MRI volume and other places in the volume are irrelevant tissues or even nothing, and such extreme class imbalance between the foreground atrial cavity and background also makes the segmentation task hard. So we divide the segmentation into two steps, the first is to locate the atria in the beginning and the second is to segment the cavity from a much smaller cropped sub-volume, which can be used to train networks on a normal PC. To make the input data uniformly sized and suitable for V-Net architecture, we firstly crop and zero-pad all volumes to with size $576 * 576 * 96$ and the predictions are transformed to the original size $576 * 576 * 88$ or $640 * 640 * 88$ in a post-precessing step. Then we use a 3D version of contrast limited adaptive histogram localization (CLAHE) [12] to enhance the contrast of GE-MRIs, and finally apply sample-wise normalization wherein each volume is subtracted with the mean value of intensity and divided by the deviation of intensity.

## 2.2   Network Architecture

The segmentation network involved in our framework is adapted from V-Net [10] as illustrated in Fig. 2. It is a fully convolutional neural network, in which convolution operations are used to both extract features in different scales from the data and reduce the resolution by applying appropriate stride. The left part of the network is an encoding path following a typical architecture of a standard convolutional network, which captures the context information in a local-to-global sense, and the right part decodes the signal to its original size and output two volumes indicating the probability of each voxel to be foreground and background respectively.

The left side of the network is divided in a few stages that operate at different resolutions, each stage consists of one or two convolutional layers, and learns a residual function, that is, the input of each stage is added to the output of the last convolutional layer of that stage. The convolutions performed in each layer use volumetric kernels with size 5 * 5 * 5, and the pooling is achieved by convolution operation with size 2 * 2 * 2 and stride 2. Moreover, the number of feature channels doubles at each stage of the encoding path while the resolutions halves. In the end of each layer, batch normalization and PRelu non linearities are used.

The right side of the network is a symmetric counterparts of the left that extracts features and expands the spatial support to output a two channel volumetric segmentation. Similar to the left part of the network, each stage of the right part contains one or two convolutional layers, and also learns a residual function. The convolutions performed in each layer also use volumetric kernels with size 5 * 5 * 5, and the up-pooling is achieved by de-convolution operation with size 2 * 2 * 2 and stride 2. The features extracted from the left part of the network are forwarded to the corresponding stage of the right part, which is

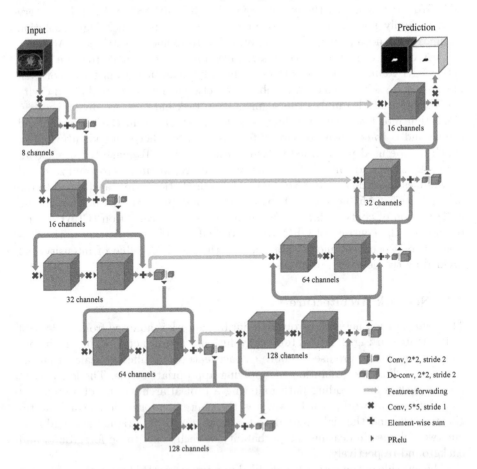

**Fig. 2.** The architecture of our segmentation network adapted from V-Net [10].

shown as horizontal connections in Fig. 2. The same as those in the left part, batch normalization and PRelu non linearities are used in the end of each layer.

### 2.3  Loss Function

The segmentation network predict two volumes of the same size of the input, which are computed after a voxel-wise softmax activation in the final layer and indicate the probability of each voxel to be foreground or background. In segmentation tasks, our aim is to train a network whose foreground prediction is as similar as the given ground truth mask. As the left atrial cavity only takes a small fraction of the volume, we adopt the dice coefficient to define the loss function that to be minimize. The dice coefficient is used to measure the similarity between two given binary data, and be expressed as

$$Dice = \frac{2|\mathbf{a} \cdot \mathbf{b}|}{|\mathbf{a}|^2 + |\mathbf{b}|^2},  \tag{1}$$

where $\mathbf{a}, \mathbf{b}$ are two binary vectors. If $\mathbf{a} = \mathbf{b}$, $Dice = 1$, and if $\mathbf{a_i} \neq \mathbf{b_i}$ for all $i$, $Dice = 0$. In our implementation, we use the foreground prediction (probability) and the given ground truth as $\mathbf{a}, \mathbf{b}$ respectively to compute the loss of the network, which is simply defined as

$$Loss = 1 - Dice.  \tag{2}$$

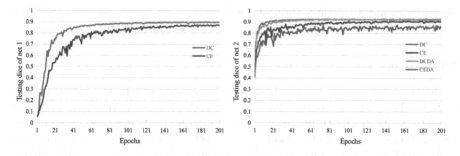

**Fig. 3.** The dice coefficients on validation data after training different epochs of the two segmentation networks in our framework using dice loss (DC) and cross-entropy loss (CE). Moreover, data augmentation (DCDA & CEDA) is applied in the second network when using both loss functions.

This formulation do not need to assign weights to samples of different classes to establish the right balance between foreground and background voxels and is very easy to understand and implement. We also compared dice loss with traditional cross-entropy loss, the results can be found in Fig. 3, wherein we can see the segmentation networks converge faster and reach higher dice coefficients when using dice loss function.

## 2.4   Training

As we mentioned in the beginning, our framework consists of two main stages, the first is to locate the target based on coarse segmentation, and the second is to segment the left atria cavity from the cropped target region. So, we need to train two segmentation networks. For the first network, we firstly down-sample the input with sampling rate $0.25 * 0.25 * 0.5$ and reduce the resolution from $576 * 576 * 96$ to $144 * 144 * 48$, which makes the network consumes much lower memory and can be trained on a normal personal computer. Here choosing sampling rate 0.5 in Z-axis instead of 0.25 is simply to avoid extreme narrow feature maps produced by pooling. Then we feed the input into the network, the weights are initialized using He initialization [5] and updated using Adam algorithm with a fixed learning rate 0.001. We choose 80 out of the 100 data as training data and the rest 20 as testing data, training is completed after 200 epochs and the model with best dice score is saved, and we use mini-batch of size 4 in the first network. For the second network, we firstly compute the barycenter of the given ground truth mask, and crop a region of size $240 * 160 * 96$ centered with the barycenter from the original data. We have calculated the bounding box of all given masks, and found that the maximum widths along x, y, z axis are 209, 128, 73 voxels respectively, so a region of size $240 * 160 * 96$ is big enough to cover the whole cavity. And then we feed the cropped input into the network and train it in the same way as that in the first stage except that batch size is 1 due to memory restriction. To further improve the generalization ability and segmentation accuracy of our framework, we also apply data augmentation in the second network. Before feeding the cropped volume into the network, we randomly choose to slightly translate, scale, rotate, or flip the input data in 3D, and 3D elastic deformation is also used to generate shape diversity.

## 2.5   Testing

In the testing phase, a previously unseen MRI volume is firstly down-sampled to $144 * 144 * 48$, and fed into the first network. The network will output the probability map for both background and foreground, we apply a simple binary test on these two volumetric map where voxels are assign to be foreground or background according which corresponding probability is higher, and this binary mask is used to locate the target region. We compute the barycenter of the predicted mask, crop a region of size $240 * 160 * 96$ centered with this barycenter and then feed it into the second network. The second network also output the probability map and we can compute a binary mask inside the target region and map it back to the original size volume, which is the final left atrial cavity segmentation result, as shown in Fig. 1.

## 3   Result

We implemented our framework using PyTorch [1] with cuDNN, and ran all experiments on a personal computer with 8 GB of memory, Intel Core i7 6700K

CPU @ 4.00 Ghz, and a Nvidia GTX 1060 6G GPU. We validated our framework on the 100 GE-MRIs for training provided by 2018 Atrial Segmentation Challenge, and conducted a 5-fold cross validation, which leaves 80 volumes for training and the rest 20 for validation.

We firstly compared between using dice loss and traditional cross-entropy loss in our framework. The detailed statistics are listed in Table 1, each row contains the segmentation dice coefficients using dice loss and cross-entropy loss of the first, second network and the entire framework in each fold and the average is shown in the bottom. The segmentation accuracy are better when using dice loss than cross-entropy loss in both two networks, and the data augmentation applied in the second network also improves the performance when using dice loss. However, when using cross-entropy loss, data augmentation leads side effects on the accuracy instead. One possible reason is that operations for data augmentation, such as scale and deformation, greatly increase the variation of atrial volume sizes which makes the problem of class-imbalance between foreground and background more serious. For example, the size proportion between atrial cavity and the background varies from sample to sample, but the class weights used in cross-entropy are usually computed as fixed averages, and dice loss do not suffer from class-imbalance problem at all. This situation can also be found in plots of Fig. 3, where the dice coefficients oscillates when applying data augmentation with cross-entropy loss while the other plots are smoother and stabler in contrast. So when using cross-entropy in our framework, we do not apply data augmentation in the second network. Moreover, we can also see from the statistics that the segmentation accuracy of the entire framework is actually the same with the second network, that means the accuracy of the first step do not affect the final segmentation accuracy of the entire framework as long as it gives relatively right target location and the cropped sub-region that covers the entire left atrial cavity. Thus, we can improve the final segmentation accuracy simply by further improving the second network's performance, this is also why we apply data augmentation only for the second network when using dice loss.

The first network takes about 4.7 h to train and consume 4.2 GB GPU memory in average (input size is 144 * 144 * 48 and bath size is 4), and the second network takes about 13.6 h to train and consume about 4 GB GPU memory in average (input size is 240 * 160 * 96 and batch size is 1). At test time, our framework can generate the entire segmentation output within 2s using about 2.6 GB GPU memory. This show our approach's great potential for practical clinical use, because of its simplicity, effectiveness, high accuracy and efficiency. 5 selected atrial segmentation results are listed in Fig. 4 comparing to the given ground truth, the first 4 are those with top dice coefficients and the last one is the worst case among all MRIs. For those MRIs with relative high equality, our frameworks works pretty well, but when facing with MRIs that are unclear and blurry, our method still struggles for higher segmentation accuracy.

**Further Improvement for the Challenge.** To reach higher segmentation accuracy for the challenge, we doubled the number of feature channels in each stage of the network in Fig. 2 and used more convolutional layers (one or two more in each stage). Then we retrained the second network using a Nvidia GTX 1080Ti 11G in each fold of the 5-fold cross validation shown in Table 1, the models with best dice scores were saved and the average dice score increased from 0.923 to 0.927 using more feature channels and deeper architecture. The final predictions on testing data are computed as the average of all predictions of these 5 models. According to the evaluation from the organizers, our method achieves an average Dice Similarity Coefficient score of 0.932 on the 54 testing data and rank 1st among all 27 participants, for more details please refer to the challenge homepage, http://atriaseg2018.cardiacatlas.org/.

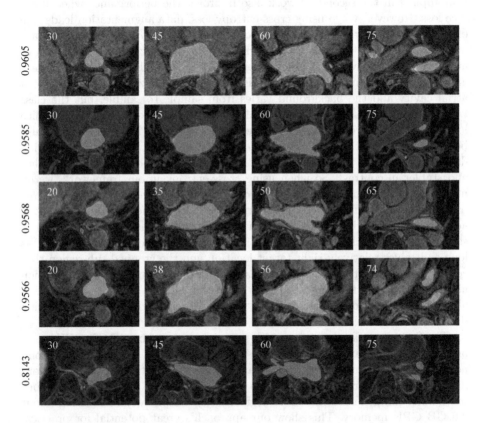

**Fig. 4.** Segmentation results of 5 patients comparing to given ground truth. The ground truths are given in red color and the predictions are colored in gray. The dice coefficients of all predictions are shown in the left most column and the slice number (from axial view) of each picture is shown in the upper left corner of itself. (Color figure online)

**Table 1.** Segmentation accuracy. From left to right: barycenter estimation error (BCE1) (in voxels), segmentation dice coefficients using dice loss (DC1) and cross-entropy loss (CE1) of the first network, dice coefficients of the second network using dice loss and cross-entropy without (DC2 & CE2) and with (DCDA2 & CEDA2) data augmentation, and the entire framework on validation data using dice loss (DC) and cross-entropy loss (CE).

| Fold | BCE1 | DC1 | CE1 | DC2 | DCDA2 | CE2 | CEDA2 | DC | CE |
|------|------|-----|-----|-----|-------|-----|-------|-----|-----|
| 1 | 0.69,0.21,0.51 | 0.885 | 0.871 | 0.917 | 0.923 | 0.904 | 0.872 | 0.923 | 0.904 |
| 2 | 0.68,0.44,0.63 | 0.864 | 0.834 | 0.902 | 0.909 | 0.888 | 0.869 | 0.909 | 0.888 |
| 3 | 0.45,0.34,0.52 | 0.883 | 0.877 | 0.913 | 0.924 | 0.917 | 0.884 | 0.924 | 0.917 |
| 4 | 0.36,0.27,0.61 | 0.889 | 0.870 | 0.906 | 0.932 | 0.911 | 0.879 | 0.932 | 0.911 |
| 5 | 0.52,0.27,0.42 | 0.894 | 0.871 | 0.920 | 0.929 | 0.908 | 0.871 | 0.929 | 0.908 |
| Avg | 0.54,0.31,0.54 | 0.884 | 0.865 | 0.912 | 0.923 | 0.906 | 0.875 | **0.923** | 0.906 |

## 4   Conclusion

This paper detailed a simple but effective approach for automatic 3D atrial segmentation from GE-MRIs, which consists of two volumetric fully convolutional networks adapted from V-Net. The first network is used to coarsely segment the atria from a low-resolution version of the input and estimate the location of the atrial cavity. The second is used to further precisely segment the atria from the cropped sub-region that covers the whole atria. This multi-resolution solution has low memory costs, allowing the network to be trained on a normal personal computer, and the high efficiency make it very easy to apply our segmentation method to clinical use, for example, to reconstruct the structure of human atria and to help researchers develop effective treatments for atrial fibrillation.

## References

1. Pytorch. http://pytorch.org/
2. Çiçek, Ö., Abdulkadir, A., Lienkamp, S.S., Brox, T., Ronneberger, O.: 3D U-Net: learning dense volumetric segmentation from sparse annotation. In: Ourselin, S., Joskowicz, L., Sabuncu, M.R., Unal, G., Wells, W. (eds.) MICCAI 2016. LNCS, vol. 9901, pp. 424–432. Springer, Cham (2016). https://doi.org/10.1007/978-3-319-46723-8_49
3. Ciresan, D., Giusti, A., Gambardella, L.M., Schmidhuber, J.: Deep neural networks segment neuronal membranes in electron microscopy images. In: Advances in Neural Information Processing Systems, pp. 2843–2851 (2012)
4. Hansen, B.J., et al.: Atrial fibrillation driven by micro-anatomic intramural re-entry revealed by simultaneous sub-epicardial and sub-endocardial optical mapping in explanted human hearts. Eur. Heart J. **36**(35), 2390–2401 (2015)
5. He, K., Zhang, X., Ren, S., Sun, J.: Delving deep into rectifiers: surpassing human-level performance on imagenet classification. In: Proceedings of the IEEE International Conference on Computer Vision, pp. 1026–1034 (2015)

6. He, K., Zhang, X., Ren, S., Sun, J.: Deep residual learning for image recognition. In: Proceedings of the IEEE Conference on Computer Vision and Pattern Recognition, pp. 770–778 (2016)
7. Litjens, G., et al.: A survey on deep learning in medical image analysis. Med. Image Anal. **42**, 60–88 (2017)
8. Long, J., Shelhamer, E., Darrell, T.: Fully convolutional networks for semantic segmentation. In: Proceedings of the IEEE Conference on Computer Vision and Pattern Recognition, pp. 3431–3440 (2015)
9. McGann, C., et al.: Atrial fibrillation ablation outcome is predicted by left atrial remodeling on MRI. Circ. Arrhythmia Electrophysiol. **7**(1), 23–30 (2014)
10. Milletari, F., Navab, N., Ahmadi, S.A.: V-net: fully convolutional neural networks for volumetric medical image segmentation. In: 2016 Fourth International Conference on 3D Vision (3DV), pp. 565–571. IEEE (2016)
11. Nishida, K., Nattel, S.: Atrial fibrillation compendium: historical context and detailed translational perspective on an important clinical problem. Circ. Res. **114**(9), 1447–1452 (2014)
12. Pizer, S.M., Johnston, R.E., Ericksen, J.P., Yankaskas, B.C., Muller, K.E.: Contrast-limited adaptive histogram equalization: speed and effectiveness. In: 1990 Proceedings of the First Conference on Visualization in Biomedical Computing, pp. 337–345. IEEE (1990)
13. Ronneberger, O., Fischer, P., Brox, T.: U-Net: convolutional networks for biomedical image segmentation. In: Navab, N., Hornegger, J., Wells, W.M., Frangi, A.F. (eds.) MICCAI 2015. LNCS, vol. 9351, pp. 234–241. Springer, Cham (2015). https://doi.org/10.1007/978-3-319-24574-4_28
14. Tobon-Gomez, C., et al.: Benchmark for algorithms segmenting the left atrium from 3D CT and MRI datasets. IEEE Trans. Med. Imaging **34**(7), 1460–1473 (2015)
15. Zhao, J., et al.: Three-dimensional integrated functional, structural, and computational mapping to define the structural "fingerprints" of heart-specific atrial fibrillation drivers in human heart ex vivo. J. Am. Heart Assoc. **6**(8), e005922 (2017)

# Automatically Segmenting the Left Atrium from Cardiac Images Using Successive 3D U-Nets and a Contour Loss

Shuman Jia[1]([✉]), Antoine Despinasse[1], Zihao Wang[1], Hervé Delingette[1],
Xavier Pennec[1], Pierre Jaïs[2], Hubert Cochet[2], and Maxime Sermesant[1]

[1] Université Côte d'Azur, Epione Research Group, Inria, Sophia Antipolis, France
shuman.jia@inria.fr
[2] IHU Liryc, University of Bordeaux, Pessac, France

**Abstract.** Radiological imaging offers effective measurement of anatomy, which is useful in disease diagnosis and assessment. Previous study [1] has shown that the left atrial wall remodeling can provide information to predict treatment outcome in atrial fibrillation. Nevertheless, the segmentation of the left atrial structures from medical images is still very time-consuming. Current advances in neural network may help creating automatic segmentation models that reduce the workload for clinicians. In this preliminary study, we propose automated, two-stage, three-dimensional U-Nets with convolutional neural network, for the challenging task of left atrial segmentation. Unlike previous two-dimensional image segmentation methods, we use 3D U-Nets to obtain the heart cavity directly in 3D. The dual 3D U-Net structure consists of, a first U-Net to coarsely segment and locate the left atrium, and a second U-Net to accurately segment the left atrium under higher resolution. In addition, we introduce a Contour loss based on additional distance information to adjust the final segmentation. We randomly split the data into training datasets (80 subjects) and validation datasets (20 subjects) to train multiple models, with different augmentation setting. Experiments show that the average Dice coefficients for validation datasets are around 0.91–0.92, the sensitivity around 0.90–0.94 and the specificity 0.99. Compared with traditional Dice loss, models trained with Contour loss in general offer smaller Hausdorff distance with similar Dice coefficient, and have less connected components in predictions. Finally, we integrate several trained models in an ensemble prediction to segment testing datasets.

**Keywords:** 3D U-Net · Segmentation · Left atrium · Loss function ·
Contour loss · Distance map · Ensemble prediction

## 1 Introduction

Atrial fibrillation (AF) is the most frequently encountered arrhythmia in clinical practice, especially in aged population [2,3]. It is characterized by uncoordinated electrical activation and disorganized contraction of the atria. This condition is

© Springer Nature Switzerland AG 2019
M. Pop et al. (Eds.): STACOM 2018, LNCS 11395, pp. 221–229, 2019.
https://doi.org/10.1007/978-3-030-12029-0_24

associated with life-threatening consequences, such as heart failure, stroke and vascular cerebral accident. AF also leads to increased public resource utilization and expense on health care.

With evolving imaging technologies, the analysis of cardiovascular diseases and computer-aided interventions has been developing rapidly. Imaging of the heart is routinely performed in some hospital centers when managing AF and prior to atrial ablation therapy, an invasive treatment to establish trans-mural lesions and block the propagation of arrhythmia. Automated segmentation from cardiac images will benefit the studies of left atrial anatomy, tissues and structures, and provide tools for AF patient management and ablation guidance.

In recent years, with the continuous development of deep learning, neural network models have shown significant advantages in different visual and image processing problems [4]. Automatic segmentation of 3D volumes from medical images by deep neural network also attracts increasing attention in the research community of medical image analysis [5,6].

In this study, we utilize 3D U-Nets with convolutional neural network (CNN), which shows clear advantages compared with traditional feature extraction algorithms [7,8]. Based on that, Ronneberger et al. proposed the original U-Net structure [9] with CNN. Traditional 2D U-Net has achieved good results in the field of medical image segmentation [9,10]. However, it performs convolution on the 2D slices of the images and cannot capture the spatial relationship between slices. Its 3D extension [11], expands the filter operator into 3D space. This extracts image features in 3D, and hence takes into account the spatial continuity between slices in medical imaging. This may better reflect shape features of the corresponding anatomy, enabling full use of the spatial information in 3D images.

Previously, Tran et al. used 3D CNNs to extract temporal and spatial features [12]. They experimented with different sets of data. Hou et al. used 3D CNN to detect and segment pedestrians in a video sequence [13]. The previous studies show that 3D CNN outperformed 2D CNN when dealing with sequences issues.

3D U-Net was used in [11] to realize semi-automatic segmentation of volumetric images. Oktay et al. segmented the ventricle from magnetic resonance (MR) images with 3D U-Net. They introduced an anatomical regularization factor into the model [14], while we choose to use loss function at pixel level.

In the following sections, we will present the two-stage network to segment the left atrium from MR images. The network consists of two successive 3D U-Nets. The first U-Net is used to locate the segmentation target. The second U-Net performs detailed segmentation from cropped region of interest. We introduce a new loss function, Contour loss for the second U-Net. Results will be shown in Sect. 3.

## 2    Method

### 2.1    Dual 3D U-Nets - Cropping and Segmenting

U-Net is a typical encoder-decoder neural network structure. The images are encoded by the CNN layers in the encoder. The output characteristics and the feature maps at different feature levels of the encoder serve as input of the decoder. The decoder is an inverse layer-by-layer decoding process. Such a codec structure can effectively extract image features of different levels so as to analyze the images in each dimension. The 3D U-Net used in this paper is a specialization of 3D U-Net proposed by Çiçek et al. [11]. The implementation of U-Net follows the work of Isensee et al. [15]. We propose a successive dual 3D U-Net architecture, illustrated in Fig. 1.

The first 3D U-Net locates and coarsely extracts the region of interest. Its input is MR images normalized and re-sized to [128, 128, 128]. Its output is preliminary predicted masks of the left atrium. We keep the largest connected component in the masks, and compute the spatial location of the left atrium. Then, we crop the MR images and ground truth masks with a cuboid centered at the left atrium.

The second network performs a secondary processing of the cropped images using the full resolution. Because the higher is the resolution, the larger is the needed memory, we keep only the region around the left atrium, so as to preserve information that is essential for left atrial segmentation. But also, this allows to put a higher resolution on the region of interest with the same amount of memory resource. The input for the second U-Net is MR images cropped around the predicted left atrium without re-sampling of size [224, 144, 96]. Its output is our prediction for the left atrial segmentation. We train the second U-Net with two kinds of ground truths, binary segmentation masks $M$ and euclidean distance maps $D(M)$, as shown in Fig. 2. Here, we introduce a new loss function based on Contour distance.

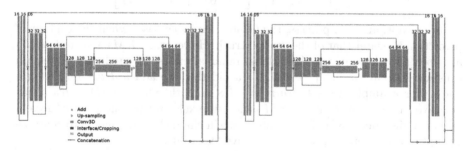

**Fig. 1.** Proposed dual 3D U-Net structure. Green blocks represent 3D features; Dark blue refers to interface operation to crop the region of interest based on first U-Net prediction. (Color figure online)

## 2.2 Loss Functions

Using Dice coefficient as loss function can reach high accuracy. However, experiments show that, because the inside of the left atrial body accounts for most pixels, the network would stop to optimize when it finds satisfying segmentation of the left atrial body. Instead of the volume inside, the contour is what we want to obtain accurately for the segmentation. It is challenging to segment accurately especially the region around the pulmonary veins and the appendage. Hence, we introduce the Contour distance into the loss function.

The distance maps $D(M)$ of the ground truth segmentation $M$ illustrate how far is each pixel from the contour of the left atrium. We compute the distance for pixels inside and outside the left atrium, based on the euclidean distance transform algorithm implemented in scipy. The definition of a Hausdorff distance is symmetric between two point sets. But to make it easy to be implemented in neural networks, we do not compute the distance map of the changing prediction $P$ in the training process, and use an unitary distance:

$$loss_{contour} = \sum_{p \in contour(P)} \min_{m \in M} \|p - m\|_2 = \sum(D(M) \circ contour(P)), \qquad (1)$$

where $\circ$ performs element-wise multiplication. $P$ is the prediction of the U-Net, after sigmoid activation of the output layer. To compute a contour of predicted left atrium, we can apply 3D Sobel filters on $P$ and add the absolute value of results in three directions:

$$contour(P) = |P * s_x| + |P * s_y| + |P * s_z|, \qquad (2)$$

where $*$ denotes 3D convolution operation and $+$ denotes pixel-wise addition. $s_x$, $s_y$ and $s_z$ are 3D Sobel–Feldman kernels in x-, y- and z-direction, for example:

$$s_z = \left[ \begin{bmatrix} +1 & +2 & +1 \\ +2 & +4 & +2 \\ +1 & +2 & +1 \end{bmatrix}, \begin{bmatrix} 0 & 0 & 0 \\ 0 & 0 & 0 \\ 0 & 0 & 0 \end{bmatrix}, \begin{bmatrix} -1 & -2 & -1 \\ -2 & -4 & -2 \\ -1 & -2 & -1 \end{bmatrix} \right].$$

In the optimization process, the Contour loss decreases when the contour of prediction is nearer that of the ground truth. However, if $D(M)$ and $contour(P)$ are both always positive, a bad global minimal exists: let prediction $P$ remain constant so that $contour(P) \sim 0$. To avoid this, we add a *drain* on distance maps. For example, we set $drain = -1.8$, and therefore $D(M) + drain$ has negative value on the contour of $M$, since all pixels on the contour of $M$ have a euclidean distance $\leqslant \sqrt{3}$. This creates a drain channel for the model to continue the optimization process towards negative loss value.

$$loss_{contour} = \sum((D(M) + drain) \circ contour(P)). \qquad (3)$$

The loss function is differentiable and converging in 3D U-Net training trials[1].

## 2.3   Ensemble Prediction

Contour loss provides a spatial distance information for the segmentation, while Dice coefficient measures the volumes inside the contour. We combine the two loss functions in a ensemble prediction model.

We visualize the process in Fig. 2.

---

[1] The code is available on https://gitlab.inria.fr/sjia/unet3d_contour.

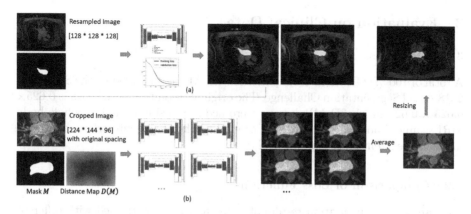

**Fig. 2.** The framework of successive U-Nets training. (a) The fist U-Net - cropping; (b) the second U-Net - segmenting, with ensemble prediction models. We show here axial slices of MR images, overlapped with manual segmentation of the left atrium in blue, our segmentation in red, intersection of the two in purple. (Color figure online)

**Fixed Experimental Setting.** In this study, we set batch size to 1. Training-validation split ratio equals to 0.8. We perform normalization of image intensity. The initial learning rate of our neural network is $5e^{-4}$. Learning rate will be reduced to half after 10 epochs if the validation loss is not improving. The early convergence is defined as no improvement after 50 epochs. Number of base filters is 16. Maximum number of epochs is 500, and 200 steps per epoch. Using large initial learning rate reduces the time to find the minimum and may also overstep the local minima to converge closer to the global minimum of the loss function. However, the accuracy of segmentation also relies on when we stop the training to avoid over-fitting.

We use computation cluster with GPU capacity 6.1 for training. The first U-Net takes around 200–300 epochs to reach convergence. We extract the largest connected components in predictions of the first U-Net to crop the original MRI images around the left atrium, without re-sampling, of size $[224, 144, 96]$. The second U-Net takes around 150–200 epochs to reach convergence. Then the prediction results of the second U-Net are re-sized without re-sampling to the original image size.

**Varying Experimental Setting.** For the second U-Net, we change on the one hand, the options for augmentation: horizontal/vertical flip set to True or False; rotation range set to 0, 7, 10°; width/height shift range set to 0.0–0.1; zoom in or not, zoom range set to 0.1, 0.2. On the other hand, we alter the loss function option, Dice coefficient and Contour loss. We choose multiple trained U-Net models with above experimental settings for ensemble prediction. We also train twice with some parameters but with different validation splitting. We make the final decision of segmentation based on the average of all predictions, similar to letting multiple agents vote for each pixel if it belongs to left atrium or not, in majority voting system.

# 3   Evaluation on Clinical Data

## 3.1   Materials

A total of 100 3D GE-MRIs from patients with AF are provided by the STACOM 2018 Atrial Segmentation Challenge. The original resolution of the data is $0.625 \times 0.625 \times 0.625$ mm$^3$. 3D MR images were acquired using a clinical whole-body MRI scanner and the corresponding ground truths of the left atrial masks were manually segmented by experts in the field.

## 3.2   Comparison of Loss Functions

We assess the segmentation results of individual models, trained with different experimental setting, as described in Sect. 2.3, to compare the prediction performance using Dice loss and Contour loss.

**Evaluation Metrics.** The evaluation metrics are Dice coefficient, Hausdorff distance (HD) and confusion matrix, as shown in Table 1. Multiple evaluation metrics provide us different views to assess the performance of models.

The Dice index for predicted segmentation in validation datasets attained 0.91–0.92. The sensitivity of prediction was around 0.90–0.94 and the specificity 0.99. The proposed method closely segmented the atrial body, with both loss functions, compared with manual segmentation. Different from traditional Dice loss, models trained with Contour loss in general offered smaller Hausdorff distance with similar Dice coefficient.

**Visualization.** We visualize the predicted segmentation results of validation datasets in Fig. 3 in a 3D view. Case 1 and Case 2 are selected to represent respectively, good scenario and bad scenario. For the two loss functions, differences lay in the boundary, the region close to the pulmonary veins and septum. With Dice loss function, there were more details and sharp edges, and therefore more disconnected spots not belonging to the left atrium. With the Contour loss function, the smoothness of the contour, and shape consistency were better maintained.

**Connected Components.** The left atrium should be a single connected component in binary segmentation mask. We present in Table 2 the number of connected components in raw predictions given by U-Nets and compare the two loss functions. Using Dice coefficient loss alone produced more disconnect components not belonging to the left atrial structures.

**Table 1.** Evaluation of validation datasets segmentations using two loss functions: Dice coefficient loss (top) and Contour loss (bottom).

|  |  | Confusion Matrix | | Dice | Contour distance (pixel) | |
|---|---|---|---|---|---|---|
|  |  | Sensitivity | Specificity |  | Average HD | HD |
| Model (Dice loss) | 1 | 0.93 ± 0.03 | 0.99 ± 0.00 | 0.910 ± 0.032 | 1.71 ± 0.53 | 34.29 ± 15.00 |
|  | 2 | 0.94 ± 0.04 | 0.99 ± 0.00 | 0.913 ± 0.029 | 1.63 ± 0.57 | 27.39 ± 12.19 |
|  | 3 | 0.91 ± 0.05 | 0.99 ± 0.00 | 0.914 ± 0.023 | 1.62 ± 0.42 | 30.04 ± 11.76 |
|  | 4 | 0.92 ± 0.04 | 0.99 ± 0.00 | 0.920 ± 0.030 | 1.41 ± 0.51 | 23.50 ± 12.34 |
|  | 5 | 0.91 ± 0.05 | 0.99 ± 0.00 | 0.918 ± 0.019 | 1.47 ± 0.47 | 21.66 ± 7.75 |
|  | 6 | 0.91 ± 0.04 | 0.99 ± 0.00 | 0.921 ± 0.024 | 1.45 ± 0.48 | 26.52 ± 16.14 |
|  | 7 | 0.92 ± 0.03 | 0.99 ± 0.00 | 0.923 ± 0.027 | 1.40 ± 0.67 | 23.38 ± 12.03 |
|  | 8 | 0.93 ± 0.04 | 0.99 ± 0.00 | 0.927 ± 0.024 | 1.28 ± 0.46 | 22.48 ± 16.35 |
| Model (Contour loss) | 1 | 0.90 ± 0.05 | 0.99 ± 0.00 | 0.911 ± 0.020 | 1.60 ± 0.41 | 19.80 ± 6.82 |
|  | 2 | 0.92 ± 0.04 | 0.99 ± 0.00 | 0.906 ± 0.045 | 1.60 ± 0.58 | 18.88 ± 7.13 |
|  | 3 | 0.91 ± 0.04 | 0.99 ± 0.00 | 0.917 ± 0.025 | 1.48 ± 0.51 | 21.57 ± 8.13 |
|  | 4 | 0.92 ± 0.04 | 1.00 ± 0.00 | 0.917 ± 0.024 | 1.40 ± 0.33 | 20.21 ± 7.53 |
|  | 5 | 0.90 ± 0.04 | 1.00 ± 0.00 | 0.919 ± 0.024 | 1.53 ± 0.51 | 24.81 ± 6.57 |
|  | 6 | 0.89 ± 0.03 | 1.00 ± 0.00 | 0.921 ± 0.019 | 1.58 ± 0.59 | 24.04 ± 11.72 |
|  | 7 | 0.91 ± 0.04 | 0.99 ± 0.00 | 0.914 ± 0.021 | 1.56 ± 0.47 | 22.65 ± 7.52 |
|  | 8 | 0.92 ± 0.04 | 1.00 ± 0.00 | 0.917 ± 0.030 | 1.41 ± 0.45 | 19.66 ± 4.52 |

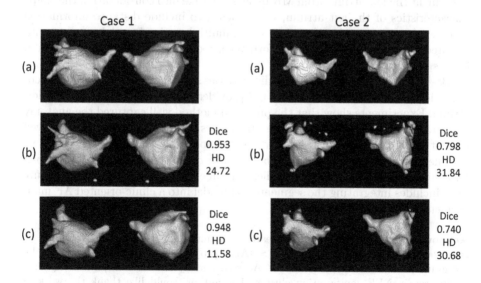

**Fig. 3.** Visualization of good and bad validation datasets segmentations, with their Dice, Hausdorff distance (HD) with respect to the ground truth. (a) manually segmented; (b) predicted with Dice coefficient loss; (c) predicted with Contour loss.

**Table 2.** Number of connected components in predicted segmentations.

|  | Mean $\pm \sigma$ | Maximum |
|---|---|---|
| Model (Dice loss) | $2.85 \pm 2.83$ | 20 |
| Model (Contour loss) | $1.28 \pm 0.65$ | 5 |

**Ensemble Prediction.** To reduce the irregular bumps and disconnected components, we choose in total 11 trained U-Net models with both loss functions, to perform an ensemble prediction for testing datasets. We add the probabilistic segmentations of all U-Nets and threshold their sum $\geq 5.5$.

With Dice loss, more details can be captured. With Contour loss, the shapes look more regular and smoother globally. The difficult regions to segment remain the septum and especially the appendage for both loss functions. The manual segmentations are usually performed slice-by-slice, there exists sudden discontinuity between two axial slices. While our segmentation is based on 3D operators, the segmented region is continuous between slices, which accounts for part of mismatch between manual segmentation and network segmentation that cannot be avoided.

## 4 Conclusion

In this paper, we proposed a deep neural network with dual 3D U-Net structure, to segment the left atrium from MR images. To take into consideration the shape characteristics of the left atrium, we proposed to include distance information and created a Contour loss function. Using multiple trained models in an ensemble prediction can improve the performance, reducing the impact of accidental factors in neural network training.

Based on previous studies on cardiac cavities segmentation, our method accurately located the region of interest and provided good segmentations of the left atrium. Experiments show that the proposed method well captured the anatomy of the atrial volume in 3D space from MR images. The new loss function achieved a fine-tuning of contour distance and good shape consistency.

The automated segmentation model can in return reduce manual work load for clinicians and has promising applications in clinical practice. Potential future work includes integrating the segmentation model into a clinic-oriented AF management pipeline.

**Acknowledgments.** Part of the research was funded by the Agence Nationale de la Recherche (ANR)/ERA CoSysMed SysAFib project. This work was supported by the grant AAP Santé 06 2017-260 DGA-DSH, and by the Inria Sophia Antipolis - Méditerranée, NEF computation cluster. The author would like thank the work of relevant engineers and scholars.

# References

1. McGann, C., et al.: Atrial fibrillation ablation outcome is predicted by left atrial remodeling on MRI. Circ.: Arrhythm. Electrophysiol. **7**(1), 23–30 (2014)
2. Zoni-Berisso, M., Lercari, F., Carazza, T., Domenicucci, S., et al.: Epidemiology of atrial fibrillation: European perspective. Clin. Epidemiol. **6**(213), e220 (2014)
3. Morillo, C.A., Banerjee, A., Perel, P., Wood, D., Jouven, X.: Atrial fibrillation: the current epidemic. J. Geriat. Cardiol.: JGC **14**(3), 195 (2017)
4. De Cecco, C.N., Muscogiuri, G., Varga-Szemes, A., Schoepf, U.J.: Cutting edge clinical applications in cardiovascular magnetic resonance. World J. Radiol. **9**(1), 1 (2017)
5. Litjens, G., et al.: A survey on deep learning in medical image analysis. Med. Image Anal. **42**, 60–88 (2017)
6. Liu, J., et al.: Applications of deep learning to MRI images: a survey. Big Data Min. Anal. **1**(1), 1–18 (2018)
7. Zeiler, M.D., Fergus, R.: Visualizing and understanding convolutional networks, 28 November 2013. arXiv:1311.2901v3 [cs.CV]. Computer Vision–ECCV 2014, vol. 8689, pp. 818–833 (2014)
8. Bengio, Y., Courville, A., Vincent, P.: Representation learning: a review and new perspectives. IEEE Trans. Pattern Anal. Mach. Intell. **35**(8), 1798–1828 (2013). https://doi.org/10.1109/TPAMI.2013.50
9. Ronneberger, O., Fischer, P., Brox, T.: U-Net: convolutional networks for biomedical image segmentation. In: Navab, N., Hornegger, J., Wells, W.M., Frangi, A.F. (eds.) MICCAI 2015. LNCS, vol. 9351, pp. 234–241. Springer, Cham (2015). https://doi.org/10.1007/978-3-319-24574-4_28
10. Li, X., Chen, H., Qi, X., Dou, Q., Fu, C., Heng, P.: H-DenseUNet: hybrid densely connected UNet for liver and tumor segmentation from CT volumes. IEEE Trans. Med. Imaging **37**(12), 2663–2674 (2018). https://doi.org/10.1109/TMI.2018.2845918
11. Çiçek, Ö., Abdulkadir, A., Lienkamp, S.S., Brox, T., Ronneberger, O.: 3D U-Net: learning dense volumetric segmentation from sparse annotation. In: Ourselin, S., Joskowicz, L., Sabuncu, M.R., Unal, G., Wells, W. (eds.) MICCAI 2016. LNCS, vol. 9901, pp. 424–432. Springer, Cham (2016). https://doi.org/10.1007/978-3-319-46723-8_49
12. Tran, D., Bourdev, L.D., Fergus, R., Torresani, L., Paluri, M.: Learning spatiotemporal features with 3D convolutional networks. CoRR, abs/1412.0 (2015)
13. Hou, R., Chen, C., Shah, M.: An end-to-end 3D convolutional neural network for action detection and segmentation in videos. arXiv preprint arXiv:1712.01111 (2017)
14. Oktay, O., et al.: Anatomically constrained neural networks (ACNNs): application to cardiac image enhancement and segmentation. IEEE Trans. Med. Imaging **37**(2), 384–395 (2018). https://doi.org/10.1109/TMI.2017.2743464
15. Isensee, F., Kickingereder, P., Wick, W., Bendszus, M., Maier-Hein, K.H.: Brain tumor segmentation and radiomics survival prediction: contribution to the BRATS 2017 challenge. In: Crimi, A., Bakas, S., Kuijf, H., Menze, B., Reyes, M. (eds.) BrainLes 2017. LNCS, vol. 10670, pp. 287–297. Springer, Cham (2018). https://doi.org/10.1007/978-3-319-75238-9_25

# Fully Automated Left Atrium Cavity Segmentation from 3D GE-MRI by Multi-atlas Selection and Registration

Mengyun Qiao[1], Yuanyuan Wang[1(✉)], Rob J. van der Geest[2], and Qian Tao[2(✉)]

[1] Biomedical Engineering Center, Fudan University, Shanghai, China
yywang@fudan.edu.cn
[2] Department of Radiology, Leiden University Medical Center, Leiden, the Netherlands
q.tao@lumc.nl

**Abstract.** This paper presents a fully automated method to segment the complex left atrial (LA) cavity, from 3D Gadolinium-enhanced magnetic resonance imaging (GE-MRI) scans. The proposed method consists of four steps: (1) preprocessing to convert the original GE-MRI to a probability map, (2) atlas selection to match the atlases to the target image, (3) multi-atlas registration and fusion, and (4) level-set refinement. The method was evaluated on the datasets provided by the MICCAI 2018 STACOM Challenge with 100 dataset for training. Compared to manual annotation, the proposed method achieved an average Dice overlap index of 0.88.

**Keywords:** Left atrium · Atlas selection · Multi-atlas segmentation

## 1 Introduction

The left atrial (LA) cavity segmentation is an important step in reconstructing and visualizing the patient-specific atrial structure for clinical use. Various important clinical parameters can be derived from accurately segmented LA structures [3,9]. With development of imaging techniques, the three-dimensional LA geometry can be non-invasively visualized by computed tomography angiography (CTA) or magnetic resonance angiography (MRA). Gadolinium-enhanced magnetic resonance imaging (GE-MRI) is typically used to study the extent of fibrosis (scars) across the LA, which provides clinically important diagnostic and prognostic information. However, manual segmentation of the atrial chambers from GE-MRI is a highly challenging and time-consuming task, due to the complex shape of LA, as well as the low contrast between the atrial tissue and background. Accurate automated computer methods to segment the LA and reconstruct it in three dimensions are highly desirable [4,8,12].

In this work, we present and evaluate a systematic workflow to segment the LA in a fully automated manner. The proposed workflow consists of four steps: (1) preprocessing to convert the original GE-MRI to a probability map, (2) atlas selection, (3) multi-atlas registration and fusion, and (4) level-set refinement.

© Springer Nature Switzerland AG 2019
M. Pop et al. (Eds.): STACOM 2018, LNCS 11395, pp. 230–236, 2019.
https://doi.org/10.1007/978-3-030-12029-0_25

## 2    Data and Method

### 2.1    Data

The majority of the dataset used in this work were provided by The University of Utah (NIH/NIGMS Center for Integrative Biomedical Computing (CIBC)), and the rest were from multiple other centers. For all clinical data, institutional ethics approval was obtained.

In total, 100 datasets were provided with annotated reference standard for training purposes. Each 3D MRI patient data was acquired using a clinical whole-body MRI scanner and contained the raw MRI scan and the corresponding ground truth labels for the LA cavity and pulmonary veins. The ground truth segmentation was performed by expert observers.

### 2.2    Probability Map

GE-MRI typically has non-quantitative signal intensity values that vary among different scans. The difference is especially pronounced if the scans are from different vendors or centers. To normalize the signal intensity value, we converted the original GE-MRI to a probability tissue map as a preprocessing step, with the signal intensity value normalized between 0 and 100%. This was realized by the Coherent Local Intensity Clustering (CLIC) algorithm [7]: the class membership functions were calculated by a non-supervised clustering algorithm, and the probability maps were derived accordingly. The preprocessing was applied on the entire 3D volume of GE-MRI. Figure 1 shows a slice of the original GE-MRI and the resulting probabilistic image.

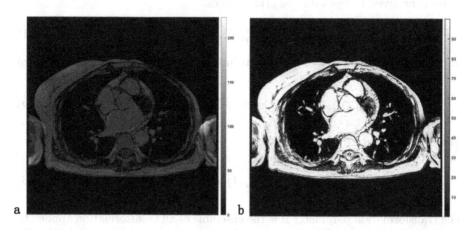

**Fig. 1.** An example of the GE-MRI slice: (a) the original GE-MRI scan (arbitrary range), (b) the preprocessed probability map (range 0–100%).

### 2.3   SIMPLE Atlas Selection

Atlas images were selected based on image similarity between the atlas and the target. We used the selective and iterative method for performance level estimation (SIMPLE) method for atlas selection based on 3D atrial anatomy [6]. Poorly performing atlases can be regarded as noise in the majority voting. These segmentations can be down-weighted by their estimated performance to reduce their influence on the result of the majority voting. We used the groupwise registration to get the propagated segmentations from atlas $L$, and then the segmentations were combined into $L_{mean}$ using a weighted majority vote. $L_{mean}$ was assumed to be reasonably similar to $L_{target}$ [1]. The estimated performance was recomputed as $f(L_i, L^0_{mean})$, where $f$ is the Dice overlapping index, $i$ is the atlas index, and 0 denotes the number of iterations. The poorly performing atlas images were discarded if is lower than a predefined threshold, while the rest was combined into a new atlas set for next iteration of groupwise registration. The iteration was stopped when the atlas set converged.

### 2.4   Multi-atlas Segmentation by Groupwise Registration

Mutli-atlas segmentation is an image-based approach to segment complex anatomical structures in medical images [5]. Define $A_i$ as the atlas image, with known label $L_i$, $i = 1, 2, ..., N$, where $N$ is the total number of atlases. Multiple images with labels were used as a base of knowledge, to segment a new image $I$ containing the same anatomical structure.

In this work, we used the groupwise strategy to perform multi-atlas registration [10,11]. We registered all the atlas images in one groupwise registration to the given image. We formulated it as optimizing a group objective function that merges the given image with the atlas images:

$$\hat{\mu} = \arg \min_{\mu} C(T_\mu; I, A_1, A_2, ..., A_N) \tag{1}$$

where the cost function $C$ is defined to minimize the variance in the group of images:

$$C(\mu) = (T_\mu(I) - \bar{I}_\mu)^2 + \sum_{i=1}^{N}(T_\mu(A_i) - \bar{I}_\mu)^2 \tag{2}$$

with $\bar{I}_\mu$ is defined as:

$$\bar{I}_\mu = \frac{1}{N+1}\left\{ T_\mu(I) + \sum_{i=1}^{N} T_\mu(A_i) \right\} \tag{3}$$

The transformation parameter $\mu$ is the ensemble of all transformation applied to $N$ atlas images $A_i$ and the given image $I$. Given the preprocessing step, the signal intensity range in the atlas images and target image is comparable. Therefore minimization of the variance indicates that all images in the groupwise registration are well registered. The propagated labels were fused by the majority-vote method to obtain the segmentation result.

## 2.5    Refinement of LA Segmentation by Level-Set

From the atlas-based segmentation, the local methods can still provide incremental improvement. In this step, we used a level-set approach to refine and grow the initial segmentation, attending to local image details. With the atlas-based segmentation as initialization, the level-set [2] was applied with the energy function formulated as:

$$F(\phi) = \mu \int_\Omega |\nabla H(\phi)| \, d\Omega + \nu \int_\Omega H(\phi) d\Omega$$
$$+ \lambda_1 \int_\Omega |I(x,y,z) - c_1|^2 H(\phi) d\Omega \qquad (4)$$
$$+ \lambda_2 \int_\Omega |I(x,y,z) - c_2|^2 (1 - H(\phi)) d\Omega$$

where $\phi$ is the level set function. The model assumes that the image $I$ is a three-dimensional image with piecewise constant values. It defines the evolving surface $\Omega = 0$ as the boundary of object to be detected in image $I$. $c_1$, $c_2$ are the average values inside and outside the boundary, respectively. $H$ is the Heaviside function. $\mu \geq 0$, $\nu \geq 0$, $\lambda_1, \lambda_2 > 0$ are weighting factors of the four terms. The first term denotes the surface area, and second term denotes the volume inside

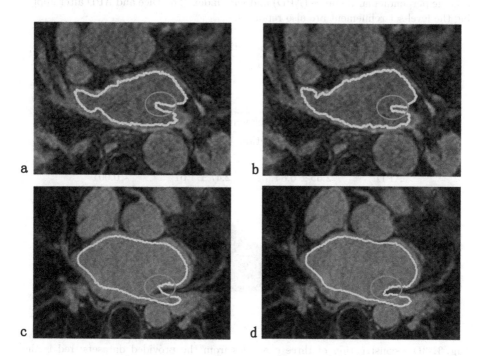

**Fig. 2.** Two examples of the level-set refinement: (a) and (c) are before refinement with dice index of 0.89, while (b) and (d) are after refinement with dice index of 0.91. The blue circle highlights the area of improvement. (Color figure online)

the surface. The last two terms are the variance inside and outside the boundary. A characteristic of the level-set method approach is that it does not have prior assumptions on the object geometry and can deal with complex shapes such as the LA. Figure 2 shows an example of the refinement step.

## 3    Experiments and Results

Our proposed method was evaluated on the training datasets (N = 100) and the testing datasets (N = 54) datasets provided by the MICCAI 2018 STACOM challenge organizers. The leave-one-out strategy was used: for one target image, we used the rest of 99 datasets as the potential atlases. We carried out preliminary experiments and compared the performance of two atlas selection methods: (1) random atlas selection, (2) SIMPLE atlas selection. In each method 10 best atlases were finally selected. The performance in terms of Dice index is reported in Table 1. The proposed method that tested on the training datasets resulted in an average Dice index of $0.88 \pm 0.03$ and average perpendicular distance (APD) of $2.47 \pm 1.01$ mm, superior to random atlas selection (P<0.05 by the Paired

**Table 1.** Performance of our proposed method on the training datasets in terms of average perpendicular distance (APD) and Dice index. The Dice and APD after applying the level-set refinement are also reported.

| Evaluation after atlas segmentation | | | |
|---|---|---|---|
| | Random-selection | SIMPLE-selection | P-value |
| Dice indices | $0.83 \pm 0.07$ | $0.87 \pm 0.04$ | <0.05 |
| APD | $2.67 \pm 0.93$ | $2.49 \pm 0.01$ | <0.001 |
| Evaluation after level-set refinement | | | |
| | Random-selection | SIMPLE-selection | P-value |
| Dice indices | $0.84 \pm 0.08$ | $0.88 \pm 0.03$ | <0.05 |
| APD | $2.65 \pm 0.84$ | $2.47 \pm 1.01$ | <0.001 |

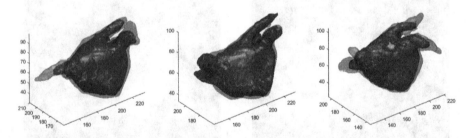

**Fig. 3.** 3D reconstruction of three examples from the provided datasets: red is the manual ground truth segmentation, and blue is the automated segmentation. (Color figure online)

Wilcoxon-test). The method tested on the testing datasets resulted in an average Dice index of 0.93. Figure 3 shows three examples of the 3D reconstruction of the LA cavity segmentation by the proposed method, compared to the ground truth manual segmentation.

## 4   Discussion and Conclusion

In this work, we have developed an atlas-based method to segment the complex LA cavity from GE-MRI scans. The method includes atlas selection, registration, and label fusion. In addition, for better performance of the atlas approach, we have normalized the GE-MRI images by the probability map, and refined the final results using the level-set approach.

Depending on similarity between the given image to the individual atlases, we used iterative rules to select suitable atlases. The atlas-selection method estimated the performance of segmentation of an individual atlas in an iterative manner, such that the best performed atlases were selected in *a posteriori* manner for progressively better segmentation performance.

We believe that the atlas approach is still interesting in face of the current trend of deep-learning-based medical image segmentation. Especially when the training data is scarce, the atlas approach preserves the geometrical characteristics of anatomical structures; by comparing a target image with the atlases, the algorithm still performs a type of "learning", although not data-intensive.

In conclusion, we have presented a complete workflow to fully automatically segment the LA cavity from GE-MRI. Trained and tested on the MICCAI 2018 STACOM Challenge database, the proposed method showed reasonably high accuracy of LA segmentation compared to the time-consuming manual annotation.

**Acknowledgements.** This research is supported by grants from National Key R&D Program of China 2018YFC0116303.

## References

1. Artaechevarria, X., Munoz-Barrutia, A., Ortiz-de Solórzano, C.: Combination strategies in multi-atlas image segmentation: application to brain mr data. IEEE Trans. Med. Imaging **28**(8), 1266–1277 (2009)
2. Chan, T.F., Vese, L.A.: Active contours without edges. IEEE Trans. Image Process. **10**(2), 266 (2001). A Publication of the IEEE Signal Processing Society
3. Haissaguerre, M., et al.: Spontaneous initiation of atrial fibrillation by ectopic beats originating in the pulmonary veins. N. Engl. J. Med. **339**(10), 659–666 (1998)
4. Hansen, B.J., et al.: Atrial fibrillation driven by micro-anatomic intramural re-entry revealed by simultaneous sub-epicardial and sub-endocardial optical mapping in explanted human hearts. Eur. Hear. J. **36**(35), 2390–2401 (2015)
5. Iglesiasa, J.E., Sabuncu, M.R.: Multi-atlas segmentation of biomedical images: a survey. Med. Image Anal. **24**(1), 205–219 (2015)

6. Langerak, T.R., van der Heide, U.A., Kotte, A.N., Viergever, M.A., Van Vulpen, M., Pluim, J.P.: Label fusion in atlas-based segmentation using a selective and iterative method for performance level estimation (simple). IEEE Trans. Med. Imaging **29**(12), 2000–2008 (2010)

7. Li, C., Xu, C., Anderson, A.W., Gore, J.C.: MRI tissue classification and bias field estimation based on coherent local intensity clustering: a unified energy minimization framework. In: Prince, J.L., Pham, D.L., Myers, K.J. (eds.) IPMI 2009. LNCS, vol. 5636, pp. 288–299. Springer, Heidelberg (2009). https://doi.org/10.1007/978-3-642-02498-6_24

8. McGann, C., et al.: Atrial fibrillation ablation outcome is predicted by left atrial remodeling on MRI. Circ. Arrhythmia Electrophysiol. **7**(1), 23–30 (2014)

9. Oral, H., et al.: Circumferential pulmonary-vein ablation for chronic atrial fibrillation. N. Engl. J. Med. **354**, 934–941 (2006)

10. Wachinger, C., Navab, N.: Simultaneous registration of multiple images: similarity metrics and efficient optimization. IEEE Trans. Pattern Anal. Mach. Intell. **35**(5), 1221–1233 (2013). https://doi.org/10.1109/TPAMI.2012.196

11. Yigitsoy, M., Wachinger, C., Navab, N.: Temporal groupwise registration for motion modeling. In: Székely, G., Hahn, H.K. (eds.) IPMI 2011. LNCS, vol. 6801, pp. 648–659. Springer, Heidelberg (2011). https://doi.org/10.1007/978-3-642-22092-0_53

12. Zhao, J., et al.: Three-dimensional integrated functional, structural, and computational mapping to define the structural fingerprints of heart-specific atrial fibrillation drivers in human heart ex vivo. J. Am. Hear. Assoc. **6**(8), e005922 (2017)

# Pyramid Network with Online Hard Example Mining for Accurate Left Atrium Segmentation

Cheng Bian[1], Xin Yang[2,3]($\boxtimes$), Jianqiang Ma[1], Shen Zheng[1], Yu-An Liu[1],
Reza Nezafat[3], Pheng-Ann Heng[2], and Yefeng Zheng[1]

[1] Tencent YouTu Lab, Shenzhen, China
[2] Department of Computer Science and Engineering,
The Chinese University of Hong Kong, Hong Kong, China
`xinyang@cse.cuhk.edu.hk`
[3] Department of Medicine (Cardiovascular Division),
Beth Israel Deaconess Medical Center and Harvard Medical School, Boston, USA

**Abstract.** Accurately segmenting left atrium in MR volume can benefit the ablation procedure of atrial fibrillation. Traditional automated solutions often fail in relieving experts from the labor-intensive manual labeling. In this paper, we propose a deep neural network based solution for automated left atrium segmentation in gadolinium-enhanced MR volumes with promising performance. We firstly argue that, for this volumetric segmentation task, networks in 2D fashion can present great superiorities in time efficiency and segmentation accuracy than networks with 3D fashion. Considering the highly varying shape of atrium and the branchy structure of associated pulmonary veins, we propose to adopt a pyramid module to collect semantic cues in feature maps from multiple scales for fine-grained segmentation. Also, to promote our network in classifying the hard examples, we propose an Online Hard Negative Example Mining strategy to identify voxels in slices with low classification certainties and penalize the wrong predictions on them. Finally, we devise a competitive training scheme to further boost the generalization ability of networks. Extensively verified on 20 testing volumes, our proposed framework achieves an average Dice of 92.83% in segmenting the left atria and pulmonary veins.

## 1 Introduction

Atrial fibrillation is the most common cardiac electrical disorder, which happens around left atrium and becomes a major cause of stroke [9]. Ablation therapy guided by MR imaging is a popular treatment solution for atrial fibrillation. Segmenting left atrium in MR scan can benefit the preoperative assessment, determine the catheter size in pulmonary veins, optimize the therapy plan and reduce fluoroscopy time. However, confined by low efficiency and reproducibility, manually segmenting the left atrium tends to be intractable [6].

© Springer Nature Switzerland AG 2019
M. Pop et al. (Eds.): STACOM 2018, LNCS 11395, pp. 237–245, 2019.
https://doi.org/10.1007/978-3-030-12029-0_26

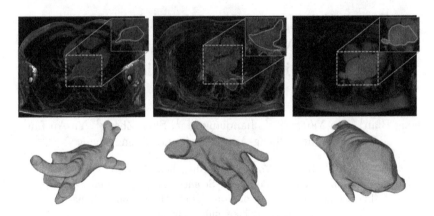

**Fig. 1.** Challenges in segmenting the left atrium and associated pulmonary veins. Green curve and surface rendering denote segmentation ground truth. (Color figure online)

As illustrated in Fig. 1, except the poor image quality, automatically segmenting left atrium with pulmonary veins in MR volume is very challenging. Firstly, left atrium only occupies a small ratio when compared with the background, which makes it difficult for algorithm to localize and recognize the boundary details. Secondly, due to the similar intensity with surrounding chambers and the thin myocardial wall with respect to the limited MR resolution, boundary ambiguity and deficiency often occur around the atrium and pulmonary veins. Thirdly, the shape and size of left atrium vary greatly across different subjects and timepoints. Also, the topological arrangement of pulmonary veins, especially the number of veins, presents weak pattern in population [12].

Automated left atrium segmentation retains intensive research interest. Deformable model [16], atlas based label fusion [11] and non-rigid registration [17] are typical methods for atrium included cardiovascular structure segmentation. However, these methods depend heavily on the hand-crafted statistical modeling and lack generalization ability against unseen cases, such as the left atrium with rare number of pulmonary veins. In the deep learning era, this segmentation task is embracing new opportunities. Stacked sparse auto-encoders [13], Fully Convolutional Network [7] and Convolutional LSTM [3] were deployed in a 2D fashion for left atrium segmentation. On the other hand, networks in 3D fashion have also been explored [14]. The superiority between 2D and 3D fashion is arguable and case-by-case [2]. Although networks in both 2D and 3D achieved remarkable performance, preserving segmentation details in varying scales and tackling classification ambiguity around boundary are overlooked in previous work.

In this paper, we propose a deep network based solution for automated left atrium segmentation in gadolinium-enhanced MR (GE-MR) volumes with promising performance. With investigations on our task, we argue that, networks with 2D architectures can easily outperform their heavy 3D versions with respect to time efficiency and segmentation accuracy. Left atrium body possesses a relative compact structure, while the associated pulmonary veins present to be branchy in varying scales. As this regard, we propose to further tailor our 2D network with pyramid pooling modules to enlarge receptive field ranges and perceive semantic cues from multiple scales. This pyramid module greatly promotes the segmentation in fine scales. To avoid the learning of non-informative negative samples and enhance the learning of hard negative examples around boundary, we propose an Online Hard Negative Example Mining strategy to identify voxels in slices with low classification certainties and penalize the wrong predictions on them. Finally, we devise a competitive training scheme in which several competitors need to compete with each other. With the scheme, the generalization ability of all competitors can be further improved. Extensively verified on 20 testing volumes, our proposed framework achieves an average Dice of 92.83% in segmenting the left atria and pulmonary veins.

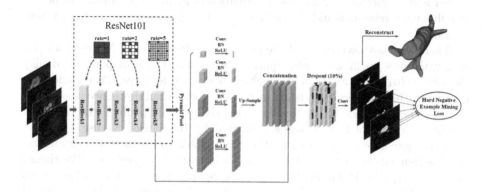

**Fig. 2.** Schematic view of our proposed framework.

## 2 Methodology

Figure 2 is the schematic view of our proposed framework. Without extra localization modules, our system takes the whole 2D slice as input, hierarchical features are firstly learned and extracted by consecutive residual blocks. Residual blocks are supervised by an auxiliary branch. Then, the adopted pyramid pooling module distills semantic cues from multiple scales. These semantic information are then merged to infer the final predictions. At the end of the network, we introduce the Online Hard Negative Example Mining loss to guide the network selectively search samples in order to learn more efficiently and effectively.

### 2.1 Choice Between 2D and 3D Architecture

For volumetric segmentation tasks, 2D and 3D architectures are two main streams and have their different strengths [2,5]. We investigate several state-of-the-art 2D and 3D architectures for the left atrium segmentation and choose the 2D fashion for following reasons. Firstly, operating in 2D endows our method with super time efficiency, which is about 4 times faster than the 3D version. Secondly, under limited GPU memory, 2D network can get more spaces to expand and deepen its layout which is crucial for the learning capacity of network. Thirdly, 2D network can get large enough and diverse receptive fields to aggregate the in-plane, multiscale context cues for more detailed segmentation.

## 2.2   Pyramid Network Design

Directly applying vanilla 2D networks to segment the left atrium performs poor. The atrium is very small when compared with the background, strong context information, such as the lung and other heart chambers, are needed to localize it. This requires the network have large receptive field and ability to extract proper feature hierarchy. Also, the deformation of atrium and branchy structure of pulmonary veins demand the network can recognize boundaries from macro to nano.

Therefore, as shown in Fig. 2, we propose to adopt the PSPNet network [15] as a workhorse to fit our task. Specifically, for the forehead part, we deploy the ResNet101 pre-trained on ImageNet to extract features. To enhance the network with large receptive fields under parameter limitations, we also utilize the dilated convolution in the ResBlock. For the pyramid pooling module, based on the intermediate feature maps generated by ResBlocks, we use 4 pooling layers to aggregate context information from global to fine scales. The bin size of these four levels are set as $1\times1$, $2\times2$, $3\times3$ and $6\times6$, respectively. We choose the Atrous Spatial Pyramid Pooling as the pooling kernels. The feature maps produced by the pyramid pooling module are then upsampled and concatenated with the original feature maps from ResBlock. A last convolution layer is applied on the concatenated feature maps to output the final predictions. Noted that, we apply the dropout on this last convolution layer as an ensemble strategy to improve the generalization ability of our network. Dropout ratio is set as 0.1.

Same with [15], we inject an auxiliary supervision branch at the last ResBlock to tackle the potential training difficulties faced by deep neural networks. As a way to tackle the class imbalance, we use the weighted cross entropy as the basic form of our objective function.

## 2.3   Online Hard Negative Example Mining

For network training, the definition of objective function directly defines the latent mapping that network needs to fit. The quality of samples counted to minimize the objective function can determine how close the network can achieve its goal. In this work, we find that there is an obvious imbalance between the amount of foreground and background samples within slices, which can bias the loss function. Also, most background samples are easy to be classified and only some voxels around the atrium boundary get high probabilities to be false alarms. Therefore, we propose an Online Hard Negative Example Mining (OHNEM) strategy to tune the objective function and selectively search informative hard negative samples to significantly improve the training efficacy.

Motivated by [10], our OHNEM strategy roots in three facts, (1) most negative samples are becoming non-informative as the training progresses and should be excluded from the loss to alleviate the foreground-background imbalance, (2) hard negative examples are those getting high prediction scores to be foreground, (3) we preserve all positive samples to learn the knowledge of foreground thoroughly. In this regard, our OHNEM consists of three major steps in each training iteration. First, determining the number of hard negative examples to be selected. We adaptively define this number as $C_{hn} = max(2 * C_p, min(5, C_n/4))$, where $C_p$ and $C_n$ are the number of foreground and background voxels obtained from the label slice, respectively. Then, all the background samples are ranked in a descending order according to their prediction scores. Finally, our loss function only counts the top $C_{hn}$ background candidates and discards the rest. Our loss function equally considers the losses from all the foreground samples and the selected background samples. Effectiveness of the proposed OHNEM is verified with ablation study in Sect. 3.

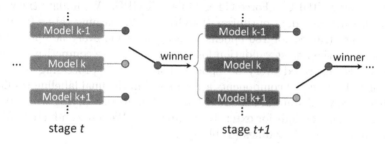

**Fig. 3.** The proposed competitive training scheme with multiple model candidates. Winner model in each stage is denoted with yellow dot. (Color figure online)

### 2.4 Competitive Training Scheme

Because of the random conditions in training, networks with the same training setting can have different generalization abilities. An ensemble of multiple networks often surpasses the single network. In this paper, we propose a new training scheme to generate an ensemble of multiple networks in a competitive way, denoted as CTS. As shown in Fig. 3, our CTS training scheme is intuitive in design. The scheme simultaneously trains several competitors with the same initial configurations. Only the winner model with the lowest validation loss can be recorded at the validation point. The winner model in stage $t$ can get the chance to broadcast its parameters to initialize all the competitors in next competitive stage. All the competitors are then be fine-tuned in stage $t + 1$ until the generation of a new winner at next validation point.

With the CTS training scheme, all competitors are facing with the more and more intensive competition. All candidates are enhanced with the local optimal model after each validation stage. They can only struggle to mine more informative examples for effective gradient update, which in turn improves the performances of all competitors. Also, different from the direct and heavy ensemble of multiple networks for testing, our network is light-weight for testing since only the winner model is kept at the end.

## 3    Experimental Results

**Experiment Materials:** We evaluated our method on the Atrial Segmentation Challenge dataset in STACOM 2018. There are two kinds of image size as 576×576×88 and 640×640×88 with the unified spacing 1.0×1.0×1.0 mm. We split the whole dataset (100 volumes with ground truth annotation) into 70/10/20 for training/validation/testing. Training dataset is further augmented with random flipping. Every volume is split into slices along the last dimension. Slices are normalized as zero mean and unit variance before training and testing.

**Implementation Details:** We implement all the compared methods with the *PyTorch* using NVIDIA GeForce GTX TITAN X GPUs. We update the weights of network with an Adam optimizer (batch size = 2, learning rate = 0.001). All the models are trained for 180 epochs with a validation every 30 epochs. Only the model with the lowest validation loss is saved. For the competitive training, the winner is generated every 30 epochs. For the post-processing in testing, small isolated connected components are removed in the final labeling result. We implemented several state-of-the-art 3D and 2D networks varying with different layouts and kernel designs for extensive comparison. We also conducted ablation study on our proposed modules (Table 1).

**Table 1.** Quantitative evaluation of our proposed framework

| Method | Metrics | | | | |
|---|---|---|---|---|---|
| | Dice[%] | Conform[%] | Jaccard[%] | Adb[mm] | Hdb[mm] |
| PSPNet [15] | 87.904 | 72.218 | 78.532 | 2.159 | 23.319 |
| PSPNetD | 92.053 | 82.536 | 85.387 | 1.564 | 21.154 |
| Unet-2D | 89.638 | 76.485 | 81.404 | 2.185 | 22.711 |
| Unet-3D | 87.020 | 68.033 | 77.785 | 2.082 | 25.243 |
| DeepLabV3 [4] | 88.456 | 72.449 | 79.846 | 2.427 | 23.099 |
| SegNet [1] | 90.762 | 78.912 | 83.422 | 2.050 | 23.362 |
| GCN [8] | 91.812 | 82.049 | 84.926 | 1.597 | 18.940 |
| PSPNetD+OHNEM | 92.688 | 84.113 | 86.434 | **1.405** | **17.224** |
| PSPNetD+OHNEM+CTS3 | **92.834** | **84.445** | **86.694** | 1.496 | 17.891 |

**Quantitative and Qualitative Analysis:** We use 5 metrics to evaluate all the compared solutions, including Dice, Conform, Jaccard, Average Distance of Boundaries (Adb) and Hausdorff Distance of Boundaries (Hdb). We first compare our core network *PSPNet* with other architectures. *PSPNet* with dilated convolution and dropout is denoted as *PSPNetD*. We can observe that, sharing the same spirit in modifying kernels to perceive broad context, both *PSPNetD* and *GCN* get better results than other architectures which use vanilla convolutions. *PSPNetD* performs even better than *GCN*, which may because of the explicit exploration on multiscale semantic cues. *Unet-3D* loses power in our task and present poor results than all 2D networks.

Based on *PSPNetD*, we conduct ablation study on OHNEM and CTS. There occurs a significant improvement (about 0.6% in Dice) when OHNEM module is involved to guide the learning of PSPNetD (denoted as *PSPNetD + OHNEM*). This proves the importance of designing proper objective function and selecting effective samples for training. For the competitive training scheme CTS, we can see that, with 3 candidates for competition, *PSPNetD + OHNEM+CTS3* brings another 0.16% improvement in Dice over *PSPNetD + OHNEM*. The learning of hard negative samples is already enhanced in *PSPNetD + OHNEM*, while CTS further pushes candidate models for more effective learning.

Finally, we visualize the segmentation results of *PSPNetD + OHNEM+CTS3* in Fig. 4 with Hausdorff distance [mm] between segmentation and ground truth. Our method presents robustness against the shape and size variations in left atrium and pulmonary veins. Most points on segmentation surface around the atrium body present low Hausdorff distances (cold color) to the ground truth.

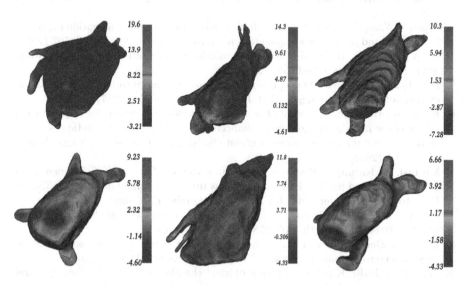

**Fig. 4.** Visualization of our segmentation results with the mapping of Hausdorff distance. Color bar is annotated with mean in the center, min and max on the ends. (Color figure online)

Large distance values (hot color) often occur around the end of pulmonary veins which are hard to define since the length of pulmonary vein varies greatly.

## 4   Conclusions

In this paper, we present a fully automatic framework for the segmentation of left atrium and pulmonary veins in GE-MR volume. We start with network architecture in 2D, and adopt kernels for large receptive field. The involvement of pyramid pooling improves our network in combing global and local cues to recognize detailed boundaries. The proposed Online Hard Negative Example Mining significantly boosts the training efficacy. Competitive training scheme pushes the network for further improvement. Verified on 20 testing cases, our method presents promising performance in segmenting left atrium and pulmonary veins across different subjects and imaging conditions.

**Acknowledgments.** The work in this paper was supported by a grant from the Research Grants Council of the Hong Kong Special Administrative Region (Project no. GRF 14225616).

## References

1. Badrinarayanan, V., et al.: Segnet: a deep convolutional encoder-decoder architecture for image segmentation. IEEE TPAMI **39**(12), 2481–2495 (2017)
2. Bernard, O., et al.: Deep learning techniques for automatic mri cardiac multi-structures segmentation and diagnosis: is the problem solved? IEEE TMI (2018)
3. Chen, J., Yang, G., Gao, Z., et al.: Multiview two-task recursive attention model for left atrium and atrial scars segmentation. arXiv preprint arXiv:1806.04597 (2018)
4. Chen, L.C., et al.: Rethinking atrous convolution for semantic image segmentation. arXiv preprint arXiv:1706.05587 (2017)
5. Dou, Q., Yu, L., et al.: 3D deeply supervised network for automated segmentation of volumetric medical images. Med. Image Anal. **41**, 40–54 (2017)
6. Karim, R., Housden, R.J., et al.: Evaluation of current algorithms for segmentation of scar tissue from late gadolinium enhancement cardiovascular magnetic resonance of the left atrium: an open-access grand challenge. J. Cardiovasc. Magn. Reson. **15**(1), 105 (2013)
7. Mortazi, A., Karim, R., Rhode, K., Burt, J., Bagci, U.: *CardiacNET*: Segmentation of left atrium and proximal pulmonary veins from MRI using multi-view CNN. In: Descoteaux, M., Maier-Hein, L., Franz, A., Jannin, P., Collins, D.L., Duchesne, S. (eds.) MICCAI 2017, Part II. LNCS, vol. 10434, pp. 377–385. Springer, Cham (2017). https://doi.org/10.1007/978-3-319-66185-8_43
8. Peng, C., Zhang, X., et al.: Large kernel matters-improve semantic segmentation by global convolutional network. arXiv preprint arXiv:1703.02719 (2017)
9. Peng, P., Lekadir, K., et al.: A review of heart chamber segmentation for structural and functional analysis using cardiac magnetic resonance imaging. Magn. Reson. Mater. Phys. Biol. Med. **29**(2), 155–195 (2016)
10. Shrivastava, A., Gupta, A., Girshick, R.: Training region-based object detectors with online hard example mining. In: CVPR, pp. 761–769 (2016)

11. Tao, Q., Ipek, E.G., et al.: Fully automatic segmentation of left atrium and pulmonary veins in late gadolinium-enhanced MRI: towards objective atrial scar assessment. J. Magn. Reson. Imaging **44**(2), 346–354 (2016)
12. Tobon-Gomez, C., Geers, A.J., et al.: Benchmark for algorithms segmenting the left atrium from 3d ct and mri datasets. IEEE TMI **34**(7), 1460–1473 (2015)
13. Yang, G., Zhuang, X., Khan, H., et al.: A fully automatic deep learning method for atrial scarring segmentation from late gadolinium-enhanced mri images. In: ISBI 2017, pp. 844–848. IEEE (2017)
14. Yang, X., Bian, C., Yu, L., Ni, D., Heng, P.-A.: Hybrid loss guided convolutional networks for whole heart parsing. In: Pop, M., Sermesant, M., Jodoin, P.-M., Lalande, A., Zhuang, X., Yang, G., Young, A., Bernard, O. (eds.) STACOM 2017. LNCS, vol. 10663, pp. 215–223. Springer, Cham (2018). https://doi.org/10.1007/978-3-319-75541-0_23
15. Zhao, H., Shi, J., Qi, X., Wang, X., Jia, J.: Pyramid scene parsing network. In: CVPR, pp. 2881–2890 (2017)
16. Zheng, Y., et al.: Multi-part modeling and segmentation of left atrium in c-arm ct for image-guided ablation of atrial fibrillation. IEEE TMI **33**(2), 318–331 (2014)
17. Zhuang, X., et al.: A registration-based propagation framework for automatic whole heart segmentation of cardiac mri. IEEE TMI **29**(9), 1612–1625 (2010)

# Combating Uncertainty with Novel Losses for Automatic Left Atrium Segmentation

Xin Yang[1,4(✉)], Na Wang[2,3], Yi Wang[2,3], Xu Wang[2,3], Reza Nezafat[4], Dong Ni[2,3], and Pheng-Ann Heng[1]

[1] Department of Computer Science and Engineering,
The Chinese University of Hong Kong, Hong Kong, China
xinyang@cse.cuhk.edu.hk
[2] National-Regional Key Technology Engineering Laboratory for Medical Ultrasound, School of Biomedical Engineering, Health Science Center, Shenzhen University, Shenzhen, China
[3] Medical UltraSound Image Computing (MUSIC) Lab, Shenzhen, China
[4] Department of Medicine (Cardiovascular Division), Beth Israel Deaconess Medical Center and Harvard Medical School, Boston, USA

**Abstract.** Segmenting left atrium in MR volume holds great potentials in promoting the treatment of atrial fibrillation. However, the varying anatomies, artifacts and low contrasts among tissues hinder the advance of both manual and automated solutions. In this paper, we propose a fully-automated framework to segment left atrium in gadolinium-enhanced MR volumes. The region of left atrium is firstly automatically localized by a detection module. Our framework then originates with a customized 3D deep neural network to fully explore the spatial dependency in the region for segmentation. To alleviate the risk of low training efficiency and potential overfitting, we enhance our deep network with the transfer learning and deep supervision strategy. Main contribution of our network design lies in the composite loss function to combat the boundary ambiguity and hard examples. We firstly adopt the *Overlap* loss to encourage network reduce the overlap between the foreground and background and thus sharpen the predictions on boundary. We then propose a novel *Focal Positive* loss to guide the learning of voxel-specific threshold and emphasize the foreground to improve classification sensitivity. Further improvement is obtained with an recursive training scheme. With ablation studies, all the introduced modules prove to be effective. The proposed framework achieves an average Dice of 92.24% in segmenting left atrium with pulmonary veins on 20 testing volumes.

## 1 Introduction

Cardiovascular diseases are keeping as the leading causes of death in the world. About half of the diagnosed cases suffer from the atrial fibrillation caused stroke [8]. MR, especially the Gadolinium Enhancement MR (GE-MR), is a dominant

X. Yang and N. Wang—Authors contribute equally to this work.

© Springer Nature Switzerland AG 2019
M. Pop et al. (Eds.): STACOM 2018, LNCS 11395, pp. 246–254, 2019.
https://doi.org/10.1007/978-3-030-12029-0_27

imaging modality to localize scars and provide guidance for ablation therapy [6]. Segmenting the left atrium with associated pulmonary veins and reconstructing its patient-specific anatomy are beneficial to optimize the therapy plan and reduce the risk of intervention. However, manual delineation tends to be time-consuming and presents low inter- and intra-expert reproducibilities.

Automatically segmenting left atrium with pulmonary veins is a nontrivial task. As shown in Fig. 1, left atrium and pulmonary veins vary greatly in shape and size across subjects. Affected by the artifacts and differences in gadolinium dose, appearance of atrium and pulmonary veins can change dramatically. Due to the thin myocardial wall of left atrium and the low contrast between left atrium and other surroundings, the boundary of left atrium and pulmonary veins are hard and ambiguous for computer to recognize. Conquering the complex anatomical variance and boundary ambiguity are the main concerns of this work.

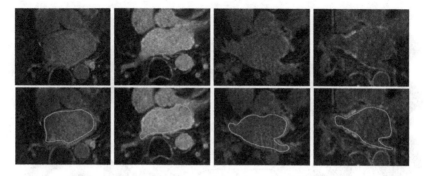

**Fig. 1.** Challenges in segmenting the left atrium and associated pulmonary veins. Yellow curve denotes segmentation ground truth. (Color figure online)

Cardiovascular image segmentation is an active research area. Deformable models were proposed to segment chambers and vessels [16]. Other popular streams are the multi-atlas [1] and non-rigid registration based methods [17]. However, those methods bear the difficulties in designing boundary descriptors and generalizing the model trained on limited data to unseen deformable cases [10]. Deep neural network rapidly emerges and pushes the upper-bound of cardiovascular image segmentation [3,7,14]. Whereas, limited by computational resources, most methods exploit 2D architectures which ignore the global spatial information in volume and the segmentation consistency across slices. Also, few works paid attention to the importance of loss function design. Hybrid loss function combining *Dice* loss and weighted cross entropy loss can address class-imbalance and preserve both concrete and branchy structures in segmentation [15]. Loss function enforcing the low overlap between foreground and background, denoted as *Overlap* loss, presents potential in promoting segmentation [13].

In this paper, we propose a fully-automated framework to segment left atrium and pulmonary veins in gadolinium-enhanced MR volumes. We firstly deploy a detection module to accurately localize the left atrium region in the raw volume. We then propose our customized segmentation network in 3D fashion. Transfer learning and dense deep supervision strategy are involved to alleviate the risk of low training efficiency and potential overfitting. To effectively guide our network combat the boundary ambiguity and hard examples, we introduce a composite loss function which consists of two complementary components: (1) the *Overlap* loss to encourage network reduce the intersection between foreground and background probability maps and thus make the predictions on boundary less fuzzy, (2) a novel *Focal Positive* loss to guide the learning of voxel-specific threshold and emphasize the foreground to improve classification sensitivity. Based on the design, we obtain further improvement by fine-tuning our network with a recursive scheme. With ablation studies, all the introduced modules prove to be effective. The proposed framework achieves an average Dice of 92.24% in segmenting left atrium with pulmonary veins on 20 testing volumes.

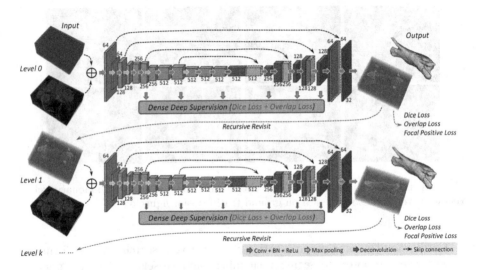

**Fig. 2.** Schematic view of our proposed framework.

## 2    Methodology

Figure 2 is the schematic illustration of our proposed framework. The localized left atrium ROI serves as the input of our customized 3D network. Our network generates an intermediate segmentation and also a single channel to represent the foreground probability map. The foreground map then merges with original ROI and goes through the next refinement level. The last recursive level outputs the final segmentation of left atrium and pulmonary veins.

## 2.1   Localization of Left Atrium Region

To exclude the noise from background and narrow down the searching area, we propose to exploit a Faster-RCNN [9] network to localize the ROI of atrium and pulmonary veins. We train our Faster-RCNN with 2D slices and bounding boxes. Testing MR volume is firstly split into slices along a canonical axis and all the localized ROI in these slices are then merged into a 3D bounding box. Our ROI detection module achieves 100% accuracy in hitting the atrium area.

## 2.2   Enhance the Training of 3D Network

Shown as Fig. 2, we customize a 3D U-net from [4] and take it as the workhorse. To fit our proposed loss function design, we discard the last softmax layer and only output a single channel of foreground probability. We enhance the training of our network from the following three aspects.

**Transfer Learning in 3D Fashion:** Leveraging knowledges of well-trained model can improve the generalization ability of networks [2]. Equipped with 3D convolutions to extract spatial-temporal features, C3D model proposed in [11] is proper for the transfer learning in 3D networks. Therefore, we apply the parameters of layers *conv1, conv2, conv3a, conv3b, conv4a and conv4b* in C3D model to initialize the downsampling path of our customized 3D U-net. All the layers are then fine-tuned for our atrium segmentation task.

**Dense Deep Supervision (*DDS*):** Introducing deep supervision mechanism is effective in addressing the gradient vanishing problem faced by deep networks [5]. As shown in Fig. 2, deep supervision adds auxiliary side-paths and thus exposes shallow layers to extra supervisions. Let $\mathcal{X}^{w \times h \times d}$ be the ROI input, $W$ be the parameters of main network, $w = (w^1, w^2, ..., w^S)$ be the parameters of side-paths and $w^s$ denotes the parameters of the $s^{th}$ side-path. $\tilde{\mathcal{L}}$ and $\mathcal{L}_s$ are main loss function and loss function in $s^{th}$ side-path, respectively. Components of $\tilde{\mathcal{L}}$ and $\mathcal{L}_s$ are further explained in following sections. In this work, we extend the deep supervision in a dense form, that is, we attach auxiliary side-paths in both down- and up-sampling branches and finally 6 losses in total. The final loss function $\mathcal{L}$ for our 3D U-net with dense deep supervision is elaborated in Eq. 1, where $\beta_s$ ($s \in (1, 2, ..., 6)$) is the weight of different side-paths.

$$\mathcal{L}(\mathcal{X}; \mathcal{W}, w) = \tilde{\mathcal{L}}(\mathcal{X}; \mathcal{W}) + \sum_{s \in S} \beta_s \mathcal{L}_s(\mathcal{X}; \mathcal{W}, w^s) + \lambda(||\mathcal{W}||^2 + \sum_{s \in S} ||w^s||^2) \quad (1)$$

**Dice Loss to Address Class Imbalance (*DCL*):** Class imbalance can bias the traditional loss function and thus make network ignore minor classes [15]. Dice coefficient based loss is becoming a promising choice in addressing the problem and present more clean predictions. In this work, we adopt Dice loss as a basic component for main loss and all auxiliary losses.

### 2.3   Composite Loss Against Classification Uncertainty

**Overlap Loss  (*OVL*):** Boundary ambiguity raises uncertainties for background-foreground classification. The uncertainty can be observed from the fuzzy areas in predicted probability maps. Enlarging the gap between background and foreground predictions can suppress this kind of uncertainty. In this work, we adopt the *Overlap* loss (OVL) [13] to measure this kind of gap. OVL loss is a basic component of main loss and all auxiliary losses. OVL loss is defined as follows,

$$\mathcal{L}_{ovl}(\mathcal{W}, w^s) = \sum_{i=1}^{|\mathcal{X}|}(P(y_i = 1|\mathcal{X}; \mathcal{W}, w^s) * P(y_i = 0|\mathcal{X}; \mathcal{W}, w^s)), \qquad (2)$$

where $P$ is the predicted probability maps for foreground and background, $*$ is basic multiplication. By minimizing the *Overlap* loss, our network is pushed to learn more discriminative features to distinguish background and foreground regions and thus gain confidence in recognizing ambiguous boundary locations.

**Focal Positive Loss  (*FPL*):** OVL loss focuses on enlarging the gap between foreground and background predictions. Whereas, accurately extracting foreground object, i.e. atrium and pulmonary veins, is our final task. Thus, emphasizing the foreground should be further considered. To this regard, we add a threshold map (TM) layer after probability map at the end of the main network to adaptively regularize the foreground probability map. Our network can learn to tune the TM layer and obtain voxel-specific thresholds. Finally, the TM layer can suppress weak predictions and only preserve strong positive predictions. To train the TM layer, we introduce a novel loss function, i.e. *Focal Positive* loss, which is derived in a differentiable form as follows:

$$\mathcal{L}_{fpl}(\mathcal{W}, w^s) = 1 - \frac{2 * |Mask(y_i = 1|\mathcal{X}; \mathcal{W}, w^{(t)}) * Y|}{|Mask(y_i; \mathcal{W}, w^s)| + |Y|}, \qquad (3)$$

$$Mask(y_i; \mathcal{W}, w^s) = 1/(1 + e^{-tmp}), \qquad (4)$$

$$tmp = \begin{cases} P(y_i = 1|\mathcal{X}; \mathcal{W}, w^s), & P(y_i) > Threshold\_Map(y_i) \\ -\infty, & else \end{cases} \qquad (5)$$

By minimizing $\mathcal{L}_{fpl}$, our network can learn to enforce strong positive predictions in foreground areas defined by ground truth. $\mathcal{L}_{fpl}$ is only attached in main network. In summary, our network is trained with DCL and OVL losses existing in main network and auxiliary paths, and also the FPL loss in main network.

### 2.4   Recursive Refinement Scheme (*RRS*)

Probability map contains more explicit context information for segmentation than the raw MR volume. Revisiting the probability map to explore context cues for refinement is a classical strategy, like Auto-Context [12]. The core idea of Auto-Context is to stack a series of models in a way that, the model at level $k$ not only utilizes the appearance features in intensity image, but also the contextual features extracted from the prediction map generated by the model at level $k - 1$. The general recursive process of an Auto-Context scheme is $\hat{y}^k = \mathcal{F}^k(\mathcal{J}(x, \hat{y}^{k-1}))$, where $\mathcal{F}^k$ is the model at level $k$, $x$ and $\hat{y}^k$ are the intensity image and prediction map from level $k - 1$, respectively. $\mathcal{J}$ is a join operator to combine information from $x$ and $\hat{y}^k$. Generally, $\mathcal{J}$ is a concatenation operator. In this work, we modify $\mathcal{J}$ as a element-wise summation operator. Summation $\mathcal{J}$ enables us to reuse all the parameters of level $k - 1$ in level $k$ for fine-tune. In addition, summarizing intensity image with prediction map is intuitive to highlight the target anatomical structures and also suppress the irrelevant background noise.

**Table 1.** Quantitative evaluation of our proposed method

| Method | Metrics | | | | |
|---|---|---|---|---|---|
| | Dice[%] | Conform[%] | Jaccard[%] | Adb[mm] | Hdb[mm] |
| CEL | 82.015 | 54.921 | 69.893 | 6.613 | 47.396 |
| DCL | 84.907 | 57.811 | 75.291 | 2.287 | 30.344 |
| DCL-DDS | 87.020 | 68.033 | 77.785 | 2.082 | 25.243 |
| DCL-DDS-OVL | 89.833 | 76.942 | 81.745 | 2.116 | 24.591 |
| DCL-DDS-OVL-FPL | 91.548 | 81.449 | 84.464 | **1.455** | 19.554 |
| DCL-DDS-OVL-FPL-RRS1 | 92.243 | 83.113 | 85.639 | 1.541 | 18.888 |
| DCL-DDS-OVL-FPL-RRS2 | **92.244** | **83.120** | **85.644** | 1.490 | **18.293** |

## 3  Experimental Results

**Experiment Materials:** We evaluated our method on the Atrial Segmentation Challenge 2018 dataset. We split the whole dataset (100 samples with ground truth annotation) into training set (80 volumes) and testing set (20 volumes). All the training and testing samples are normalized as zero mean and unit variance before inputting into network. We augmented the training dataset with random flipping, rotation and 3D elastic deform.

**Implementation Details:** We implemented our framework in *Tensorflow*, using 4 NVIDIA GeForce GTX TITAN Xp GPUs. We update the parameters of network with an Adam optimizer (batch size=1, initial learning rate is 0.001). Randomly cropped $144 \times 64 \times 144$ sub-volumes serve as input to our network. To avoid shallow layers being over-tuned during fine-tuning, we set smaller initial learning rate for *conv1, conv2, conv3a, conv3b, conv4a and conv4b* as *1e−6, 1e−6, 1e−5, 1e−5, 1e−4, 1e−4*. We adopt sliding window with proper overlap ratio and overlap-tiling stitching strategies to generate predictions for the whole volume, and remove small isolated connected components in final segmentation result.

**Quantitative and Qualitative Analysis:** We use 5 metrics to evaluate the proposed framework, including Dice, Conform, Jaccard, Average Distance of Boundaries (Adb) and Hausdorff Distance of Boundaries (Hdb). We take the customized 3D U-net with basic cross entropy loss (CEL) as a baseline. We conduct intensive ablation study on our introduced modules, including DCL, DDS, OVL, FPL and RRS. All of the compared methods share the same basic 3D U-net architecture. Table 1 illustrate the detailed quantitative comparisons among different module combinations. For simplicity, all compared methods are denoted with the main module names. Both the DCL and DDS brings improvement over the traditional CEL. Significant improvements firstly occurs as we

**Fig. 3.** Visualization of probability maps generated by networks with different losses. From left to right, CEL, DCL-DDS, DCL-DDS-OVL, DCL-DDS-OVL-FPL and DCL-DDS-OVL-FPL-RRS1.

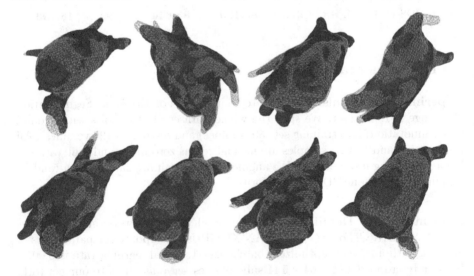

**Fig. 4.** Visualization of our segmentation results in testing datasets. Green mesh denotes ground truth, while blue surface denotes our segmentation result. Our segmentation presents high overlap ratio with the ground truth. (Color figure online)

inject the *Overlap* loss. FPL further boosts the segmentation about 2% in Dice. Improvements brought by OVL and FPL are also verified as we visualize the foreground probability maps in Fig. 3. The foreground probability map becomes more sharp and noise-free as OVL and FPL involves. Mining context cues in probability maps with RRS contributes about 1% in Dice as two RRS levels (RRS2) are utilized. We only adopt two RRS levels to balance the performance gain and computation burden. Finally, DCL-DDS-OVL-FPL-RRS2 achieves the best performance in almost all metrics. We also visualize the segmentation results with ground truth in Fig. 4. Our method conquers complex variance of left atrium and pulmonary veins and achieves promising performance.

## 4    Conclusion

In this paper, we present a fully automatic framework for left atrium segmentation and GE-MR volumes. Originating with a network in 3D fashion to better tackle complex variances in shape and size of left atrium, we present our main contributions in introducing and verifying the composite loss which is effective in combating the boundary ambiguity and hard examples. We also propose a modified recursive scheme for successive refinement. As extensively validated on a large dataset, our proposed framework and modules prove to be promising for the left atrium segmentation in GE-MR volumes.

**Acknowledgments.** The work in this paper was supported by a grant from the Research Grants Council of the Hong Kong Special Administrative Region (Project no. GRF 14225616), a grant from National Natural Science Foundation of China under Grant 61571304, Grant 81771598, and Shenzhen Peacock Plan under Grant KQTD2016053112051497.

## References

1. Bai, W., Shi, W., Ledig, C., Rueckert, D.: Multi-atlas segmentation with augmented features for cardiac MR images. Med. Image Anal. **19**(1), 98–109 (2015)
2. Chen, H., Ni, D., Qin, J., et al.: Standard plane localization in fetal ultrasound via domain transferred deep neural networks. IEEE JBHI **19**(5), 1627–1636 (2015)
3. Chen, J., Yang, G., Gao, Z., et al.: Multiview two-task recursive attention model for left atrium and atrial scars segmentation. arXiv preprint arXiv:1806.04597 (2018)
4. Çiçek, Ö., Abdulkadir, A., et al.: 3d u-net: Learning dense volumetric segmentation from sparse annotation. arXiv preprint arXiv:1606.06650 (2016)
5. Dou, Q., Yu, L., et al.: 3D deeply supervised network for automated segmentation of volumetric medical images. Med. Image Anal. **41**, 40–54 (2017)
6. McGann, C., Akoum, N., et al.: Atrial fibrillation ablation outcome is predicted by left atrial remodeling on MRI. Circ. Arrhythm. Electrophysiol. **7**, 23–30 (2013)
7. Mortazi, A., Karim, R., Rhode, K., Burt, J., Bagci, U.: *CardiacNET*: segmentation of left atrium and proximal pulmonary veins from MRI using multi-view CNN. In: Descoteaux, M., Maier-Hein, L., Franz, A., Jannin, P., Collins, D.L., Duchesne, S. (eds.) MICCAI 2017, Part II. LNCS, vol. 10434, pp. 377–385. Springer, Cham (2017). https://doi.org/10.1007/978-3-319-66185-8_43

8. Peng, P., Lekadir, K., et al.: A review of heart chamber segmentation for structural and functional analysis using cardiac magnetic resonance imaging. Magn. Reson. Mater. Phys. Biol. Med. **29**(2), 155–195 (2016)
9. Ren, S., He, K., Girshick, R., Sun, J.: Faster R-CNN: towards real-time object detection with region proposal networks. In: NIPS, pp. 91–99 (2015)
10. Tobon-Gomez, C., Geers, A.J., et al.: Benchmark for algorithms segmenting the left atrium from 3D CT and MRI datasets. IEEE TMI **34**(7), 1460–1473 (2015)
11. Tran, D., Bourdev, L., Fergus, R., Torresani, L., Paluri, M.: Learning spatiotemporal features with 3D convolutional networks. In: ICCV, pp. 4489–4497 (2015)
12. Tu, Z., Bai, X.: Auto-context and its application to high-level vision tasks and 3D brain image segmentation. IEEE TPAMI **32**(10), 1744–1757 (2010)
13. Wang, Z., et al.: Automated detection of clinically significant prostate cancer in mp-MRI images based on an end-to-end deep neural network. IEEE TMI **37**, 1127–1139 (2018)
14. Yang, G., Zhuang, X., Khan, H., et al.: A fully automatic deep learning method for atrial scarring segmentation from late gadolinium-enhanced MRI images. In: ISBI 2017, pp. 844–848. IEEE (2017)
15. Yang, X., Bian, C., Yu, L., Ni, D., Heng, P.-A.: Hybrid loss guided convolutional networks for whole heart parsing. In: Pop, M., et al. (eds.) STACOM 2017. LNCS, vol. 10663, pp. 215–223. Springer, Cham (2018). https://doi.org/10.1007/978-3-319-75541-0_23
16. Zheng, Y., et al.: Multi-part modeling and segmentation of left atrium in C-arm CT for image-guided ablation of atrial fibrillation. IEEE TMI **33**(2), 318–331 (2014)
17. Zhuang, X., et al.: A registration-based propagation framework for automatic whole heart segmentation of cardiac MRI. IEEE TMI **29**(9), 1612–1625 (2010)

# Attention Based Hierarchical Aggregation Network for 3D Left Atrial Segmentation

Caizi Li[1,2], Qianqian Tong[1], Xiangyun Liao[2(✉)], Weixin Si[2(✉)], Yinzi Sun[2],
Qiong Wang[2], and Pheng-Ann Heng[2,3]

[1] School of Computer Science, Wuhan University, Wuhan 430072, China
[2] Shenzhen Key Laboratory of Virtual Reality and Human Interaction Technology,
Shenzhen Institutes of Advanced Technology, Chinese Academy of Sciences,
Shenzhen 518000, China
xyunliao@gmail.com, wxsics@gmail.com
[3] Department of Computer Science and Engineering,
The Chinese University of Hong Kong, Hong Kong, China

**Abstract.** Atrial fibrillation (AF) is the most common type of cardiac arrhythmia. The atrial segmentation is essential for the understanding of the human atria structure which is vital to the AF treatment. In this paper, we propose a novel three-dimensional (3D) segmentation network combining hierarchical aggregation and attention mechanism for 3D left atrial segmentation, named attention based hierarchical aggregation network (HAANet). In our network, the shallow and deep feature fusion capability of encoder-decoder convolutional neural networks is enhanced through hierarchical aggregation. Besides, attention mechanism is adopted to promote the ability of extracting efficient features. Experimental results demonstrate the HAANet can produce good results for 3D left atrial segmentation and the dice score of our HAANet reaches 92.30.

**Keywords:** Hierarchical aggregation · Attention mechanism ·
3D left atrial segmentation

## 1 Introduction

Atrial fibrillation (AF) has been the most common type of cardiac arrhythmia, and the lack of understanding of the structure of the human atria results in the poor performance of current AF treatment. Nowadays, gadolinium contrast agencies are applied to MRI scans to improve the clarity of the images of a patient's internal structure. Gadolinium-enhanced magnetic resonance imaging (GE-MRI) is an important tool to evaluate the atrial fibrosis [1]. The fingerprints of atrial structure may hold the key to understanding and reversing AF according to recent studies on human atria imaged with GE-MRI [2,3]. Due to the low contrast between the atrial tissue and background, manually segmenting atrial chambers from GE-MRIs is prone to generate errors. Thus, fully automatic atrial segmentation of the left atrial (LA) is being desperately needed to reconstruct and visualize the atrial structure in the clinical usage.

© Springer Nature Switzerland AG 2019
M. Pop et al. (Eds.): STACOM 2018, LNCS 11395, pp. 255–264, 2019.
https://doi.org/10.1007/978-3-030-12029-0_28

It is very challengeable to segment LA from 3D GE-MRIs. Firstly, the poor contrast between LA and its background reduces the visibility of LA boundaries. Besides, the image quality may be impaired due to the patients' irregular breath rhythm and heart rate variability during the scan. With the rapid development of deep learning in recent years, several fully automatic methods have been proposed for the LA segmentation. Mortazi et al. [4] designed a multi-view convolutional neural network (CNN) with an adaptive fusion strategy. In their work, 3D data were respectively parsed into 2D components from axial, sagittal and coronal, and then a separate CNN for each component was utilized. Chen et al. [5] extended the multi-view strategy by adopting dilated residual network and sequential learning network composed by ConvLSTM [6].

In this paper, given by the inspiration of 3D U-Net [7] which is extended from 2D U-Net [8] and deep layer aggregation strategy [9] which has been used for computer vision tasks, we propose a hierarchical aggregation encoder-decoder network for the 3D atrial segmentation. By using the hierarchical aggregation strategy, successive layers with different depth at the same stage are being aggregated iteratively to obtain better fuse information. Recently, attention mechanism has been employed as a feature selector during the forward inference in computer vision. Wang et al. [10] proposed a residual attention network built by stacking residual attention learning modules to generate attention-ware features. Inspired by [10], we attach a mask branch as a residual branch to each stage in the encoder part of our network, thus spatial information from shallow layers can be used to gradually reinforce the contour features of deeper layers which own abundant semantic information. Through integrating hierarchical aggregation and attention mechanism, our proposed attention based hierarchical aggregation network (HAANet) achieves better performance than 3D U-Net.

## 2    Method

In our work, a 3D U-Net with two convolutional layer within its each stage is firstly utilized as a region of interest (ROI) detection network to segment the ROI which contains the LA from the entire 3D volume of each MRI, which helps to alleviate the impact from the surrounding tissues. After the ROIs are being detected, they are all cropped out from the raw MRIs and then fed into the proposed HAANet to obtain accurate prediction of the LA segmentation. More details about ROI are demonstrated in Sect. 2.1. As shown in Fig. 1(a), our HAANet is based on 3D U-Net and it composes of an encoder path and a decoder path: the encoder path includes a series of stacked attention based hierarchical aggregation modules (HAAM) which comprise a hierarchical aggregation unit (HAU) as a trunk branch and an attention unit (AU) as a mask branch; the decoder path is the same as U-Net which consists of several repeated convolutional layers followed by Batch Normalization (BN) and rectified linear unit (ReLU) to reconstruct the spatial information from the deepest layer. Skip connection is adopted to combine the spatial information and the semantic information.

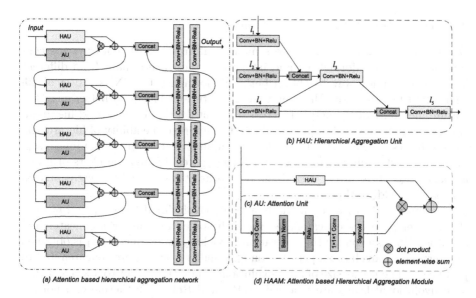

**Fig. 1.** An overview of the proposed attention based hierarchical aggregation network (HAANet). (a) The architecture of our HAANet. (b) Components of the Hierarchical Aggregation Unit (HAU). (c) Components of the Attention Unit (AU). (d) The Attention based Hierarchical Aggregation Module (HAAM) consists of HAU and AU.

## 2.1   ROI Detection

ROI detection is the first step in our segmentation architecture to generate the inputs of HAANet. There are two reasons that the ROI detection is performed. Firstly, for the training data, every LA in each MRI only takes a small percentage, which means most of the volume data in a MRI is useless for our task. Besides, letting so much useless data participates in the computation is a huge waste for computing resources. Secondly, detecting the ROI is an efficient way to promote the performance of our method because the surrounding tissues around the LA have a great impact for our segmentation task due to the very low contrast between LA and its background.

In our work, we first resize all of the raw training samples and their corresponding labels into a fixed shape (64, 128, 128) which is ordered by (z-axis, height, width) that be adopted throughout the paper, and then these resized training samples are fed into 3D U-Net to obtain coarse predictions. After these coarse predictions being inferred, we resize them into their respective origin shape and crop the ROIs out. In case there may be errors when predicting the coarse results, we expand the targets' areas to make sure the entire raw targets be definitely cropped out. As a result, the shape of each raw ROI is (88, 160, 256).

## 2.2 Hierarchical Aggregation Unit

While deeper layers contain more semantic information, shallower layers contain more spatial information, and integrating the feature maps from different layers is a common way to obtain better performance for CNNs. Yu et al. [9] aggregated hierarchically through a tree structure of blocks to better span the feature hierarchy of the network across different depths. In our work, hierarchical aggregation depicted in [9] are adopted to improve the ability of hierarchical feature representation. Due to the huge computation cost in the 3D image segmentation task, different from [9] which aggregates different numbers of layers at each stage, three layers with different depths at each stage are aggregated in this paper, yielding HAU.

As illustrated in Fig. 1(b), the first two convolutional layers $l_1$ and $l_2$ from the encoder path are concatenated for each stage, and then the concatenated result is fed into a convolution operation to yield $l_3$. Subsequently, $l_4$ is obtained by exerting a convolution operation on $l_3$. Finally, $l_3$ and $l_4$ are concatenated to produce $l_5$. The kernel size of all convolutional operations is set to $3 \times 3 \times 3$, and all convolutional layers are followed by BN and ReLU. Furthermore, a normal convolutional layer with $3 \times 3 \times 3$ kernel size and $(2, 2, 2)$ strides followed by BN and ReLU is employed for downsampling among each stage.

## 2.3 Attention Mechanism

There are mainly two factors which may affect the precision of our segmentation model. One results from the reality that the shape of the LA varies a lot among different patients, and the other results from that the downsampling operation in CNN can easily lose spatial information as depth going deeper. In order to alleviate these two problems and force our network to focus on the targets, attention mechanism is integrated into our network as one mask branch at each stage. As depicted in Fig. 1(c), the attention unit in our network consists of two convolutional layers and a sigmoid layer, and the former convolutional layer is followed by BN and ReLU. By passing the attention unit, the value of feature maps are normalized into the range of [0, 1] for each channel and each spatial position so as to obtain the attention mask. The sigmoid operation can be defined as follows,

$$f(x_{i,c}) = \frac{1}{1 + \exp(-x_{i,c})}, \tag{1}$$

where exp denotes the exponential function, and $x_{i,c}$ denotes the $i^{th}$ value of a feature map at the $c^{th}$ channel.

## 2.4 Attention Based Hierarchical Aggregation Module

As shown in Fig. 1(d), once the attention mask is obtained, the dot production operation will be performed between the mask branch and the trunk branch. However, due to values of the attention mask ranging from zero to one, repeated dot production with mask branch will degrade the value of feature maps in deep layers, and the attention mask will potentially break good property of the trunk branch [10]. To tackle this issue, the residual learning mode is adopted to make identical mapping between the input and output of the trunk branch. The output $H(x_{i,c})$ of our attention based hierarchical aggregation module including the trunk branch and mask branch is defined below:

$$H(x_{i,c}) = (M_{i,c}(x) \otimes F_{i,c}(x)) \oplus F_{i,c}(x), \tag{2}$$

where $M_{i,c}(x)$ indicates the output of the mask branch with range of $[0, 1]$, $F_{i,c}(x)$ denotes the trunk branch generated by our hierarchical aggregation unit, $\otimes$ denotes dot product, and $\oplus$ denotes element-wise sum.

## 3  Experiments and Results

### 3.1  Dataset

There are 100 training data for the LA cavity segmentation provided by the 2018 atrial segmentation challenge. The original resolution of the data is $0.625 \times 0.625 \times 0.625\,\mathrm{mm}^3$, and each sample contains 88 slices in the Z-axis. Because there is no validation data released, we randomly split out 10 patient data from the training dataset as our validation data to test the proposed model, thus there are 90 patient data for training and 10 patient data for validation. During the training process, we resize the ROIs to the fixed size of (88, 80, 128) for the input of our model.

Data augmentation is applied for expanding our training set to alleviate overfitting during the training. On a random basis, the data is rotated between 0 to $2\pi$ degree along the Z axis, scaled between 0.8 to 1.2 range, mirror and translation transformation along Z axis are also adopted. Besides, $\gamma$ transformation ranges from 0.8 to 1.3 is utilized. Finally, Z-score standardization is employed after all the above transformation being performed. It is worth noting that all transformations are not performed by the probability of 50%. By applying these random data augmentation methods, theoretically we have unlimited data for training. In our work, these random data augmentation methods are being performed during the training stage in each iteration.

### 3.2  Training Protocols

During our experiment, for both 3D U-Net and our network, we set the number of channels of convolutional layers at the first stage in our network to 8, and double increase the number of channels with each downsampling operation being exerted until the fifth stage. The training batch size is set to 8, and our network is optimized by the Adam algorithm [11]. There are 100 iterations are run in every epoch. Learning rate is initialized to 1e−3 and decayed with the factor 0.1 when no update occurs within 5 epochs. Dice loss is adopted as the loss function and it is denoted as

$$dice\_loss = 1 - 2 * \frac{|y_{true} \bigcap y_{pred}|}{|y_{true}| \bigcup |y_{pred}|}, \tag{3}$$

where $y_{true}$ denotes the training labels, $y_{pred}$ denotes the output of our network.

Our network was implemented by Keras 2.1.5 which uses tensorflow 1.4 as its backend, and our model was trained and tested on an Nvidia Geforce GTX 1080Ti GPU, developed on a 64-bit ubuntu 16.04 platform with Intel®Core™ i5-7640X CPU @ 4.00 GHZ × 4, 32 GB memory (RAM).

### 3.3    Experimental Settings

The goal of our experiments is to assess the validity of both HAU and AU, the evaluation criteria is the dice coefficient which demonstrates coincidence degree between the predicted value and the true value. We set six experiments for the LA segmentation, that is UNet-2, HANet-2, HAANet-2, UNet-3, HANet-3, HAANet-3, and the postfix number denotes the number of convolutional layers at each stage in the network, by the way, UNet-2 is also used in ROI detection but with different input shape. HANet indicates our architecture without attention unit, and HAANet means our attention based hierarchical aggregation network. In order to equally compare the differences of these models, all of the six networks are set with the same hyperparameters and the same training protocols mentioned in Sect. 3.2.

### 3.4    Quantitative Results

All of the six networks with the same experiment settings mentioned above are applied for LA segmentation. Among the six networks, UNet-2 represents our baseline with two convolutional layers at each stage, UNet-3 represents there are three convolutional layers relative to UNet-2 at each stage. HANet-2 and HANet-3 represent our proposed network illustrated by Fig. 1(a) but without attention module, there are respectively two and three convolutional layers for HAU in HANet-2 and HANet-3, and HAU with two convolutional layers is a submodule including $l_1$, $l_2$ and $l_3$ which indicated in Fig. 1(b). We obtain HAANet-2 and HAANet-3 by integrating the AU into HANet-2 and HANet-3 to further optimize the HAU. We progressively test the six networks so as to demonstrate the effectiveness of HAU and AU.

Table 1 shows the comparison results of the six networks on the validation data. The results of six experiments illustrate that integrating HAU and AU can obtain a better performance than U-Net which is the classic medical image segmentation architecture. The tendency of the results for the six networks with both two and three convolutional layers shows that our attention based aggregation module is a promising strategy for LA segmentation. The discussions of respective effectiveness of HAU and AU are represented as follows.

**Table 1.** Comparison results of six networks on validation data.

| Method | Dice coefficient | Method | Dice coefficient |
|---|---|---|---|
| UNet-2 | 92.46 | UNet-3 | 92.61 |
| HANet-2 | 92.75 | HANet-3 | 92.81 |
| **HAANet-2** | **92.86** | **HAANet-3** | **93.00** |

**Performance of Hierarchical Aggregation Unit.** As shown in Table 1, the dice score of HANets is higher than that of UNets whether there are 2 or 3 layers at each stage. Hierarchical aggregation module makes the traditional stacked convolutional layers merge as a tree structure, and it can promote the performance of the network by learning richer features. Concatenating the feature maps from different layers can preserve the shallow features and integrate convolutional layers with different size of receptive field which is important for semantic segmentation.

**Performance of Attention Unit.** Attention mechanism is an effective way to force the network to focus the target that is the LA cavity in our work. Through generating the normalized intermediate mask by sigmoid function, our proposed HAANets adopt residual attention learning strategy to attach attention maps of shallow convolutional layers to the output of the whole block at each stage, which

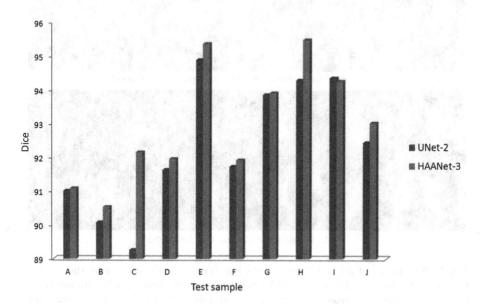

**Fig. 2.** Comparison results between UNet-2 and our HAANet-3 for ten patient data. The vertical axis denotes the dice score and the horizontal axis denotes ten patient samples (A–J).

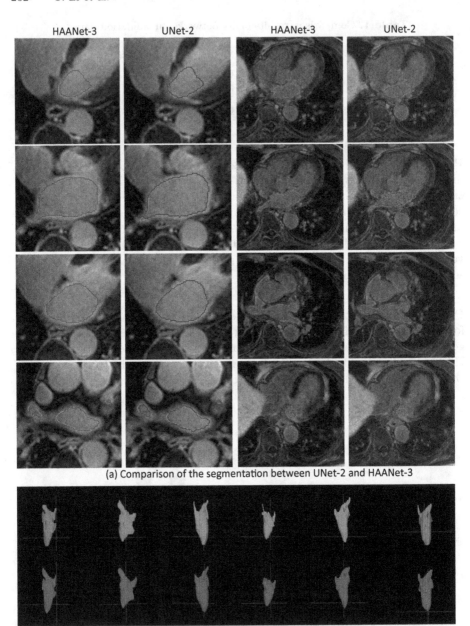

(a) Comparison of the segmentation between UNet-2 and HAANet-3

(b) 3D segmentation results of HAANet-3

**Fig. 3.** Illustration of 2D and 3D segmentation results. (a) Comparison of the segmentation results between Unet-2 and HAANet-3 (the green curve denotes ground truth, the red curve denotes automated segmentation results of HAANet-3, and the blue curve denotes automated segmentation results of UNet-2). (b) 3D segmentation results (the green ones denote ground truth and the red ones denote 3D segmentation results of HAANet-3). (Color figure online)

can not only avoid the potential of breaking good property of trunk branch but also further improve the performance of our network. In Table 1, our HAANets perform better than HANets, showing the effectiveness of our residual attention unit.

### 3.5 Segmentation Results

By integrating HAU and AU together, the HAANet-3 achieves the highest dice of 93.00. The dice score of our ten validation patient data for UNet-2 and our HAANet-3 are shown in Fig. 2. It can be observed that our HAANet-3 achieves higher dice than UNet-2 for almost all validation data except that the dice score of HAANet-3 is a little lower than that of UNet-2 for the sample $I$. This comparison results further illustrate that the proposed HAANet is promising for 3D left atrial segmentation.

Figure 3(a) illustrates automatic 2D LA segmentation results of the proposed HAANet-3 and UNet-2. We can see that HAANet-3 has a good advantage over UNet-2 especially for hard part in a MRI. Figure 3(b) depicts the 3D LA segmentation results reconstructed by ITK-SNAP for six different patients, and the green ones denote ground truth and the red ones denote 3D segmentation results of HAANet-3.

## 4   Conclusion

This paper presents an attention based hierarchical aggregation network (HAANet) which can be trained end-to-end from scratch for fully automatic LA segmentation. The HAAM in our network contains a trunk branch and a mask branch, the trunk branch is an aggregation unit which merges shallow layers and deep layers at each stage of the network, while the mask branch is an attention unit which serves the network as a feature selector during the forward inference. The experimental results shows the proposed HAANet can obtain better LA segmentation results contrasted to U-Net. Additionally, we submit our segmentation results of the testing data online, and obtain the final dice score of 92.30 from the organizers.

**Acknowledgment.** This work is supported by grants from Shenzhen Key Laboratory of Virtual Reality and Human Interaction Technology, Shenzhen Institutes of Advanced Technology, Chinese Academy of Sciences, China (No. ZDSYS201605101739178), Natural Science Foundation of Guangdong (No. 2018A030313100), Research Grants Council of the Hong Kong Special Administrative Region (No. GRF 14225616), Guangdong Province Science and Technology Plan Project (No. 2016A020220013), the Science and Technology Plan Project of Guangzhou (No. 201704020141) and China Posdoctoral Science Foundation (No. 2017M622831).

# References

1. McGann, C., et al.: Atrial fibrillation ablation outcome is predicted by left atrial remodeling on MRI. Circ. Arrhythm. Electrophysiol. **7**, 23–30 (2013)
2. Hansen, B.J., et al.: Atrial fibrillation driven by micro-anatomic intramural re-entry revealed by simultaneous sub-epicardial and sub-endocardial optical mapping in explanted human hearts. Eur. Heart J. **36**(35), 2390–2401 (2015)
3. Zhao, J., et al.: Three-dimensional integrated functional, structural, and computational mapping to define the structural "fingerprints" of heart-specific atrial fibrillation drivers in human heart ex vivo. J. Am. Heart Assoc. **6**(8), e005922 (2017)
4. Mortazi, A., Karim, R., Rhode, K., Burt, J., Bagci, U.: *CardiacNET*: segmentation of left atrium and proximal pulmonary veins from MRI using multi-view CNN. In: Descoteaux, M., Maier-Hein, L., Franz, A., Jannin, P., Collins, D.L., Duchesne, S. (eds.) MICCAI 2017, Part II. LNCS, vol. 10434, pp. 377–385. Springer, Cham (2017). https://doi.org/10.1007/978-3-319-66185-8_43
5. Chen, J., et al.: Multiview two-task recursive attention model for left atrium and atrial scars segmentation. arXiv preprint arXiv:1806.04597 (2018)
6. Xingjian, S.H., Chen, Z., Wang, H., Yeung, D.-Y., Wong, W.-K., Woo, W.-C.: Convolutional LSTM network: a machine learning approach for precipitation nowcasting. In: Advances in Neural Information Processing Systems, pp. 802–810 (2015)
7. Çiçek, Ö., Abdulkadir, A., Lienkamp, S.S., Brox, T., Ronneberger, O.: 3D U-Net: learning dense volumetric segmentation from sparse annotation. In: Ourselin, S., Joskowicz, L., Sabuncu, M.R., Unal, G., Wells, W. (eds.) MICCAI 2016, Part II. LNCS, vol. 9901, pp. 424–432. Springer, Cham (2016). https://doi.org/10.1007/978-3-319-46723-8_49
8. Ronneberger, O., Fischer, P., Brox, T.: U-Net: convolutional networks for biomedical image segmentation. In: Navab, N., Hornegger, J., Wells, W.M., Frangi, A.F. (eds.) MICCAI 2015, Part III. LNCS, vol. 9351, pp. 234–241. Springer, Cham (2015). https://doi.org/10.1007/978-3-319-24574-4_28
9. Yu, F., Wang, D., Darrell, T.: Deep layer aggregation. arXiv preprint arXiv:1707.06484 (2017)
10. Wang, F., et al.: Residual attention network for image classification. arXiv preprint arXiv:1704.06904 (2017)
11. Kingma, D.P., Adam, J.B.: A method for stochastic optimization. arXiv preprint arXiv:1412.6980 (2014)

# Segmentation of the Left Atrium from 3D Gadolinium-Enhanced MR Images with Convolutional Neural Networks

Chandrakanth Jayachandran Preetha[1], Shyamalakshmi Haridasan[1], Vahid Abdi[1], and Sandy Engelhardt[2,3]([⊠]) [iD]

[1] Faculty of Electrical Engineering and Information Technology,
University of Magdeburg, Magdeburg, Germany
[2] Faculty of Computer Science and Research Campus STIMULATE,
University of Magdeburg, Magdeburg, Germany
[3] Department of Computer Science, Mannheim University of Applied Sciences,
Mannheim, Germany
s.engelhardt@hs-mannheim.de

**Abstract.** Gadolinium contrast agents are used in a third of all magnetic resonance scans to study the extent of fibrosis across the left atria in patients with atrial fibrillation. Direct segmentation of the atrial heart chambers from these 3D gadolinium-enhanced magnetic resonance images (GE-MRI) is very demanding due to the low contrast between the atrial tissue and background. Current automatic segmentation methods delineate the left atrium on auxiliary *bright-blood* MRI scans, which need to be registered to GE-MRI in an additional, potentially error-prone step. Yet, it could render extremely useful to obtain direct segmentation on GE-MRI to simultaneously estimate the left atrium anatomy and the extent of fibrosis. In this work, we present a deep learning approach which is able to segment the left atrium from 3D GE-MRI. 100 data sets provided by the MICCAI 2018 Atrial Segmentation Challenge have been used to train and test deep convolutional neural networks (CNN), which follow a 2D architecture with deep supervision (train-validation-test 70-5-25). After performing a four fold cross validation, the network achieved a mean dice of 0.8945. Within the scope of the test phase of the challenge, we trained the network on 100 data sets, predicted novel segmentations on the official test set and, according to the leaderboard, achieved a final score of 0.888.

**Keywords:** Heart segmentation · Atrial segmentation ·
Deep learning · Convolutional neural networks

## 1 Introduction

Atrial fibrillation (AF) is the most common sustained heart rhythm disturbance that causes considerable increase in morbidity and mortality among adults. It has been observed that ectopic beats from the pulmonary veins (PVs) commonly

© Springer Nature Switzerland AG 2019
M. Pop et al. (Eds.): STACOM 2018, LNCS 11395, pp. 265–272, 2019.
https://doi.org/10.1007/978-3-030-12029-0_29

266 C. J. Preetha et al.

initiate AF. Ablation therapies attempt to disrupt electrical reentry pathways that cause the arrhythmia. Minimally invasive catheter ablation (CA) using radio-frequency energy has become one of the most common treatments for AF patients refractory to drug treatment [10]. Therefore, CA strategies attempt to electrically isolate the PVs from the left atrial (LA) body. Gadolinium-enhanced cardiac MRI [5], with histologic validation, can resolve the underlying 3D structure and fibrosis distribution in regions of AF drivers in LA from human hearts. They produce hyperenhancement of the myocardial tissues in suspicious regions, such as myocardial infarction or fibrosis. GE-MRI in AF patients has recently emerged as a promising tool for detecting atrial scars to stratify patients, guide ablation therapy and predict treatment success or failure [15]. For example, among patients with AF undergoing CA, atrial tissue fibrosis estimated by GE-MRI is associated with likelihood of post-ablation recurrent arrhythmia [10].

Quantification of scar tissue regions require a segmentation of both the left atrium and the high intensity scar area from GE-MRI images. However, clinical interpretation and segmentation of these GE-MRI scans for AF patients is challenging due to poor image quality, attributed to the variability in heart rate, very low signal-to-noise ratio (SNR), canceling of healthy tissue signal, and contrast agent wash-out during image acquisition. Moreover, the boundaries of the thin LA wall (3–4 mm) is mostly captured unclear, making it hard to differentiate it from surrounding structures. Furthermore, the LA has a small size and a high anatomical variability. For example, the arrangement of the PV varies greatly between subjects. Topological variants include four veins (74%), five veins (17%) or three veins (9%). Because of the named obstacles, previous studies had to heavily rely on manual delineation for the PVs and LA geometry from GE-MRI [1]. Besides manual supervision, a second *bright-blood* MRI acquisition for anatomical segmentation of the LA often needs to be acquired, necessitate prolonged scan times and error-prone image fusion. Hence, segmenting the LA anatomy automatically from a single GE-MRI acquisition is highly in demand and has not been focus of much related work so far.

The segmentation approaches for other MRI modalities, such as magnetic resonance angiography (MRA) or 3D balanced steady state free precession MRI have evolved from purely data driven methods, like region growing [2] and graph cuts [3] to other advanced methods using prior information. Peters et al. [4] implemented segmentation of whole heart anatomy using shape-constrained deformable models. In 2013, Zhu et al. [6] proposed an automatic segmentation technique that is robust to the large variability of the LA datasets. The technique adopted a variational region-growing framework, searching locally for a seed region of the LA from a slice of a MR image. One of the major cornerstones to benchmark existing algorithms for atrial segmentation from CT and MRI datasets has been the MICCAI STACOM Challenge 2013[1]. Tobon-Gomez et al. [9], published the corresponding manuscript comparing several existing segmentation algorithms. Most participants obtained good accuracy on the middle of the LA body for MRI segmentation. However, the PVs were often missing

---

[1] http://www.cardiacatlas.org/challenges/left-atrium-segmentation-challenge/.

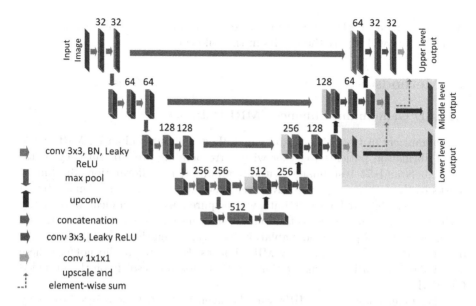

**Fig. 1.** 2D U-Net shaped architecture of the network. Deep supervision operations are highlighted in yellow. (Color figure online)

and the lower part of the LA body (closer to the mitral valve) was frequently oversegmented.

Convolutional neural networks (CNN) have recently shown outstanding performance in medical image segmentation [11], where they typically take the form of U-Net-like architectures for 2D [8] or 3D input [16]. These architectures have proven very successful within the scope of the MICCAI STACOM ACDC Challenge 2017[2] for multistructure cardiac MRI segmentation [13]. Another CNN-based approach has been used by Mortazi et al. [12] for atrial segmentation: They employed a multi-view CNN with an adaptive fusion strategy and achieved state-of-the-art sensitivity (90%), specificity (99%), precision (94%) on the STACOM 2013 challenge data set. Yang et al. [7] developed a deep learning based approach for atrial scar segmentation on GE-MRI via stacked sparse auto-encoders. However, their approach did not segment the atrial structure directly on this modality, as this has been stated to be too challenging.

We introduce a convolutional neural network segmentation procedure to solve for automatic delineation of the left atrial geometry directly on GE-MRI data sets. The networks are based on popular approaches suggested by the group around Ronneberger [8,16] that are capable of yielding precise segmentations by introducing skipping connections between encoder-decoder paths. Furthermore, we employ deep supervision techniques, as proposed by Kayalibay et al. [14].

---

[2] https://www.creatis.insa-lyon.fr/Challenge/acdc/index.html.

We evaluated our methods for segmentation on the MICCAI Atrial Segmentation data set[3] and will report the results in the following.

## 2   Methods

### 2.1   3D Gadolinium-Enhanced MRI Data Set

The organizers of the challenge provided 3D gadolinium-enhanced MRI of 100 atrial fibrillation patients together with corresponding ground-truth segmentation labels and 54 test images without ground-truth information. 3D binary masks of the left atrial cavity have been made available as labels made by an expert. Each 3D MRI patient data was acquired using a clinical whole-body MRI scanner. Additionally to the images, information on whether the data set has been acquired pre- or post-ablation has been given. For some patients, the data cohort was accompanied by MRA images. However, we did not use these additional information to ensure that segmentation relies solely on the provided GE-MRI.

The dimensions of the MRIs vary between $640 \times 640$ or $576 \times 576$ in xy and contain exactly 88 slices in the z-direction. The provided data sets had an isotropic spacing of $1 \times 1 \times 1$ mm$^3$. Therefore, preprocessing in terms of re-sampling was not necessary.

### 2.2   Segmentation with 2D U-Net

The proposed model generally follows the architecture of the U-Net proposed by Ronneberger et al. [8]. It consists of an encoding path and a decoding path as shown in Fig. 1. The basic block in the encoding path consists of two $3 \times 3$ convolutional layers, each followed by batch normalization and Leaky ReLU activation. Down-sampling is performed using $2 \times 2$ max pooling which reduces the size of the feature maps to half its size at each block while the convolutional layers in the encoding path increase the number of feature maps from 1 to 32 to 512. In the decoding path, first a deconvolution operation is performed using a $2 \times 2$ filter with stride of $2 \times 2$, which doubles the dimensions of the feature maps in both directions but reduces the number of feature maps by a factor of two. The model also employs skip connections from layers of equal resolution in the encoding path in order to provide high-resolution features to the decoding path. The $3 \times 3$ convolutional layers in the decoding path also reduce the number of feature maps obtained by the concatenation of equal resolution feature maps in the encoding and decoding paths. In the last layer a $1 \times 1 \times 1$ convolutional filter decreases the number of feature maps to one and sigmoid activation is used to obtain the prediction for the segmentation maps. One of the main problems encountered during the training of deep neural networks is that the gradient backpropagation becomes weak towards the shallow layers of the network. The proposed network employs deep supervision (marked in

[3] http://atriaseg2018.cardiacatlas.org/.

yellow in Fig. 1) to alleviate gradient vanishing and the loss of information during forward propagation [14]. The segmentation maps from two intermediate layers of lower resolution immediately before the final output layer are upsampled using $2 \times 2$ transpose convolution and combined with the output of the highest resolution segmentation map using element-wise summation. In addition to this, these upsampled lower level features are also used to generate predictions by applying softmax function on them. These predictions are also compared with the ground truth to calculate auxiliary losses, which are combined with the loss from the final layer, resulting in more effective gradient backpropagation.

## 3    Evaluation

For the 2D U-Net approach, individual 2D-slices in axial direction were resized to $512 \times 512$. Grey level information of every image was normalized to zero-mean and unit-variance. The model was trained for 100 epochs. We split the dataset into 70-5-25 for training, validation and testing and computed a four fold cross validation. The network was trained using ADAM solver and a pixel-wise dice loss with a initial learning rate of $10^{-4}$. The training was done using a batch size of 10 with 5325 images and validated on 592 images in each epoch. Learning rate decay was done by monitoring the dice coefficient value on the validation data. Slices where no mask was visible have been omitted in the training process to decrease the class imbalances.

## 4    Results

The 2D U-Net performs 2D slice-by-slice prediction for every slice of the MRI, which are then stacked together into a 3D volume. The ground truth for each patient which was resized to $512 \times 512 \times 88$ was used to evaluate the output predictions of the model. With regard to the expert segmentations on the original data set, a mean dice score of 0.8945 was achieved for the 2D model with deep supervision after performing a four fold cross validation. The model performed well in the center region of the atrium, as shown in Fig. 2. The main mode of failure were the regions of the pulmonary veins, where the model struggled to fully include them in the predicted masks. The predictions in 2D consist of non-atrial regions also.

The training took 10 h to complete for 100 epochs on 70 patients on a GeForce GTX 1080 Ti GPU model. Training for 100 epochs with deep supervision witnessed an advantage of steady increase in the training dice coefficient followed by increase in validation dice coefficient. A score of 0.88 was achieved in the final test dataset of the 2018 Atrial segmentation challenge according to the leaderboard.

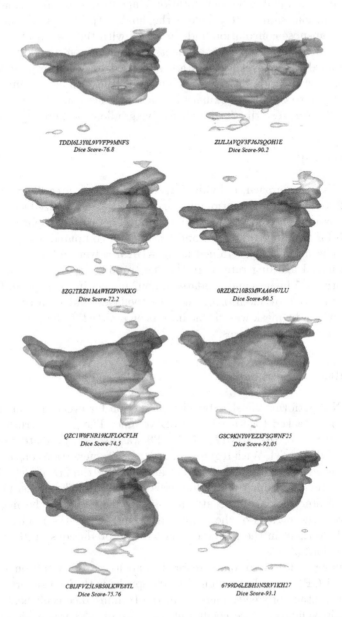

**Fig. 2.** Best (right column) and worst (left column) result in each cross validation set. Prediction are shown as green surfaces while ground truth is shown as white surfaces. (Color figure online)

# 5    Discussion

In this paper, we presented an automatic atrial segmentation method based on a 2D U-Net for GE-MRI data sets. The predictions in 2D consist of non-atrial regions, which could be filtered away in a simple postprocessing step. Deep supervision has turned out to be especially beneficial for the 2D U-Net. However, using the 3D version of the U-Net might be beneficial to obtain better consistency in z-direction.

**Acknowledgements.** The work was supported by the European Regional Development Fund under the operation number 'ZS/2016/04/78123' as part of the initiative "Sachsen-Anhalt WISSENSCHAFT Schwerpunkte" and by the German Research Foundation (DFG) project 398787259, EN1197/2-1.

# References

1. Ravanelli, D., et al.: A novel skeleton based quantification and 3-D volumetric visualization of left atrium fibrosis using late gadolinium enhancement magnetic resonance imaging. IEEE Trans. Med. Imag. **33**(2), 566–576 (2014)
2. Karim, R., Mohiaddin, R., Rueckert, D.: Left atrium segmentation for atrial fibrillation ablation. In: Proceedings of the SPIE 6918, Medical Imaging 2008: Visualization, Image-Guided Procedures, and Modeling, 69182U, 15 March 2008. https://doi.org/10.1117/12.771023
3. Lombaert, H., Sun, Y., Grady, L., Xu, C.: A multilevel banded graph cuts method for fast image segmentation. In: 10th IEEE International Conference Computer Vision, vol. 1, pp. 259–265 (2005)
4. Peters, J., Ecabert, O., Meyer, C., Schramm, H., Kneser, R., Groth, A., Weese, J.: Automatic whole heart segmentation in static magnetic resonance image volumes. In: Ayache, N., Ourselin, S., Maeder, A. (eds.) MICCAI 2007. LNCS, vol. 4792, pp. 402–410. Springer, Heidelberg (2007). https://doi.org/10.1007/978-3-540-75759-7_49
5. Kim, R.J., et al.: The use of contrast-enhanced magnetic resonance imaging to identify reversible myocardial dysfunction. N. Engl. J. Med. **343**, 1445–1453 (2000). https://doi.org/10.1056/NEJM200011163432003
6. Zhu, L., Gao, Y., Yezzi, A., Tannenbaum, A.: Automatic segmentation of the left atrium from MR images via variational region growing with a moments-based shape prior. IEEE Trans. Image Process. **22**(12), 5111–5122 (2013)
7. Yang, G., et al.: A fully automatic deep learning method for atrial scarring segmentation from late gadolinium-enhanced MRI images. In: 2017 IEEE 14th International Symposium on Biomedical Imaging
8. Ronneberger, O., Fischer, P., Brox, T.: U-Net: convolutional networks for biomedical image segmentation. In: Navab, N., Hornegger, J., Wells, W.M., Frangi, A.F. (eds.) MICCAI 2015. LNCS, vol. 9351, pp. 234–241. Springer, Cham (2015). https://doi.org/10.1007/978-3-319-24574-4_28
9. Tobon-Gomez, C., et al.: Benchmark for algorithms segmenting the left atrium from 3D CT and MRI datasets. IEEE Trans. Med. Imag. **34**(7), 1460–1473 (2015)
10. Marrouche, N.F., et al.: Association of atrial tissue fibrosis identified by delayed enhancement MRI and atrial fibrillation catheter ablation: the DECAAF study. JAMA **311**(5), 498–506 (2014). https://doi.org/10.1001/jama.2014.3

11. Litjens, G., et al.: A survey on deep learning in medical image analysis Medical Image Analysis. Med. Image Anal. **42**, 60–88 (2007)
12. Mortazi, A., Karim, R., Rhode, K., Burt, J., Bagci, U.: *CardiacNET*: segmentation of left atrium and proximal pulmonary veins from MRI using multi-view CNN. In: Descoteaux, M., Maier-Hein, L., Franz, A., Jannin, P., Collins, D.L., Duchesne, S. (eds.) MICCAI 2017. LNCS, vol. 10434, pp. 377–385. Springer, Cham (2017). https://doi.org/10.1007/978-3-319-66185-8_43
13. Isensee, F., Jaeger, P.F., Full, P.M., Wolf, I., Engelhardt, S., Maier-Hein, K.H.: Automatic cardiac disease assessment on cine-MRI via time-series segmentation and domain specific features. In: Pop, M., et al. (eds.) STACOM 2017. LNCS, vol. 10663, pp. 120–129. Springer, Cham (2018). https://doi.org/10.1007/978-3-319-75541-0_13
14. Kayalibay, B., Jensen, G., van der Smagt, P.: CNN-based segmentation of medical imaging data, arXiv preprint arXiv:1701.03056 (2017)
15. McGann, C., Akoum, N., Patel, A., et al.: Atrial fibrillation ablation outcome is predicted by left atrial remodeling on MRI. Circ. Arrhythmia Electrophysiol. **7**(1), 23–30 (2014). https://doi.org/10.1161/CIRCEP.113.000689
16. Çiçek, Ö., Abdulkadir, A., Lienkamp, S.S., Brox, T., Ronneberger, O.: 3D U-Net: learning dense volumetric segmentation from sparse annotation. In: Ourselin, S., Joskowicz, L., Sabuncu, M.R., Unal, G., Wells, W. (eds.) MICCAI 2016. LNCS, vol. 9901, pp. 424–432. Springer, Cham (2016). https://doi.org/10.1007/978-3-319-46723-8_49

# V-FCNN: Volumetric Fully Convolution Neural Network for Automatic Atrial Segmentation

Nicoló Savioli[1(✉)], Giovanni Montana[1,2(✉)], and Pablo Lamata[1(✉)]

[1] Department of Biomedical Engineering, King's College London,
London SE1 7EH, UK
{nicolo.l.savioli,giovanni.montana,pablo.lamata}@kcl.ac.uk
[2] WMG, University of Warwick, Coventry CV4 71AL, UK
g.montana@warwick.ac.uk

**Abstract.** Atrial Fibrillation (AF) is a common electro-physiological cardiac disorder that causes changes in the anatomy of the atria. A better characterization of these changes is desirable for the definition of clinical biomarkers. There is thus a need for its fully automatic segmentation from clinical images. This work presents an architecture based on 3D-convolution kernels, a Volumetric Fully Convolution Neural Network (V-FCNN), able to segment the entire atrial anatomy in a one-shot from high-resolution images ($640 \times 640$ pixels). A loss function based on the mixture of both Mean Square Error (MSE) and Dice Loss (DL) is used, in an attempt to combine the ability to capture the bulk shape as well as the reduction of local errors caused by over-segmentation.

Results demonstrate a good performance in the middle region of the atria along with the challenges impact of capturing the pulmonary veins variability or valve plane identification that separates the atria to the ventricle. Despite the need to reduce the original image resolution to fit into Graphics Processing Unit (GPU) hardware constraints, 92.5% and 85.1% were obtained respectively in the 2D and 3D Dice metric in 54 test patients (4752 atria test slices in total), making the V-FCNN a reasonable model to be used in clinical practice.

**Keywords:** Cardiac imaging · Segmentation · FCNN · Atria · Fibrillation · Shape · Clinical biomarkers · Anatomy · Deep learning · MRI scanner

## 1   Introduction

Atrial Fibrillation (AF) is a common electro-physiological cardiac disorder with a large worldwide prevalence [1] that causes changes in the anatomy of the atria. A better characterization of these changes, which are known to further promote and sustain fibrillation, is desirable for the definition of clinical biomarkers. These biomarkers can be directly started from the image (i.e. the shape of the atria [2],

© Springer Nature Switzerland AG 2019
M. Pop et al. (Eds.): STACOM 2018, LNCS 11395, pp. 273–281, 2019.
https://doi.org/10.1007/978-3-030-12029-0_30

or the fibrotic burden by late gadolinium enhanced (LGE) magnetic resonance imaging (MRI) [3]) rather than mechanistic simulations of the function (i.e. computation of the risk of arrhythmia perpetuation [4]).

There is thus a need for a full atrium automatic segmentation from clinical images, especially in LGE studies. The current state of the art is based on tedious along with error-prone manual procedures, and the main difficulty is the lack of atrial tissue contrast. Fully automated solutions are desirable to speed up the process and remove inter- and intra-observer variability. In this direction, a combination of multi-atlas registration within 3D level-set has been proposed, reporting a reasonable performance in the main atrial body and pulmonary vein regions [5]. The large computational burden of this multi-atlas approach can be alleviated by the use of Convolutional Neural Networks (CNNs), as it has been illustrated in [6] for the analysis of 2D MRI slices.

The first idea followed in this work is the use of 3D CNNs as the effective solution for the 3D LGE datasets segmentation. The starting point is the v-net [7], which takes into consideration the spatial redundancy naturally present on the entire volumetric stack within 3D-kernels, showing good benefits in different cardiac segmentation problems [8,9]. This architecture is modified to reduce the memory burden and speed-up its training. The last idea explored in this work is the sensible choice of a loss function, the joint combination of both the mean squared error (MSE) and Dice Loss, which has been reported to be beneficial

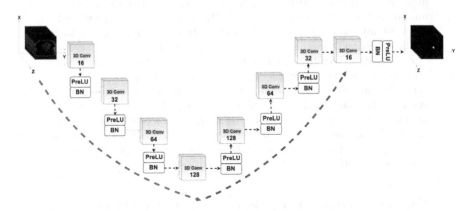

**Fig. 1.** V-FCNN architecture. Input is the (XYZ) 3D MRI volume of size (127 × 127 × 88), also passed through the down-sampling path (blue arrow), represented by a 3D kernels Convolution Neural Network (CNN) able to progressively reduce the input volume slices. Then, the hidden features, at the end of it, are restored within 3D up-sampling kernels (red-arrow), ending in an output being a 3D mask of size (127 × 127 × 88). Both down-sampling and up-sampling paths consist of four 3D-convolutions blocks (blue boxes) followed by PreLU plus 3D-Batch Normalisation (BN). The number of feature maps for each convolution layers are 16, 32, 64, 128 both in down and up-sampling. (Color figure online)

as the MSE minimizes global image details while the Dice Loss reduces local over-segmentation errors [7].

## 2   Atrial Datasets

A population of 100 3D LGE-MRIs, and masks, were made available through the [2018 Atrial Segmentation Challenge] and used in this work. Images had a acquisition resolution of $0.625 \times 0.625 \times 0.625 \, \text{mm}^3$.

## 3   Method

A volumetric fully convolutional network, V-FCNN, is designed with two main paths (see Fig. 1): the volumetric down-sampling path as well as the volumetric up-sampling path. The volumetric down-sampling path has four 3-D convolutions blocks, each following by PreLU along with 3-D Batch Normalisation (BN) layers. This takes as input the entire (XYZ) volume and progressively reduces the size of each slice (XY) together with the number of spatial slices stack (Z). In this phase, the volume is compressed and presents both (XY) and (Z) reduction. In a complementary fashion, the up-sampling volumetric restores the compressed volume to its initial size, with every 3-D up-sampling convolution blocks being followed by PreLU and 3-D Batch Normalisation (BN). Each sampling directions of V-FCNN contains four blocks within 16,32,64,128 -3D kernels respectively (with the size fix to $3 \times 3$).

Image down-sampling, during a segmentation task, presents the problem of feature map reduction, followed by a strong spatial information loss. This problem has been addressed in v-net [7,8] by adding skips layers between down-sampling and up-sampling layers in order to fuse low-level features within high level features.

Our work explores an alternative approach to the skip path connections. In particular, we boost our model to capture fine details in two ways. First, the use of max-pooling operations is avoided in order to prevent the loss of spatial resolution (i.e. if pools do not overlap well, pooling operation loses appreciable information where the objects are located in the image). The second idea is to use a loss layer that combines the $MSE_{loss}$ and $DICE_{loss}$ metrics as an attempt to recover details in the image. The rationale is that the $DICE_{loss}$ term searches for local details in the volume data, as a consequence, the $MSE_{loss}$ can be seen as a regularization that instead focuses on global features on the MRI volume. This bimodal loss also prevents the V-FCNN to fall in a local minimum; especially within small atrial regions. Specifically, the loss is defined as:

$$Loss = MSE_{loss} + \lambda \cdot DICE_{loss}$$

$$= \frac{1}{Z} \sum_{s=0}^{Z} (y[s] - \hat{y}[s])^2 + \lambda \cdot \frac{1}{Z} \sum_{s=0}^{Z} \frac{2 \sum_i^N y[s]_i \hat{y}[s]_i}{\sum_i^N y[s]_i + \sum_i^N \hat{y}[s]_i}, \tag{1}$$

where $y[s]$ are the Ground Truth (GT) binary slices, as well as $\hat{y}[s]$ are the correspondent's prediction masks. $s$ is an index through the spatial slices in the Z (z-axis). Then, the sums for $i$ in $DICE_{loss}$ runs over all N pixels of the prediction masks $\hat{y}[s]$ and the $y[s]$ GT masks. Finally, $\lambda$ controls the amount of $DICE_{loss}$ during the optimisation training process (set to $1e-3$).

The size of each MRI slice is reduced to $127 \times 127$ pixels using down-sampling bi-cubic interpolation. Please note this is a necessary operation linked to the vRAM (6–12 GB) of current conventional GPUs, not enough for storing high-resolution volumetric images. Finally, an up-sampling bi-cubic interpolation is finally used for restoring the mask to the size of $640 \times 640$ pixels.

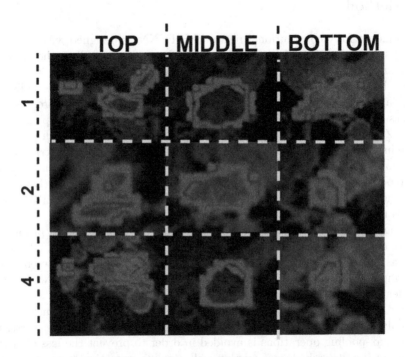

**Fig. 2.** Visual comparison of the segmentations obtained from V-FCNN (green line) vs clinical ground truth (red line) in three different test patients (number 1, 2 or 4). The comparison is made at three different sections of the atrium: top, middle and bottom. Note how the V-FCNN is able to segment not only visually simpler slices (middle section) but also more complex cases (top and bottom sections). (Color figure online)

## 3.1   Implementation Details

For our experiments, we train the network with a number of epochs of 1000 up to convergence. The Stochastic Gradient Descent (SDG) was used with a learning rate of $1e-4$, while the momentum and weight decay are 0.9, $1e-5$ respectively. Furthermore, to increase the generalization of the network, data argumentation was used, finding particularly effective the random vertical and horizontal flip in close combination with plane translation. Input sequences were equalized in grayscale intensity with CLAHE (Contrast Limited Adaptive Histogram Equalization) [10], and the noise was minimized with a combination of High-Pass Filters and Gaussian blurring filters.

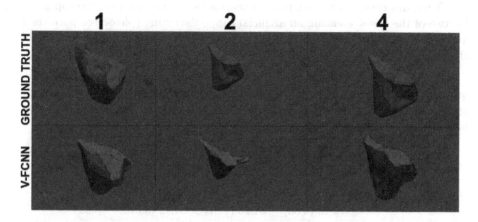

**Fig. 3.** Visual comparison between ground truth (red) compared with those obtained by proposed V-FCNN (green). Note that, the mesh coarse resolution is related to the low number of triangles used. (Color figure online)

## 4   Experiments

The experimentation phase is divided into two phases. In the first phase (preparation phase) 5 patients are used for validation (440 slices in total), and the rest (90) is used for training. In the second phase (competition), our algorithm was evaluated in the test-set with 54 patients (4752 atria slices in total).

Segmentation accuracy was measured with the 2D Dice Metric (DM) [11], which was subdivided into 3 regions of the atria: top (including the pulmonary veins), middle and bottom (including the valve plane that divides the atria and ventricle). Besides, 2D DM and the surface Hausdorff Distance (HD) [12] are computed for the entire atrial anatomy. The 3D Dice of the full volume is finally also measured at the second phase (please see [Dice]).

Particularly, given a point $p$ and a surface $S$, the distance $\varepsilon(p.S)$ is defined as:

$$\varepsilon(p, S) = min_{p' \in S} d(p, p')  \tag{2}$$

whereas the $d(\cdot)$ is the Euclidean distance between two points in a Euclidean space. Then, the HD between mesh surfaces $S_{GroundTruth}$ and $S_{vfcnn}$ is given as:

$$HD(S_{GroundTruth}, S_{vfcnn}) = max_{p \in S_{GroundTruth}} \varepsilon(p, S_{vfcnn})  \tag{3}$$

Proposed V-FCNN achieved in the first phase an average 2D DM of 69.6 ± 16.1, 82.1±1.4 and 78.0±6.0 in the top, middle and bottom regions, respectively (see Table 1 and an illustrative example in Fig. 2). The HD ranged from 0.31 to 0.86 mm. The average 2D DM and HD were 76.58(7.87) and 0.59 respectively. The 2D and 3D DM in the competition phase were 92.5% and 85.1% respectively.

Visual inspection of the contours reveals a loss of accuracy on the top slices for two of the cases, creating an artificial shape flattening (please see patients 1 and 2 of Fig. 3), and on the top of the atria with the more variable anatomy of the pulmonary veins.

**Table 1.** Automatic segmentation results (2D Dice Metric and Hausdorff Distance) for all five test patients. Results report the mean and standard deviation, and are divided into three different atrium sections: top, middle and bottom.

| Patient number | Top | Mid | Bottom | 3D-mesh |
|---|---|---|---|---|
| | DM (%) | DM (%) | DM (%) | HD |
| 1 | 77.74 (8.46) | 84.62 (2.40) | **76.99 (10.89)** | 0.86 |
| 2 | 72.26 (9.50) | 81.25 (1.57) | 54.18 (33.70) | 0.66 |
| 3 | 74.10 (8.92) | 77.75 (1.36) | 68.27 (14.75) | **0.31** |
| 4 | 81.60 (0.74) | 81.54 (1.20) | 72.32 (14.32) | 0.55 |
| 5 | **84.48 (2.58)** | **85.36 (0.66)** | 76.33 (7.05) | 0.58 |

## 5    Discussion and Conclusion

The exploitation of spatial coherence across different volumetric slices is an important resource for improving the accuracy of fully automated segmentation. Proposed V-FCNN achieves a good segmentation performance, mostly in the middle atrial section. Limitations occur in the top and bottom sections, caused by the presence of the pulmonary veins and the difficulty to identify where the atria and ventricle split. In the preparation phase the V-FCNN achieved a 2D DM between 82.05(3.43) (top case) and 69.23(14.92) (worst case), along with a 2D DM of 92.5% and 3D DM of 85.1% at the competition phase, satisfactory for using V-FCNN model in clinical practice.

A cardiac imaging study has a huge space coherence that many segmentation algorithms have exploited within different deep-learning techniques [7–9,13,14]. The use of 3D convolutional kernels [7–9] has the advantage of directly capturing the spatial information in each convolutional layer without adding extra parameters (i.e. the addition of a recurring network). This is the main reason that has driven us to use them in our optimized V-FCNN (Fig. 4).

Proposed V-FCNN simplifies the v-net [7] by removing the skip-paths, in an effort to achieve a much quicker training convergence. The constitutive advantage of the skip-paths is to increase the localization, but at the cost of a considerable slow down of the network speed (i.e. GPU global memory leak) while propagating

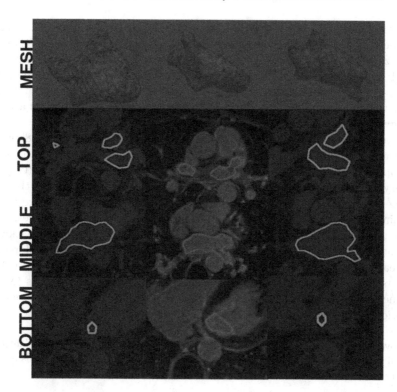

**Fig. 4.** Exemplary result in 3 cases of the competition cases, illustrated with the 3D reconstruction (red, top) and 3 different sections of atrium segmentation: top, middle, and bottom. (Color figure online)

the gradient, at every iteration, forward and back through those paths [15]. We were not able to directly compare with the default v-net architecture, with the skip-paths, but our results suggest that the speed-up in training did not cause a significant loss in segmentation accuracy.

The choice of the loss function is an important consideration in any CNN, and the joint minimization of MSE and Dice Loss has been adopted in our solution. The interpretation is that the MSE looks at global volumetric features, while the Dice Loss (DL) regularize it trying to fit local details. This choice allowed the learning to avoid local minima and the corresponding slow convergence of using only MSE. The optimal weight between them, and the inclusion of further criteria such as an L1/L2 loss or a statistical distance to a library of existing cases, should be addressed in future extensions of this work.

The limitation of the 3D convolutional kernels is the large memory burden, and this is the reason why the original detail of the images was reduced (from 640 to 127 pixels). This image reduction is required due to the current hardware limitations (i.e. memory of the GPU) and obviously caused a loss in performance.

The concatenation of two CNNs, the second working at the full resolution cropping the region of interests containing the atria, is also a good strategy that will minimize the impact of this choice. An alternative is to work with the information at two levels of detail, which has shown to achieve good performance in this problem of atrial segmentation [16]. Recent evidence also suggests that not simply the concatenation, but the combination of two fields of view is a useful strategy to maximize the segmentation performance of MRIs [17].

An alternative strategy to 3D convolution is recurrent units [13,14,18], where recurrence is used to capture the redundancy between adjacent slices. The use of this approach is more memory efficient (i.e. fewer parameters in the GPU global memory to capture recurring partners), also reporting improvements at the challenging apical slices of the left ventricle [13]. These solutions are limited by the vanishing gradient that unfortunately occurs within very long sequences, what can be partially avoided by imposing upper bound constraints on the backward gradient (i.e. gradient clipping) or with a regularisation term [19]. Moreover, the creation of Long short-term memory (LSTM) co-processor, for future MRI scanners inference, requires several parallel Multiply–Accumulate units (MAC) (i.e. due to multiple linear LSTM gates) which require huge amounts of memory bandwidth [20].

**Acknowledgements.** This work was supported by the Wellcome/EPSRC Centre for Medical Engineering at King's College London [g.a. 203148/Z/16/Z]. PL holds a Wellcome Trust Senior Research Fellowship [g.a. 209450/Z/17/Z].

# References

1. Prystowsky, E.N., Benson, D.W., Fuster, V.: Management of patients with atrial fibrillation?: a statement for healthcare professionals from the subcommittee on electrocardiography and electrophysiology. American Heart Association. Circulation **93**, 15 (1996)
2. Varela, M., et al.: Novel computational analysis of left atrial anatomy improves prediction of atrial fibrillation recurrence after ablation. Front. Physiol. **8**, 68 (2017)
3. Kim, R.J., Wu, E., Rafael, A., et al.: The use of contrast-enhanced magnetic resonance imaging to identify reversible myocardial dysfunction. N. Engl. J. Med. **343**, 1445–1453 (2000)
4. Boyle, P.M., Zahid, S., Trayanova, N.A.: Towards personalized computational modelling of the fibrotic substrate for atrial arrhythmia. EP Europace **18**, 136–145 (2016)
5. Tao, Q., Shahzad, R., Ipek, E.G., Berendsen, F.F., Nazarian, S., van der Geest, R.J.: Fully automated segmentation of left atrium and pulmonary veins in late gadolinium enhanced MRI. J. Cardiovasc. Magn. Reson. **18**(1), O84 (2016)
6. Mortazi, A., Karim, R., Rhode, K.S., Burt, J., Bagci, U.: CardiacNET: segmentation of left atrium and proximal pulmonary veins from MRI using multi-view CNN. CoRR, abs/1705.06333 (2017)
7. Milletari, F., Navab, N., Ahmadi, S.-A.: V-Net: fully convolutional neural networks for volumetric medical image segmentation. CoRR, abs/1606.04797 (2016)

8. Isensee, F., Jaeger, P., Full, P.M., Wolf, I., Engelhardt, S., Maier-Hein, K.H.: Automatic cardiac disease assessment on cine-MRI via time-series segmentation and domain specific features. CoRR, abs/1707.00587 (2017)

9. Winther, H.B., et al.: $\nu$-Net: deep learning for generalized biventricular cardiac mass and function parameters. CoRR, abs/1706.04397 (2017)

10. Kaur, H., Sharmila Rani, J.: MRI brain image enhancement using histogram equalization techniques. In: 2016 International Conference on Wireless Communications, Signal Processing and Networking (WiSPNET), pp. 770–773 (2016)

11. Avendi, M.R., Kheradvar, A., Jafarkhani, H.: A combined deep-learning and deformable-model approach to fully automatic segmentation of the left ventricle in cardiac MRI. Med. Image Anal. **30**, 108–119 (2016)

12. Cignoni, P., Rocchini, C., Scopigno, R.: Metro: measuring error on simplified surfaces. CoRR (1996)

13. Poudel, R.P.K., Lamata, P., Montana, G.: Recurrent fully convolutional neural networks for multi-slice MRI cardiac segmentation. In: Zuluaga, M.A., Bhatia, K., Kainz, B., Moghari, M.H., Pace, D.F. (eds.) RAMBO/HVSMR -2016. LNCS, vol. 10129, pp. 83–94. Springer, Cham (2017). https://doi.org/10.1007/978-3-319-52280-7_8

14. Chen, J., Yang, L., Zhang, Y., Alber, M.S., Chen, D.Z.: Combining fully convolutional and recurrent neural networks for 3D biomedical image segmentation. CoRR, abs/1609.01006 (2016)

15. Paszke, A., Chaurasia, A., Kim, S., Culurciello, E.: ENet: a deep neural network architecture for real-time semantic segmentation. CoRR, abs/1606.02147 (2016)

16. Xiong, Z., Fedorov, V.V., Fu, X., Cheng, E., Macleod, R., Zhao, J.: Fully automatic left atrium segmentation from late gadolinium enhanced magnetic resonance imaging using a dual fully convolutional neural network. IEEE Trans. Med. Imaging, 1 (2018)

17. Savioli, N., Vieira, M.S., Lamata, P., Montana, G.: A generative adversarial model for right ventricle segmentation. arXiv:1810.03969, 27 September 2018

18. Savioli, N., Vieira, M.S., Lamata, P., Montana, G.: Automated segmentation on the entire cardiac cycle using a deep learning work-flow. arXiv:1809.01015, 31 August 2018

19. Pascanu, R., Mikolov, T., Bengio, Y.: Understanding the exploding gradient problem. CoRR, abs/1211.5063 (2012)

20. Chang, A.X.M., Martini, B., Culurciello, E.: Recurrent neural networks hardware implementation on FPGA. CoRR, abs/1511.05552 (2015)

# Ensemble of Convolutional Neural Networks for Heart Segmentation

Wilson Fok$^{(\boxtimes)}$, Kevin Jamart, Jichao Zhao, and Justin Fernandez

Auckland Bioengineering Institute, Auckland, New Zealand
wfok007@aucklanduni.ac.nz

**Abstract.** Training an ensemble of convolutional neural networks requires much computational resources for a large set of high-resolution medical 3D scans because deep representation requires many parameters and layers. In this study, 100 3D late gadolinium-enhanced (LGE)-MRIs with a spatial resolution of 0.625 mm × 0.625 mm × 0.625 mm from patients with atrial fibrillation were utilized. To contain cost of the training, down-sampling of images, transfer learning and ensemble of network's past weights were deployed. This paper proposes an image processing stage using down-sampling and contrast limited adaptive histogram equalization, a network training stage using a cyclical learning rate schedule, and a testing stage using an ensemble. While this method achieves reasonable segmentation accuracy with the median of the Dice coefficients at 0.87, this method can be used on a computer with a GPU that has a Kepler architecture and at least 3 GB memory.

**Keywords:** Convolutional neural network · LGE-MRI · Human atria · Pre-trained model · Cyclical learning rate · Ensemble of networks · Dice loss

## 1 Introduction

In computer vision, major advances in the performance of neural networks comes empirically from an increase in their number of learnable parameters and in floating point operations per seconds [1]. In 2015, an ensemble of two ResNets yielded 3.57% test error in the ImageNet classification challenge. The ensemble is computationally expensive to train as one ResNet, the version of the architecture with 152 parameters layers, contains 60-million parameters [2] and uses $11.3 \times 10^9$ floating point operations per second in a single forward pass [3]. Another example is a winning entry for brain tumor segmentation challenge (BRATS 2017); it deployed an ensemble of multiple neural networks with unique architectures that were separately trained and then put together [4].

Segmentation of images of diseased hearts with atrial fibrillation is useful scientifically and clinically. The left atrium can be imaged using 3D late Gadolinium-Enhanced Magnetic Resonance Imaging [5]. However, manual segmentation requires expert knowledge and is laborious and time-consuming. Thus, the task of the segmentation of the left atria can be benefited from the use of automated deep neural networks.

© Springer Nature Switzerland AG 2019
M. Pop et al. (Eds.): STACOM 2018, LNCS 11395, pp. 282–291, 2019.
https://doi.org/10.1007/978-3-030-12029-0_31

However, these advanced networks are computationally expensive to train and deploy. Because of resource constraint, one should not ignore the computational overhead during training and inference. To this end, this paper aims to emphasize the importance of the concept of reuse and to demonstrate its practical value in getting good segmentation performance at a fraction of the overhead. Reuse appears in two ways. One is transfer learning; the other is an ensemble of networks. The goal is to strike a balance between memory usage, floating point operations, and segmentation performance.

Transfer learning is commonly used in object localization/detection and segmentation [6–9]. The theory is that network's weights after training on a large image dataset such as the ImageNet dataset are useful in related tasks such as localization and segmentation. These weights can then be further fine-tuned for other imaging tasks. The advantage of transfer learning is that these weights' values provide better convergence to the objective function of a related task than those that are randomly sampled from a normal distribution. During training, these weights can be tuned, albeit at a lower learning rate, or be frozen altogether. If these weights are held fixed, one eliminates the need to perform operations to update them and to store their error gradient during backpropagation. Furthermore, ensembles of networks typically require separate training on a number of different neural networks with various architectures or different weight initialization [10]. To reduce the computational cost of training such an ensemble, [10] observed that improvement can also be gained by averaging a single network with a fixed architecture but different values of weights that were captured at different points in time during training. It has been shown that DenseNet performance can be boosted by this ensemble technique [10]. During the training phase, the learning rate was perturbed periodically at regular intervals. This is very different from the method of learning rate decay or of setting the learning rate to be dependent on the accumulation of the magnitude of the error gradients over epochs [11, 12]. [13] appears to support cyclical learning rate as it reduces training time for GoogLeNet with inception modules and AlexNet in ImageNet challenge without sacrificing the accuracy. In addition, Cyclic Cosine Annealing, a recently proposed learning rate schedule, may help to move network weight values from one local minimum of the parameter space to the next [14]. An ensemble can improve accuracy for different architectures such as DenseNet and ResNet by averaging its members [10].

This paper proposes an algorithm that combines (1) ideas from transfer learning to use weights from ResNet (networks trained on ImageNet dataset for image classification task), and (2) techniques to create an ensemble of convolutional neural networks. The method section provides information on the data used in the heart segmentation challenge, and training and evaluation methods of the ensemble of neural networks. The result section delineates the progress of training, validation and testing. The discussion section summarizes the key findings, and it shows that the lessons learned from classification tasks can be carried forward nicely to segmentation tasks. The conclusion describes future directions of this paper.

## 2   Methods

### 2.1   Data Source

The dataset contains 100 cases of 3D late gadolinium-enhanced (LGE)-MRIs with a spatial resolution of 0.625 mm × 0.625 mm × 0.625 mm. These images have 88 slices in the z direction and either 640 × 640 or 576 × 576 in the x and y directions respectively. The edges of the 640 × 640 were trimmed down to 576 × 576. No additional data source was used except this MRI dataset. These images have detailed structural information of the diseased hearts, showing atria and its scar. Heart functions are linked to their structures [15, 16]. All the images of the hearts are in grey-scale, and come with a ground truth mask that outlines the left atria.

### 2.2   Data Processing

As the intensity contrast of the MRIs was low, contrast limited adaptive historgram equalization technique was used to enhance its contrast [17]. This technique worked by locally enhancing the contrast of the images. It used a pixel neighborhood in which the intensity was adjusted. One of the crucial parameters was thus the size of this neighborhood/ kernel. As the kernel size varied, the output image differed. A range of kernel sizes was tested from ¼ down to 1/128 of the image's height. The final selection included sizes of 1/8 and 1/32 on an ad hoc basis. These two kernels operated on different scales and thus revealed different boundaries of the part to be segmented.

### 2.3   Image Processing and Augmentation

The original resolution of the MRI scans was down-sampled to 88 by 112 by 112 using a bi-cubic interpolation at each depth to save memory usage. Out of these 100 cases, training took 70%; validation 10%; testing 20%.

To augment the dataset, image transformations were performed, including histogram matching and various image transformations (rotation, cropping and translation, zooming, skewing, shearing, randomly erasing patches, tilting, flipping and reversing stack orders). The rotation operation randomly rotated the images at an angle between 0-360°. The cropping and translation operation randomly cut out 2/3 of the stack of images using a square bounding box that was randomly placed inside the scans. The zooming operation randomly zoomed in and out based on a ratio between 1.5 and 0.5. The perspective skewing calculated a transformation plane that, out of four corners (upper left, upper right, lower left or lower right), skewed toward one corner. The tilting operation calculated the transformation plane that, out of four sides (top, bottom, left and right), tilted toward one side. The shearing operation transformed the image in either x or y direction at an angle that was chosen randomly between ±25°. The erasing operation erased a rectangular patch whose width and height ranged from 10% to 100% of the input's width and height. As the width and height were chosen randomly and independently, the shape of the patch was usually a rectangle, even though a square was possible. The reverse of stack order meant the slices in the depth direction were sorted back to front. The flipping operation flipped the images either sideway or upside down. The histogram matching first normalized the cumulative distribution of pixel intensity of a source and a template, both sampled at random from the cases. A source was to be matched to a template. The quantiles of the source's pixel intensity were mapped and adjusted to the corresponding values dictated by the template's distribution. If the amount of free hard disk space is of concern, one can opt for augmenting dataset during training on demand rather than storing the transformed images and masks on hard disk. As a result of the augmentation procedure, the training dataset increased to 11250 cases while the validation and the testing datasets both increased to 3750.

## 2.4 Network Training, Validation and Testing

Figure 1 is the schematic of the proposed architecture of our deep convolutional neural network. The inputs were two stacks of the contrast-enhanced MRI scans from an identical case. The 2D feature maps produced by pre-trained ResNet34 were represented by flat squares. The layers of ResNet that could extract highly expressive features were kept in the original sequential or hierarchical order, and the rest (classification layer and the pooling layer) were discarded. Specifically, these layers were layer 0–2, 4, 5, 6, 7 in Pytorch implementation. The feature maps were then concatenated to show the two input stacks at each level. Subsequently, 3D convolution was performed on them to account for the fact that the object to-be-segmented was 3D.

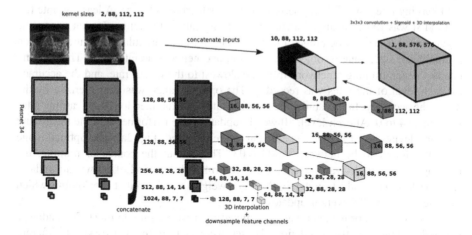

**Fig. 1.** An overview of the neural network architecture. The input is contrast-enhanced with 1/8 and 1/32 kernels. The weights from the ResNet34 process the stacks slice by slice in 2D. The resultant feature maps are then convolved with 3D filters at their corresponding levels (horizontal arrows). The feature cubes are interpolated to match the dimension of the upper levels and then concatenated with those higher up. Inputs are reused and concatenated to the highest level. The network predicts the mask by $3 \times 3 \times 3$ convolution, followed by sigmoid activations and interpolation. The size of the squares and cubes alludes to the change in dimension. The detailed dimension (number of feature channels, depth, height, width) is written in bold near to the layers.

3D Convolution was followed by batch-norm [18] whose outputs were passed through a rectified linear unit [19]. The network resembled U-Net or V-Net on the ascending arm [20, 21]. It expanded its resolution and joined the lower level to the upper level level-by-level. In order to match the dimensions between the upper and lower levels, interpolation was used to obtain the exact size and to minimize misalignment. Finally, the features went through a 3D $3 \times 3 \times 3$ convolution with sigmoid activation units for binary classification or segmentation. An additional interpolation pushed the predicted mask to the dimension back to the original scans. Unlike U-Net, this network's architecture is a hybrid with the descending arm composed of 2D convolutions from a pre-trained ResNet34 and ascending arm composed of 3D convolutions.

Dice loss as in [21] was used, for it seemed to be more suitable than cross-entropy loss for segmentation. It penalized both false positive and false negative and thus was more similar to that of Dice coefficient, the scoring criterion. The total loss was calculated as a sum of the Dice losses on all individual slices. On segmenting, this automatically put more weights on the smaller parts than the larger parts. Misclassification of small regions as appeared on the MR images became costlier mathematically because of the ratios among true positive, false positive and false negative in Dice loss. As the region shrank, any misclassified voxel would give rise to an increase in the loss.

Training began by randomly shuffling the training set. The learnable weights' values were initialized as in [22]. The mini-batch size was 5. Adam algorithm [11] optimized the network.

Learning rate, one of the hyper-parameters, is hard to optimize. A learning rate test as in [13] was run to estimate a suitable learning rate value range. In that paper, the suggestion is to try a wide range of learning rates for any particular configuration (such as minibatch size, dataset, objective function, and network architecture). The learning rate test began training a network from the slowest to the fastest rate, and the accuracy for each level of the rates was recorded. The optimal range was characterized by the rise in the accuracy. In this study, the learning rate test examined log10 of learning rates spanning −4 to 0. At each level, 10 weight updates were performed and the means and standard deviations of the Dice coefficients were recorded (Fig. 2a). The optimal range was later found to be between 0.003–0.03. Throughout the course of training, the learning rate schedule was a triangular waveform with a period of 50 weight updates and a peak of 0.03 and a trough of 0.003. The weights were saved at checkpoints which were set at every 200 weight updates.

Cessation of improvement was determined by a paired sample t-test. 20 randomly selected samples in the validation dataset were fed to the networks at different checkpoints. Training was terminated when t-tests showed no improvement in the validation loss. While t-test may not be necessary for determining whether difference in losses exists, it can still be useful as it quantifies the lack of difference by its p-value.

Amongst the checkpoints, six were selected based on their low validation losses. Afterwards, an ensemble of networks was formed by selecting a number of members up to all the six checkpoints. Once the ensemble was created, the ensemble's prediction was a simple average of all the predictions made by the members. To identify an ensemble that yielded low validation loss, several combinations were tried and the one with the lowest loss was kept aside. The ensemble's prediction, by nature, was

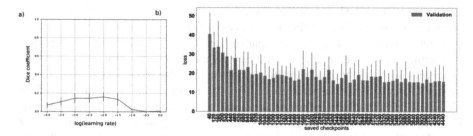

**Fig. 2.** (a) shows the results of a learning rate test. The optimal range for the learning rate spans over the hump (rise in Dice coefficient). The range of −1.5−−2.5 was used in training in the form of a cyclical learning rate scheme. (b) shows validation loss at different checkpoints. The validation loss rose and fell, like a ripple, as a consequence of the learning rate scheme, and it had shown little or no improvement after the ripple settled down. Those checkpionts, with lower losses, were selected in a list of candidates for an ensemble.

continuous, between 0–1, owing to the sigmoid activation function. To obtain a binary mask, the prediction was thresholded at a cutoff value. By comparing changes in Dice coefficients of predictions given by the earlier ensemble at different cutoffs, the optimal threshold was the one that yielded the highest Dice coefficient.

Testing was conducted on 63 combinations of ensembles of neural networks using up to the six aforementioned checkpoints. The smallest ensemble included a single network that was loaded with weights at a single checkpoint; the largest ensemble included six networks with weights from all the six checkpoints. Each test ran on 50 randomly selected samples in the augmented test set. The performance was quantified by the median of the Dice coefficients.

## 3   Experimental Results

The results are delineated in the order of training, validation and testing. Figure 2a plots the results of the learning rate test. Between –4–0, an optimal range of values sits between −1.5−−2.5 (log values)/0.003–0.03 because of higher average Dice coefficients. This optimal range was set to be the peak and tough of the cyclical learning rate schedule. Figure 2b shows the Dice loss of validation dataset at various checkpoints. The rhythmic rise and fall of the loss mirrors the cyclical learning rate. The peaks of the learning rate waveform have likely assisted the network in exploring the possible parameters' space, thereby dislodging the weights from one local minimum to the next. The price tag is a temporary increase in loss. Paired sample t-test reveals no significant evidence that the loss of checkpoint 4440 is different from that of checkpoint 3920 (p-value: 0.72). This suggests further training may not be needed. The binary mask was obtained by thresholding the network's prediction. The optimal threshold is at 0.37, giving a Dice coefficient of 0.85.

The evaluation of the performance of the ensembles was performed in two ways, using the test set. The first is a graphic method in which the network's mistake and correctness were plotted against the ground truth. The second is a quantitative way in which the median of Dice coefficients was calculated.

Figure 3 shows a side-by-side comparison between a ground truth and a predicted mask estimated by an ensemble of a test case. While the region of true positive highly aligns with that of the ground truth, the regions of the false negative and false positive are small and very small respectively. Figure 4 summarizes the top 16 medians of the Dice coefficients produced by ensembles of different combinations of neural networks. The performance of the ensembles is sensitive to group size and combinations of networks with a specific selection of weights. It is worth pointing out that only the first and the seventh columns represent the score given by a single network. The rest are ensembles of two or more networks.

## False negative  False positive  True positive  Ground truth

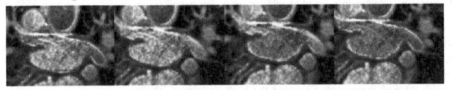

**Fig. 3.** A side-by-side comparison between the ground truth mask and a predicted mask estimated by an ensemble during testing. This case obtained a Dice coefficient of 0.87. Yellow region highlights the false positive, the false negative, and the true positive of the prediction as well as the ground truth. The shape formed by true-positive pixels resembles the shape of the ground truth. (Color figure online)

## 4 Discussion

The findings of this paper align with the existing view of the behavior, in general, of ensembles of neural networks and with the effect of using cyclical learning rate schedules in training. However, it is important to emphasize the difference between this work and the existing literature. The existing view is that ensemble outperforms individuals in terms of robustness and accuracy [10, 23]. The performance does not guarantee to improve by a merely increase in size. The optimal combinations for an ensemble are normally found by trial and error. The finding in this paper confirms this view. The ensemble of networks in Fig. 4 often produces slightly better results than its individual members because, amongst the 16, only two were single networks. As the Dice coefficient rises, it is hard to identify a clear pattern of combination of networks that would work best. The performance gain is hypothesized to be a result of averaging predictions by various sets of weights at saddle points or local minimums on the path of parameter optimization during training.

The cyclical learning rate schedule has been observed to produce an oscillating pattern [13, 14]. [10] deployed similar schedules for training and had found that the oscillation was a consistent hallmark. The exploration using fast learning rate can be beneficial in the long run despite temporary increase in loss as observed in this work.

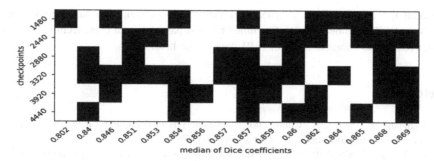

**Fig. 4.** The size and membership of an ensemble can often produce variation in the level of accuracy. The above plot highlights the median of Dice coefficients produced by ensembles of different combinations of neural networks at different checkpoints. A black square means the inclusion of that checkpoint. A white square means the opposite. The series of black squares down each column represents the configuration of that ensemble. The x axis shows the Dice coefficient in an ascending order. The y axis shows which checkpoints were used in the ensemble of networks' weights.

What makes this work novel is that the ensemble's objective is to segment images whereas the existing work in the literature concerns image classification. Although segmentation can be regarded as a classification task between yes or no, it is still non-identical to classifying many classes of objects such as 1000 different classes in ImageNet Classification Challenge. Nonetheless, the finding of this work indicates that the existing tricks for training neural networks also apply in segmentation tasks. The benefit of using transfer learning, cyclical learning rate schedule and ensembles is reduced computational overhead. In this work, while training needs only 2.4 GB of memory on GPU, testing requires no more than 3.0 GB. Consequently, it is possible to implement this work on a computer with a GPU that has a Kepler architecture and at least 3 GB memory.

## 5  Conclusions and Future Work

This work demonstrates a feasible pipeline that helps train an ensemble with limited computational resources. It suggests that the tricks used in image classification can be carried forward to segmentation. The future work will include evaluations of different learning rate schedules, and investigations of the benefits and harms of various image augmentation techniques. As learning rate is often set to monotonically decay during training, future studies should compare how the generalization of ensembles improves under various learning rate decay schedules and other cyclical learning rate schedules with different waveforms, periods, and peak and tough values. The speed of network training will also be explored under the influence of different schedules. Moreover, this preliminary work deploys many image transformation techniques, but their potential impacts on ensemble's performance are unclear. In other words, how the transformation techniques work to boost the ensemble's ability to achieve good segmentation in unseen patient cases should be considered. Future work will investigate the balance between the computational overhead on the image augmentation and performance gains, and will consider ways to fine tune or adjust parameters of these techniques as to better reflect the characteristics of MRI data of the hearts.

# References

1. Canziani, A., Paszke, A., Culurciello, E.: An analysis of deep neural network models for practical applications. arXiv preprint arXiv:1605.07678 (2016)
2. Boulch, A.: ShaResNet: reducing residual network parameter number by sharing weights. arXiv preprint arXiv:1702.08782 (2017)
3. He, K., Zhang, X., Ren, S., Sun, J.: Deep residual learning for image recognition. In: Proceedings of the IEEE Conference on Computer Vision and Pattern Recognition, pp. 770–778 (2016)
4. Kamnitsas, K., et al.: Ensembles of multiple models and architectures for robust brain tumour segmentation. arXiv preprint arXiv:1711.01468 (2017)
5. McGann, C., et al.: Atrial fibrillation ablation outcome is predicted by left atrial remodeling on MRI. Circ. Arrhythmia Electrophysiol. 7(1), 23–30 (2014)
6. Girshick, R.: Fast R-CNN, arXiv preprint arXiv:1504.08083 (2015)
7. He, K., Gkioxari, G., Dollár, P., Girshick, R.: Mask R-CNN. In: 2017 IEEE International Conference on Computer Vision (ICCV), pp. 2980–2988 (2017)
8. Iandola, F., Moskewicz, M., Karayev, S., Girshick, R., Darrell, T., Keutzer, K.: Densenet: implementing efficient convnet descriptor pyramids. arXiv preprint arXiv:1404.1869 (2014)
9. Redmon, J., Divvala, S., Girshick, R., Farhadi, A.: You only look once: Unified, real-time object detection. In: Proceedings of the IEEE Conference on Computer Vision and Pattern Recognition, pp. 779–788 (2016)
10. Huang, G., Li, Y., Pleiss, G., Liu, Z., Hopcroft, J.E., Weinberger, K.Q.: Snapshot ensembles: train 1, get M for free. arXiv preprint arXiv:1704.00109 (2017)
11. Kingma, D.P., Ba, J.: Adam: a method for stochastic optimization. arXiv preprint arXiv: 1412.6980 (2014)
12. Tieleman, T., Hinton, G.: Lecture 6.5 - RMSProp, COURSERA: Neural Networks for Machine Learning. Technical report (2012)
13. Smith, L.N.: Cyclical learning rates for training neural networks. In: 2017 IEEE Winter Conference on Applications of Computer Vision (WACV), pp. 464–472 (2017)
14. Loshchilov, I., Hutter, F.: SGDR: Stochastic Gradient Descent with Warm Restarts. arXiv preprint arXiv:1608.03983 (2016)
15. Hansen, B.J., et al.: Atrial fibrillation driven by micro-anatomic intramural re-entry revealed by simultaneous sub-epicardial and sub-endocardial optical mapping in explanted human hearts. Eur. Heart J. 36(35), 2390–2401 (2015)
16. Zhao, J., et al.: Three-dimensional integrated functional, structural, and computational mapping to define the structural 'Fingerprints' of heart-specific atrial fibrillation drivers in human heart ex vivo. J. Am. Heart Assoc. 6(8), e005922 (2017)
17. Zuiderveld, K.: Contrast limited adaptive histogram equalization. In: Heckbert, P.S. (eds.) Graphics Gems, pp. 474–485. Academic Press, Cambridge (1994)
18. Ioffe, S., Szegedy, C.: Batch normalization: accelerating deep network training by reducing internal covariate shift. arXiv preprint arXiv:1502.03167 (2015)
19. Nair, V., Hinton, G.E.: Rectified linear units improve restricted boltzmann machines. In: Proceedings of the 27th International Conference on Machine Learning (ICML-10), pp. 807–814 (2010)

20. Çiçek, Ö., Abdulkadir, A., Lienkamp, S.S., Brox, T., Ronneberger, O.: 3D U-Net: learning dense volumetric segmentation from sparse annotation. In: International Conference on Medical Image Computing and Computer-Assisted Intervention, pp. 424–432 (2016)
21. Milletari, F., Navab, N., Ahmadi, S.-A.: V-net: fully convolutional neural networks for volumetric medical image segmentation. In: 2016 Fourth International Conference on 3D Vision (3DV), pp. 565–571 (2016)
22. He, K., Zhang, X., Ren, S., Sun, J.: Delving deep into rectifiers: surpassing human-level performance on imagenet classification. In: Proceedings of the IEEE International Conference on Computer Vision, pp. 1026–1034 (2015)
23. Pereira, S., Pinto, A., Alves, V., Silva, C.A.: Brain tumor segmentation using convolutional neural networks in MRI images. IEEE Trans. Med. Imag. **35**(5), 1240–1251 (2016)
24. Xiong, Z., Fedorov, V., Fu, X., Cheng, E., Macleod, R., Zhao, J.: Fully automatic left atrium segmentation from late gadolinium enhanced magnetic resonance imaging using a dual fully convolutional neural network. IEEE Trans. Med. Imag. (2018) (in Press)

# Multi-task Learning for Left Atrial Segmentation on GE-MRI

Chen Chen$^{(\boxtimes)}$, Wenjia Bai, and Daniel Rueckert

Biomedical Image Analysis Group, Imperial College London, London, UK
cc215@ic.ac.uk

**Abstract.** Segmentation of the left atrium (LA) is crucial for assessing its anatomy in both pre-operative atrial fibrillation (AF) ablation planning and post-operative follow-up studies. In this paper, we present a fully automated framework for left atrial segmentation in gadolinium-enhanced magnetic resonance images (GE-MRI) based on deep learning. We propose a fully convolutional neural network and explore the benefits of multi-task learning for performing both atrial segmentation and pre/post ablation classification. Our results show that, by sharing features between related tasks, the network can gain additional anatomical information and achieve more accurate atrial segmentation, leading to a mean Dice score of 0.901 on a test set of 20 3D MRI images. Code of our proposed algorithm is available at https://github.com/cherise215/atria_segmentation_2018/.

**Keywords:** Multi-task learning · Atrial segmentation · Fully convolution neural network

## 1 Introduction

Atrial fibrillation (AF) is a condition of the heart that causes an irregular and often abnormally fast heart rate [1]. This can cause blood clots to form, which can restrict blood supply to vital organs, and further leads to a stroke and heart failure [2]. One of the most common treatments for AF is called ablation which can isolate the pulmonary veins (PVs) from the Left Atrium (LA) electrically by inducing circumferential lesion and destroying abnormal tissues. During this procedure, a good understanding of the patient atrial anatomy is very vital for planning and guiding the surgery, and further improving the patient outcome [2].

A good way to learn the anatomical structure of the LA is by performing LA segmentation on medical images, such as computed tomography (CT) scans and magnetic resonance images (MRI). With the development of imaging techniques and computer science, many automatic or semi-automatic algorithms [3] have been proposed for atrial segmentation. However, this is still a challenging problem and many traditional methods may fail to segment due to several reasons. For example, intensity-based methods such as region growing may fail to segment those atria with extremely thin myocardial walls, especially when

© Springer Nature Switzerland AG 2019
M. Pop et al. (Eds.): STACOM 2018, LNCS 11395, pp. 292–301, 2019.
https://doi.org/10.1007/978-3-030-12029-0_32

their surroundings have very similar intensity to their blood pool [3]. In addition, there is large shape variation among the LA of different individuals, such as atrial sizes and pulmonary vein structures [3]. These variations will make it too complex for model-based segmentation methods to impose shape prior. An alternative way is to use atlas-based methods that can be robust to the LA with high anatomical variations. However, this kind of approach is time-consuming which typically takes 8 min around [4]. Most recently, with the increase of computing hardware performance and more data becoming available, deep learning has become the state-of-the-art method due to its efficiency and effectiveness on computer vision tasks, and has been widely used in the medical domain [5].

In this paper, we focus on the segmentation of the LA from gadolinium-enhanced magnetic resonance images (GE-MRI). These images can either be taken before or after ablation treatment. Noticing that there might be contextual difference between the pre-ablation and post-ablation images (e.g. ablation will cause scars in the LA [6] and may influence the quality of images), we proposed a multi-task convolutional neural network (CNN) that could segment a patient left atrium from GE-MRI and detect whether this patient is pre- or post-ablation. In this way, our network could not only learn structural information from segmentation masks, but also retrieve contextual information through the classification task. Our network was trained simultaneously for the two tasks, using a stack of 2D slices extracted from each MRI scan along with its corresponding segmentation masks and a pre/post ablation label. In addition, in order to improve the robustness of segmentation on images with various image contrast and sizes, we employed a contrast augmentation method to augment our training set and trained our network with images in different sizes. In order to produce a fixed-length vector to classify input images in multiple sizes, spatial pyramid pooling [7] was used in this network.

The proposed framework was trained and evaluated on the data set of the Atrial Segmentation Challenge 2018. Our experimental results showed that by sharing features between related tasks, our network achieved better segmentation performance compared to a variant of U-Net trained with a single task. During the test phase, our network can directly inference the segmentation mask from a scan of GE-MRI without taking extra pre-processing steps for image contrast enhancement. In total, our method is very efficient as one 3D segmentation result for each individual was obtained in 6 s on a Nvidia Titan Xp GPU using our model, plus 3 or 4 s for post-processing on the whole volume, which is far more faster than general atlas-based methods that usually take minutes.

## 2   Methods

In this section, we present the architecture of our proposed multi-task network and how we post-process the network output to get the final 3D segmentation mask. Our proposed network is adapted from a U-Net architecture [8] where we increase the depth of the network and add a classification branch. The input to the network is a stack of 2D images. The output is a one shot prediction of the atrial segmentation mask and pre/post ablation classification score.

## 2.1   Network Architectures

In order to explore the benefit of multi-task learning, the proposed network is designed to conduct both the atrial segmentation task and an auxiliary pre/post ablation classification task with images of multiple sizes.

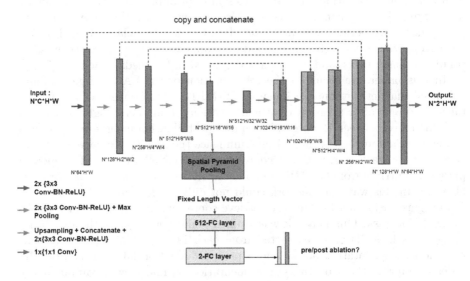

**Fig. 1.** The architecture of our proposed multi-task Deep U-Net. Best viewed in color (Color figure online)

The core of our method, named 'Deep U-Net' and shown in Fig. 1, is derived from the 2D U-Net [8] for semantic segmentation. Since the largest size of images in our dataset is $640 \times 640$ in x-y planes, we increased the receptive field of U-Net by adding more pooling layers. The modified network now consists of five down-sampling blocks and five up-sampling blocks. Each down-sampling layer contains two $3 \times 3$ convolutions, with Batch Normalizations (BN) [9] and Rectified Linear Unit (ReLU) activations, as well as a $2 \times 2$ max pooling operation with stride 2 for down-sampling. The up-sampling path is symmetric to the down-sampling path. By aggregating both coarse and fine features learned at different scales from the down-sampling path and up-sampling path, our network is supposed to achieve better segmentation performance than those networks without the aggregation operations.

Our classification task is performed by utilizing image features learned from the down-sampling path. Features after the $4^{th}$ max pooling layer are extracted for classification, which is a common practice for many existing classification networks [10,11]. In order to generate fix-length feature vectors learned from input images with different sizes and scales, spatial pyramid pooling [7] is applied. These fixed-length vectors are then processed through Fully-Connected layers

(FC) followed by a softmax layer to calculate class probabilities (pre/post-ablation) for each image. Dropout [12] is applied to the output of FC layers with a probability of 0.5 during the training process, which functions as regularizer to encourage our network to better generalize [12].

The loss function $L$ for our multi-task network is $L = L_S + \lambda L_C$, where $L_S$ is segmentation score, $L_C$ is classification score, and $\lambda = 1$ for our experiments. For segmentation part, pixel-wise cross-entropy loss was employed. For classification part, we used sigmoid cross-entropy to measure pre/post ablation classification loss. The classification ground truth of every 2D image is labeled as one if this slice is extracted from a post-ablation object. Otherwise, its ground truth is zero. Our network was trained jointly with the combined loss. The classification loss works as a regularization term, enabling the network to learn the high-level representation that generalizes well on both tasks.

Our network was optimized using Stochastic Gradient Descent (SGD) [13] with a momentum of 0.99 and weight decay of 0.0005. The initial learning rate is 0.001, which will be decreased at a rate of 0.5 after every 50 epochs.

### 2.2 Post-processing

During the inference time, axial slices extracted from a 3D image are fed into the network slice by slice. The segmentation branch predicts pixel-wise probability score for both background and atrium classes. A 2D segmentation mask is then generated by finding the class with the highest probability for each pixel on the slice. By concatenating these segmentation results slice by slice, a rough 3D mask for each patient is produced. In order to refine the boundary of those masks, we performed 3D morphological dilation and erosion, and kept the largest connected component for each volume.

## 3 Experiments and Results

### 3.1 Data

In this paper, our algorithm was trained and evaluated on the dataset of the 2018 Atrial Segmentation Challenge[1]. This dataset contains a training set of 100 3D gadolinium-enhanced magnetic resonance imaging scans (GE-MRIs) along with corresponding LA manual segmentation mask and pre/post ablation labels for training and validation. In addition, there is a set of 54 images without labels provided for testing. For model training and evaluation, we randomly split the training set into 80 : 20. We did not use any external data for training or pre-training of our network.

Images in this dataset have been resampled and preprocessed by the organizers. So there is no need to do re-sampling procedure in the pre-processing stage. Despite the consistency observed in the resolution of the data, this dataset exhibits large differences in images sizes and image contrast. For example, in our

---

[1] http://atriaseg2018.cardiacatlas.org/.

MRI dataset, there are two sizes of images: 576 × 576 and 640 × 640 on the axial planes. Apart from that, atria, in different images, can also have various shapes and sizes. These phenomena may arise due to the fact that these scans were collected from multiple sites which may have different scanners and imaging protocols. Hence, it is important to build a robust method for those images. Figure 2 visualizes the difference in image contrast of different images in different views. In this paper, we use data augmentation to increase data variety with the aim of improving the model's generalization ability on different images, which will be discussed in Sect. 3.3.

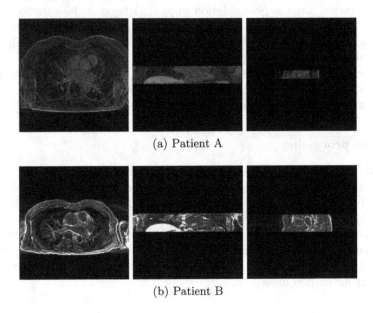

(a) Patient A

(b) Patient B

**Fig. 2.** Original images slices from different views. Despite the homogeneity in image resolution, there are significant differences in the image contrast and quality among different individuals, which can be challenging to segment the Left Atrium (red) from MRI images. Best viewed in color (Color figure online)

## 3.2  Data Pre-processing

In order to preserve the resolution of images, image re-scaling was not performed in the data pre-processing stage. Instead, multi-scale cropping was used to increase the data variety, so that network can analyze images with different contexts. More details will be described in the next section as it actually happens in the data augmentation process. For testing, images can be directly fed into the network provided that its length is a factor of 32 due to the architecture of the network. Otherwise, reflect padding is required. The only necessary step in our pre-processing stage in both training and testing stage is to normalize image intensity to zero mean and unit variance, which has been widely accepted in common practice.

### 3.3   Data Augmentation

Our training data was augmented via a composition of image transformations, including random horizontal/vertical flip with a probability of 50%, random rotation with degree range from $-10$ to $+10$, random shifting along X and Y axis within the range of 10 percent of its original image size, and zooming with a factor between 0.7 and 1.3.

In addition, we also employed random gamma correction as a way of contrast augmentation. Image contrast was adjusted based on a point-wise nonlinear transformation: $G(x, y) = F(x, y)^{1/\gamma}$ where $F(x, y)$ is the original value of each pixel in an image, and $G(x, y)$ is the transformed value for each pixel. The value of $\gamma$ is randomly chosen from the range of $(0.8, 2.0)$ for each image. By applying gamma correction randomly, the variety of image contrast in the training set was significantly increased.

In order to process images at multiple sizes with objects at multiple scales, we centrally cropped 2D images at various image scales. The cropped sizes include $256 \times 256, 384 \times 384, 480 \times 480, 512 \times 512, 576 \times 576, 640 \times 640$. If the cropped size was larger than the image's original size, zero padding was performed instead. Motivated by Curriculum Learning [14], we trained our network firstly with cropped images where the left atrium taking a large portion of the image and then we gradually increased the image size. In this way, our network learns to segment from easy scenarios to hard scenarios and this helps the model to quickly converge in the beginning [15]. Despite the change in input images sizes, our network could still output a fixed length feature vector for classification since we employed spatial pyramid pooling [7]. In practice, we found this could help the network focus on learning task-specific structural features for organ segmentation regardless of the contextual changes in sizes and scales. It is also beneficial for quantitative analysis based on medical image segmentation since we do not use rescaling nor resizing operations which have the risk of introducing scaling/shifting artifacts during prediction.

**Contrast Augmentation:** We found that there exists a diversity of image contrast in the dataset, where low-contrast effects can reduce the visual quality of an image [16] and thus affect segmentation accuracy. To solve this issue, traditional machine learning methods often require image contrast enhancement methods during image pre-processing. Here, we proposed a contrast augmentation method based on gamma correction instead, to generate a variety of images with different levels of contrast during training. In this way, our CNN could gain the ability to segment images regardless of the difference of image contrast. And there is no need to do any contrast adjustment during testing. Therefore, our method is more efficient than those general traditional methods which require those adjustments.

To show our contrast augmentation method is superior to the traditional contrast enhancement methods, we compared it with two image contrast enhancement methods: contrast limited adaptive histogram equalization (CLAHE) [17] and automatic gamma correction [16]. Both of them have been widely used in the pre-processing of CT image and MRI image applications [18–20] in order to improve medical image quality for visual tasks. For CLAHE, we divided each image into $8 \times 8$ regions and performed contrast enhancement on each region by default.

The above experiments were performed based on a simple Deep U-Net (without multi-task) for comparison. From Table 1, it can be seen that our proposed data augmentation method could significantly improve the robustness of our network for processing images with various image contrast and outperformed the traditional image pre-processing methods which may have the risk of amplifying noises and take extra processing time. Therefore, in the following sections, we would like to employ contrast augmentation as our default experimental setting.

**Table 1.** Segmentation accuracy using a single-task Deep U-net with different contrast processing strategies

| Base model | Method | Need extra time | Dice |
|---|---|---|---|
| Deep U-Net | Standard normalization | No | 0.847 (0.18) |
| Deep U-Net | Automatic gamma correction | Yes | 0.854 (0.15) |
| Deep U-Net | CLAHE | Yes | 0.876 (0.09) |
| Deep U-Net | Gamma augmentation | No | 0.883 (0.08) |

### 3.4   Results

To evaluate our segmentation accuracy for different experimental settings, we use four measurements: the Dice score (also known as Dice similarity coefficient score), the Jaccard Similarity Coefficient (JC) score, the Hausdorff Distance (HD) and the Average Symmetric Surface Distance (ASSD).

To show the advancement of our deep network with additional pooling/max-pooling layers, we compared our modified networks with the vanilla 2D U-Net. The results are shown in Table 2. It can be seen that the segmentation performance was greatly improved by increasing the depth of the network, especially in Mitral Valve(MV) planes. Our best results were achieved by using the multi-task Deep U-Net followed by post-processing, producing a Dice score of 0.901. From the visualization plots in Fig. 3, we could see that our multi-task U-Net is more robust than the other two with only one segmentation goal. One reason could be that by sharing features with segmentation and related pre/post ablation classification, the network is forced to learn better representation on images taken before the ablation treatment and those after the treatment, which could further improve segmentation performance. Figure 4 showed that our model achieved high overlap ratio between our 3D segmentation result and the ground truth in different subjects. However, one significant failure mode can be observed around the region of pulmonary veins. One possible reason might be that the number and the length of pulmonary veins vary from person to person, making it too hard for the network to learn from limited cases.

**Table 2.** Segmentation accuracy results based on different measurements for different networks and methods

|  | Dice | JC | HD | ASSD |
|---|---|---|---|---|
| Vanilla U-Net | 0.855 (0.11) | 0.760 (0.14) | 21.81 (19.35) | 1.577 (1.07) |
| Deep U-Net | 0.883 (0.08) | 0.798 (0.11) | 21.18 (21.00) | 1.199 (0.47) |
| Deep U-Net + multi-task | 0.896 (0.04) | 0.815 (0.07) | 15.40 (6.39) | 1.11 (0.35) |
| Deep U-Net + multi-task + post-proc | **0.901 (0.03)** | **0.822 (0.06)** | **14.23 (4.83)** | **1.04 (0.32)** |

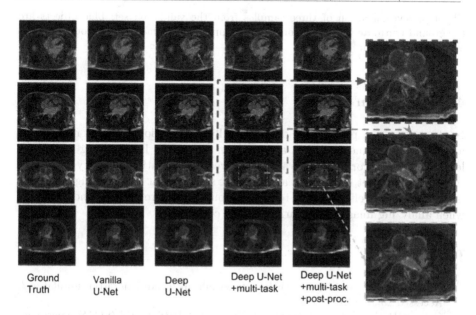

| Ground Truth | Vanilla U-Net | Deep U-Net | Deep U-Net +multi-task | Deep U-Net +multi-task +post-proc. |

**Fig. 3.** Example segmentations for axial slices from our MRI set using different methods. Each column shows axial slices from the mitral to the PVs plane from different individuals (top to bottom).

The total processing procedure (inference + post-processing) for each whole 3D MRI predicted by our network took approximately 10 s on average on one Nvidia Titan Xp GPU. It is therefore much more efficient than those atlas-based methods which typically take eight minutes [4].

For the Atrial Segmentation Challenge 2018, we adopted an ensemble method called Boostrap Aggregating (Bagging) [21] to improve our model's performance in the test phase. We noticed that samples in the dataset were collected from multiple sites while a large portion is from The University of Utah. In that case, domain shift or domain bias may exist when we use a model trained on one certain subset from a limited dataset to predict data from another subset as they may have different intensity distributions. Therefore, we trained the same model 5 times, each with a random subset and then averaged the class probabilities produced by these five models for prediction. Our ensembled results on a set of

54 test cases given by the organizers improved from an averaged Dice score of 0.9197 to 0.9206.

**Fig. 4.** 3D visualization of three samples from the validation set: blue objects are the ground truth, and the green ones are the predicted segmentation of our proposed method. (Color figure online)

## 4    Conclusion

In this paper, we proposed a deep 2D fully convolutional neural network to automatically segment the left atrium from GE-MRIs. By applying multi-task learning, our network demonstrated improved segmentation accuracy compared to a baseline U-Net method. In addition, we showed that contrast augmentation is an efficient and effective way to enhance our model's robustness and efficiency when analyzing images with various image contrast.

## References

1. NHS.    Atrial    fibrillation.    https://www.nhs.uk/conditions/atrial-fibrillation/. Accessed 25 June 2018
2. Calkins, H., Kuck, K.H., Cappato, R., et al.: 2012 HRS/EHRA/ECAS expert consensus statement on catheter and surgical ablation of atrial fibrillation: recommendations for patient selection, procedural techniques, patient management and follow-up, definitions, endpoints, and research trial design: a report of the heart rhythm society (HRS) task force on catheter and surgical ablation of atrial fibrillation. Heart Rhythm **9**(4), 632–696 (2012)
3. Tobon-Gomez, C., Geers, A.J., Peters, J., et al.: Benchmark for algorithms segmenting the left atrium from 3D CT and MRI datasets. IEEE Trans. Med. Imaging **34**(7), 1460–1473 (2015)
4. Depa, M., Sabuncu, M.R., Holmvang, G., Nezafat, R., Schmidt, E.J., Golland, P.: Robust atlas-based segmentation of highly variable anatomy: left atrium segmentation. In: Camara, O., Pop, M., Rhode, K., Sermesant, M., Smith, N., Young, A. (eds.) STACOM 2010. LNCS, vol. 6364, pp. 85–94. Springer, Heidelberg (2010). https://doi.org/10.1007/978-3-642-15835-3_9
5. Litjens, G., Kooi, T., et al.: A survey on deep learning in medical image analysis. Med. Image Anal. **42**, 60–88 (2017)
6. Malcolme-Lawes, L., et al.: Automated analysis of atrial late gadolinium enhancement imaging that correlates with endocardial voltage and clinical outcomes: a 2-center study. Heart Rhythm: Off. J. Heart Rhythm Soc. **10**, 1184–1191 (2013)

7. He, K., Zhang, X., Ren, S., Sun, J.: Spatial pyramid pooling in deep convolutional networks for visual recognition. In: Fleet, D., Pajdla, T., Schiele, B., Tuytelaars, T. (eds.) ECCV 2014. LNCS, vol. 8691, pp. 346–361. Springer, Cham (2014). https://doi.org/10.1007/978-3-319-10578-9_23

8. Ronneberger, O., Fischer, P., Brox, T.: U-net: convolutional networks for biomedical image segmentation. CoRR, abs/1505.04597 (2015)

9. Ioffe, S., Szegedy, C.: Batch normalization: accelerating deep network training by reducing internal covariate shift. CoRR, abs/1502.03167 (2015)

10. Krizhevsky, A., Sutskever, I., Hinton, G.E.: Alexnet. Advances in Neural Information Processing Systems, pp. 1–9 (2012)

11. He, K., Zhang, X., Ren, S., Sun, J.: Deep residual learning for image recognition. arXiv e-prints, December 2015

12. Srivastava, N., Hinton, G., Krizhevsky, A., Sutskever, I., Salakhutdinov, R.: Dropout: a simple way to prevent neural networks from overfitting. J. Mach. Learn. Res. **15**(1), 1929–1958 (2014)

13. Rumelhart, D.E., Hinton, G.E., Williams, R.J.: Learning representations by back-propagating errors. Nature **323**(6088), 533 (1986)

14. Bengio, Y., Louradour, J., Collobert, R., Weston, J.: Curriculum learning. In: Proceedings of the 26th Annual International Conference on Machine Learning, pp. 41–48. ACM (2009)

15. Lieman-Sifry, J., Le, M., Lau, F., Sall, S., Golden, D.: FastVentricle: cardiac segmentation with ENet. In: Pop, M., Wright, G.A. (eds.) FIMH 2017. LNCS, vol. 10263, pp. 127–138. Springer, Cham (2017). https://doi.org/10.1007/978-3-319-59448-4_13

16. Somasundaram, K., Kalavathi, P.: Medical image contrast enhancement based on gamma correction. Int. J. Knowl. Manag. e-Learn. **3**, 15–18 (2011)

17. Zuiderveld, K.: Contrast limited adaptive histogram equalization. In: Graphics Gems IV, pp. 474–485. Academic Press Professional Inc., San Diego (1994)

18. Pizer, S.M., Johnston, R.E., Ericksen, J.P., Yankaskas, B.C., Muller, K.E.: Contrast-limited adaptive histogram equalization: speed and effectiveness. In: Proceedings of the First Conference on Visualization in Biomedical Computing, pp. 337–345. IEEE (1990)

19. Selvy, P.T., Palanisamy, V., Radhai, M.S.: A proficient clustering technique to detect CSF level in MRI brain images using PSO algorithm. WSEAS Trans. Comput. **7**, 298–308 (2013)

20. Ghose, S., et al.: A random forest based classification approach to prostate segmentation in MRI. In: MICCAI Grand Challenge: Prostate MR Image Segmentation (2012)

21. Breiman, L.: Bagging predictors. Mach. Learn. **24**(2), 123–140 (1996)

# Left Atrial Segmentation Combining Multi-atlas Whole Heart Labeling and Shape-Based Atlas Selection

Marta Nuñez-Garcia[1]([✉]), Xiahai Zhuang[2], Gerard Sanroma[3], Lei Li[2], Lingchao Xu[2], Constantine Butakoff[1], and Oscar Camara[1]

[1] Physense, Department of Information and Communication Technologies, Universitat Pompeu Fabra, Barcelona, Spain
marta.nunez@upf.edu
[2] School of Data Science, Fudan University, Shanghai, China
[3] German Center for Neurodegenerative Diseases, Bonn, Germany

**Abstract.** Segmentation of the left atria (LA) from late gadolinium enhanced magnetic resonance imaging (LGE-MRI) is challenging since atrial borders are not easily distinguishable in the images. We propose a method based on multi-atlas whole heart segmentation and shape modeling of the LA. In the training phase we first construct whole heart LGE-MRI atlases and build a principal component analysis (PCA) model able to capture the high variability of the LA shapes. All atlases are clustered according to their LA shape using an unsupervised clustering method which additionally outputs the most representative case in each cluster. All cluster representatives are registered to the target image and ranked using conditional entropy. A small subset of the most similar representatives is used to find LA shapes with similar morphology in the training set that are used to obtain the final LA segmentation. We tested our approach using 80 LGE-MRI data for training and 20 LGE-MRI data for testing obtaining a Dice score of $0.842 \pm 0.049$.

**Keywords:** Left atrial segmentation · Multi-atlas segmentation · Atlas selection · LGE-MRI segmentation

## 1 Introduction

Atrial fibrillation (AF) is the most frequent type of arrhythmia affecting around 1% to 2% of the worldwide population (approximately 33 million in 2015). Prevalence of AF doubles with each decade of age and the AF incidence is rapidly growing with the increasing life expectancy in developed countries [1,13].

Inspection of LA anatomy in clinical routine often involves LA segmentation which is typically done by fusing several manual segmentations. This is the most accurate method nowadays but automatic segmentation is desirable in order to reduce the high workload and to eliminate inter- and intra-observer variability attributable to manual segmentation. However, accurate automatic

© Springer Nature Switzerland AG 2019
M. Pop et al. (Eds.): STACOM 2018, LNCS 11395, pp. 302–310, 2019.
https://doi.org/10.1007/978-3-030-12029-0_33

segmentation of the LA is challenging due to the high variability of atrial shapes: global features such as size and sphericity, and local features such as number, position and orientation of the pulmonary veins (PVs), and size and shape of the left atrial appendage (LAA) greatly differ between patients. Most of patients have 4 PVs (up to 70%) but it is relatively frequent to have a common left trunk (around 15 %) or one or even two extra right PVs [8]. The LAA size and shape could be even more arbitrary [5].

Some methods have been able to successfully segment the LA in an automatic or semi-automatic way using different image modalities [2,3,11,12]. LA automatic segmentation is especially challenging in LGE-MRI data due to the intrinsic characteristics of this image modality. LGE-MRI is used to visualize and quantify the extent of fibrosis (or scar) in the chambers of the heart. The contrast agent appears enhanced (white) in fibrotic areas while the myocardium is nulled in the rest facilitating the detection of scar but at the same time complicating the delineation of the cardiac substructures. In the case of the LA, the healthy (non-fibrotic) parts of the thin atrial wall exhibit a very similar intensity profile to the surrounding structures (e.g. the left ventricle).

Multi-atlas segmentation (MAS) has been widely used in different clinical contexts [6]. The outline of MAS involves registering N atlases (labeled images) to a given target image and derive the final segmentation by applying label fusion. Many methods have been proposed for the different steps (mainly registration and label fusion) and when the number of available atlases is increased the pipeline often involves atlas selection, i.e. selecting the most appropriate set of atlases and discarding the rest [10]. Ideally, atlas selection needs to be done before the registration step to avoid unnecessary computational costs of registering bad atlases that marginally contribute to the final segmentation [7]. On the other hand, this strategy is not optimal neither since registration quality should have a role on atlas selection (e.g. better registration quality for a given atlas). Most MAS schemes rely on registration to a common reference space were similarity metrics can be computed. However, when the population of images shows large variability in appearance and shape it becomes arduous to come up with a unique template that is representative of all the rest. As mentioned before, different LAs often do not even have the same parts (i.e. different number of PVs) and for example, a LA with an extra right PV will not be well represented using a template image with 4 PVs. Thus registration between LAs with varying topology becomes problematic.

In the absence of a suitable common reference space and to avoid a high number of pairwise registrations, we propose to group the training dataset by the corresponding LA morphology, determined from the LA shape. When a new target image has to be segmented, the images from the training set representative of each morphology cluster are registered to the target to identify the correct morphology of the target. Once the morphology is known, the segmentation is carried out via multi-atlas framework using only the atlases corresponding to this morphology. We tested our method using the dataset provided by the organizers

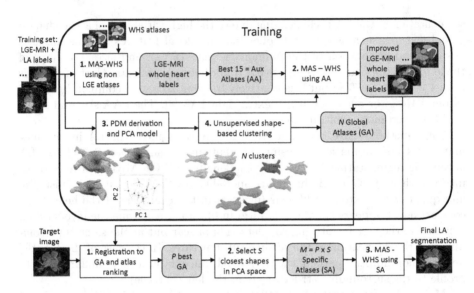

**Fig. 1.** Scheme of the proposed pipeline. The cluster representative is depicted enhanced. MAS-WHS = Multi-atlas whole heart segmentation; PDM = Point distribution model; PCA = Principal component analysis; AA = Aux atlases; GA = Global atlases; SA = Specific atlases.

of the 2018 Atrial Segmentation Challenge (STACOM'18[1]). It comprises 100 3D MRI patient data acquired using a clinical whole-body MRI scanner and the corresponding LA manual segmentations.

## 2   Methodology

The pipeline of the proposed method can be seen in Fig. 1. We built our method from the multi-atlas whole-heart segmentation (MAS-WHS) technique presented in [14] and [16] since the joint segmentation of surrounding structures may help in identifying the structure limits. Briefly, in the registration step of the MAS-WHS strategy all atlases were registered to the target image using a hierarchical scheme designed for cardiac images: first, a global affine registration was performed followed by a three-level locally affine registration (LARM) [15] to initialize substructure alignment, and by a non-rigid registration based on free-from deformations (FFD) [9] to refine local details. We considered seven substructures: left and right ventricular cavities, right atrial cavity, left atria (considering the cavity, the PVs and the LAA), the left ventricular myocardium, the ascending aorta and the pulmonary artery. For the label fusion step we applied atlas ranking based on conditional entropy [14] and a patch based strategy [16].

---

[1] http://atriaseg2018.cardiacatlas.org/.

To obtain labels of the remaining substructures (i.e. derive whole heart LGE-MRI atlases) we first performed MAS-WHS using the 30 high-resolution balanced steady state free precession (bSSFP) volumetric MRI atlases distributed by the Left Atrial Segmentation Challenge (STACOM'13) [12]. Due to the intrinsic differences between the two image modalities (LGE-MRI and bSSFP MRI for target and atlases, respectively) the automatic MAS-WHS results were poor for many cases. For that reason we re-segmented all cases applying again MAS-WHS using a new atlas set, Aux Atlas (AA), built selecting the best 15 results obtained in the initial step and replacing the LA automatic segmentation by the ground truth provided by the STACOM'18 challenge.

To group together images with similar LA shapes we first extracted all LA surfaces from the ground truth segmentations by applying the marching cubes algorithm. After that, we characterized each atrial mesh using a point distribution model (PDM) obtained by consistently sampling the LA surface meshes defining rays from the LA center of mass in all directions. We used spherical coordinates, $(r, \phi, \theta)$, to define the rays with $r = 100$, high enough to intersect the LA shape, $\phi \epsilon [0, \pi]$ with $\phi/70$ step and $\theta \epsilon [0, 2\pi]$ with a variable number of samples for each $\phi$ increasing from 1 to 70 and decreasing again to 1. We proceed like that to have more points in the areas where LA shapes are wider. In that way, the final number of points representing each shape was 2452. The LA shapes were then characterized by the intersection points between the rays and the surface. When the ray intersected more than once (because of the PVs) only the furthest intersecting point was considered. We computed the mean PDM and aligned all the PDMs to the mean using iterative closest point (ICP). Finally, we applied PCA to reduce the dimensionality of the data and clustered the shapes using the Affinity Propagation method [4]. This method clusters data by iteratively identifying a subset of representative examples and assigning the remaining elements to a specific representative when the method converges. Let us call the set of $N$ cluster representatives global-atlases (GA). Ideally, the GA set contains a good representation of highly variable atrial shapes.

To segment an unseen target image we proceeded as follows: we first registered the GA to the target image using the hierarchical registration scheme commented before [14]. We then ranked all atlases according to the negative conditional entropy of the target image $(I)$ given the warped label image $(L_n)$ which corresponds to:

$$-H(I|L_n) = \sum_{i \epsilon I, l \epsilon L_n} p(i, l) log \frac{p(i, l)}{p(l)} \qquad (1)$$

where H is entropy, $p(i, l)$ is the joint probability of the intensity value $i$ and label $l$, and $p(l)$ is the marginal probability of label $l$. We then selected the $P$ global atlases with highest negative conditional entropy and found the $S$ closest shapes in PCA space to each of the $P$ GA. The set of $M = P$ x $S$ atlases defined a target-specific set of atlases (SA) used to derive the final LA segmentation in a MAS framework.

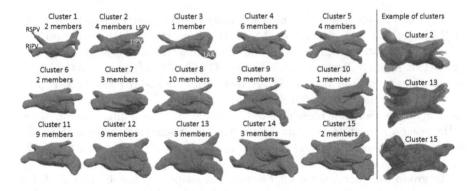

**Fig. 2.** All cluster representatives (left) and examples of small clusters showing all pertaining shapes superimposed (right). Representatives are sorted according to the first principal component that was identified as size/sphericity. RSPV = Right Superior PV, RIPV = Right Inferior PV, LSPV = Left Superior PV, LIPV = Left Inferior PV, LAA = Left Atrial Appendage.

## 3   Experiments and Results

We used data from the 2018 Atrial Segmentation Challenge (STACOM'18) to test our method. The dataset was randomly split into training and testing sets (80 and 20 cases, respectively) and, consequently, 80 LGE-MRI whole heart atlases were constructed. PDMs were built using 2452 rays, which was experimentally found appropriate to capture atrial anatomy, and a PCA model describing the LA shape variability in the training set was obtained. We decided to keep the 34 first principal components since they correspond to the 95% of cumulative variance. Fifteen representative shapes were identified by applying the Affinity Propagation clustering method and the GA was built using the corresponding atlases. The mean number of samples in each cluster was $5.33 \pm 3.94$ ranging from only 1 to 14. The 15 shape representatives can be seen in Fig. 2 as well as some of the clusters found. Our shape-based clustering method was able to identify a duplicated sample which was eliminated from the dataset. We decided to keep the $P = 3$ best GA as shape representatives and search for the $S = 5$ samples closest to each representative. We therefore constructed the SA with only 15 atlases.

To test segmentation accuracy we used three metrics: Dice similarity defined as

$$Dice(X, Y) = 2\frac{|X \cap Y|}{|X| + |Y|} \tag{2}$$

surface to surface distance, defined as

$$S2S(S, T) = mean(d(S, T), d(T, S)) \tag{3}$$

and Hausdorff distance defined as

$$HD(S, T) = max(d(S, T), d(T, S)) \tag{4}$$

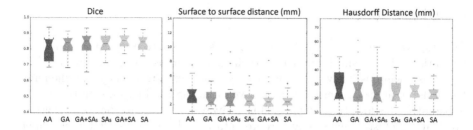

**Fig. 3.** From left to right: Dice score, surface to surface distance and Hausdorff distance. All distances are reported in mm. Results corresponding to the different sets of atlases considered are shown: AA = Aux Atlases, GA = Global Atlases, $SA_s$ = Specific Atlases found using the preliminary shape, SA = Specific Atlases selected using the proposed approach.

**Table 1.** LA segmentation results: Dice, surface to surface distance (S2S) and Hausdorff distance (HD) corresponding to the different experiments.

|      | $AA$ | $GA$ | $GA + SA_s$ | $SA_s$ | $GA + SA$ | $SA$ |
|------|------|------|-------------|--------|-----------|------|
| Dice | $0.807 \pm 0.076$ | $0.797 \pm 0.122$ | $0.819 \pm 0.094$ | $0.828 \pm 0.065$ | $0.838 \pm 0.075$ | $\mathbf{0.842\pm0.049}$ |
| S2S  | $3.675 \pm 1.892$ | $3.553 \pm 3.014$ | $3.209 \pm 2.231$ | $2.820 \pm 1.262$ | $2.843 \pm 1.768$ | $\mathbf{2.630\pm1.124}$ |
| HD   | $30.456 \pm 16.114$ | $27.442 \pm 12.831$ | $26.806 \pm 12.215$ | $24.704 \pm 8.383$ | $26.014 \pm 9.561$ | $\mathbf{24.041\pm8.141}$ |

where $X$ and $Y$ are the automatic and reference segmentation, respectively, $S$ and $T$ the set of voxels in the boundary of $X$ and $Y$, respectively, and $d$ is the Euclidean distance.

We compared our shape-based atlas selection approach with the performance obtained using other atlas sets. As baseline we used the results from the MAS-WHS performed using AA. We also derived a preliminary segmentation using the GA since those atlases were already fully registered to the target image in a previous step. We applied label fusion and obtained a preliminary LA segmentation whose related LA surface mesh was then sampled and projected to PCA space. The closest shapes were selected and a new set of target-specific atlases ($SA_s$) was established. We decided to keep the 15 closest shapes to have same atlas size as in our proposed approach. We investigated if including the highly variable (in terms of LA shape) global atlases, i.e. GA, was beneficial or disadvantageous.

As can be seen in Fig. 3 and Table 1 the best configuration corresponded to solely using the SA. The worse performance corresponded to the use of GA, and whenever the GA was used jointly with the specific atlases SA or $SA_s$ the results got worse than using only the latter.

## 4    Discussion

Results show that the highly variable GA was not able to produce accurate segmentations. This could be understood since if the atlas set is very diverse only few atlases will positively contribute to segment each target image, i.e. the number of useful atlases is actually small. Even if in the label fusion step the weight of bad atlases is low, they would still negatively contribute to the segmentation. We found 2 clusters with only 1 representative which means that their shape was so specific that could not be grouped together with other samples. Due to the relatively low accuracy of segmentations with GA, the preliminary shapes were inaccurate and quite similar between them. This was also revealed by the fact that after sampling, all corresponding PDMs were very similar, projections to PCA space fell close together and the $SA_s$ was (apart from the order) the same in most cases.

However, we found that registering GA to the target image and ranking the atlases according to the negative conditional entropy helped to position the target in PCA space and in that way infer shape features such as size or number of PVs. Then, for refinement, we decided to keep $P = 3$ best GA and look for similar shapes using Euclidean distance in PCA space. We decided to keep several cluster representatives to reduce the influence of atlas ranking errors and to relatively increase the shape variability of the final atlas set. Restricting ourselves to one cluster may increase the risk of failing if the atlas ranking or the shape-based clustering is not accurate. We could use other criteria to include similar shapes in SA such as all samples in the P clusters but we decided to use Euclidean distance in PCA space because of the different cluster size (from 1 to 14) and because we aimed to have the same number of atlases for all targets to have comparable computational time for all cases. We also based our decision of having a final set of 15 SA on the computational time. Using 3 representatives and 5 related atlases was experimentally found to provide the best compromise between accuracy and computational time but it still needs validation and would probably need to be adjusted depending on the training set (i.e. number of clusters found and samples in each cluster).

Most atlas selection methods require non rigid registration between the target image and the atlases. To reduce the computational cost of such high number of registrations some methods perform offline atlas to atlas registration in order to cluster similar atlases together. Compared to those methods our shape based atlas ranking reduces the number of non-rigid registrations required.

## 5    Conclusions

In this paper we have shown that atlas selection is an important step in MAS and we have proposed a two-step strategy to identify the most appropriate set of atlases for segmenting highly variable structures such as the LA. We showed that a simple sampling approach was able to capture LA morphology and could be used to automatically group together similar shapes. Shape information could be therefore used to discard bad atlases prior to registration avoiding a high number of computationally expensive registrations and improving segmentation results.

**Acknowledgments.** This study was partially funded by the Spanish Ministry of Economy and Competitiveness (DPI2015-71640-R) and by the "Fundació La Marató de TV3" (n° 20154031).

# References

1. Andrade, J., Khairy, P., Dobrev, D., Nattel, S.: The clinical profile and pathophysiology of atrial fibrillation: relationships among clinical features, epidemiology, and mechanisms. Circ. Res. **114**(9), 1453–1468 (2014)
2. Chen, J., et al.: Multiview two-task recursive attention model for left atrium and atrial scars segmentation. arXiv preprint arXiv:1806.04597 (2018)
3. Depa, M., Sabuncu, M.R., Holmvang, G., Nezafat, R., Schmidt, E.J., Golland, P.: Robust atlas-based segmentation of highly variable anatomy: left atrium segmentation. In: Camara, O., Pop, M., Rhode, K., Sermesant, M., Smith, N., Young, A. (eds.) STACOM 2010. LNCS, vol. 6364, pp. 85–94. Springer, Heidelberg (2010). https://doi.org/10.1007/978-3-642-15835-3_9
4. Frey, B.J., Dueck, D.: Clustering by passing messages between data points. Science **315**(5814), 972–976 (2007)
5. García-Isla, G., et al.: Sensitivity analysis of geometrical parameters to study haemodynamics and thrombus formation in the left atrial appendage. Int. J. Numer. Meth. Biomed. Eng. **34**, e3100 (2018)
6. Iglesias, J.E., Sabuncu, M.R.: Multi-atlas segmentation of biomedical images: a survey. Med. Image Anal. **24**(1), 205–219 (2015)
7. Langerak, T.R., Berendsen, F.F., Van der Heide, U.A., Kotte, A.N., Pluim, J.P.: Multiatlas-based segmentation with preregistration atlas selection. Med. Phys. **40**(9), 091701 (2013)
8. Prasanna, L., Praveena, R., D'Souza, A.S., Kumar, M.: Variations in the pulmonary venous ostium in the left atrium and its clinical importance. J. Clin. Diagn. Res. **8**(2), 10 (2014)
9. Rueckert, D., Lorenzo-Valdes, M., Chandrashekara, R., Sanchez-Ortiz, G., Mohiaddin, R.: Non-rigid registration of cardiac MR: application to motion modelling and atlas-based segmentation. In: Proceedings of 2002 IEEE International Symposium on Biomedical Imaging, pp. 481–484 (2002)
10. Sanroma, G., Wu, G., Gao, Y., Shen, D.: Learning-based atlas selection for multiple-atlas segmentation. In: Proceedings of the IEEE Conference on Computer Vision and Pattern Recognition, pp. 3111–3117 (2014)
11. Tao, Q., Ipek, E.G., Shahzad, R., Berendsen, F.F., Nazarian, S., van der Geest, R.J.: Fully automatic segmentation of left atrium and pulmonary veins in late gadolinium-enhanced MRI: towards objective atrial scar assessment. J. Magn. Reson. Imaging **44**(2), 346–354 (2016)
12. Tobon-Gomez, C., et al.: Benchmark for algorithms segmenting the left atrium from 3D CT and MRI datasets. IEEE Trans. Med. Imaging **34**(7), 1460–1473 (2015)
13. Vos, T., et al.: Global, regional, and national incidence, prevalence, and years lived with disability for 310 diseases and injuries, 1990–2015: a systematic analysis for the global burden of disease study 2015. Lancet **388**(10053), 1545–1602 (2016)

14. Zhuang, X., et al.: Multiatlas whole heart segmentation of CT data using conditional entropy for atlas ranking and selection. Med. Phys. **42**(7), 3822–3833 (2015)
15. Zhuang, X., Rhode, K.S., Razavi, R.S., Hawkes, D.J., Ourselin, S.: A registration-based propagation framework for automatic whole heart segmentation of cardiac MRI. IEEE Trans. Med. Imaging **29**(9), 1612–1625 (2010)
16. Zhuang, X., Shen, J.: Multi-scale patch and multi-modality atlases for whole heart segmentation of MRI. Med. Image Anal. **31**, 77–87 (2016)

# Deep Learning Based Method for Left Atrial Segmentation in GE-MRI

Yashu Liu, Yangyang Dai, Cong Yan, and Kuanquan Wang[✉]

Harbin Institute of Technology, Harbin 150001, China
{yashuliu,yangyangdai815,yanc}@stu.hit.edu.cn, wangkq@hit.edu.cn

**Abstract.** Understanding the anatomical structure of left atrial (LA) is crucial for clinical treatment of atrial fibrillation (AF). Gadolinium Enhanced Magnetic Resonance Imaging (GE-MRI) provides clarity images of LA structure. However, the most of LA structure analysis on GE-MRI studies are based on subjective manual segmentation. An efficient and objective segmentation method in GE-MRI is highly demanded. Although deep learning based method has achieved great success on some medical image segmentations, solving LA segmentation through deep learning is still an unsatisfied field. In this paper, we handle this unmet clinical need by exploring two convolutional neural networks (CNNs) structures, fully convolutional network (FCN) and U-Net, to improve the accuracy and efficiency of LA segmentation. Both models were trained and evaluated on GE-MRI dataset provided by 2018 atrial segmentation challenge. The results show that FCN-based LA automatic segmentation method achieves Dice score over 82%; U-Net method achieves Dice score over 83%.

**Keywords:** Left atrial segmentation · Deep learning · GE-MRI

## 1 Introduction

Atrial fibrillation (AF) is a common condition of heart arrhythmia, which can lead to blood clots, stroke, heart failure and other heart-related diseases. It is associated with structural remodeling, including fibrotic changes in the left atrial (LA) and the prevalent risk increases along with the age of the population [1–3]. Catheter ablation treatment aims to cure AF by electrically isolating the pulmonary veins. However, the success rate for a single catheter ablation procedure is just 30–50% at 5-year follow-up and multiple ablations are frequently required [4]. Accurate knowledge of LA's anatomical structure can guide the catheter ablation procedure and improve the single success rate [1,5]. Gadolinium-Enhanced Magnetic Resonance Imaging (GE-MRI) is a proven noninvasion technology to obtain clarity images of the inner structure and has broad application on studying the extent of fibrosis across the LA. However, LA segmentation challenge in GE-MRI is the low contrast between the LA and background due to the canceling of healthy tissue signal. Most existing research on LA

© Springer Nature Switzerland AG 2019
M. Pop et al. (Eds.): STACOM 2018, LNCS 11395, pp. 311–318, 2019.
https://doi.org/10.1007/978-3-030-12029-0_34

structure based on GE-MRI is subjective manual segmentation [6,7]. An effective and objective LA segmentation method in GE-MRI is heavily demanded.

Deep learning method constructs semantic information through sliding windows on appearance features. It can nearly fit any polynomial functions and achieves a great success in computer vision field. Recently deep learning based methods have been widely used in medical image segmentation. While, the fields of application is mostly in lung, brain, left ventricle (LV) and other organs [8–10]. For LA segmentation, there are rarely methods based on deep learning. [11] proposed a combined FCN and active shape model (ASM) method for LA and LAA segmentation in CT volumes. They trained a FCN to infer LA structure and an ASM model of LV to determine the fuzzy boundary between LA and LV. CardiacNET [5] was a multi-view CNN for LA and proximal pulmonary veins segmentation from MRI. They trained three similar CNNs in three views of LA to infer the LA structure and fused them with an adaptive fusion strategy. This is the first multi-view CNN for LA segmentation with our knowledge. [12] proposed an automatic two-task recursive attention model to segment LA and delineate the atrial scars simultaneously on LGE-MRI.

For further exploring the potential of deep learning on handling LA segmentation task, we researched the LA segmentation ability on GE-MRI of two CNN architectures, FCN [13] and U-Net [14]. FCN is widely used for segmentation and achieves a great success with its skip architecture. U-Net is an extension of FCN and the first CNN architecture for medical image segmentation. In this paper, we constructed a FCN model and U-Net model based on [13] and [14]. For minimizing the human intervention, we just center cropped the image into $300 \times 300$ due to limited computer memory. Both models were trained in one-stage and without transfer learning. Compared to the former complex architectures our models are more concise and efficient.

## 2   Method

The work-flow of our method is shown in Fig. 1, taking FCN as an example. We first separated the 3D volumes into 2D slices. Then the 2D slices were center cropped into $300 \times 300$ as the input of the model. The cropped size was statistical from the training data. The model was trained and tested on the cropped slices. Finally, the results were padding into the original size and compared with the ground truth.

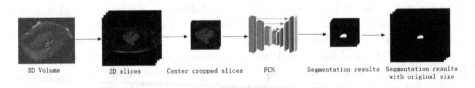

3D Volume     2D slices     Center cropped slices     FCN     Segmentation results     Segmentation results with original size

**Fig. 1.** Work-flow of our method.

FCN's excellent contribution is the skip architecture, which combines semantic information from a deep, coarse layer with appearance information from a shallow, fine layer to produce accurate and detailed segmentations [13]. In our FCN model, we used three skip architectures to maximum use the shallow appearance information. For tackling the data limited problem, U-Net extends FCN and concatenates the entire shallow feature maps to the upsampling layers correspondingly. This modification made it like a symmetric u-shape and obtained a precise segmentation result with a few training images. While, it has more parameters than FCN. Our U-Net architecture is based on [14]. We replaced its loss function by dice_coef_loss [15]. More details of both models are introduced in the following.

## 2.1  Network Architectures

Basic components in the two architectures are identical in order to compare them equitably. Following parameters' setting is the default set. Kernel size of the convolutional layers is $3 \times 3$ except the prediction layer. Prediction layer's kernel size is $1 \times 1$. Every convolutional layer is followed by rectified linear unit (ReLU) and mean-variance normalization (MVN) for accelerating model convergence and avoiding gradient diffusion. Max-pooling layer with stride 2 reduces the image dimensions by half and enlarges the receptive field. Dropout layer with drop rate improves the model's generalization. Up-sampling layer enlarges the feature map into target size. We adopted the polynomial decay, which is defined in Eq. 1, as our learning rate decay strategy. In Eq. 1, init_lr is the initialized learning rate, iter is the current iteration number, max_iter is the maximum number of iteration during training and power is the decay rate during each iteration. We set the power as 0.5 in both networks. The other parameters in the equation were depended on each network. Both networks were implemented in Keras framework on Ubuntu 16.04 at NVIDIA Titan X GPU and initialized randomly without transfer learning.

$$iter\_lr = init\_lr \times (1 - \frac{iter}{max\_iter})^{power} \tag{1}$$

**FCN.** FCN has three versions based on the multiples of upsampling layer, FCN32, FCN16 and FCN8 from coarse to fine. The smaller upsampling multiples will get the finer segmentation result due to more shallow features are "skipped" into the deep layer. Our model is FCN8-based architecture in order to obtain a fine segmentation result. The proposed FCN architecture is shown in Fig. 2. It has 19 convolutional layers, three max-pooling layers, two dropout layers with drop rate 0.5, and three upsampling layers. We utilized Adamax optimizer to minimize the binary cross entropy loss function. Merge process is the skip architecture, and implemented by averaging the former shallow layer and deep layer's feature maps. The model was trained 80 epochs with mini-batch 4 and initial learning rate 0.0001.

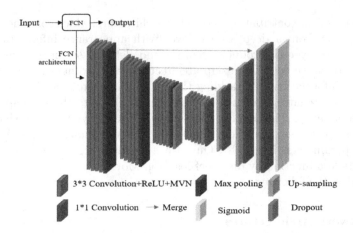

**Fig. 2.** FCN architecture: input is the cropped images and labels; output is a binary segmentation result of LA. At the bottom is the description of the basic components.

**U-Net.** Original U-Net is proposed for cell segmentation. The difficulty in cell segmentation is the blurry boundary between the cells. However, except for the blurry boundary problem, LA segmentation still has the imbalance problem of positive and negative samples. Background occupies most of the LA image. If we leave this problem, the model will not learn the sufficient useful features. To tackle this imbalance problem, we modified the original U-Net by utilizing the dice_coef_loss.

Dice_coef_loss is based on the dice score. Dice score and the dice_coef_loss are defined in the following,

$$Dice = \frac{2|X \bigcap Y|}{|X| + |Y|} \tag{2}$$

$$dice\_coef\_loss = 1 - \frac{2|X \bigcap Y| + smooth}{|X| + |Y| + smooth} \tag{3}$$

where X, Y corresponding to the ground truth and prediction map, smooth is the smoothing coefficient which is a positive number to speed up model convergence. And we set it as 10. Dice score calculates the overlap between ground truth and prediction result. Dice_coef_loss is the opposite of dice score, minimizes the loss function will maximum the dice score. The smoothing coefficient will enlarge the dice score, result in accelerating the model convergence. We also introduced MVN layer after ReLU to further accelerate the convergence. Figure 3 is the U-Net architecture. It has 18 convolutional layers, four max-pooling layers, two dropout layers with drop rate 0.2, and four upsampling layers. We utilized Adamax optimizer to minimize the dice_coef_loss function. Merge process implemented by concatenating the former shallow layer and deep layer's feature maps. The model was trained 40 epochs with mini-batch 2 and initial learning rate 0.001.

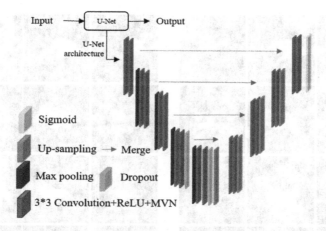

**Fig. 3.** U-Net architecture, the description of each basic component is shown in the left bottom of the network.

## 3 Experiment

We utilized 100 volumes 3D GE-MRI, which are provided by 2018 Atrial Segmentation Challenge, training and evaluating our models. We divided them into 95 training volumes and 5 test volumes randomly. Dimensions of each volume are dependent on the patient, but all volumes have 88 sequences in the Z axis. We separated the 3D volumes into 2D slices along Z axis and then center cropped these slices into $300 \times 300$. Selected 10% of training slices as validation set. A horizontal flip, vertical flip and 180° rotation augmentation process was used during training stage. Ultimately we got 22,572 training slices, 836 validation slices and 440 test slices.

Parameters setting were mentioned in Sect. 2.1. Both models were evaluated on the 440 test images from the 5 test volumes. Segmentation results of our models are shown in Fig. 4. These sub-pictures are cropped surround the heart in order to show the result clearly. FCN results display in the first three columns, the last three columns are the U-Net results. Each row represents an individual test volume. From left to right are three segmented slices of each test volume. The three slices represent the LA bottom, LA middle and LA apex, respectively. Red, green and yellow correspondingly represent the ground truth, prediction result and coincide region. We can see that, FCN prediction results have a better performance on inferring the LA boundary. But it cannot predict the LA bottom and apex perfectly (e.g. the first and third columns in Fig. 4). U-Net has a worse performance than FCN. It can infer most of the LA position correctly but tend to leakage into near structures.

Figure 5 illustrates the dice scores' distribution of the five test volumes. Blue and orange are correspondingly to the FCN and U-Net's results. The medians of individual dice ($D_i$, $i$ is the patient number) of FCN are as follows: $D_1 = 0.93, D_2 = 0.96, D_3 = 0.96, D_4 = 0.94, D_5 = 0.96$. And the medians of individual

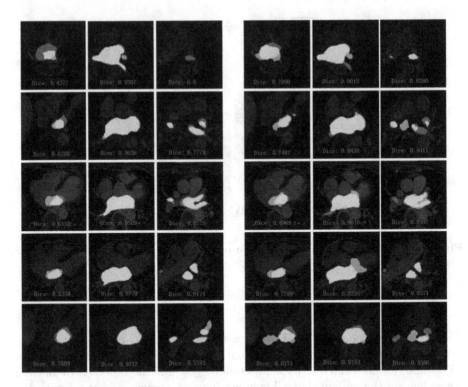

**Fig. 4.** Segmentation results of FCN and U-Net. FCN results are shown in the first three columns, the last three columns are the U-Net results. From top to bottom is three segmented slices of the five test volumes. The red text is the dice score of each image. Color representation: yellow is the coincide region, red is the ground truth, green is the prediction result. (Color figure online)

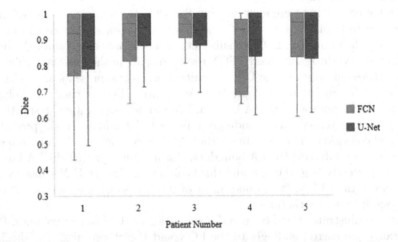

**Fig. 5.** Dice score on each volume of FCN (blue) and U-Net (orange). (Color figure online)

dice of U-Net are as follows: $D_1 = 0.87, D_2 = 0.95, D_3 = 0.93, D_4 = 0.89, D_5 = 0.93$. Mean dice score of the five test volumes is 0.8204 in FCN and 0.8307 in U-Net. Figure 5 represents that FCN got a wider range dice distribution than U-Net. However, FCN got a higher median in the all test volumes and a lower mean dice score. This distribution is consistent with Fig. 4 and LA images' distribution in the volume. LA middle occupies the most of LA images in a volume. And FCN prefers a perfect performance on inferring the LA middle with some inaccurate predictions of LA bottom and apex. These properties result in the LA middle in FCN with a higher dice and the other parts with a lower dice score. On the contrary, U-Net is expert in recalling the most parts of LA but it tends to leakage into near structures. Leading to the dice distribution in U-Net is more compact than FCN. Our FCN model was tested by the challenge test set which contains 54 volumes without public ground truth. The final score is 0.903, and ranks 6th out of 18 conference attending teams. The score is a little higher than our expect. We thought this is because the different calculation method. For a non-object slice, we set the score is zero if the model predicts a object. But there still a score for the background class in the organizer's calculate method. This is a different balance of accuracy and recall rate.

## 4   Conclusion

Most of previous LA segmentation methods are based on traditional methods. There are only few deep learning based methods for LA segmentation with different image modalities. LA segmentation in GE-MRI still a time consuming and tedious task due to its low contrast. In this paper, we evaluated two deep learning architectures, FCN and U-Net, to tackle LA segmentation problem in GE-MRI. In FCN model, we adopted a three skip architecture to infer LA and obtained a 0.8204 mean dice score. In U-Net model, we modified the loss function with dice_coef_loss due to the imbalance positive and negative samples and add MVN layer to speed up model convergence. We got a 0.8307 mean dice score in U-Net model. Segmentation result indicates that FCN and U-Net infer LA structure correctly. FCN can directly predict the LA without leakage but not good at LA bottom structure prediction. Though U-Net sometimes leakage into near structure, it can recall the most of LA parts. Our future work will further improve performance of both models with more tricks and explore more architectures.

**Acknowledgment.** This work was supported by the National Natural Science Foundation of China, grant no. 61571165. Thanks to Jichao Zhao and Zhaohan Xiong who organized the 2018 atrial segmentation challenge.

# References

1. Yang, G., et al.: Fully automatic segmentation and objective assessment of atrial scars for longstanding persistent atrial fibrillation patients using late gadoliniu-menhanced MRI. Med. Phys. **45**(4), 1562–1576 (2018)
2. McGann, C., Akoum, N., et al.: Atrial fibrillation ablation outcome is predicted by left atrial remodeling on MRI. Circ.: Arrhythm. Electrophysiol. **7**, 23–30 (2014)
3. Kuppahally, S.S., et al.: Left atrial strain and strain rate in patients with paroxysmal and persistent atrial fibrillation relationship to left atrial structural remodeling detected by delayed-enhancement MRI. Circ.: Arrhythm. Electrophysiol. **3**, 231–239 (2010)
4. Ouyang, F., et al.: Long-term results of catheter ablation in paroxysmal atrial fibrillation lessons from a 5-year follow-up. Circulation **122**, 2368–2377 (2010)
5. Mortazi, A., Karim, R., Rhode, K., Burt, J., Bagci, U.: *CardiacNET*: segmentation of left atrium and proximal pulmonary veins from MRI using multi-view CNN. In: Descoteaux, M., Maier-Hein, L., Franz, A., Jannin, P., Collins, D.L., Duchesne, S. (eds.) MICCAI 2017. LNCS, vol. 10434, pp. 377–385. Springer, Cham (2017). https://doi.org/10.1007/978-3-319-66185-8_43
6. Karim, R., Housden, R.J., et al.: Evaluation of current algorithms for segmentation of scar tissue from late gadolinium enhancement cardiovascular magnetic resonance of the left atrium: an open-access grand challenge. J. Cardiovasc. Magn. Reson. **15**(1), 105–122 (2013)
7. McGann, C., et al.: Atrial fibrillation ablation outcome is predicted by left atrial remodeling on MRI. Circ. Arrhythm. Electrophysiol. **7**, 23–30 (2014)
8. Setio, A., Ciompi, F., et al.: Pulmonary nodule detection in CT images: false positive reduction using multi-view convolutional networks. IEEE Trans. Med. Imaging **35**, 1160–1169 (2016)
9. Zhao, X., et al.: A deep learning model integrating FCNNs and CRFs for brain tumor segmentation. Med. Image Anal. **43**, 98–111 (2018)
10. Avendi, M.R., Kheradvar, A., Jafarkhani, H.: A combined deep-learning and deformable-model approach to fully automatic segmentation of the left ventricle in cardiac MRI. Med. Image Anal. **30**, 101–119 (2016)
11. Liu, H., Feng, J., Feng, Z., Lu, J., Zhou, J.: Left atrium segmentation in CT volumes with fully convolutional networks. In: Cardoso, M.J., et al. (eds.) DLMIA/ML-CDS -2017. LNCS, vol. 10553, pp. 39–46. Springer, Cham (2017). https://doi.org/10.1007/978-3-319-67558-9_5
12. Chen, J., Yang, G., et al.: Multiview two-task recursive attention model for left atrium and atrial scars segmentation. arXiv:1806.04597, (2018)
13. Shelhamer, E., Long, J., Darrell, T.: Fully convolutional networks for semantic segmentation. IEEE Trans. Pattern Anal. Mach. Intell. **39**(4), 640–651 (2016)
14. Ronneberger, O., Fischer, P., Brox, T.: U-Net: convolutional networks for biomedical image segmentation. In: Navab, N., Hornegger, J., Wells, W.M., Frangi, A.F. (eds.) MICCAI 2015. LNCS, vol. 9351, pp. 234–241. Springer, Cham (2015). https://doi.org/10.1007/978-3-319-24574-4_28
15. Milletari, F., Navab, N., Ahmadi, S.: V-Net: fully convolutional neural networks for volumetric medical image segmentation. In: Fourth International Conference on 3D Vision, pp. 565–571 (2016). https://doi.org/10.1109/3DV.2016.79

# Dilated Convolutions in Neural Networks for Left Atrial Segmentation in 3D Gadolinium Enhanced-MRI

Sulaiman Vesal[✉], Nishant Ravikumar, and Andreas Maier

Pattern Recognition Lab, Friedrich-Alexander-Universität Erlangen-Nürnberg,
Erlangen, Germany
sulaiman.vesal@fau.de

**Abstract.** Segmentation of the left atrial chamber and assessing its morphology, are essential for improving our understanding of atrial fibrillation, the most common type of cardiac arrhythmia. Automation of this process in 3D gadolinium enhanced-MRI (GE-MRI) data is desirable, as manual delineation is time-consuming, challenging and observer-dependent. Recently, deep convolutional neural networks (CNNs) have gained tremendous traction and achieved state-of-the-art results in medical image segmentation. However, it is difficult to incorporate local and global information without using contracting (pooling) layers, which in turn reduces segmentation accuracy for smaller structures. In this paper, we propose a 3D CNN for volumetric segmentation of the left atrial chamber in LGE-MRI. Our network is based on the well known U-Net architecture. We employ a 3D fully convolutional network, with dilated convolutions in the lowest level of the network, and residual connections between encoder blocks to incorporate local and global knowledge. The results show that including global context through the use of dilated convolutions, helps in domain adaptation, and the overall segmentation accuracy is improved in comparison to a 3D U-Net.

## 1 Introduction

Atrial fibrillation (AF) is the most common type of cardiac arrhythmia, with higher rates of incidence among aging populations [1]. AF is caused by impaired electrical activity within the atria, which causes cardiac muscle fibers to contract rapidly, in an irregular fashion. The poor performance of current AF treatment strategies results from a lack of adequate understanding of the electrical and structural remodeling characteristics of the human atria [2,3]. AF is associated with structural remodeling in the form of fibrosis/scars within the LA and an associated reduction in myocardial voltage. The degree of reduction in voltage and the extent of scar tissue, are indicative of the severity of the pathology [4].

Current clinical protocol for assessing the left atrium (LA) is electro-anatomical mapping, performed during an electro physiological study. However, this is an invasive technique which uses ionizing radiation, with sub-optimal accuracy, resulting in large localization errors for LA tissue. Gadolinium-enhanced

© Springer Nature Switzerland AG 2019
M. Pop et al. (Eds.): STACOM 2018, LNCS 11395, pp. 319–328, 2019.
https://doi.org/10.1007/978-3-030-12029-0_35

magnetic resonance imaging (GE-MRI) has been shown to improve the visibility of a patient's internal structures, such as the left atria. Consequently, GE-MRI has been widely adopted in recent years, to assess the extent of scar tissue/fibrosis within the LA wall, as a result of AF. While histological analyses have confirmed scar tissue within the LA to be localized to regions of low voltage, quantification of the extent and precise location of the same in a clinical setting, requires invasive techniques. Consequently, the effect of such structural remodeling of cardiac tissue on patient outcome following treatment, and the risk of AF recurrence, are yet to be well understood [4]. Accurate segmentation of the LA is thus of considerable clinical interest, for diagnosis, treatment planning and patient prognosis. However, manual segmentation of the LA in 3D for structural analysis, is very time consuming, susceptible to manual errors, and is subject to inter-rater differences. Consequently, an automatic approach to LA segmentation is imperative for improved diagnosis and clinical decision making.

Segmentation of the left atrium (LA) in GE-MRI images is a challenging task. Firstly, the LA is hard to distinguish even for expert cardiologists specialized in cardiac MRI. Secondly, respiratory motion, heart rate variability, low signal-to-noise ratio (SNR), and contrast agent wash-out during the long acquisition times, frequently result in poor image quality. [5] addressed this challenge using two segmentation steps, namely; an initial global segmentation step using multi-atlas registration; and a subsequent local refinement step based on 3D level-set propagation. However, more recently, convolutional neural networks (CNNs) have demonstrated state-of-the-art performance in various medical image segmentation tasks. [6] introduced the U-Net architecture, which is trained end-to-end to process the entire image and perform a pixel-wise classification. The V-Net [7] architecture, a 3D extension to the U-net, was introduced for volumetric images and enables 3D segmentation, as opposed to processing volumes slice-by-slice. A multi-view CNN was proposed in [8], with an adaptive fusion strategy to segment the LA and proximal pulmonary veins in MRI. While, in [9] the LA was segmented in computed tomographic angiography (CTA) data, using fully convolutional neural networks (FCNs) with 3D conditional random fields (CRFs), achieving state-of-the-art segmentation accuracy. Other studies [10] have also investigated automatic segmentation of the left ventricle, in GE-MRI, to assess the extent of scarring resulting from myocardial infarction.

In this paper, we propose a fully automatic 3D segmentation approach for the LA, in AF patients scanned using GE-MRI. Accurate organ segmentation requires incorporation of both local and global information. For this purpose, we constructed a modified version of the 3D U-Net, using dilated convolutions [11] in the lowest layer of the encoder-branch, to extract features spanning a wider spatial range. Additionally, we added residual connections between convolution layers in the encoder branch of the network, to better incorporate multi-scale image information and ensure smoother flow of gradients in the backward pass. A schematic of the proposed network is presented in Fig. 1. The main contribution of this study is a modified 3D U-Net based segmentation approach (henceforth referred to as 3D U-Net + DR), that incorporates global context via dilated convolutions in the deepest part of the network. We evaluated the LA chamber segmentation accuracy of our network using a data set provided as part of the STACOM MICCAI 2018 challenge, and compared it with a 3D U-Net.

## 2 Methodology

Incorporating both local and global information is beneficial for many segmentation tasks. In a conventional U-Net however, the lowest level of the network has a small receptive field which prevents the network from extracting features that capture non-local information. Hence, the network would have no understanding that there is only one left atrial cavity within the image, leading to misclassifications of areas with similar appearance. Dilated convolutions [11] provide a suitable solution to this problem. They introduce an additional parameter, called the dilation rate, to convolution layers, which defines the spacing between values in a kernel. This helps dilate the kernel such that a $3 \times 3 \times 3$ kernel with a dilation rate of 2 will have a receptive field size equal to that of a $7 \times 7 \times 7$ kernel. Additionally, this is achieved without any increase in complexity, as the number of parameters associated with the kernel remains the same.

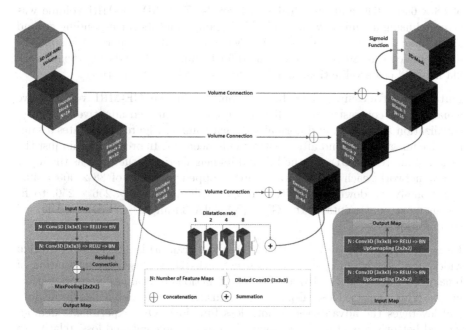

**Fig. 1.** 3D U-Net + DR network comprising residual connections in the encoder branch and a summation of 4 dilated convolutions layers (with various dilation rates) in the bottleneck. (Color figure online)

The proposed 3D U-Net + DR network (refer to Fig. 1) comprises three down-sampling and upsampling convolution blocks within the encoder and decoder branches, respectively. We use two convolutions with a kernel size of $3 \times 3 \times 3$

voxels per block, with batch normalization, rectified linear units (ReLUs) as activation functions, and a subsequent 3D-max pooling operation, as illustrated in Fig. 1. Image dimensions are preserved between the encoder-decoder branches following convolutions, by zero-padding the estimated feature maps. This enabled corresponding feature maps to be concatenated between the branches. A sigomid activation function was used in the last layer to produce a value between 0 and 1, to distinguish the background from the foreground. Furthermore, to improve the flow of gradients in the backward pass of the network, the conventional convolution layers in the encoder branch were replaced with residual convolution layers. In each encoder-convolution block, the input to the first convolution layer is concatenated with the output of second convolution layer (red line in Fig. 1), and the subsequent 3D max-pooling layer reduces volume dimensions by half. The bottleneck between the branches employs four dilated convolutions, with dilation rates $1 - 4$. The outputs of each were subsequently summed up and provided as input to the decoder branch.

**Data Acquisition:** A total of 100 3D GE-MRIs from patients with AF were provided as part of the STACOM 2018 challenge, for atrial segmentation[1]. The resolution of the provided data is $0.625 \times 0.625 \times 0.625$ mm$^3$, with dimensions of $88 \times 640 \times 640$ and $88 \times 576 \times 576$ voxels. Each 3D GE-MRI volume was acquired using a clinical whole-body MRI scanner and its corresponding ground truth binary mask for the LA cavity, was annotated by experts. We split the data set such that 80 volumes were used for training and validating the network via five-fold cross validation, and 20 volumes were used for testing.

**Data Pre-processing:** Due to low contrast in the GE-MRI volumes, we enhanced the contrast slice-by-slice, using contrast limited adaptive histogram equalization (CLAHE), and normalized each volume. Figure 2 illustrates a sample slice image before and after contrast enhancement. In order to retain just the region of interest (ROI), i.e. the LA and its neighboring structures, as the input to our network, each volume was center cropped to a size of $88 \times 400 \times 400$. Additionally we downsampled the cropped volumes to $80 \times 256 \times 256$, to fit the memory constraints of the GPU (NVIDIA Titan XP 12 GB) used for all computations.

**Loss Function:** The dice coefficient (DC) loss (Eq. 1a) is a measure of overlap widely used for training segmentation networks [7]. We used a combination of binary cross entropy (Eq. 1b) and DC loss functions to evaluate network performance. This combined loss (Eq. 1c) is less sensitive to class imbalance in the data and leverages the advantages of both loss functions. Our experiments demonstrated better segmentation accuracy when using the combined loss, relative to employing either individually.

$$\zeta_{dc}(y, \hat{y}) = 1 - \sum_k \frac{\sum_n y_{nk}\hat{y}_{nk}}{\sum_n y_{nk} + \sum_n \hat{y}_{nk}} \tag{1a}$$

---

[1] http://atriaseg2018.cardiacatlas.org/.

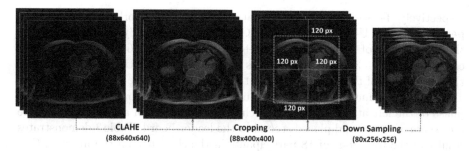

**Fig. 2.** Pre-processing workflow: (1) contrast in GE-MRI volumes is enhanced using CLAHE; (2) resulting volumes are cropped to retain just the ROI; and eventually the cropped volume is down sampled.

$$\zeta_{bce}(y, \hat{y}) = -\sum_{k}[\hat{y}_{nk}log(\hat{y}_{nk}) + (1 - y_{nk})(1 - \hat{y}_{nk})] \tag{1b}$$

$$\zeta(y, \hat{y}) = \zeta_{dc}(y, \hat{y}) + \zeta_{bce}(y, \hat{y}) \tag{1c}$$

In Eqs. (1a–1b) $\hat{y}_{nk}$ denotes the output of the model, where $n$ represents the pixels and $k$ denotes the classes. The ground truth labels are denoted by $y_{nk}$ and we use the two-class version of the DC loss $\zeta_{dc}(y, \hat{y})$ proposed in [7].

## 3 Results and Discussion

The challenge data set was split into 80% for training and validation and 20% for testing. We further split the former into 64 volumes for training and 16 volumes for validation, within a five-fold cross-validation scheme. Networks trained in all experiments were evaluated using the held out test data (20 volumes), and employed the Adam optimizer with a learning rate of 0.001 and 150 epochs.

Three evaluation metrics were used namely: Dice Similarity Coefficient (DC), Jaccard Index (JI) and Accuracy (AC) for the final evaluation of our network. DC and JI measure the degree of overlap between the estimated segmentations and the ground truth masks, while AC measures the proportion of correctly classified pixels. Jaccard is numerically more sensitive to mismatch when there is reasonably strong overlap than DC and AC. The average DC, JI, and AC measures achieved by both 3D U-Net and 3D U-Net + DR across five-fold cross validation experiments, using different data normalization techniques, are summarized in Tables 1 and 2, respectively. The DC scores achieved by the 3D U-Net with min/max normalization for the validation and test sets were 81.8% and 81.8%, respectively. Our approach, on the other hand, achieved DC scores of 85.8% and 84.8% for the validation and test sets, respectively. Similarly, we also trained both models with mean and standard deviation-based normalization, which drastically improved the segmentation accuracy. The DC scores achieved by 3D U-Net with mean/std normalization for the validation and test sets were 87.9% and 89.8%, while our approach achieved DC scores of 90.2% and 90.1%,

respectively. These results indicate that the proposed network significantly outperforms the 3D U-Net, in terms of DC and JI. Additionally, as summarized in
the first column of Table 1, 3D U-Net + DR requires just 2.6 million parameters
compared to the 3.3 million parameters required by the 3D U-Net. This is due to
the summation of the feature maps in the lowest level of the network. Additionally, we also evaluated our model on unseen test data, provided in the second
phase of the left atrial segmentation challenge. Our model achieved an average
dice score of 92.6% and placed second in the leader-board. Table 3 demonstrates
results of the top 6 out of 18 participants in the challenge on the unseen LGE-
MRI dataset.

**Table 1.** Segmentation accuracy evaluated in terms of DC, JI and AC metrics, for 3D
U-Net and 3D U-Net + DR with min/max normalization.

| Methods | Train data | | | Validation data | | | Test data | | |
|---|---|---|---|---|---|---|---|---|---|
| | AC | DC | JI | AC | DC | JI | AC | DC | JI |
| 3D U-Net (3.3 M) | 0.805 | 0.896 | 0.844 | 0.805 | 0.818 | 0.728 | 0.808 | 0.818 | 0.740 |
| 3D U-Net + DR (2.6 M) | 0.806 | **0.923** | **0.873** | 0.806 | **0.858** | **0.768** | 0.808 | **0.848** | **0.770** |

**Table 2.** Segmentation accuracy evaluated in terms of DC, JI and AC metrics, for 3D
U-Net and 3D U-Net + DR with mean/std subtraction.

| Methods | Train data | | | Validation data | | | Test Data | | |
|---|---|---|---|---|---|---|---|---|---|
| | AC | DC | JI | AC | DC | JI | AC | DC | JI |
| 3D U-Net (3.3 M) | 0.807 | 0.968 | 0.943 | 0.806 | 0.879 | 0.797 | 0.804 | 0.898 | 0.819 |
| 3D U-Net + DR (2.6 M) | 0.807 | **0.975** | **0.951** | 0.807 | **0.902** | **0.826** | 0.803 | **0.901** | **0.825** |

**Table 3.** The final rankings based on overall dice coefficient score for the top 6 out of
18 participants on 54 unseen LGE-MRI test dataset.

| Rank | Methods | Dice score |
|---|---|---|
| 1 | Xia et al. [12] | 0.932 |
| 2 | Bian et al. [13] | 0.926 |
| **2** | **U-Net + DR (ours)** | **0.926** |
| 3 | Li et al. [14] | 0.923 |
| 3 | Puybareau et al. [15] | 0.923 |
| 3 | Yang et al. [16] | 0.923 |

The images presented in Fig. 3 help visually assess the segmentation quality
of the proposed method on three test volumes. Here, the green contour represents
the ground truth segmentation, while the red and blue contours represent the

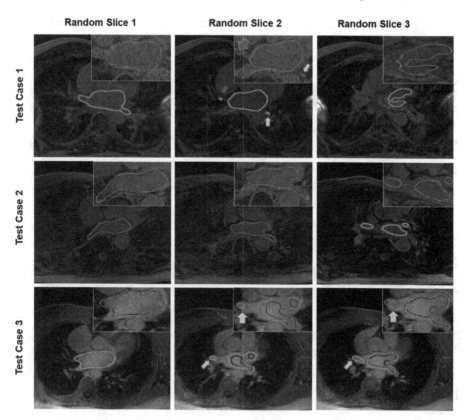

**Fig. 3.** Each row shows 3 GE-MRI slices chosen at random from three patients in the test data set. Their corresponding segmentation contours are overlaid: Green represents the ground-truth, while the red and blue contours are estimated by 3D U-Net+DR and 3D U-Net, respectively. The red arrow in the top row, middle slice highlights the true positive tissue areas missed by both methods. The yellow arrows highlight false negatives of 3D U-Net that are captured by 3D U-Net + DR. (Color figure online)

outputs of our method, and 3D U-Net, respectively. These images indicate that the proposed network produces more accurate segmentations, with fewer false negatives than the 3D U-Net. The red arrows in the top row, middle slice highlight atrial regions that both methods missed, while the yellow arrows outline regions where our method succeeded but the 3D U-Net did not. These false negatives of the 3D U-Net were confirmed by visualizing the surface meshes of the estimated segmentations and the ground truth masks for these patient volumes, depicted in Fig. 4. 'Test Case 3' presented in the bottom row of Fig. 4, clearly highlights the advantage offered by the use of dilated convolutions in the bottleneck as with our approach, relative to the 3D U-Net. The additional global context imbued within the former, aids in capturing the LA more accurately than the latter.

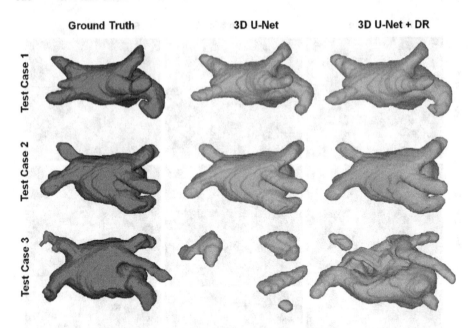

**Fig. 4.** 3D surface visualization for the ground-truth and the output generated by 3D U-Net and 3D U-Net + DR respectively.

## 4    Conclusion

In this study we proposed a 3D CNN, called 3D U-Net + DR, for automatic segmentation of the LA in GE-MRI volumes of patients diagnosed with AF. Accurate and precise segmentation of the LA is essential for diagnosis, ablation therapy planning and patient prognosis. We leveraged the advantage of training a segmentation network on the complete 3D cardiac MRI volume, rather than 2D slices, to prevent loss of inter-slice information. The proposed method utilizes both local and global information, by expanding the receptive-field in the lowest level of the network, using dilated convolutions. Five-fold cross validation experiments using 100 GE-MRI volumes, revealed significant improvements in segmentation accuracy, evaluated using the DC and JI metrics, were achieved using our approach, compared to a conventional 3D U-Net. In the future we will look to extend our network to incorporate an automatic ROI detection component, within a multi-task learning framework, for improved robustness.

**Acknowledgements.** This study was partially supported by the project BIG-THERA: Integrative 'Big Data Modeling' for the development of novel therapeutic approaches for breast cancer, and the BMBF grant: Using big data to advance breast cancer research. Radiogenomics in breast cancer: Combining imaging and genomic data to address breast cancer complexity. The authors would also like to thank NVIDIA for donating a Titan XP GPU.

# References

1. Guang, Y., et al.: Fully automatic segmentation and objective assessment of atrial scars for long-standing persistent atrial fibrillation patients using late gadolinium-enhanced MRI. Med. Phys. **45**(4), 1562–1576 (2018)
2. McGann, C., et al.: Atrial fibrillation ablation outcome is predicted by left atrial remodeling on MRI. Circ.: Arrhythm. Electrophysiol. **7**(1), 23–30 (2014)
3. Zhao, J., et al.: Three-dimensional integrated functional, structural, and computational mapping to define the structural "fingerprints" of heart-specific atrial fibrillation drivers in human heart ex vivo. J. Am. Heart Assoc. **6**(8), e005922 (2017)
4. Oakes, R.S., et al.: Detection and quantification of left atrial structural remodeling with delayed-enhancement magnetic resonance imaging in patients with atrial fibrillation. Circulation **119**(13), 1758–1767 (2009)
5. Tao, Q., Ipek, E.G., Shahzad, R., Berendsen, F.F., Nazarian, S., van der Geest, R.J.: Fully automatic segmentation of left atrium and pulmonary veins in late gadolinium-enhanced MRI: towards objective atrial scar assessment. J. Magn. Reson. Imaging **44**(2), 346–354 (2016)
6. Ronneberger, O., Fischer, P., Brox, T.: U-Net: convolutional networks for biomedical image segmentation. In: Navab, N., Hornegger, J., Wells, W.M., Frangi, A.F. (eds.) MICCAI 2015. LNCS, vol. 9351, pp. 234–241. Springer, Cham (2015). https://doi.org/10.1007/978-3-319-24574-4_28
7. Milletari, F., Navab, N., Ahmadi, S.A.: V-Net: fully convolutional neural networks for volumetric medical image segmentation. In: 2016 Fourth International Conference on 3D Vision (3DV), pp. 565–571, October 2016
8. Mortazi, A., Karim, R., Rhode, K., Burt, J., Bagci, U.: *CardiacNET*: segmentation of left atrium and proximal pulmonary veins from MRI using multi-view CNN. In: Descoteaux, M., Maier-Hein, L., Franz, A., Jannin, P., Collins, D.L., Duchesne, S. (eds.) MICCAI 2017. LNCS, vol. 10434, pp. 377–385. Springer, Cham (2017). https://doi.org/10.1007/978-3-319-66185-8_43
9. Jin, C., et al.: Left atrial appendage segmentation using fully convolutional neural networks and modified three-dimensional conditional random fields. IEEE J. Biomed. Health Inform. 1 (2018)
10. Kurzendorfer, T., Forman, C., Schmidt, M., Tillmanns, C., Maier, A., Brost, A.: Fully automatic segmentation of left ventricular anatomy in 3-D LGE-MRI. Comput. Med. Imaging Graph. **59**, 13–27 (2017)
11. Yu, F., Koltun, V.: Multi-scale context aggregation by dilated convolutions. In: ICLR (2015)
12. Xia, Q., Hao, Y., Hu, Z., Hao, A.: Automatic 3D atrial segmentation from gadolinium-enhanced MRI using volumetric fully convolutional networks. In: Statistical Atlases and Computational Models of the Heart, STACOM 2018. Springer, Heidelberg (2018)
13. Bian, C., et al.: Pyramid network with online hard example mining for accurate left atrium segmentation. In: Statistical Atlases and Computational Models of the Heart, STACOM 2018. Springer, Heidelberg (2018)
14. Li, C., et al.: Attention based hierarchical aggregation network for 3D left atrial segmentation. In: Statistical Atlases and Computational Models of the Heart, STACOM 2018. Springer, Heidelberg (2018)

15. Puybareau, E., Zhou, Z., Khoudli, Y., Xu, Y., Lacotte, J., Géraud, T.: Left atrial segmentation in a few seconds using fully convolutional network and transfer learning. In: Statistical Atlases and Computational Models of the Heart, STACOM 2018. Springer, Heidelberg (2018)
16. Yang, X., et al.: Combating uncertainty with novel losses for automatic atrium segmentation. In: Statistical Atlases and Computational Models of the Heart, STACOM 2018. Springer, Heidelberg (2018)

# A Semantic-Wise Convolutional Neural Network Approach for 3-D Left Atrium Segmentation from Late Gadolinium Enhanced Magnetic Resonance Imaging

Davide Borra, Alessandro Masci, Lorena Esposito, Alice Andalò,
Claudio Fabbri, and Cristiana Corsi[✉]

Department of Electrical and Information Engineering,
University of Bologna, Cesena, Italy
cristiana.corsi3@unibo.it

**Abstract.** Several studies suggest that the assessment of viable left atrial (LA) tissue is a relevant information to support catheter ablation in atrial fibrillation (AF). Late gadolinium enhanced magnetic resonance imaging (LGE MRI) is a new emerging technique which is employed for the non-invasive quantification of LA fibrotic tissue. The analysis of LGE MRI relies on manual tracing of LA boundaries. This procedure is time-consuming and prone to high inter-observer variability given the different degrees of observers' experience, LA wall thickness and data resolution. Therefore, an automatic approach for the LA wall detection would be highly desirable. This work focuses on the design and development of a semantic-wise convolutional neural network based on the successful architecture U-Net (U-SWCNN). Batch normalization, early stopping and parameter initializers consistent with the activation functions chosen were used; a loss function based on the Dice coefficient was employed. The U-SWCNN was trained end-to-end with the 3-D data available from the 2018 Atrial Segmentation Challenge. The training was completed using 80 LGE MRI data and a post-processing step based on the 3-D morphology was then applied. After the post-processing step, the average Dice coefficient on the validation set (20 LGE MRI data) was 0.911, while on the test set (54 LGE MRI data) was 0.898.

**Keywords:** Left atrium segmentation ·
Convolutional neural networks ·
Late gadolinium enhanced magnetic resonance imaging

## 1 Introduction

Atrial fibrillation (AF) is the most common arrythmia worldwide. It has been estimated that the prevalence of AF in US is about 2.2 million including paroxysmal or persistent AF [1]. There are about 160,000 new AF cases each year in the US and in the European countries alone. Consequences of AF could imply

© Springer Nature Switzerland AG 2019
M. Pop et al. (Eds.): STACOM 2018, LNCS 11395, pp. 329–338, 2019.
https://doi.org/10.1007/978-3-030-12029-0_36

a notable reduction in quality of life, poor mental health, disability, dementia and mainly an increment of the stroke risk five-fold [2]. Radio frequency ablation (RFA) of the left atrium (LA) represents the therapy for AF patients where anti-arrhythmic drugs and direct current cardio-version show no efficacy. Despite huge improvements for targeting and delivery of AF ablation, the long-term durable restoration of sinus rhythm is achieved only for a part of AF patients. Indeed, AF-free rates after a single ablation vary between 30 and 50% at 5 years follow-up [3,4]. The low success of the current AF treatment could be related to the incapability to define a personalized approach for ablation, also including atrium specific anatomy and fibrotic tissue location [5].

Late gadolinium enhanced magnetic resonance imaging (LGE MRI) is a new emerging non-invasive imaging acquisition which might be employed for the assessment of LA myocardial tissue in patients affected by AF. With this technique, healthy and scar tissues are differentiated: scar tissue is visualized as a region of enhanced or high signal intensity while healthy tissue is characterized by low signal intensity. For this purpose, several clinical studies suggested that the information on LA scar tissue can provide relevant information for the assessment of the appropriate strategy in catheter ablation; moreover fibrotic changes in the LA substrate have been proposed to explain the persistence and sustainability of AF [6,7]. Through LGE MRI, the detection of the fibrotic tissue to identify native and post-ablation atrial scarring is provided and this might imply an improvement of the success rate of the RFA [8,9]. However, in clinical practice the detection of the LA anatomy from LGE MRI is a very challenging task, given the complexity of the atrial and pulmonary veins (PVs) structures and the limited contrast of this imaging acquisition technique. Indeed, the thickness of the LA wall is very low. Other potential sources of error are the residual motion due to patient breathing, heart rate variability, low signal-to-noise ratio, and contrast agent washout during the long acquisition which results in a reduction of the image quality.

Several studies aimed at LA anatomy and fibrotic tissue assessment from LGE MRI showed promising results. However, most of them were based on manual segmentation of the LA wall and PVs [10–12]. This implies a time consuming subjective task, resulting in a poor reproducibility between multicenter studies. Therefore, the availability of a fully automatic algorithm for LA chamber segmentation would be very useful in order to accurately reconstruct and visualize the atrial structure for clinical use. To this purpose, the use of convolutional neural networks (CNNs) represents a suitable approach for the LA + PVs segmentation. Recently, Mortazi et al. [13] developed a 2-D convolutional neural network (CNN) approach for the LA + PVs segmentation from cine MRI SSFP sequences and CT data (STACOM 2013 Cardiac Segmentation Challenge [14]) and Baumgartner et al. [15] proposed a fully automated framework, combining U-Net [16] and batch normalization, for the segmentation of the left and right ventricles from short-axis cardiac cine MRI data, training the network in 2-D and 3-D.

The aim of this study was the design and development of an automatic image segmentation algorithm of the LA cavity from LGE MRI based on the use of CNNs, exploiting a deep learning pipeline based on the successful architecture U-Net. The network was trained end-to-end from scratch using 3-D data available in the training phase of the STACOM 2018 Atrial Segmentation Challenge.

## 2 Methods

### 2.1 Dataset

One hundred cardiac LGE MRI data with ground truth labels and 54 LGE MRI data without ground truth labels were provided by the organizers of the STACOM 2018 Atrial Segmentation Challenge during the training and testing phases, respectively. The original resolution of the data is $0.625 \times 0.625 \times 0.625\,\text{mm}^3$. In order to train the model looking at the performance on unseen data during the optimization, we manually split the dataset in a training (80%) and a validation set (20%).

### 2.2 Pre-processing

Let $X^{(i)}$ be the i-th MRI data and $Y^{(i)}$ its true binary segmentation (0 background, 1 LA) $i \in [1, 100]$. $X^{(i)}$ and $Y^{(i)}$ have the same size of $m \times n \times d$, where $m$ and $n$ are 576 or 640 and $d$ is 88. Then, a 3-D crop containing LA of $m1 \times n1 \times d1 = 320 \times 384 \times 88$ for each image and mask is extracted, generating $X_c^{(i)}$ and $Y_c^{(i)}$ (cropped data and masks). The assessment of the coordinates of this crop is obtained by applying a rough segmentation based on Otsu's algorithm to each middle axial slice of the entire dataset (Fig. 1).

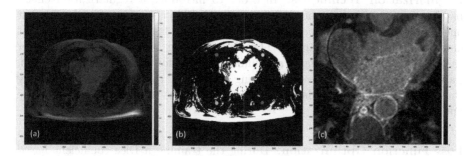

**Fig. 1.** Otsu's algorithm for the assessment of the atrial location. (a) Initial image; (b) binary image resulting from Otsu's segmentation and points automatically selected to crop the image - starting from the center of the binary region in the center of the image (red point), the limits of this region on the x direction are automatically computed (green and pink points); the mid point between them is computed and the image crop is automatically obtained considering a fixed window size; (c) final crop. (Color figure online)

Considering (1) that LGE MRI images are acquired in the axial plane apply-ing a standard protocol which always considers the LA chamber in the center of the acquired image and (2) the results of the rough segmentation, LA is easily located and the images are consequently cropped. Once the crop is extracted, a subsampling procedure is performed in order to reduce the computational cost according to the available hardware; in this way $X_{sc}^{(i)}$ and $Y_{sc}^{(i)}$ are generated (subsampled cropped data and masks). The subsampled region of interest size is $144 \times 176 \times 32$ and the network is trained with these 3-D data.

### 2.3   3-D Semantic Segmentation with U-SWCNN

In the following the CNN architecture, regularization techniques and loss func-tion optimization are described.

**CNN Architecture.** In order to perform the LA segmentation task, we intro-duced a deep learning approach based on the U-Net architecture (Fig. 2). In the convolutional layers, kernel size of $3 \times 3 \times 3$, stride size of $1 \times 1 \times 1$ and Recti-fied Linear Units (ReLUs) activation functions in the hidden layers or sigmoidal activation function in the output layer are used. In the max pooling layers, a pooling size of $2 \times 2 \times 2$ and stride size of $2 \times 2 \times 2$, halving the shape of hidden activations, are employed. Lastly, in the transposed convolutional layers, kernel size of $2 \times 2 \times 2$ and stride size of $2 \times 2 \times 2$. For both convolutional and transposed convolutional layers, padding size is such that the output shape of the layer is the same of the input shape. Furthermore, biases and weights are randomly ini-tialized from a truncated normal distribution and using the initialization scheme proposed by He et al. [17] for ReLUs, respectively.

**Regularization Techniques.** In order to introduce a regularization effect, batch normalization [18] and early stopping techniques are employed. The first one is an adaptive reparametrization introduced to reduce the covariance shift and it also acts as regularizer. These normalizing layers are included after each hidden convolutional layer and immediately before the activation function. Lastly, we employed the early stopping technique, that consists in returning the model with lowest validation set error during the training and stopping the optimization.

**Loss Function Optimization.** During the training a Dice coefficient based loss function, named Dice loss, is computed. Let $dc = \frac{2N_{tp}}{2N_{tp}+N_{fp}+N_{fn}}$ be the Dice coefficient computed between the true mask and the predicted mask, expressed in terms of true positives ($N_{tp}$), false positives ($N_{fp}$) and false negatives ($N_{fn}$). Then, the Dice loss is defined as $dl = 1 - dc$. In order to solve the optimization problem, the Adam adaptive learning rate optimization algorithm [19] is used.

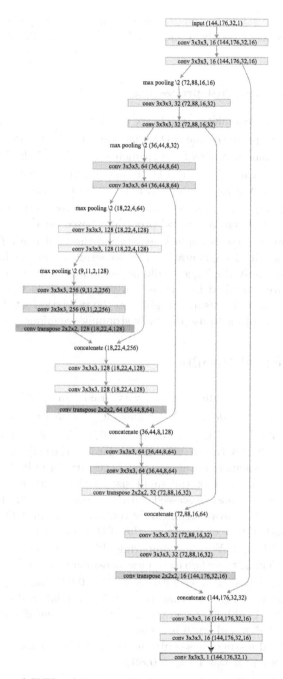

**Fig. 2.** The proposed CNN architecture. The last convolutional layer (red boxed) differs from others in the activation function and the absence of the normalization layer. In each item are specified the kind of layer, the kernel size, the number of filters and the output shape. (Color figure online)

Exponential decay rates $\beta_1$ and $\beta_2$ are 0.9 and 0.999 respectively, while the learning rate $\epsilon$ is 1e−4. Lastly, a max number of epochs of 255 with a batch size of 2 is used.

## 2.4   Prediction and Post-processing

Once the training is completed, the CNN is fed with an unseen input $X_{sc}^{(i)}$ and the model directly produces the 3-D segmentation $P_{sc}^{(i)}$. During the prediction step, at first the model predicts the segmentation $P_{sc}^{(i)}$ of the i-th image sub-sampled crop $X_{sc}^{(i)}$, then the final 3-D predicted segmentation $P^{(i)}$ is obtained by reshaping the subsampled predicted segmentation to the size $m1 \times n1 \times d1$ and by padding this intermediate result $P_c^{(i)}$ with zeros to the i-th input original size $m \times n \times d$. Then, a removal procedure based on the evaluation of the connected-regions volume is applied, generating the final mask $P_{la}^{(i)}$. In particular, since the obtained segmentation $P^{(i)}$ sometimes contains LA and various spurious elements, only the biggest volume associated with the LA is kept.

The procedure described in this section was applied on the validation and test sets. Furthermore, it was also applied on the validation set during the optimization process in order to use the early stopping technique correctly.

## 3   Experimental Results

For the performance evaluation $dc$ was used. In particular, $DC_{sc}^{(i)} = dc\left(P_{sc}^{(i)}, Y_{sc}^{(i)}\right)$, $DC^{(i)} = dc\left(P^{(i)}, Y^{(i)}\right)$ and $DC_{la}^{(i)} = dc\left(P_{la}^{(i)}, Y^{(i)}\right)$ were computed and we obtained $DC_{sc} = 0.916 \pm 0.027$, $DC = 0.907 \pm 0.029$ and $DC_{la} = \mathbf{0.911 \pm 0.028}$ (mean value ± standard deviation) on the validation set. The spurious regions removal step was necessary for only 3 of 20 predicted validation masks. Considering the small size of the spurious regions and the low number of images post-processed with this procedure, after this step the performance did not improve significantly (on average from 0.907 to 0.911).

In Figs. 3 and 4 two examples of the 3-D validation data segmentation obtained with 3-D U-SWCNN, comparing $P_{la}^{(i)}$ (red) and $Y^{(i)}$ (blue), are reported. In particular, these two examples corresponds to the best and the worst predictions of the neural network with $DC_{la}^{(best)} = 0.964$ and $DC_{la}^{(worst)} = 0.850$, respectively. Lastly, in Fig. 5 an example of a 3-D validation data segmentation in which the CNN is able to reconstruct all the PVs even if some of them are not included in the ground truth is reported.

During the final testing phase only the average Dice coefficient was provided by the organizers and it was $DC_{la} = \mathbf{0.898}$.

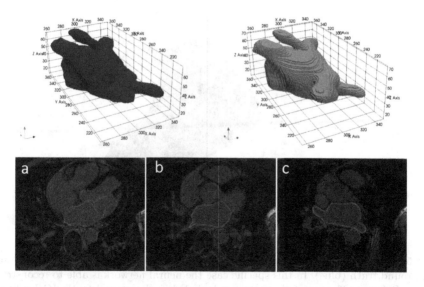

**Fig. 3.** 3-D segmentation (top) obtained with the proposed CNN (red) and the ground truth (blue) of the best prediction ($DC_{la}^{(best)} = 0.964$). Contours (bottom) are extracted at 20 (a), 40 (b) and 60% (c) of the LA extension along the longitudinal axis. (Color figure online)

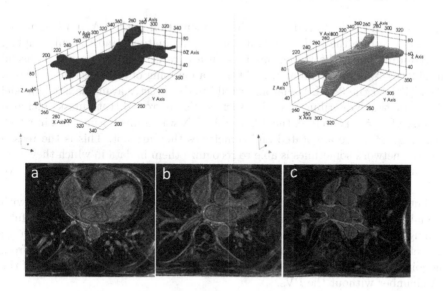

**Fig. 4.** 3-D segmentation (top) obtained with the proposed CNN (red) and the ground truth (blue) of the worst prediction ($DC_{la}^{(worst)} = 0.850$). Contours (bottom) are extracted at 20 (a), 40 (b) and 60% (c) of the LA extension along the longitudinal axis. (Color figure online)

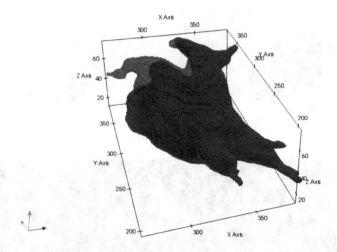

**Fig. 5.** Example of a 3-D segmentation obtained with the proposed CNN (red) and the ground truth (blue). In this specific case the neural network is able to reconstruct all the PVs even if some of them are not included in the ground truth. (Color figure online)

## 4    Discussion and Conclusions

The proposed method produces a joint segmentation of the LA and PVs in AF patients. Despite the high variability of the LA anatomy and the subsampling procedure applied, the model provides an accurate mask that could be useful for ablation therapy planning. In addition, a mapping of gray level intensities of LGE MRI on the 3-D LA anatomy would make directly available a 3-D model of fibrotic tissue distribution on LA surface. Some ground truth masks included PVs (see Fig. 5). Therefore, the 3-D U-SWCNN was trained using data in which sometimes PVs are annotated and sometimes they are not. This is the reason why the network sometimes is able to recognize them in data in which they were not annotated in the ground truth. This leads to an additional source of error in the evaluation metric.

Future developments include: (1) training the network with 3-D data with a smaller subsampling ratio in order to exploit original information, hopefully leading to a higher $DC$ and in order to reduce the gap between $DC$ and $DC_{sc}$; (2) definition of a new loss function; (3) separation of the PVs structures from the joint segmentation of LA + PVs and evaluation of the Dice coefficient of the LA chamber without the PVs.

**Acknowledgement.** Thanks to Dr. Marilisa Cortesi for lending the needed hardware. The experiments have been conducted using Keras [20] with TensorFlow [21] backend on CPU (Intel Xeon E5-2640) of a workstation equipped with 64 GB of RAM.

# References

1. Chugh, S.S., Blackshear, J.L., Shen, W.K., Hammill, S.C., Gersh, B.J.: Epidemiology and natural history of atrial fibrillation: clinical implications. J. Am. Coll. Cardiol. **37**(2), 371–378 (2001)
2. Wolf, P.A., Abbott, R.D., Kannel, W.B.: Atrial fibrillation as an independent risk factor for stroke: the Framingham study. Stroke **22**(8), 983–988 (1991)
3. Ouyang, F., et al.: Long-term results of catheter ablation in paroxysmal atrial fibrillationclinical perspective: lessons from a 5-year follow-up. Circulation **122**(23), 2368–2377 (2010)
4. Weerasooriya, R., et al.: Catheter ablation for atrial fibrillation: are results maintained at 5 years of follow-up? J. Am. Coll. Cardiol. **57**(2), 160–166 (2011)
5. Akoum, N., et al.: Atrial fibrosis helps select the appropriate patient and strategy in catheter ablation of atrial fibrillation: a DE-MRI guided approach. J. Cardiovasc. Electrophysiol. **22**(1), 16–22 (2011)
6. Oakes, R.S., et al.: Detection and quantification of left atrial structural remodeling with delayed-enhancement magnetic resonance imaging in patients with atrial fibrillation. Circulation **119**(13), 1758–1767 (2009)
7. Daccarett, M., et al.: Association of left atrial fibrosis detected by delayed-enhancement magnetic resonance imaging and the risk of stroke in patients with atrial fibrillation. J. Am. Coll. Cardiol. **57**(7), 831–838 (2011)
8. Malcolme-Lawes, L., et al.: Automated analysis of atrial late gadolinium enhancement imaging that correlates with endocardial voltage and clinical outcomes: a 2-center study. Heart Rhythm **10**(8), 1184–1191 (2013)
9. Marrouche, N.F., et al.: Association of atrial tissue fibrosis identified by delayed enhancement MRI and atrial fibrillation catheter ablation: the DECAAF study. Jama **311**(5), 498–506 (2014)
10. Spragg, D.D., et al.: Initial experience with magnetic resonance imaging of atrial scar and co-registration with electroanatomic voltage mapping during atrial fibrillation: success and limitations. Heart Rhythm **9**(12), 2003–2009 (2012)
11. Sohns, C., et al.: Quantitative magnetic resonance imaging analysis of the relationship between contact force and left atrial scar formation after catheter ablation of atrial fibrillation. J. Cardiovasc. Electrophysiol. **25**(2), 138–145 (2014)
12. Karim, R., et al.: Evaluation of current algorithms for segmentation of scar tissue from late gadolinium enhancement cardiovascular magnetic resonance of the left atrium: an open-access grand challenge. J. Cardiovasc. Magn. Reson. **15**(1), 105 (2013)
13. Mortazi, A., Karim, R., Rhode, K.S., Burt, J., Bagci, U.: CardiacNET: segmentation of left atrium and proximal pulmonary veins from MRI using multi-view CNN. CoRR abs/1705.06333 (2017). http://arxiv.org/abs/1705.06333
14. Tobon-Gomez, C., Geers, A.J., et al.: Benchmark for algorithms segmenting the left atrium from 3D CT and MRI datasets. IEEE Trans. Med. Imaging **34**(7), 1460–1473 (2015). https://doi.org/10.1109/TMI.2015.2398818
15. Baumgartner, C.F., Koch, L.M., Pollefeys, M., Konukoglu, E.: An exploration of 2D and 3D deep learning techniques for cardiac MR image segmentation. CoRR abs/1709.04496 (2017). http://arxiv.org/abs/1709.04496
16. Ronneberger, O., Fischer, P., Brox, T.: U-net: convolutional networks for biomedical image segmentation. CoRR abs/1505.04597 (2015). http://arxiv.org/abs/1505.04597

17. He, K., Zhang, X., Ren, S., Sun, J.: Delving deep into rectifiers: surpassing human-level performance on imagenet classification. CoRR abs/1502.01852 (2015). http://arxiv.org/abs/1502.01852
18. Ioffe, S., Szegedy, C.: Batch normalization: accelerating deep network training by reducing internal covariate shift. CoRR abs/1502.03167 (2015). http://arxiv.org/abs/1502.03167
19. Kingma, D.P., Ba, J.: Adam: a method for stochastic optimization. CoRR abs/1412.6980 (2014). http://arxiv.org/abs/1412.6980
20. Chollet, F., et al.: Keras (2015). https://github.com/keras-team/keras
21. Abadi, M., Agarwal, A., et al.: TensorFlow: large-scale machine learning on heterogeneous systems (2015). Software available from tensorflow.org. https://www.tensorflow.org/

# Left Atrial Segmentation in a Few Seconds Using Fully Convolutional Network and Transfer Learning

Élodie Puybareau[1]([✉]), Zhou Zhao[1,2], Younes Khoudli[1], Edwin Carlinet[1], Yongchao Xu[1,2], Jérôme Lacotte[3], and Thierry Géraud[1]

[1] EPITA Research and Development Laboratory (LRDE),
Le Kremlin-Bicêtre, France
elodie.puybareau@lrde.epita.fr
[2] Huazhong University of Science and Technology, Wuhan, China
[3] Institut Cardiovasculaire Paris Sud, Hôpital Privé Jacques Cartier, Massy, France

**Abstract.** In this paper, we propose a fast automatic method that segments left atrial cavity from 3D GE-MRIs without any manual assistance, using a fully convolutional network (FCN) and transfer learning. This FCN is the base network of VGG-16, pre-trained on ImageNet for natural image classification, and fine tuned with the training dataset of the MICCAI 2018 Atrial Segmentation Challenge. It relies on the "pseudo-3D" method published at ICIP 2017, which allows for segmenting objects from 2D color images which contain 3D information of MRI volumes. For each $n^{\text{th}}$ slice of the volume to segment, we consider three images, corresponding to the $(n-1)^{\text{th}}$, $n^{\text{th}}$, and $(n+1)^{\text{th}}$ slices of the original volume. These three gray-level 2D images are assembled to form a 2D RGB color image (one image per channel). This image is the input of the FCN to obtain a 2D segmentation of the $n^{\text{th}}$ slice. We process all slices, then stack the results to form the 3D output segmentation. With such a technique, the segmentation of the left atrial cavity on a 3D volume takes only a few seconds. We obtain a Dice score of 0.92 both on the training set in our experiments before the challenge, and on the test set of the challenge.

**Keywords:** 3D heart MRI · Atrial segmentation ·
Fully convolutional network

## 1 Introduction

*Motivation.* Atrial fibrillation (AF) is the most common heart rhythm disease, corresponding with the activation of an electrical substrate within the atrial myocardium. AF is already an endemic disease, and its prevalence is soaring, due to both an increasing incidence of the arrhythmia and an age-related increase in its prevalence [3,10,18]. Indeed, 1–2% of the population suffer from AF at

© Springer Nature Switzerland AG 2019
M. Pop et al. (Eds.): STACOM 2018, LNCS 11395, pp. 339–347, 2019.
https://doi.org/10.1007/978-3-030-12029-0_37

present, and the number of affected individuals is expected to double or triple within the next two to three decades both in Europe and in the USA [5].

Due to the limited effects of anti-arrhythmic drugs, AF can only be cured by percutaneous radiofrequency catheter ablation (CA) targeting triggers and critical areas responsible for AF perpetuation in left atrium (LA). Identification and quantification of AF electrical substrate prior to AF ablation remains an unsolved issue as the number of targets remains unpredictable using clinical criterias. AF CA is still a challenging intervention requiring a perioperative 3D mapping to identify AF substrate to select the best ablation strategy [5].

Exploration of LA substrate has suggested that AF may be a self-perpetuating disease with a voltage or electrogram (EGM) amplitude reduction which is an indicator of the severity of tissue corresponding with collagen deposition in the myocardial interstitial space. Non-invasive assessment of myocardial fibrosis has proved useful as a diagnostic, prognostic, and therapeutic tool to understand and reverse AF [3,18]. Visualization and quantification of gadolinium in late gadolinium-enhanced cardiac magnetic resonance (LGE-CMR) sequences estimate the extracellular matrix volume and have been used as a LA fibrosis surrogate, and over the last years, several groups tested the ability of LGE-CMR to detect pre-existing fibrosis [1,10].

Although these reports suggested that the extent of fibrosis may predict recurrences after ablation procedures, the lack of 3D automated LA reconstruction, the lack of reference values for normality has prompted the publication of several image acquisition and post-processing protocols and thresholds to identify fibrosis, eventually limiting the external validation and reproducibility of this technique [9,11,13]. Last, let us note that, because of these technical limits, the assessment of LA fibrosis has not yet been widely adopted in the clinical practice [2].

*Context.* The method presented in this paper has been developed in the context of the MICCAI 2018 Atrial Segmentation Challenge[1]. The aim was to provide a fully automated pipeline for the segmentation of the LA cavity from 3D GE-MRIs without any manual assistance.

Despite the relevance of LA segmentation on GE-MRIs, this task remains manual, tedious and user-dependant. This segmentation is challenging due to the low contrast between atrial tissue and background. The development of an algorithm that can perform fully automatic atrial segmentation for the LA cavity, reconstruct and visualize the atrial structure for clinical usage would be an important improvement for patients and practitioners.

We received 100 volumes with associated masks to develop our method. A set of 25 new volumes, released 2 weeks before the end of the challenge, has been used to evaluate our method by comparing our segmentation results with the manual segmented masks (these masks will not be released). This challenge will establish a fair comparison to state-of-the-art methods, while providing the community with a large dataset of 3D GE-MRI heart images.

---

[1] http://atriaseg2018.cardiacatlas.org/.

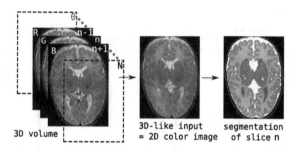

**Fig. 1.** Illustration of the main idea used in [15]: a segmentation of a 3D medical image is performed on a set of 2D color images. For the $n^{\text{th}}$ slice, a 2D color image (middle) is formed from the triplet of slices $n-1, n, n+1$ (left) of the original volume, and the segmentation (right) is performed on this 2D color image. Actually, such a **2D** image contains **3D** information thanks to the 3 color channels (R, G, B); so, this approach is called "**pseudo-3D**". (Color figure online)

*Main Related Work.* As machine learning really improved the results of some segmentation tasks, the use of such strategy seems meaningful in the context of medical image segmentation. In a work published in the IEEE Intl. Conf. on Image Processing (ICIP) in 2017 [15], 3D brain MR volumes are segmented using fully convolutional network (FCN) and transfer learning. The network used for transfer learning is VGG (Visual Geometry Group) [14], pre-trained on the ImageNet dataset. It takes as input a 2D color image, formed from 3 consecutive slices of the 3D volume (see Fig. 1). Thus, we have a 2D image containing 3D information; this approach is called "pseudo-3D". This method only relies on a single modality, and obtains good results for brain segmentation. For the left atrial segmentation challenge, we use the same approach, leveraging the power of a fully convolutional network pre-trained on a large dataset and later fine-tuned on the training set provided for the challenge.

## 2 Description of the Proposed Method

An overview of the proposed method is given in Fig. 2. The method is fully automatic, and takes 2D color images as input. The method is very computationally efficient, as it processes each complete scan volume in less than 2 s.

### 2.1 Pre-processing

The study of the histograms of the training volumes shows a high variability among the different volumes, as depicted in Fig. 3(a). A clue to improve the network results is to normalize the input volumes according to their histograms. To that aim, for each volume, we compute the histogram of the central sub-volume (1/4 of the pixels in all three dimensions, so containing $4^3$ times less voxels than the whole volume; see Fig. 4(c)). Let $m$ be the maximum gray-level having a non-null histogram value. We requantize all voxel values using a

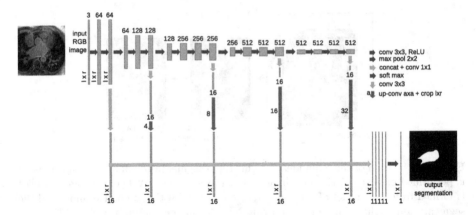

**Fig. 2.** Architecture of the proposed network. We fine tune it and combine linearly fine to coarse feature maps of the pre-trained VGG network [14]. Note that each input 2D color image is built from the slice $n$ and its neighbouring slices $n-1$ and $n+1$. (Color figure online)

linear function so that the gray-level range $[0, m]$ is mapped to $[0, 255]$. For our application, the question amounts to how to prepare appropriate inputs, RGB input images, given that a heart MR image is a 3D volume. To that aim, we propose to stack successive 2D slices. Precisely, to form an input artificial color image for the pre-trained network to segment the $n^{\text{th}}$ slice, we use the slice $n$ of the volume, its predecessor, the $(n-1)^{\text{th}}$ slice, and its successor, the $(n+1)^{\text{th}}$ slice, as respectively the green, red and blue channels (note that the order is not relevant). This process is depicted in Fig. 2 (left). Each 2D color image thus forms a representation of a part (a small volume) of the MR volume.

## 2.2 Deep FCN for Left Atrial Segmentation

Fully convolutional network (FCN) and transfer learning has proved their efficiency for natural image segmentation [7]. The paper [15] proposed to rely on a FCN and transfer learning to segment 3D brain MR images, although those images are very different from natural images. As it was a success, we adapted it to LA segmentation. We rely on the 16-layer VGG network [14], which was pre-trained on millions of natural images of ImageNet for image classification [6]. For our application, we keep only the 4 stages of convolutional parts called "*base network*", and we discard the fully connected layers at the end of VGG network. This base network is mainly composed of convolutional layers: $z_i = w_i \times x + b_i$, Rectified Linear Unit (ReLU) layers for non-linear activation function: $f(z_i) = \max(0, z_i)$, and max-pooling layers between two successive stages, where $x$ is the input of each convolutional layer, $w_i$ is the convolution parameter, and $b_i$ is the bias term. The three max-pooling layers divide the base network into four stages of fine to coarse feature maps. Inspired by the work in [7,8], we add specialized convolutional layers (with a $3 \times 3$ kernel size) with

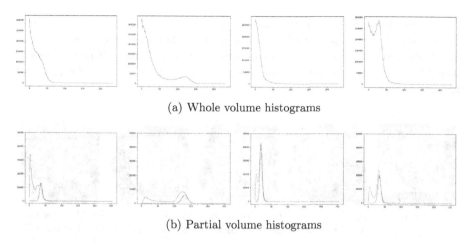

(a) Whole volume histograms

(b) Partial volume histograms

**Fig. 3.** (a): The histograms of the original volumes have various shapes; (b): to normalize the gray-level scale of each volume, we consider the histogram of their central sub-volume (in orange; see also Fig. 4(c)), which has the same dynamic than the one of the left atrial region given by the ground-truth (in green). (Color figure online)

$K$ (e.g. $K = 16$) feature maps after the convolutional layers at the end of each stage. All the specialized layers are then rescaled to the original image size, and concatenated together. We add a last convolutional layer with kernel size $1 \times 1$ at the end. This last layer combine linearly the fine to coarse feature maps in the concatenated specialized layers, and provide the final segmentation result. The proposed network architecture is schematized in Fig. 2.

The architecture described above is very similar with the one used in [8] for retinal image analysis, where the retinal images are already 2D color images. Using such a 2D representation avoids the expensive computational and memory requirements of fully 3D FCN.

For the training phase, we use the multinomial logistic loss function for a one-of-many classification task, passing real-valued predictions through a softmax to get a probability distribution over classes. We rely on the ADAM optimization procedure [4] (AMSGrad variant [12]) to minimize the loss of the network. The relevant parameters of the methods are the following: the learning rate is set to 0.002 (we did not use learning rate decay), the beta_1 and beta_2 are respectively set to 0.9 and 0.999, and we use a fuzz factor (epsilon) of 0.001. Each batch is composed of 12 centered images (for each channel we subtract 127 to values to ensure that input values are within the $[-127, 127]$ range). We trained the network for 10 epochs.

At test time, after having pre-processed the 3D volume (requantization), we prepare the set of 2D color images. Then we subtract 127 for each channel, and pass every image through the network.

We run the train and test phases on an NVIDIA GPU card. The training phase lasts about 50 min, while the testing lasts less than 2 s, 1.78 s in average. These reasonable computational times allow a re-training if new scans are available, and allow clinical use. To this running time, 1 s is added for pre- and post-processing.

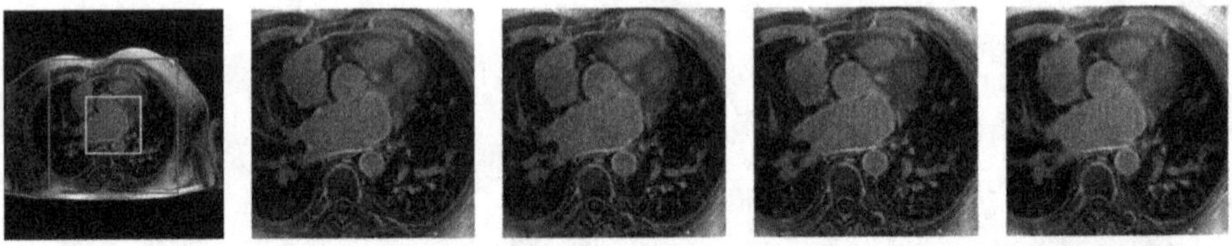

(a) Initial image  (b) Slice $n-1$ as red channel  (c) Slice $n$ as green channel  (d) Slice $n+1$ as blue channel  (e) 2D color image for the $n^{\text{th}}$ slice

**Fig. 4.** The yellow box in (a) depicts the sub-volume used for the gray-level normalization stage, and the red box depicts the boundary of the cropped input image. Three successive cropped slices (b–d) are used to build a 2D color image (e). (Color figure online)

### 2.3 Post-processing

The output of the network for one slice during the inference phase is a 2D segmented slice. After treating all the slices of the volume, all the segmented slices are stacked to recover a 3D volume with the same shape as the initial volume, and containing only the segmented lesions. We clean up the segmentation keeping only the largest component. We regularized the 3D segmentation result using a $5 \times 5$ median filter in the sagittal axis. This step smoothes the segmentation and erases potential inconsistencies.

## 3    Experiments and Results

### 3.1    Experiments and Results for the Training Phase

To experiment, we only relied on the 100 training scans provided by the challenge.

Our model has been trained on 80 scans, chosen randomly, using the parameters described in the previous section. Then we have tested our model on the 20 remaining scans. This procedure have been repeated 5 times to cross-validate the results.

Our preliminary results on the training dataset without any normalization showed a dice score of 0.86 after 10 epochs. We performed two 5-fold. With the help of normalization, the dice score reaches $0.9 \pm 0.02$ (under cross-validation), after 10 epochs, and with an average of 0.92. Note that these results were obtained without any training data augmentation.

The evolution of the dice score according to the number of epochs (Fig. 5) shows that the network converges quickly and, actually, 4 epochs are enough to reach a dice score of 0.90.

Some qualitative results can be found in Fig. 6.

## 3.2   Results of the Challenge

Our method reaches a top rank during the challenge, with a Dice score of 0.923. The best score was 0.932 and the second best 0.926.

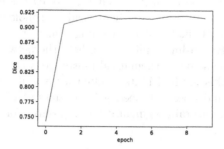

**Fig. 5.** Evolution of the dice score with the number of epochs.

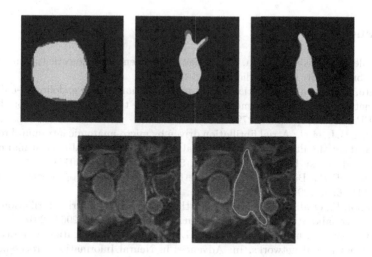

**Fig. 6.** Examples of segmentation results. First row: superposition between our result and GT (wrongly labeled pixels are in green); second row: an MRI slice (left) and our segmentation superimposed over it (right). (Color figure online)

## 4 Conclusion

In this article, we proposed a method to accurately segment the left atrial cavity in few seconds based on transfer learning from VGG-16, a pre-trained network used to classify natural images. This method takes the advantage of keeping 3D information of the MRI volume and the speed of processing only 2D images, thanks to the pseudo-3D concept. An entire 3D volume is processed in less than 3 s. We were ranked $3^{rd}$ during the challenge.

This method can also deal with multi-modality, and can be applied to other segmentation problems, such as in [17], where a similar method is proposed to segment white matter hyperintensities, but pseudo-3D has been replaced by an association of multimodality and mathematical morphology pre-processing to improve the detection of small lesions. We might also try to modify our inputs thanks to some highly non-linear filtering to help the network segment the LA, more specifically using some mathematical morphology operators [16].

The strength of this method is its modularity and its simplicity. It is easy to implement, fast, and does not need a huge amount of annotated data for training (in our work on brain segmentation [15], only 2 or 4 images were used for training).

**Acknowledgments.** The authors want to thank the organizers of the Atrial Segmentation Challenge. We gratefully acknowledge the support of NVIDIA Corporation with the donation of the Quadro P6000 GPU used for this research.

## References

1. Ambale-Venkatesh, B., Lima, J.: Cardiac MRI: a central prognostic tool in myocardial fibrosis. Nat. Rev. Cardiol. **12**, 18–29 (2015)
2. Benito, E.M., et al.: Left atrial fibrosis quantification by late gadolinium-enhanced magnetic resonance: a new method to standardize the thresholds for reproducibility. EP Eur. **19**(8), 1272–1279 (2016)
3. Hansen, B.J., et al.: Atrial fibrillation driven by micro-anatomic intramural re-entry revealed by simultaneous sub-epicardial and sub-endocardial optical mapping in explanted human hearts. Eur. Heart J. **36**(35), 2390–2401 (2015)
4. Kingma, D.P., Ba, J.: Adam: a method for stochastic optimization. CoRR abs/1412.6980 (2014)
5. Kirchhof, P., et al.: ESC guidelines for the management of atrial fibrillation developed in collaboration with EACTS. Eur. Heart J. **37**, 2893–2962 (2016)
6. Krizhevsky, A., Sutskever, I., Hinton, G.E.: Imagenet classification with deep convolutional neural networks. In: Advances in Neural Information Processing Systems, pp. 1097–1105 (2012)
7. Long, J., Shelhamer, E., Darrell, T.: Fully convolutional networks for semantic segmentation. In: Proceedings of IEEE International Conference on Computer Vision and Pattern Recognition, pp. 3431–3440 (2015)
8. Maninis, K.-K., Pont-Tuset, J., Arbeláez, P., Van Gool, L.: Deep retinal image understanding. In: Ourselin, S., Joskowicz, L., Sabuncu, M.R., Unal, G., Wells, W. (eds.) MICCAI 2016. LNCS, vol. 9901, pp. 140–148. Springer, Cham (2016). https://doi.org/10.1007/978-3-319-46723-8_17

9. Marrouche, N., et al.: Association of atrial tissue fibrosis identified by delayed enhancement MRI and atrial fibrillation catheter ablation: the DECAAF study. JAMA **311**, 498–506 (2014)
10. McGann, C., et al.: Atrial fibrillation ablation outcome is predicted by left atrial remodeling on MRI. Circ.: Arrhythmia Electrophysiol. **7**(1), 23–30 (2013)
11. Oakes, R.S., et al.: Detection and quantification of left atrial structural remodeling with delayed-enhancement magnetic resonance imaging in patients with atrial fibrillation. Circulation **119**(13), 1758–1767 (2009)
12. Reddi, S.J., Kale, S., Kumar, S.: On the convergence of adam and beyond. In: International Conference on Learning Representations (2018)
13. Seitz, J., Horvilleur, J., Lacotte, J., et al.: Correlation between AF substrate ablation difficulty and left atrial fibrosis quantified by delayed-enhancement cardiac magnetic resonance. Pacing Clin. Electrophysiol. **34**(10), 1267–1277 (2011)
14. Simonyan, K., Zisserman, A.: Very deep convolutional networks for large-scale image recognition. CoRR abs/1409.1556 (2014)
15. Xu, Y., Géraud, T., Bloch, I.: From neonatal to adult brain MR image segmentation in a few seconds using 3D-like fully convolutional network and transfer learning. In: Proceedings of the 23rd IEEE International Conference on Image Processing (ICIP), Beijing, China, pp. 4417–4421, September 2017
16. Xu, Y., Géraud, T., Najman, L.: Connected filtering on tree-based shape-spaces. IEEE Trans. Pattern Anal. Mach. Intell. **38**(6), 1126–1140 (2016)
17. Xu, Y., Géraud, T., Puybareau, É., Bloch, I., Chazalon, J.: White matter hyper-intensities segmentation in a few seconds using fully convolutional network and transfer learning. In: Crimi, A., Bakas, S., Kuijf, H., Menze, B., Reyes, M. (eds.) BrainLes 2017. LNCS, vol. 10670, pp. 501–514. Springer, Cham (2018). https://doi.org/10.1007/978-3-319-75238-9_42
18. Zhao, J., et al.: Three dimensional integrated functional, structural, and computational mapping to define the structural "fingerprints" of heart specific atrial fibrillation drivers in human heart EX vivo. J. Am. Heart Assoc. **6**(8), 5922 (2017)

# Convolutional Neural Networks for Segmentation of the Left Atrium from Gadolinium-Enhancement MRI Images

Coen de Vente[1,2], Mitko Veta[2], Orod Razeghi[1], Steven Niederer[1], Josien Pluim[2], Kawal Rhode[1], and Rashed Karim[1(✉)]

[1] School of Biomedical Engineering & Imaging Sciences,
King's College London, London, UK
rashed.karim@kcl.ac.uk
[2] Department of Biomedical Engineering,
Eindhoven University of Technology, Eindhoven, The Netherlands

**Abstract.** This paper introduces a left atrial segmentation pipeline that utilises a deep neural network for learning segmentations of the LA from Gadolinium enhancement magnetic resonance images (GE-MRI). The trainable fully-convolutional neural network consists of an encoder network and a corresponding decoder network followed by a pixel-wise classification layer. The entire network has 17 convolutional layers, with the encoder network containing 5 convolutional layers, and the decoder network containing 11 convolution layers with 1 additional convolution layer in between. The training image database consisted of manually annotated GE-MRI images ($n = 75$). Dice scores of $0.87 \pm 0.07$ and $0.80 \pm 0.12$ were achieved on our test set ($n = 25$) and a multi-centre independent set using transfer learning, respectively. On the test set that was provided by the challenge ($n = 54$), a Dice score of 0.897 was achieved. We experimentally demonstrated the robustness of the proposed method as a segmentation pipeline for potential use in clinical research.

**Keywords:** Convolutional neural networks · U-Net ·
Image segmentation · Left atrium

## 1 Introduction

Atrial fibrillation (AF) is the most common cardiac arrhythmia causing chaotic activation and compromised contraction in the atria. The contraction function tends to stop with AF. Nearly 1.3 million people in the UK have been diagnosed with AF, and it is estimated that there are at least 500,000 people living with undiagnosed AF (BHF-UK Factsheet 2018). For AF patients, pulmonary vein (PV) isolation is often the first procedure performed in patients referred for catheter ablation of AF. Recent advances in catheter technology and MRI sequence development are now enabling catheter ablations to be performed within an MRI.

© Springer Nature Switzerland AG 2019
M. Pop et al. (Eds.): STACOM 2018, LNCS 11395, pp. 348–356, 2019.
https://doi.org/10.1007/978-3-030-12029-0_38

This opens the door to imaging lesion formation and ensuring that no gaps are formed during ablations. However, analysis of complex 3D real time images to identify gaps manually is not easy or reliable during procedures. This requires an automatic workflow to analyse MRI images to identify gaps in near real time.

Lesions are imaged with Gadolinium enhancement magnetic resonance imaging (GE-MRI) and left atrium (LA) segmentations from these images is challenging due to contrast and image quality. In the workflow for identifying lesion gaps, the atria is normally segmented from contrast enhanced magnetic resonance angiography (CE-MRA) or steady state free precessing (SSFP) MRI. However, it must then be registered to the GE-MRI [1] so lesions can be visualised. Unlike GE-MRI, both CE-MRA and SSFP provide with good contrast in the LA, allowing it to be distinguished from its surrounding structures. The left atrium segmentation challenge (LASC) 2013 benchmark [2] for LA segmentation on SSFP images compared algorithms that demonstrated excellent agreement (Dice $> 0.9$) in some techniques, but not as good agreement (Dice $< 0.8$) near the PV region of the LA. The results of this study could not be directly compared with the results of the current work, since only manual annotations of the entire LA were available for our datasets, instead of separate segmentations of the PVs and LA body. The registration between CE-MRA or SSFP to GE-MRI is non-rigid and a sub-optimal registration introduces further errors into the lesion image analysis workflow.

In this work we utilised a deep neural network that learns to generate dense volumetric segmentations of the LA from 2D image tiles for training. It segmented directly form GE-MRI images, removing the need for image registration steps in the atrial image processing pipeline. The network was based on a previous U-Net architecture [3], consisting of a contractor encoder part to analyse the entire image and a successive expanding decoder part generated segmentation of the same size as the input. 2D tiles of the GE-MRI images were used as input for the model. The proposed U-Net is deeper than the original U-Net architecture. Moreover, it was observed that reducing the number of convolutional layers in each convolutional group of the U-Net from two to one, reduced the required memory substantially, while not affecting the performance significantly. This work was in response to the 2018 Atrial Segmentation Challenge held in conjunction with MICCAI 2018.

## 2 Methods

### 2.1 Imaging Data

The challenge dataset consisted of 100 GE-MRI images from multiple institutions. The slice sizes were either $576 \times 576$ or $640 \times 640$ voxels. Each image contained 88 slices. The image resolution was $0.625 \times 0.625 \times 0.625$ mm$^3$. This dataset was randomly split up into a train, validation and test set of 50, 25 and 25 images, respectively. An independent dataset of 18 GE-MRI images from The University of Utah and King's College London was also used to evaluate the trained model. Moreover, this dataset was also split into a train, validation and test set, of 8, 5, and 5 images, respectively. The sizes of the images in the x- and y-direction varied: $480^2, 576^2, 640^2, 768^2$ or $864^2$ voxels. The number of slices also varied: 40, 44, 50, 80 or 100. The voxel spacing differed as well: 0.35, 0.39, 0.63 or 0.66 mm in the x- and y-direction, and 1.30, 2.00 or 2.05 mm in the z-direction. Our model was also tested on an additional test set which was provided by the challenge organizers. The ground truth of this set was left invisible to us. The left atrium was manually delineated by experts in the field for each GE-MRI image in all datasets. All data that was used received institutional ethics approval.

## 2.2  Network Architecture

The network architecture is depicted in Fig. 1. The network was an adaptation of the original U-Net architecture [3]. The input of this fully convolutional network was a $448 \times 448$ image and the output was a probability map of the same size as the input. The network consisted of convolutional groups. These groups contained two $3 \times 3$ convolutions in the original U-Net [3]. In this work, however, we made one of these layers optional, i.e. the presence of these layers was a hyperparameter during model selection. In Fig. 1, these layers can be seen coloured in yellow. The number of feature maps $k$ in the convolutions of the first convolutional group was varied, compared to a constant number of maps ($k = 64$) in the original U-Net architecture. These convolutional groups were connected to each other by $2 \times 2$ max pooling layers in the contracting path and up-convolutions ($2 \times 2$ upsampling followed by a $2 \times 2$ convolution) in the expanding path.

The number of max pooling layers, which is equal to the number of up-convolutions, was 4 in the original U-Net architecture [3]. However, in this work we generalised this number to the depth of the network $m$ as a hyperparameter during model selection. The number of feature maps in the output of the convolutions doubled after every max pooling layer, and this number halved after every up-convolution. The convolutions in these convolutional groups and those in the up-convolutions used padding, such that the output of the convolution was of the same size as the input of this convolution. The convolutions in the convolutional blocks were followed by rectified linear unit (ReLU) activation and batch normalisation. After the final convolutional block, there was a $1 \times 1$ convolution with sigmoid activation function. The output of this layer was considered to be the output of the model. The total number of convolutional layers $N_{conv}$ in the network could be expressed in terms of network depth $m$:

$$N_{conv} = \begin{cases} 5m + 3, & \text{if the optional convolutions are used} \\ 3m + 2, & \text{if the optional convolutions are not used.} \end{cases} \tag{1}$$

## 2.3  Training

Although the number of unique slices in the training data was 4,400, this dataset contained many slices that were highly similar, since consecutive slices in the

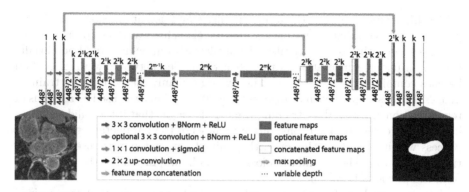

**Fig. 1.** General network architecture with $m$ concatenations between contracting and expanding paths. $k$ is the number of feature maps in the output of the first convolution. The optional layers were used in some of the trained models, which was a hyperparameter during model selection. (Color figure online)

volume images could be nearly identical. Since the training data contained data of only 50 patients, data augmentation was used to artificially generate more training data. Data augmentation was applied to 2D slices. The total number of unique slices after augmentation was increased to 44,000. The types of augmentation were rotation between $-20°$ and $+20°$, scaling between $-20\%$ and $+20\%$, shearing between $-10\%$ and $+10\%$, additive Gaussian noise with a mean of 0 and a standard deviation between 1 and 15 pixels, and contrast changing through the power law transformation.

For model selection, six different models were trained and tested on the validation set $(n = 25)$ using grid search on $m$ and setting of optional convolutions. The six models were all permutations of our hyperparameters: varying network depth $m \in \{4, 5, 6\}$, presence or absence of optional convolutions and number of feature maps $k \in \{32, 64\}$. For these models, we chose a larger number ($k = 64$) of feature maps, whenever the graphical processing unit (GPU) memory permitted, otherwise, a smaller number of $k = 32$ feature maps were used. An NVIDIA TitanX GPU with 3072 CUDA cores was used.

The optimisation algorithm used was the Adaptive Moment Estimation [4] also known as ADAM, with an initial learning rate of 0.0001. The loss function $L$ that was used was an adaptation to the Dice score, which was proposed by Milletari *et al.* [5]. $L$ was defined as follows:

$$L = 1 - \frac{2\sum_{i,j}(A \cdot B)_{ij} + 1}{\sum_{i,j} A_{ij} + \sum B_{ij} + 1},$$
(2)

where $A$ is the ground truth tensor, and $B$ the output of the neural network with values between 0 and 1. Since $B$ is not a binary image, we did not use the intersection between $A$ and $B$ in the numerator as one does when calculating the Dice score. Furthermore, in both the numerator and the denominator, a $+1$ was included to prevent division by 0. The fraction in the equation was subtracted from 1, such that minimisation of $L$ actually leads to an increased Dice overlap.

During training, 50% of the input 2D tiles contained at least one foreground pixel and 50% contained only background pixels. This was done to prevent the network from seeing too many tiles without left atrium, which could potentially lead to an undersegmentation of the left atrium. The networks were trained on batches containing four tiles each. Tiles were randomly picked from all training images. Early stopping with a *patience* of 4000 iterations was used as a means of regularisation and to reduce training time. This means that if the validation loss had not decreased for 4000 consecutive updates of the parameters, the training process was interrupted. Moreover, the trained parameters were used from the training iteration where validation loss was minimum, instead of the parameters from the previous iteration.

The independent dataset was used to evaluate the model that was trained using the challenge dataset. As a means of transfer learning, the trained parameters from the best performing model were used to continue training with the training images from the independent dataset. Training on the independent dataset was done with the same model settings as the training on the challenge dataset. The images were resized to $576 \times 576$ as the image sizes varied quite substantially in this dataset. Using the validation set of the independent dataset, the model that was solely trained on the challenge set was compared to the model that was trained on both sets.

## 2.4   Processing Pipeline

The tiles were normalised between 0 and 1 before being fed into the network, both during training and prediction. When a volume image was used to make a prediction, the 3D image was split up into tiles with an overlap of $32 \times 32$ voxels. In the z-direction, there was no overlap, since the tiles were 2D. Overlapping tiles were used, since objects that were very close to the border of a tile have little context. The final prediction for voxels which were located at overlapping tiles, was calculated by averaging the predictions from these different tiles. As a post-processing step, only the largest connected component in the 3D image was kept. All other connected components were removed from the prediction. Before testing on the independent dataset, all slices in this set were resized to $576 \times 576$, since the training set also consisted of images with this size.

## 3   Results

The results of the hyperparameter optimisation are given in Table 1. The model with the highest Dice score on the validation set ($n = 25$) was Model 2. Thus, this model was selected as the final model and was therefore used to predict on the test images. This model has no optional convolutional layers, $m = 5$ and $k = 64$. This model had 17 convolutional layers and 61,477,057 trainable parameters in total.

Figure 3 shows the smoothed loss curves of all trained models. Even though the loss was calculated in a very similar fashion as the Dice score, the model

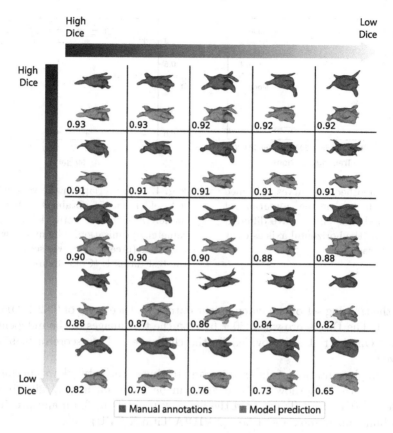

High
Dice

Low
Dice

High
Dice

Low
Dice

0.93  0.93  0.92  0.92  0.92

0.91  0.91  0.91  0.91  0.91

0.90  0.90  0.90  0.88  0.88

0.88  0.87  0.86  0.84  0.82

0.82  0.79  0.76  0.73  0.65

■ Manual annotations    ■ Model prediction

**Fig. 2.** 3D surface plots of atrial segmentations for each test image ($n = 25$) are shown and ordered according to descending Dice scores going left to right, with wrap around. In the bottom left corner of each image, the Dice score of the corresponding volume image is displayed.

with the lowest Dice score was not necessarily the model with lowest loss. This was due to the fact that the points on the loss curve were not determined by the entire validation set, but by four random tiles from the validation set.

In Fig. 4, boxplots of every trained model were drawn. The performance of the models on one image was consistently below 20%. This sample could be seen in Fig. 4 as an outlier in the boxplots of all models. For Model 1 and 2, this was the sole outlier. The other models had either two or three outliers.

The Dice score of the predictions from the final model on the test set is $0.87 \pm 0.07$ (mean ± SD). The middle slices of all images from the test set ($n = 25$) were given in Fig. 5, ordered by their Dice scores. Manual delineations were plotted in red, and the delineations which were predicted by the model were coloured in green. Figure 2 shows the predictions on all test images as 3D surface plots, ordered by their Dice scores.

The Dice score of the model that was solely trained on the challenge dataset was $0.78 \pm 0.14$ on the validation set. When training of this model was continued

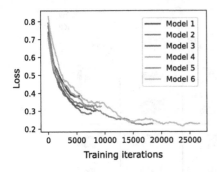

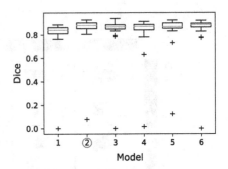

**Fig. 3.** Smoothed loss curves of trained models. The loss on the validation set is displayed. The curves have different domains on the horizontal axis as early stopping was used.

**Fig. 4.** Dice overlaps in the validation dataset ($n = 25$) obtained with each model ($n = 6$) for network hyperparameter optimisation. The model with best Dice (green circle) was selected for the test dataset. (Color figure online)

using the training set of the independent dataset, a Dice score of $0.82 \pm 0.05$ was achieved. The Dice score was $0.80 \pm 0.12$ on the test images of the independent dataset. On the test set provided by the challenge, a Dice score of $0.897$ was achieved.

The implementation of the training and testing of the network was performed on Keras with Tensorflow backend and can be found on https://github.com/coendevente/LAUnet. Training of the network took about 3.5 h and predicting one volume image took about 50 s (NVIDIA TitanX GPU).

**Table 1.** Hyperparameter optimisation process summary

| Models | Optional convolutions | $m$ | $k$ | Trainable parameters | $N_{conv}$ | Dice |
|---|---|---|---|---|---|---|
| 1 | No | 4 | 64 | 15,329,473 | 14 | $0.80 \pm 0.17$ |
| **2** | **No** | **5** | **64** | **61,477,057** | **17** | **$0.85 \pm 0.16$** |
| 3 | No | 6 | 32 | 61,522,017 | 20 | $0.83 \pm 0.17$ |
| 4 | Yes | 4 | 64 | 31,042,369 | 23 | $0.82 \pm 0.17$ |
| 5 | Yes | 5 | 32 | 31,105,953 | 28 | $0.84 \pm 0.15$ |
| 6 | Yes | 6 | 32 | 124,448,673 | 33 | $0.84 \pm 0.18$ |

Hyperparameter selection explored in possible models with grid search. The main hyperparameters were the presence or absence of optional convolutions and network depth $m$. Model 5 and 6 were too large to fit into GPU memory when setting $k = 64$. Therefore, $k$ was set to 32 for these models.

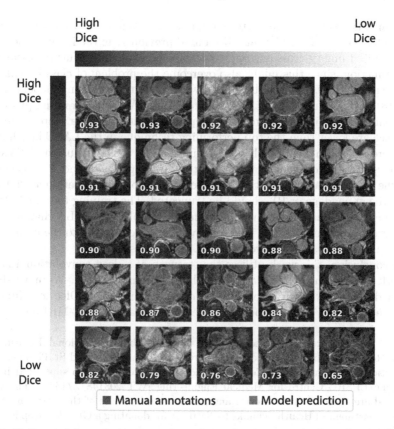

**Fig. 5.** Delineated boundaries from network prediction compared to their manually annotated boundaries. The middle slice in each test image ($n = 25$) is shown and ordered according to descending Dice scores going left to right, with wrap around. In the bottom left corner of each surface plot, the Dice score is displayed. (Color figure online)

## 4  Discussion and Conclusions

This work demonstrated that the original U-Net implementation required modifications by removing optional convolutions and an increased depth ($m = 5$ versus $m = 4$). The best performing model had more parameters than the original U-net. More complex feature could have been learned because of this, which could explain why the model performed better than the original U-Net and the other architectures with less parameters. The two models that were evaluated which had a higher depth did not further improve the performance. A possible explanation of this could be that the models with more parameters overfitted to the training data. Also, the loss function needed to be adapted as U-Net's cross entropy loss was assumed not to be suited to our specific application. However, the neural networks from this work could in further research still be trained with binary cross entropy as loss, to support this assumption. The Dice loss function was thought to be better suited as the ground truth data has a large class imbalance, and Dice has also been used previously in the LASC benchmark [2]. Weighted binary cross entropy could also be explored in future research as a loss function, since this might also overcome the problem of class imbalance.

The network was found to be robust when it was tested on an independent set ($n = 18$) that was not part of the challenge data. When the training procedure was continued on images from the independent set, the Dice score increased. The Dice are lower than the top performing technique in the LASC benchmark, but there is good indication that this could be due to the inclusion of PVs and appendage in the challenge labelled data. The appendage and PVs generally pose the greatest variation in the LA. There was also no information available on where the appendage or PV should be truncated. The training data included at least one outlier image where validation failed completely, the LA chamber's coefficient of variation ($\sigma/\mu$) for its intensity was 0.41 compared to the population, where quartiles Q1, Q2, Q3 were 0.12, 0.15, 0.17 respectively.

In conclusion, this work utilises the strength of a deep convolutional neural network for learning segmentations of the LA from GE-MRI. We have shown that a deeper U-Net with no optional convolutions and a Dice loss function could be optimised for good performance on the challenge GE-MRI data.

**Acknowledgements.** This research was supported by the National Institute for Health Research (NIHR) Biomedical Research Centre at Guy's and St Thomas' NHS Foundation Trust and King's College London. This work was also supported by the Wellcome/EPSRC Centre for Medical Engineering [WT 203148/Z/16/Z]. The views expressed are those of the author(s) and not necessarily those of the NHS, the NIHR or the Department of Health. Thanks to NVIDIA for donating a GPU for deep learning experiments.

# References

1. Karim, R., et al.: Automatic segmentation of left atrial geometry from contrast-enhanced magnetic resonance images using a probabilistic atlas. In: Camara, O., Pop, M., Rhode, K., Sermesant, M., Smith, N., Young, A. (eds.) STACOM 2010. LNCS, vol. 6364, pp. 134–143. Springer, Heidelberg (2010). https://doi.org/10.1007/978-3-642-15835-3_14
2. Tobon-Gomez, C., et al.: Benchmark for algorithms segmenting the left atrium from 3D CT and MRI datasets. IEEE Trans. Med. Imaging **34**(7), 1460–1473 (2015)
3. Ronneberger, O., Fischer, P., Brox, T.: U-Net: convolutional networks for biomedical image segmentation. In: Navab, N., Hornegger, J., Wells, W.M., Frangi, A.F. (eds.) MICCAI 2015. LNCS, vol. 9351, pp. 234–241. Springer, Cham (2015). https://doi.org/10.1007/978-3-319-24574-4_28
4. Kingma, D.P., Ba, J.: Adam: a method for stochastic optimization. arXiv preprint arXiv:1412.6980 (2014)
5. Milletari, F., Navab, N., Ahmadi, S.A.: V-Net: fully convolutional neural networks for volumetric medical image segmentation. In: Fourth International Conference on 3D Vision (3DV), pp. 565–571. IEEE (2016)

# Mixture Modeling of Global Shape Priors and Autoencoding Local Intensity Priors for Left Atrium Segmentation

Tim Sodergren[1], Riddhish Bhalodia[1], Ross Whitaker[1], Joshua Cates[2], Nassir Marrouche[2], and Shireen Elhabian[1(✉)]

[1] Scientific Computing and Imaging Institute, School of Computing,
University of Utah, Salt Lake City, UT, USA
{tsodergren,riddhishb,whitaker,shireen}@sci.utah.edu
[2] Comprehensive Arrhythmia Research and Management Center,
Division of Cardiovascular Medicine, School of Medicine,
University of Utah, Salt Lake City, UT, USA
josh.cates@gmail.com, nassir.marrouche@carma.utah.edu

**Abstract.** Difficult image segmentation problems, e.g., left atrium in MRI, can be addressed by incorporating *shape* priors to find solutions that are consistent with known objects. Nonetheless, a single multivariate Gaussian is not an adequate model in cases with significant nonlinear shape variation or where the prior distribution is multimodal. Nonparametric density estimation is more general, but has a ravenous appetite for training samples and poses serious challenges in optimization, especially in high dimensional spaces. Here, we propose a maximum-a-posteriori formulation that relies on a generative image model by incorporating both local intensity and global shape priors. We use deep autoencoders to capture the complex intensity distribution while avoiding the careful selection of hand-crafted features. We formulate the shape prior as a mixture of Gaussians and learn the corresponding parameters in a high-dimensional shape space rather than pre-projecting onto a low-dimensional subspace. In segmentation, we treat the identity of the mixture component as a latent variable and marginalize it within a generalized expectation-maximization framework. We present a conditional maximization-based scheme that alternates between a closed-form solution for component-specific shape parameters that provides a global update-based optimization strategy, and an intensity-based energy minimization that translates the global notion of a nonlinear shape prior into a set of local penalties. We demonstrate our approach on the left atrial segmentation from gadolinium-enhanced MRI, which is useful in quantifying the atrial geometry in patients with atrial fibrillation.

**Keywords:** Statistical shape models · Autoencoders · Segmentation · Mixture of Gaussians · Generalized expectation-maximization

M. Pop et al. (Eds.): STACOM 2018, LNCS 11395, pp. 357–367, 2019.
https://doi.org/10.1007/978-3-030-12029-0_39

# 1    Introduction

Automatic image segmentation is an important enabling technology in most medical imaging applications that involve soft tissue imaging, e.g., neurology, cardiology, and oncology. In particular, the preoperative anatomical representation of the left atrium (LA) is important for ablation guidance, fibrosis quantification, and biophysical modeling in artial fibrillation patients [1–4]. *Difficult* segmentation problems typically have two aspects. First, *image features*, such as intensity and texture are noisy and unreliable. Second, *anatomical boundaries* are ill-defined, often in the presence of low contrast, background clutter, and partial volumes, while irregular shapes with high variability limit the ability to find invariant features. Nonetheless, if one operates in a context with expectations of particular classes of anatomies (e.g., LA), such challenges can be addressed by means of *shape prior information* to guide and constrain the segmentation process; motivating a *Bayesian formulation*. In LA segmentation context, shape-driven methods were found to be the most appropriate in addressing inherent challenges [5]; thin myocardial wall, surrounding anatomical structures with similar image intensities, and topological variants pertaining to pulmonary veins arrangements [6]. In this paper, we propose a Bayesian *surface-based* segmentation framework that is based on a *mixture-based global shape prior* along with a *feature-based local intensity prior*.

Traditionally, when dealing with low-quality image segmentation, statistical shape information has proved to be helpful in delineating correct object boundaries (e.g., [7,8]). For example, active shape models [9] and their variants incorporate over-restrictive shape constraints in the form of statistical shape models (SSM) by limiting the solution to some low-dimensional linear subspace defined via training shape exemplars. However, these approaches are limited in their ability to accommodate shapes that are not represented in a low-dimensional description; a typical situation arises in applications with small and large-scale shape variability (e.g., [6]). Further, these methods only handle unimodal Gaussian-like shape densities. When it comes to modeling complex shape distributions, a single Gaussian can not adequately model cases of nonlinear shape variation where the probability distribution is multimodal [10].

On the other end of the spectrum, multiatlas-based segmentation (MAS) approaches require a large database of atlases to capture wide range of shape variation where label maps are propagated to the testing image through registration. In essence, MAS can be viewed as a nonparametric estimate of the prior probabilities that converges to the true densities with the number of atlases [11]. To reduce the computational burden introduced by registration, atlas selection is usually performed to exclude irrelevant atlases that might misguide the segmentation process [12]. Generally, kernel-based methods (e.g., [8,13]) are popular strategies for dealing with complex distributions in explicitly or implicitly represented training data. Under mild assumptions, they converge to the true distribution in the limit of infinite sample size [14]. For example, Cremers et al. [8] used kernel density estimate (KDE) to derive non-linear shape priors which are based on a shape distance between implicitly embedded training shapes. Nonetheless, such methods are prone to over-fitting due to small sample size in high-dimensional shape spaces, limiting the generality of the resulting models to fit unseen examples. Specifically, in multivariate density estimation, KDE requires larger kernel width to accommodate more exemplars as the dimension of a variable increases. This will eventually result in model under-fit due to high bias [15]. Further, for kernel based methods, the dimension of the parameter space is proportional to the training sample size.

A finite *mixture* of Gaussians can approximate a full kernel density, to capture nonlinearities in the distribution or subpopulations in the underlying shape subspace [16] where the expectation-maximization (EM) framework is usually used to find the maximum likelihood estimate of mixture parameters. However, modeling shape distribution after being projected onto low-dimensional subspace (e.g., [14,16]) will often collapse or mix the subpopulations, which would derail learning of the mixture structure of the underlying shape space [17]. Nonlinearity of shape statistics can also be modeled by lifting training shapes to a higher, probably infinite, dimensional feature (a.k.a. kernel) space where the shape distribution is assumed to be Gaussian distributed [10], yet this approach results in an infinite-dimensional optimization scheme while sacrificing the efficiency of optimizing in low-dimensional subspaces [14]. Further, one can settle for only an approximate solution for the reverse mapping from feature space to shape space.

The proposed bayesian formulation relies on a maximum-a-posteriori estimation from a generative statistical model that incorporates global, nonlinear shape priors modeled as a mixture of Gaussian components, and local, nonlinear intensity prior as automatically learned image features via *deep autoencoders*. The Gaussian mixture model (GMM) takes into account the linear subspace spanned by each mixture component to avoid the classical problem of model over fitting in high-dimensional spaces [18], and can adequately model non-linear shape variations. To use these shape priors in segmentation, we treat the identity of the mixture component as a latent variable while marginalizing it within a *generalized expectation-maximization* framework (GEM) [19]. We present a conditional maximization-based scheme that alternates between a closed-form solution for component-specific shape parameters that provides a global update-based optimization strategy, and an intensity-based energy minimization that translates the global notion of a nonlinear shape prior into a set of local penalties. Preliminary results show improved accuracy from traditional shape-based approaches.

## 2   Methods

Given an image $\mathcal{I} \in \mathbb{R}^D$, where $D$ is the total number of voxels and $\mathcal{I}(\mathbf{x})$ is the intensity value of the voxel located at $\mathbf{x} \in \mathbb{R}^3$, the Bayesian formulation of the segmentation problem amounts to finding the optimal surface $\mathcal{S}^*$ that maximizes the log-posterior probability $p(\mathcal{S}|\mathcal{I})$, where a shape's surface is represented by a dense set of geometrically consistent $M$-points (landmarks) $\{\mathbf{x}_i^{\mathcal{S}}\}_{i=1}^M$. In order to obtain segmentations that preserve the global shape characteristics of the shape population of interest, the segmentation process is influenced by the prior, in the form of the shape probability distribution. We model such a prior distribution as a finite mixture of Gaussians, parameterized by the mixture component identity $\mathcal{Z}$, which is treated as a latent variable, and the model parameters $\{\Theta_z\}_{z=1}^K$ for $K$-components. Hence, given a latent variable $\mathcal{Z} = z$, the log-posterior can be written as[1],

$$\log p(\mathcal{S}|\mathcal{I}) = \log p(\mathcal{S}, \mathcal{Z} = z|\mathcal{I}) - \log p(\mathcal{Z} = z|\mathcal{S}, \mathcal{I}), \tag{1}$$

where $p(\mathcal{Z}|\mathcal{S}, \mathcal{I})$ defines a conditional probability of the latent variable $\mathcal{Z}$ given the input image $\mathcal{I}$ and the estimated surface $\mathcal{S}$.

## 2.1 Generalized EM for Shape-Driven Segmentation

The use of log-posterior allows us to marginalize the over the latent variable $\mathcal{Z}$. Using *generalized expectation-maximization* (GEM) [19, p. 318], we iterate to find the mode of the marginal posterior $p(\mathcal{S}|\mathcal{I})$ by averaging over the latent variable $\mathcal{Z}$. GEM starts with an initial estimate $\mathcal{S}^o$ of the surface and iteratively refines it by marginalizing over $\mathcal{Z}$. Within each iteration, the posterior $p(\mathcal{S}|\mathcal{I})$ is guaranteed to increase, i.e., converge to a local maxima. Given the current surface guess $\mathcal{S}^{(t)}$, GEM consists of two steps: (1) *E-step*, which computes the latent conditional probability distribution $p(\mathcal{Z}|\mathcal{S}^{(t)}, \mathcal{I}) \; \forall \mathcal{Z} \in \{1, .., K\}$ and (2) *M-step*, which finds a new surface $\mathcal{S}^{(t+1)}$ such that $p(\mathcal{S}^{(t+1)}|\mathcal{I}) \geq p(\mathcal{S}^{(t)}|\mathcal{I})$. The details of the algorithm are as follows.

**E-step:** The conditional probability of the latent variable $\mathcal{Z}$ given the input image and its labeling function can be computed as,

$$p(\mathcal{Z}|\mathcal{S}^{(t)}, \mathcal{I}) = \frac{p(\mathcal{I}|\mathcal{S}^{(t)}, \mathcal{Z})p(\mathcal{S}^{(t)}|\mathcal{Z})p(\mathcal{Z})}{p(\mathcal{I}|\mathcal{S}^{(t)})p(\mathcal{S}^{(t)})} \tag{2}$$

where the intensity model $p(\mathcal{I}|\mathcal{S}^{(t)}, \mathcal{Z})$ in the vicinity of the current surface guess $\mathcal{S}^{(t)}$ is assumed to be conditionally independent of the shape generating component, leaving the E-step to be fully shape-driven. Note the effect of the prior probability $p(\mathcal{Z})$ on the expectation step, where the conditional component distribution would favor mixture components with higher support/proportion, the M-step afterwards would pull the new surface towards the most probable mixture component(s) which generated the shape to be segmented as being compatible with the local intensity and global shape priors.

---

[1] For notational simplicity, we will refer to $p(\mathcal{Z} = z) = p(\mathcal{Z})$ hereafter.

**M-step:** Taking the expectations on both sides of (1) while treating $\mathcal{Z}$ as a random variable with the distribution $p(\mathcal{Z}|\mathcal{S}^{(t)}, \mathcal{I})$ yields,

$$\log p(\mathcal{S}|\mathcal{I}) = \mathbb{E}_{p(\mathcal{Z}|\mathcal{S}^{(t)}, \mathcal{I})} \left[ \log p(\mathcal{S}, \mathcal{Z}|\mathcal{I}) \right] - \mathbb{E}_{p(\mathcal{Z}|\mathcal{S}^{(t)}, \mathcal{I})} \left[ \log p(\mathcal{Z}|\mathcal{S}, \mathcal{I}) \right] \quad (3)$$

where the left side of (1) does not depend on $\mathcal{Z}$. The second term in (3) is maximized when $\mathcal{S} = \mathcal{S}^{(t)}$ (the key result of EM [19]). Hence, it is sufficient for the new surface estimate $\mathcal{S}^{(t+1)}$ to maximize the first term of (3), which is the *expected complete-data log-likelihood.*

$$\mathcal{S}^{(t+1)} = \arg\max_{\mathcal{S}} \sum_{\mathcal{Z}=1}^{K} p(\mathcal{Z}|\mathcal{S}^{(t)}, \mathcal{I}) \log p(\mathcal{S}, \mathcal{Z}|\mathcal{I})$$

$$\propto \arg\max_{\mathcal{S}} \sum_{\mathcal{Z}=1}^{K} p(\mathcal{Z}|\mathcal{S}^{(t)}, \mathcal{I}) \left\{ \log p(\mathcal{I}|\mathcal{S}, \mathcal{Z}) + \log p(\mathcal{S}|\mathcal{Z}) + \log p(\mathcal{Z}) \right\} (4)$$

The intensity term $p(\mathcal{I}|\mathcal{S}, \mathcal{Z})$ gives rise to an intensity model in the vicinity of an estimated surface that resembles the $\mathcal{Z}$–th component. Intensity features (learned via a deep autoencoder [20]) $\mathcal{F}_i^{\mathcal{I}}$ are assumed to follow a normal distribution in the learned feature space with parameters $\Theta_i^{\mathcal{F}} = \{\mu_i^{\mathcal{F}}, \Sigma_i^{\mathcal{F}}\}$ that are estimated from the training data, where $\mu_i^{\mathcal{F}}$ is the mean intensity feature at the $i$–th point and $\Sigma_i^{\mathcal{F}}$ is the corresponding covariance matrix. Notice that $\mathcal{S}^{(t+1)}$ increases the first term of (3) since it is obtained based on its maximization. In addition, $\mathcal{S}^{(t+1)}$ decreases the second term of (3), because such a term is maximized only when $\mathcal{S} = \mathcal{S}^{(t)}$. Hence, if $\mathcal{S}^{(t+1)} = \mathcal{S}^{(t)}$, then the GEM method has converged to a local maxima of the posterior $p(\mathcal{S}|\mathcal{I})$.

## 2.2   Mixture Modeling of Global Shape Priors

The combination of high dimensional shape space and a small number of example shapes usually makes modeling of shape priors subject to *curse-of-dimensionality*, which hinders the effectiveness of mixture learning and density estimation in high dimensional spaces. To avoid over fitting, we can parameterize the covariance matrix of a component, in a manner similar to [18], through its eigen decomposition. Hence we can learn the Gaussian mixture model taking into consideration the dominant linear subspace spanned by each mixture component, where the noise variance is assumed to be isotropic and contained in a subspace orthogonal to the component's subspace.

We first model a dense set of
$M$ homologous landmarks via par-
ticle based modeling. Assume that
the anatomy-specific shape space
of all shapes defined using the
$M$−landmarks representation is com-
prised of $K$−mixture components.
Each component is parameterized by
$\Theta_k = \{\pi_k, \mu_k, \Lambda_k, \Psi_k, d_k, \sigma_k\}$ and
learned via high-dimensional EM [18]
where $\pi_k = p(\mathcal{Z} = k)$ is the component probability/proportion, $\mu_k$ is
the mean vector, $d_k$ is the intrinsic dimension of the $k$−th mixture
component, i.e., the number of dominant modes of shape variation, $\Lambda_k$
is a diagonal matrix containing the
largest $d_k$ eigenvalues of the component's full covariance matrix $\Sigma_k$, $\Psi_k$

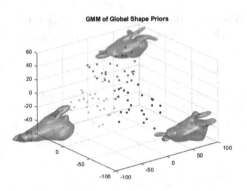

**Fig. 1.** Low-dimensional subspace representation of high-dimensional shape data. Each
sample in the training set is represented by a
single point with color representing different
components of the GMM. An example segmentation from each component is included
to illustrate differences.

is the orthonormal matrix containing the corresponding $d_k$ eigenvectors, and $\sigma_k$
is the standard deviation of the off-subspace noise. Hence the component-specific
subspace $\mathbb{S}_k$ can be defined as follows:

$$\mathbb{S}_k = \left\{ \mathcal{S}_k^{\beta} = \mu_k + \Psi_k \beta_k \mid \mathcal{S}_k^{\beta} \in \mathbb{R}^{M \times 3}, \beta_k \in \mathbb{R}^{d_k} \right\} \tag{5}$$

where $\beta_k \in \mathbb{R}^{d_k}$ encodes the shape parameters w.r.t. the $k$−th component subspace $\mathbb{S}_k$. The shape probability distribution $p(\mathcal{S})$ is defined as a weighted sum
of component-conditional probabilities $\{p(\mathcal{S}|\Theta_k)\}_{k=1}^{K}$ using the mixture weights
$\{\pi_k\}$. Notice, that the component-wise model beyond the first $d_k$ modes consists
of an isotropic variance in the directions orthogonal to the subspace, which has
the effect of penalizing point wise differences from the learned, $d_k$-dimensional
subspace. Hence, the surface probability conditional on the $k$−th mixture component becomes a product of two marginal and independent Gaussians; within-
subpace and off-subspace [18].

We illustrate this training step in Fig. 1. These data are from the samples
selected for training and illustrate a low-dimensional subspace embedding of
every sample colored according to their class membership in a mixture model
of 3 components along with example LA segmentations from each. One can
observe key differences between each component such as the number and shape
of pulmonary veins.

## 2.3   Autoencoding Local Intensity Priors

Similar to the active shape model (ASM) approach of [16], we model local intensity profiles at the object surface consisting of the normalized first derivative of image intensities oriented along the approximate normal to the surface and centered at each landmark location, with radius of $\ell$ voxels with total length of $L = 2\ell + 1$ (Fig. 2). Corresponding profiles (i.e., those that share the same landmarks) are combined across all training images but are treated independently of one another. The traditional approach to segmentation is to use the Mahalanobis distance of candidate profiles to the training data as a cost function to be minimized, however, it assumes

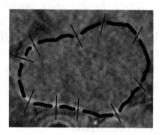

**Fig. 2.** Intensity gradients across the segmentation surface are taken at each point location.

the feature space to be first order linear and normally distributed. In order to overcome this limitation, we employ deep autoencoders to adaptively learn the higher order non-linear features. In this case, we use a 2 layer × 10 hidden units sparse autoencoder to generate pseudolinear representations of the feature space for each landmark. Due to the computationally expensive nature of this step we determined these parameters through a qualitative empirical analysis but there is potential for improvement via more rigorous cross-validation. To improve the statistics and capture lateral intensity changes, we append additional profiles which are parallel to the main profile and, offset by a subvoxel distance, with interpolated intensity values, to form "thick" profiles. We repeat this for multiple image resolutions and store the autoencoder as well as the encoded output.

## 2.4   Segmentation via Conditional Maximization

The marginal posterior optimization problem in (4) involves two types of latent variables, the labeling function to be estimated $\mathcal{S}$ and the component-specific shape parameters $\beta_k$. To seek an optimal solution for $\mathcal{S}$, we propose a *conditional maximization*-based scheme [19, p. 312] that alternates between two optimization phases. In the first phase, we optimize for $\{\beta_k\}_{k=1}^K$ given $\mathcal{S}$, and in the second phase we optimize for $\mathcal{S}$ given individual shape parameters $\{\beta_k\}_{k=1}^K$. Starting with the coarsest image resolution, we take the mean landmark positions from the training data as the initial model and iterate through these steps.

**Localized feature-based optimization:** For each landmark position, we generate a series of overlapping profiles of $L-$dimension that extend above and below the original profile by some specified search length $s$. For each of these candidate profiles, we generate features using pre-trained autoencoders. We then take the Mahalanobis distance of each candidate profile to the encoded training features and update the landmark position to the center of the optimal profile, oen with the smallest distance. This step ensures that landmarks are locally optimal, but they may no longer by globally optimal as there is no shape information encoded into these features.

**Component-specific shape parameters optimization:** For a given surface $\mathcal{S}$ (from the intensity-based step), component-specific shape parameters $\beta_k$ can be obtained by projecting the surface onto the component-specific subpace, leading a closed-form solution $\beta_k = \Psi_k^T(\mathcal{S} - \mu_k)$. By limiting the model parameters to what is considered a normal contour with respect to shape, the landmark positions become globally optimal. We repeat this a predetermined number of times at each resolution, coarse to fine.

## 3 Results

We tested these methods on 100 3D MRI images, separated into 80 for training and 20 for testing. Figure 3 shows a sample MRI image with the provided segmentation. The segmented volume is also separately displayed in blue along with modeled landmark positions. We first develop a shape model of 2048 homologous landmark positions modeled with a 3 component Gaussian mixture model. We then model local intensity gradients for each position across all samples using an autoencoder on profiles 11 voxels in length (optimized through cross-validation). In order to improve model statistics, we encode additional parallel information to create "thick" profiles. This model was then used to segment the 20 test images according to the methods described above. To compare the effectiveness of our method we evaluate results both with and without autoencoded local intensity profiles as well as with and without the mixture model. Results are in Table 1. While there appears to be some improvement in median Dice coefficient, this may not be the best metric for evaluating the efficiency of the proposed method [21]. We also evaluated the euclidean distance between final landmark positions from our segmentations to those of the ground truth. This also allows for a more detailed point by point analysis of error as shown in Fig. 4. Here we see two examples from the test data set showing clear improvement in segmentation results when using autoencoded profiles. We truncate the color scale at 10 (1 cm) and flag any higher error as an extreme outlier. Error is generally limited to image edges around the pulmonary veins when using autoencoded intensity

(a) MRI test image            (b) Original segmentation

**Fig. 3.** MRI image (a) of test sample next to LA segmentation with landmarks (b).

profiles as opposed to nearly universally high error without. These results could be further improved by initializing the model with the component mean shapes as opposed to the global mean which is how we are presently doing it.

**Table 1.** Dice scores for test results.

| Autoencoder | yes | yes | no | no |
|---|---|---|---|---|
| Mixture model | yes | no | yes | no |
| Median Dice coefficient | 0.743 | 0.747 | 0.729 | 0.739 |

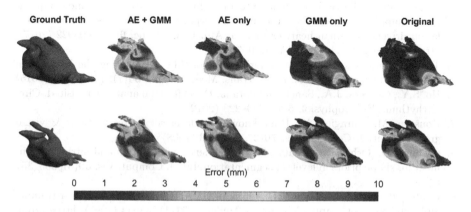

**Fig. 4.** Example results for 2 different samples with point by point error mapped onto segmentation surface. AE = autoencoder, GMM = mixture modeling, Original = neither (original ASM method).

## 4   Conclusion

This paper proposed a shape/feature-based generative model for left atrium segmentation by modeling non-Gaussian global shape priors as mixture of Gaussians on landmark-based representations of training data and learning feature-based representations for local intensity priors. The mixture parameters were learned in the high dimensional shape space by taking into account component-specific subspaces to avoid over fitting. The method used a variant of ASM-based segmentation framework that relies on a maximum a-posteriori estimation with a marginalization over class membership within a generalized EM framework. While these results are preliminary, they suggest that ASM's for LA segmentation can be improved through optimized prior modeling. Local optimization through autoencoding of non-linear intensity features and global optimization through mixture modeling of shape parameters increases the accuracy of the original method.

**Acknowledgment.** This work was supported by the National Institutes of Health [grant numbers R01-HL135568-01 and P41-GM103545-19].

# References

1. Calkins, H., et al.: 2017 hrs/ehra/ecas/aphrs/solaece expert consensus statement on catheter and surgical ablation of atrial fibrillation. Hear. Rhythm **14**(10), e275–e444 (2017)
2. McGann, C., et al.: Atrial fibrillation ablation outcome is predicted by left atrial remodeling on mri. Circ. Arrhythmia Electrophysiol. **7**(1), 23–30 (2014)
3. Hansen, B.J., et al.: Atrial fibrillation driven by micro-anatomic intramural re-entry revealed by simultaneous sub-epicardial and sub-endocardial optical mapping in explanted human hearts. Eur. Hear. J. **36**(35), 2390–2401 (2015). https://doi.org/10.1093/eurheartj/ehv233
4. Zhao, J., et al.: Three-dimensional integrated functional, structural, and computational mapping to define the structural "fingerprints" of heart-specific atrial fibrillation drivers in human heart ex vivo. J. Am. Hear. Assoc. **6**(8), e005922 (2017). https://doi.org/10.1161/JAHA.117.005922
5. Tobon-Gomez, C., et al.: Benchmark for algorithms segmenting the left atrium from 3d ct and mri datasets. IEEE Trans. Med. Imaging **34**(7), 1460–1473 (2015)
6. Ho, S.Y., Cabrera, J.A., Sanchez-Quintana, D.: Left atrial anatomy revisited. Circ. Arrhythmia Electrophysiol. **5**(1), 220–228 (2012)
7. Rousson, M., Paragios, N.: Prior knowledge, level set representations & visual grouping. Int. J. Comput. Vis. **76**(3), 231–243 (2008)
8. Cremers, D., Osher, S.J., Soatto, S.: Kernel density estimation and intrinsic alignment for shape priors in level set segmentation. Int. J. Comput. Vis. **69**(3), 335–351 (2006)
9. Cootes, T., Taylor, C., Cooper, D., Graham, J.: Active shape models-their training and application. Comput. Vis. Image Underst. **61**(1), 38–59 (1995). https://doi.org/10.1006/cviu.1995.1004
10. Cremers, D., Kohlberger, T., Schnörr, C.: Shape statistics in kernel space for variational image segmentation. Pattern Recognit. **36**(9), 1929–1943 (2003)
11. Awate, S., Whitaker, R.: Multiatlas segmentation as nonparametric regression. IEEE Trans. Med. Imaging **33**(9), 1803–1817 (2014)
12. Aljabar, P., Heckemann, R.A., Hammers, A., Hajnal, J.V., Rueckert, D.: Multi-atlas based segmentation of brain images: atlas selection and its effect on accuracy. Neuroimage **46**(3), 726–738 (2009)
13. Wimmer, A., Soza, G., Hornegger, J.: A generic probabilistic active shape model for organ segmentation. In: Yang, G.-Z., Hawkes, D., Rueckert, D., Noble, A., Taylor, C. (eds.) MICCAI 2009, Part II. LNCS, vol. 5762, pp. 26–33. Springer, Heidelberg (2009). https://doi.org/10.1007/978-3-642-04271-3_4
14. Rousson, M., Cremers, D.: Efficient kernel density estimation of shape and intensity priors for level set segmentation. In: Duncan, J.S., Gerig, G. (eds.) MICCAI 2005, Part II. LNCS, vol. 3750, pp. 757–764. Springer, Heidelberg (2005). https://doi.org/10.1007/11566489_93
15. Friedman, J.H., Stuetzle, W., Schroeder, A.: Projection pursuit density estimation. J. Am. Stat. Assoc. **79**(387), 599–608 (1984). https://doi.org/10.1080/01621459.1984.10478086
16. Cootes, T.F., Taylor, C.J.: A mixture model for representing shape variation. Image Vis. Comput. **17**(8), 567–573 (1999)
17. Dasgupta, S.: Learning mixtures of Gaussians. In: 40th Annual Symposium on Foundations of Computer Science, pp. 634–644. IEEE (1999)

18. Bouveyron, C., Girard, S., Schmid, C.: High-dimensional data clustering. Comput. Stat. Data Anal. **52**(1), 502–519 (2007)
19. Gelman, A., Carlin, J.B., Stern, H.S., Rubin, D.B.: Bayesian Data Analysis. Chapman & Hall/CRC texts in statistical science, Boca Raton (2003)
20. Vincent, P., Larochelle, H., Lajoie, I., Bengio, Y., Manzagol, P.A.: Stacked denoising autoencoders: learning useful representations in a deep network with a local denoising criterion. J. Mach. Learn. Res. **11**, 3371–3408 (2010)
21. Taha, A.A., Hanbury, A.: Metrics for evaluating 3D medical image segmentation: analysis, selection, and tool. BMC Med. Imaging **15**, 29 (2015). https://doi.org/10.1186/s12880-015-0068-x. http://www.ncbi.nlm.nih.gov/pmc/articles/PMC4533825/, 68[PII]

# Left Ventricle Full Quantification Challenge

# Left-Ventricle Quantification Using Residual U-Net

Eric Kerfoot, James Clough, Ilkay Oksuz$^{(\boxtimes)}$, Jack Lee, Andrew P. King,
and Julia A. Schnabel

School of Biomedical Engineering & Imaging Sciences,
King's College London, London, UK
{eric.kerfoot,ilkay.oksuz}@kcl.ac.uk

**Abstract.** Estimating dimensional measurements of the left ventricle
provides diagnostic values which can be used to assess cardiac health and
identify certain pathologies. In this paper we describe our methodology
of calculating measurements from left ventricle segmentations automatically generated using deep learning. We use a U-net convolutional neural
network architecture built from residual units to segment the left ventricle and then process these segmentations to estimate the area of the
cavity and myocardium, the dimensions of the cavity, and the thickness of the myocardium. Determining if an image is part of the diastolic
or systolic portion of the cardiac cycle is done by analysing the cavity
volume. The quality of our results are dependent on our training regime
where we have generated a large derivative dataset by augmenting the
original images with free-form deformations. Our expanded training set,
in conjunction with simple affine image transforms, creates a sufficiently
large training population to prevent over-fitting of the network while
still creating an accurate and robust segmentation network. Assessing
our method on the STACOM18 LVQuan challenge dataset we find that
it significantly outperforms the previously published state-of-the-art on
a 5-fold validation all tasks considered.

**Keywords:** Cardiac MR · Cardiac quantification ·
Convolutional neural networks

## 1 Introduction

Dimensional measurements of the left ventricle (LV) can be used to quantify cardiac health and identify some pathologies. For example, measuring the cavity volume throughout the cardiac cycle allows the calculation of the ejection fraction
and stroke volume biomarkers, and measuring the thickness of the myocardium
aids in diagnosing hypertrophic cardiomyopathy. Calculating these measurements automatically and accurately is an open challenge which, if addressed,
would provide a time and cost saving tool for clinical diagnostic use.

In this paper we describe our method for left ventricle quantification based
on the automatic segmentation of the LV using deep learning. We trained a

© Springer Nature Switzerland AG 2019
M. Pop et al. (Eds.): STACOM 2018, LNCS 11395, pp. 371–380, 2019.
https://doi.org/10.1007/978-3-030-12029-0_40

network to segment the myocardium of full cycle cardiac MR images (cMRI) captured at three positions on the LV. These segmentations are then processed to calculate, for each image, cavity and myocardium volumes, cavity dimensions, regional wall thicknesses, and assignment to diastolic or systolic portion of the cycle. Our method relies on an accurate and robust segmentation network and so we use data augmentation to create an expanded input dataset from an original smaller set, and apply further augmentations during the training process.

This work is our entry for the STACOM Left Ventricle Full Quantification Challenge for MICCAI 2018. The challenge involves the calculation of cavity and myocardial areas, three dimensional values of the cavity, six regional wall thickness values, and assigning each image to diastole or systole phase. We will discuss in this paper our method for inferring a myocardial segmentation from the training input images, and then the calculation of these metrics from the segmentations using image processing routines.

## 2  Background

The quantification of cardiac indices from cardiac MRI has been investigated with a variety of approaches in literature [7]. Early works relied on manual delineation of the myocardial borders and used these manual contours to calculate the cardiac indices [10]. Recent methods have bypassed the segmentation step and posed the problem as a direct regression of cardiac indices from cMRI images. The early efforts at regression focused on a two step strategy, where hand-crafted features are generated and used in a regression setting for cardiac indices estimation [14]. With the development of deep learning techniques more advanced models have been proposed to circumvent the problems of hand-crafted features by using end-to-end trained networks [16]. More recently, Xue et al. [17] have incorporated the temporal information from different phases of the cardiac cycle using a recurrent neural network.

Machine learning techniques to automate myocardial segmentation have achieved considerable success [7]. The U-net approach of Ronneberger et al. [8] has been widely adopted in the context of medical image analysis. In the context of cardiac MR segmentation, Avendi et al. [1] combined CNNs and a deformable model to perform LV segmentation, though only for the endocardial wall and only in the end-diastole and end-systole phases.

Tran [12] used a 15-layer CNN to perform complete left and right ventricular segmentation. Tan et al. [11] proposed to use radial distances calculated with a polar transform rather than per-pixel binary image segmentations. Most recently, Oktay et al. [6] incorporate global shape information into neural networks, and Bai et al. [2] showcases human-level performance for segmentation using a fully convolutional neural network.

# 3  Methods

The proposed framework consists of three stages: (1) image preprocessing and augmentation, (2) image segmentation with our CNN architecture, and (3) estimation the cardiac indices using the output segmentation information.

## 3.1  Image Preprocessing

An accurate and robust segmentation network requires a large dataset so that it learns a general solution which can correctly segment images not seen in training, and does not become over-fitted to the input data. The key concept is variety since the network is being trained to identify geometry embedded in varying contexts. If some ancillary feature in this context is often present with important features the network will correlate these features and produce a poor result for inputs lacking the ancillary feature. To provide this varied data, we first create an expanded input dataset from the original challenge data and then apply data augmentation during the training process.

The training dataset for the STACOM challenge consists of 145 subjects each with 20 image and segmentation frames, resulting in an input image matrix of dimensions $2900 \times 80 \times 80$ pixels and a binary segmentation image of the same size. As a preprocessing step we apply a non-rigid deformation to each image/segmentation pair twenty times to produce an expanded image matrix and a segmentation matrix of dimensions $2900 \times 21 \times 80 \times 80$. This processing step deforms the pixels of the image/segmentation pair near the centre in randomised directions using a smooth interpolation function, thus producing an image which has slightly different geometry to its original but still has an accompanying segmentation which segments the myocardium.

During training, we use data augmentation [5,9] when creating an input batch of images at each step. Images from the expanded dataset are selected at random, then a random selection of transpose, flip, 90/180/270-degree rotation, and shift operations are applied to each image/segmentation pair. Applying these transformations essentially produces a further expanded dataset which contains increased image variation although not further geometric variation. This prevents the network from fixating on features in specific regions of its perceptive field since the random transformations move such features around the field during training. Augmentation is the only technique we use to prevent over-fitting, other techniques like dropout were found not to improve performance and so omitting them contributed to a simpler network architecture.

## 3.2  Segmentation Network Architecture

Our network architecture (Fig. 1) is based on U-net [8] with the layers of the encoding and decoding stages defined using residual units [4,18]. Each box of the encoding portion is implemented with a residual unit and labelled with the output volume shape. The decoding portion boxes are implemented with an upsample unit and have the same output volume dimensions as the encoding layer on the same level.

Parametric rectifying linear units [3] are used to allow the network to learn a better activation which improves segmentation. Instance normalization is used to prevent contrast shifting, ensuring input image contrast is not skewed by being batched with images having significantly different contrast ranges [13].

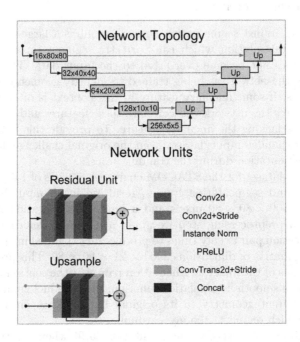

**Fig. 1.** U-net network architecture

As Fig. 1 shows, we use convolutions with a stride of 2 for downsampling and transpose convolutions with a stride of 2 for upsampling, rather than pooling and unpooling layers. This allows the network to learn optimal downsample/upsample operations while also reducing the number of layers in the network's units. Note that the first residual unit, marked $16 \times 80 \times 80$, uses a stride of 1 in its first convolution layer so that the original input image is not immediately downsampled.

### 3.3 Cardiac Indices Estimation

All indices are calculated from myocardial segmentations derived from the CNN using conventional image processing routines. The myocardial area is estimated by counting the number of pixels in the segmentation image and scaling by the image dimensions. Similarly, the cavity area is estimated by counting pixels in the enclosed area of the segmentation image. Phase estimation is derived from the cavity areas, where the image having the smallest area is considered the end systole image, and the largest is the end diastole image.

Calculating the three cavity dimension values (IS-AL, I-A, and IL-AS) and the six regional wall thickness values (IS, I, IL, AL, A, and AS) requires measuring thickness values over an area of the segmentation image. Our technique is to unravel the segmentation by converting each pixel coordinate into a polar coordinate then projecting those coordinates onto a Cartesian plane. Figure 2 illustrates this process by unravelling a shaded annulus into an image where, for each pixel, the row index represents the distance from the original image centre and the column index represents the angle in degrees around the centre clockwise starting from the top.

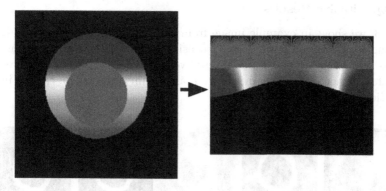

**Fig. 2.** Segmentation polar projection showing transformation of original image to unravelled equivalent image (colouration chosen to highlight transformation). (Color figure online)

By unravelling the segmentation images in this way, the cavity dimensions can be calculated by averaging the maximal row index of non-zero pixels over each column corresponding to the angular areas of the segmentation to be considered. For example, to calculate the IS-AL cavity dimension, the average of the last row index of non-zero pixels is calculated over columns 30 to 90 and 210 to 270 then multiplied by two. Similarly the region wall thicknesses are calculated by averaging the difference between the last row index of non-zero pixels and the first over the columns to consider for each measurement zone.

### 3.4 Implementation Details

We trained our network using the PyTorch 0.4.0 library on nVidia TITAN Xp and P6000 GPUs. Segmentation analysis routines are all implemented in Python 3.6 using the Numpy, SciPy, and Numba libraries. The training takes approximately 3 h for 10000 steps with a batch size of 1200. Our code is available to download at https://github.com/ericspod/STACOM18.

## 4    Experimental Results

To demonstrate the robustness of our network and analysis routines, we trained the network on three different dataset folds where the set of subjects is divided into three, five, and seven groups. We separately trained fifteen networks on each dataset instance with one of these groups reserved as the validation set, each network trained for 10000 steps with a batch size of 1200.

This batch size was chosen as the largest we could fit on one of our GPUs. The folds were created using the recommended process of the STACOM LVQuan18 challenge, ensuring that data from one subject is not split between training and test sets, and facilitating comparison with other methods in the challenge.

### 4.1    Qualitative Results

In Fig. 3, we show an example image from the training dataset with the corresponding segmentation mask generated using our algorithm. The segmentation mask is very accurate and has only few over- or under-segmented regions. The successful segmentations generated by our algorithm is the key for cardiac indices estimation.

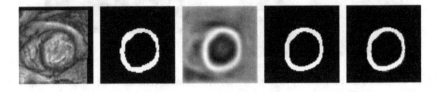

**Fig. 3.** Example training images: first and second images are input image and associated ground truth segmentation from training dataset, third is output logits from network, fourth is estimated segmentation from thresholding logits, fifth illustrates difference between ground truth and estimated segmentation, where red represents the under segmented regions and blue the over-segmented regions. (Color figure online)

### 4.2    Quantitative Results

Table 1 lists the results of applying the trained networks to the validation set and then applying our metric routines to calculate the area, dimension, wall thickness, and phase values for each image. The table lists, for each metric type, the mean absolute error (MAE) and Pearson correlation coefficient ($\rho$) when comparing the ground truth metrics against the calculated values. The last column lists the intersection-over-union error between the ground truth myocardial segmentation images versus the calculated segmentations. The last row includes the results of applying our methodology to the challenge's test dataset, omitting IoU as the ground truth segmentations are not yet available.

The conclusion from this study is that our network is generally stable and insensitive to varying training input datasets, exhibiting a loss of accuracy correlated with fold size. In particular the IoU values for fold three are greater than

**Table 1.** Fold validation results. The area, cavity dimension, and region wall thickness metrics were calculated for the validation portion of the fold (e.g. for fold 3,0 the first third of images from the dataset were validation). The mean absolute error (MAE, in mm$^2$ units for area and mm units for dim/RWT) and Pearson correlation coefficient ($\rho$) were calculated for each metric as a group. The IoU column is the intersection-over-union difference between the segmentations inferred from the fold network and the ground truth segmentations.

| Fold | | Area | | Cavity Dim. | | RWT | | Phase | IoU |
|---|---|---|---|---|---|---|---|---|---|
| | | MAE | $\rho$ | MAE | $\rho$ | MAE | $\rho$ | % Error | |
| 3 | 0 | 93.184 | 0.9631 | 1.0112 | 0.9808 | 0.8864 | 0.9355 | 6.758 | 0.1218 |
| | 1 | 71.872 | 0.9827 | 0.8712 | 0.9864 | 0.7352 | 0.9565 | 7.379 | 0.1214 |
| | 2 | 79.744 | 0.9659 | 0.8784 | 0.9836 | 0.7880 | 0.9416 | 6.931 | 0.1274 |
| Mean | | 81.600 | 0.9706 | 0.9203 | 0.9836 | 0.8032 | 0.9445 | 7.023 | 0.1235 |
| 5 | 0 | 64.576 | 0.9841 | 0.7840 | 0.9875 | 0.7008 | 0.9597 | 6.482 | 0.1059 |
| | 1 | 58.368 | 0.9873 | 0.7808 | 0.9883 | 0.6800 | 0.9621 | 6.689 | 0.1078 |
| | 2 | 56.512 | 0.9879 | 0.7728 | 0.9885 | 0.6456 | 0.9641 | 6.965 | 0.1101 |
| | 3 | 68.928 | 0.9793 | 0.8488 | 0.9861 | 0.7056 | 0.9551 | 7.172 | 0.1125 |
| | 4 | 63.104 | 0.9837 | 0.7712 | 0.9903 | 0.6776 | 0.9614 | 6.310 | 0.1100 |
| Mean | | 62.298 | 0.9845 | 0.7915 | 0.9881 | 0.6819 | 0.9605 | 6.724 | 0.1093 |
| 7 | 0 | 59.648 | 0.9863 | 0.7912 | 0.9878 | 0.6640 | 0.9635 | 6.344 | 0.1031 |
| | 1 | 55.488 | 0.9887 | 0.7496 | 0.9896 | 0.6352 | 0.9662 | 6.482 | 0.1053 |
| | 2 | 64.768 | 0.9870 | 0.8336 | 0.9900 | 0.6560 | 0.9703 | 6.758 | 0.1038 |
| | 3 | 64.704 | 0.9879 | 0.7720 | 0.9900 | 0.6456 | 0.9694 | 6.137 | 0.1057 |
| | 4 | 54.080 | 0.9867 | 0.7416 | 0.9871 | 0.6328 | 0.9615 | 7.379 | 0.1032 |
| | 5 | 54.912 | 0.9879 | 0.7096 | 0.9905 | 0.6112 | 0.9689 | 6.724 | 0.1015 |
| | 6 | 62.528 | 0.9886 | 0.7360 | 0.9913 | 0.6576 | 0.9678 | 6.517 | 0.1048 |
| Mean | | 59.447 | 0.9875 | 0.7619 | 0.9895 | 0.6432 | 0.9668 | 6.620 | 0.1039 |
| Test | | 301.811 | 0.9384 | 2.3252 | 0.9742 | 2.1531 | 0.7789 | 10.0 | N/A |

those for fold seven as one would expect given that the former had fewer training images and more unseen validation images. However the IoU values within folds are roughly equivalent, demonstrating the training was not sensitive to the presence of absence of particular input images.

The results from the test dataset correlate in general with this conclusion although there is a meaningful loss of accuracy across all metrics. For a challenge of this type it is expected that the test set would be particularly difficult and the results show this, in particular we observed some images of the test set could not be segmented correctly and so our algorithms had to compensate for an incomplete annular segmentation to generate a result.

## 5    Discussion and Conclusion

In this paper we have proposed our CNN-based approach to calculating left-ventricle metrics for the STACOM 2018 challenge. The central component to this approach is a U-net based network for segmenting the myocardium of the left ventricle. Our modifications to the classic U-net architecture and our training process are critical components to producing an accurate segmentation. Segmentations generated by this network are then processed to calculate the metrics in question.

We chose a method which relies on segmentation, in contrast to [15] which proposes a regression approach for calculating metrics directly with a RNN, for a number of practical and quantitative reasons. U-net and other architectures as applied to cardiac segmentation in particular are well understood and known to be robust when trained with a varied dataset. In particular residual units and data augmentation are now common practice when training such networks and have obvious benefits in terms of training speed and producing a network which is robust against input dissimilar from the training set.

More quantitatively, using segmentations allows the numeric and visual assessment of the network's outputs. Since the segmentations are used as input to routines for calculating the metrics, which are accurate when applied to the ground truth segmentations, the accuracy of the final metrics relies entirely on that of the segmentations. To determine if the network is producing accurate segmentations, metrics such as intersection-over-union can be used to assess accuracy, or image processing routines can be used to ensure each output image represents a single correct annulus. Most easily a simple visual inspection can be made of the outputs as overlaid on the source images.

In our experience it was clear when subjects were segmented poorly by inspecting the produced myocardial segmentations and it was equally clear why input image quality caused the network to fail. If images dissimilar from the training data produce poor results when applied to a network then visual analysis can help in diagnosing why, for example determining if other image structures were identified as the LV, if image noise made the myocardium indistinct, or if the network suffers from overfitting and so produced indistinct and jumbled segmentations. This contributes to a methodology which is less of a black-box approach compared to others, and which permits visual inspection and analysis when failure occurs.

Our scheme of image augmentations was designed to prevent overfitting to the set of training images, and so make our method more generalisable to the images in the test set (for which we do not have ground-truth segmentations due to the nature of the STACOM 2018 challenge). We assessed the effect of these augmentations by training the network with just the original training data and applying it to the test set images to produce segmentations.

Although we do not have ground-truth segmentations to compare with, visual inspection of the results, in Fig. 4 demonstrates the usefulness of the augmentation scheme. Furthermore, we counted the number of non-annular segmentations as a rough measure of segmentation quality. Overall, the test set yielded

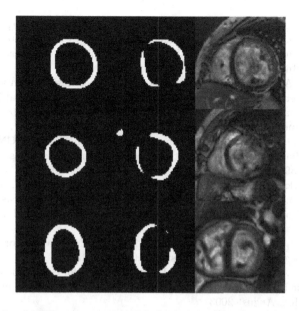

**Fig. 4.** Example of bad segmentations. Mask on left is produced by network trained with augments, mask on right without. Right-hand side is source image from test dataset.

6 non-annular segmentations when the network was trained with our augmentations, and 158 such segmentations when trained without the augmentations, demonstrating that they are key to training a network which generalises well.

Corollary to this is the observation that training without augmentations produces a much smaller final loss value as compared to training with augmentations. Having trained on the whole dataset for 5000 iterations with a batch size of 600 images, the final loss value without augmentations is 0.0092 but with augmentations is 0.0575. In conjunction with the observation of how manybreak non-annular segmentations are produced, this clearly indicates significant over-fitting to the training dataset when augmentations are not used.

**Acknowledgments.** This work was supported by an EPSRC programme Grant (EP/P001009/1) and the Wellcome EPSRC Centre for Medical Engineering at the School of Biomedical Engineering and Imaging Sciences, King's College London (WT 203148/Z/16/Z). The GPUs used in this research was generously donated by the NVIDIA Corporation.

# References

1. Avendi, M., Kheradvar, A., Jafarkhani, H.: A combined deep-learning and deformable-model approach to fully automatic segmentation of the left ventricle in cardiac MRI. Med. Image Anal. **30**, 108–119 (2016)
2. Bai, W., et al.: Automated cardiovascular magnetic resonance image analysis with fully convolutional networks. J. Cardiovasc. Magn. Reson. **20**, 65 (2018)

3. He, K., Zhang, X., Ren, S., Sun, J.: Delving deep into rectifiers: surpassing human-level performance on imageNet classification. CoRR abs/arXiv:1502.0185 (2015)
4. He, K., Zhang, X., Ren, S., Sun, J.: Identity mappings in deep residual networks. CoRR abs/1603.05027 (2016)
5. Krizhevsky, A., Sutskever, I., Hinton, G.E.: Imagenet classification with deep convolutional neural networks. In: Proceedings of the 25th International Conference on Neural Information Processing Systems, NIPS 2012, vol. 1, pp. 1097–1105. Curran Associates Inc., USA (2012)
6. Oktay, O., et al.: Anatomically constrained neural networks (ACNNs): application to cardiac image enhancement and segmentation. IEEE Trans. Med. Imaging 37(2), 384–395 (2018)
7. Peng, P., Lekadir, K., Gooya, A., Shao, L., Petersen, S.E., Frangi, A.F.: A review of heart chamber segmentation for structural and functional analysis using cardiac magnetic resonance imaging. Magn. Reson. Mater. Phys. Biol. Med. 29(2), 155–195 (2016)
8. Ronneberger, O., Fischer, P., Brox, T.: U-Net: convolutional networks for biomedical image segmentation. CoRR abs/1505.04597 (2015)
9. Simard, P.Y., Steinkraus, D., Platt, J.: Best practices for convolutional neural networks applied to visual document analysis. Institute of Electrical and Electronics Engineers, Inc. August 2003
10. Suinesiaputra, A., et al.: Quantification of LV function and mass by cardiovascular magnetic resonance: multi-center variability and consensus contours. J. Cardiovasc. Magn. Reson. 17(1), 63 (2015)
11. Tan, L.K., Liew, Y.M., Lim, E., McLaughlin, R.A.: Convolutional neural network regression for short-axis left ventricle segmentation in cardiac cine MR sequences. Med. Image Anal. 39, 78–86 (2017)
12. Tran, P.V.: A fully convolutional neural network for cardiac segmentation in short-axis MRI. arXiv preprint arXiv:1604.00494 (2016)
13. Ulyanov, D., Vedaldi, A., Lempitsky, V.S.: Instance normalization: The missing ingredient for fast stylization. CoRR abs/1607.08022 (2016). http://arxiv.org/abs/1607.08022
14. Wang, Z., Salah, M.B., Gu, B., Islam, A., Goela, A., Li, S.: Direct estimation of cardiac biventricular volumes with an adapted bayesian formulation. IEEE Trans. Biomed. Eng. 61(4), 1251–1260 (2014)
15. Xue, W., Brahm, G., Pandey, S., Leung, S., Li, S.: Full left ventricle quantification via deep multitask relationships learning. Med. Image Anal. 43, 54–65 (2018)
16. Xue, W., Islam, A., Bhaduri, M., Li, S.: Direct multitype cardiac indices estimation via joint representation and regression learning. IEEE Trans. Med. Imaging 36(10), 2057–2067 (2017)
17. Xue, W., Nachum, I.B., Pandey, S., Warrington, J., Leung, S., Li, S.: Direct estimation of regional wall thicknesses via residual recurrent neural network. In: Niethammer, M., et al. (eds.) IPMI 2017. LNCS, vol. 10265, pp. 505–516. Springer, Cham (2017). https://doi.org/10.1007/978-3-319-59050-9_40
18. Zhang, Z., Liu, Q., Wang, Y.: Road extraction by deep residual U-Net. CoRR abs/1711.10684 (2017)

# Left Ventricle Full Quantification Using Deep Layer Aggregation Based Multitask Relationship Learning

Jiahui Li[1] and Zhiqiang Hu[2(⊠)]

[1] Beijing University of Post and Telecommunication, Beijing, China
jarveelee@gmail.com
[2] Peking University, Beijing, China
huzq@pku.edu.cn

**Abstract.** Left ventricle full quantification is important in the assessment of cardiac functionality and diagnosis of cardiac diseases, but is also challenging due to the sample variability and label correlations. In this paper, we propose a deep-learning based approach for left ventricle full quantification, including 11 indices regression and cardiac phase recognition. We utilize Deep Layer Aggregation as backbone, perform 11 indices regression simultaneously supervised by multitask relationship loss, and then derive the cardiac phase by searching maximum and minimum frame from polynomial-fitted cavity area. Experiments demonstrate the superiority of the proposed method in performance.

**Keywords:** Left ventricle full quantification ·
Deep Layer Aggregation · Multitask relationship learning ·
Polynomial fitting

## 1 Introduction

Left ventricle (LV) full quantification is important in the assessment of cardiac functionality and diagnosis of cardiac diseases. Full quantification aims to simultaneously quantify all LV indices, including two areas (the cavity area and the myocardium area), six regional wall thicknesses (along different directions, abbr. RWTs), three LV dimensions (also along different directions), and the cardiac phase (diastole or systole), as shown in Fig. 1. However, the task is also challenging, not only because of the sample variability across subjects, such as the difference between subjects with or without cardiac diseases, but also due to the complex correlations between the LV indices, such as the relationships between the cavity area and three LV dimensions, the relationships between the cavity area and the cardiac phase, et al.

In this paper, we propose a deep-learning based approach for the task of left ventricle full quantification. The model consists of two parts for indices regression and cardiac phase recognition, respectively. In the part of indices regression, we

© Springer Nature Switzerland AG 2019
M. Pop et al. (Eds.): STACOM 2018, LNCS 11395, pp. 381–388, 2019.
https://doi.org/10.1007/978-3-030-12029-0_41

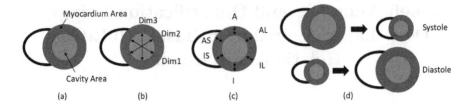

**Fig. 1.** Illustration of LV indices, including (a) the cavity area and the myocardium area, (b) three LV dimensions, (c) six regional wall thicknesses and (d) the cardiac phase (diastole or systole).

target at the 11 indices including two areas, six RWTs and three LV dimensions. The cardiac phase recognition task is handled in the second part.

For indices regression, we utilize Deep Layer Aggregation [8] as backbone. Deep Layer Aggregation extends the relatively shallow and simple aggregations in neural networks with their deep counterparts. By deep aggregations neural networks are able to fuse information from different layers and scales in a more effective way. The architecture yields promising performance in classification and segmentation tasks across multiple datasets [8].

On top of the backbone, we add heads to perform 11 indices regression simultaneously. The indices regression heads share the backbone, which makes it possible to explore the relationships among the indices implicitly. On the other hand, we supervise the regression with multitask relationship loss [5] to handle the correlations explicitly. The multitask relationship loss formulates the correlations with a symmetric positive semidefinite matrix that is jointly optimized together with parameters of the model. Experiments demonstrated that the learned correlation matrix is very close to its ground truth counterpart [5].

For cardiac phase recognition, it is natural to resort to Recurrent Neural Networks (RNNs). Although widely used in sequence modeling tasks, we found RNNs are prone to make un-continuous predictions like "diastole-systole-diastole-systole". As a result, in this paper we apply another strategy for cardiac phase recognition: We first smooth out the cavity area predictions by polynomials, then search the maximum and minimum frame with respect to the smoothed cavity area, and finally derive the cardiac phase based on the maximum and minimum frame. Compared with RNNs, the proposed method has far less parameters and will never make un-continuous predictions.

We evaluate the proposed model in dataset provided in Left Ventricle Full Quantification Challenge MICCAI2018 (LVQuan18) with five fold cross validation. Experiments demonstrate the superiority of the proposed method in performance.

## 2    Methodology

Figure 2 presents the overview of our model. Given input images, we firstly perform data preprocessing, including image normalization and augmentation, then we utilize deep neural networks for 11 indices regression and polynomial fitting for cardiac phase recognition.

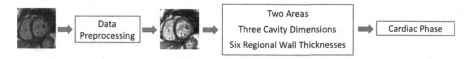

**Fig. 2.** Overview of the proposed model. Given input images, we firstly perform data preprocessing, then we utilize deep neural networks for 11 indices regression and polynomial fitting for cardiac phase recognition.

## 2.1 Data Preprocessing

It is observed that cardiac images vary a lot in brightness, contrast, et al., as shown in upper part of Fig. 3. The variability results from different machine configurations, and makes it harder for neural networks to process. As a result, we apply histogram equalization to reduce variability among images. In the lower part of Fig. 3 we present the same images after histogram equalization normalization. It is clear to see that brightness and contrast variability are largely diminished after normalization.

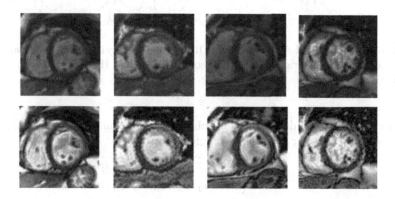

**Fig. 3.** Image samples. Upper part: original images vary a lot in brightness, contrast, et al. Lower part: after normalization, brightness and contrast variability are largely diminished.

Although images in the LVQuan18 dataset are manually labeled the landmarks, and rotated, cropped, resized to be approximately aligned, small subjective errors are inevitable in the alignment. Therefore, we apply data augmentations during training, including up-to-5-pixel random shift in horizontal/vertical directions, up-to-10-degree random rotation, and [0.8, 1.2] random scaling. Center crop or zero padding are applied after random scaling if necessary. Also, the labels are updated accordingly after image augmentations: the areas vary quadratically with respect to the scaling factor, while dimensions and RWTs vary linearly.

## 2.2   Indices Regression

**Deep Layer Aggregation.** We utilize Deep Layer Aggregation (DLA) [8] as backbone. Figure 4 illustrates the architecture. DLA extends neural networks by two types of layer aggregations: Iterative Deep Aggregation (IDA) and Hierarchical Deep Aggregation (HDA). As shown in Fig. 4, IDA connects layers across scales, and progressively aggregates and deepens the representation. IDA aggregates from the shallowest scale and iteratively merges deeper scales. Shallow feature maps get aggregated and refined through multiple stages, compared with UNet [3] architecture where shallowest feature maps get aggregated the least. On the other hand, HDA aggregates representations within one scale. The aggregation proceeds in a tree-like manner. HDA allows shallow and deep layers to aggregate and form a feature hierarchy, and still keeps the depth in control by the branching structure.

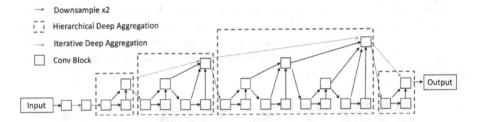

**Fig. 4.** Illustration of DLA architecture. DLA extends neural networks by two types of layer aggregations: Iterative Deep Aggregation (IDA) and Hierarchical Deep Aggregation (HDA).

For clarity we provides formal formulations for IDA and HDA. IDA function $I$ for a series of layers $x_1, x_2, ..., x_n$ is defined as

$$I(x_1, x_2, ..., x_n) = \begin{cases} x_1 & \text{if } n = 1 \\ I(N(x_1, x_2), x_3, ..., x_n) & \text{otherwise} \end{cases} \tag{1}$$

where $N$ denotes some aggregation function. HDA function $T_n$ for input layer $x$ with branching depth $n$ is defined as

$$T_n(x) = N(R_{n-1}^n(x), R_{n-2}^n(x), ..., R_1^n(x), L_1^n(x), L_2^n(x)) \tag{2}$$

where $N$ denotes the aggregation function, and $R, L$ are defined as

$$L_2^n(x) = \text{ConvBlock}(L_1^n(x)), L_1^n(x) = \text{ConvBlock}(R_1^n(x)) \tag{3}$$

$$R_m^n(x) = \begin{cases} T_m(x) & \text{if } m = n - 1 \\ T_m(R_{m+1}^n(x)) & \text{otherwise} \end{cases} \tag{4}$$

The aggregation function $N$ is this paper is defined as

$$N(x_1, x_2, ..., x_n) = \text{ReLU}(\text{BatchNorm}(\sum_i w_i x_i + b)) \qquad (5)$$

**Multitask Relationship Loss.** On top of the backbone, we add heads to perform 11 indices regression simultaneously. The indices regression heads share the backbone, which makes it possible to explore the relationships among the indices implicitly. On the other hand, we supervise the regression with multitask relationship loss [5] to handle the correlations explicitly. The loss function is formally formulated as

$$l_1 = \frac{1}{N_I N_T} \sum_{i,t} \|y_{i,t} - \hat{y}_{i,t}\|_2^2 \qquad (6)$$

$$l_2 = \|W_{\text{DLA}}\|_2^2 + \|W_{\text{head}}\|_2^2 \qquad (7)$$

$$l_3 = \text{trace}(W_{\text{head}} \Omega^{-1} W_{\text{head}}^T) \qquad (8)$$

$$l = l_1 + \alpha l_2 + \beta l_3 \qquad (9)$$

where $y_{i,t}, \hat{y}_{i,t}$ denote ground truth and predictions for LV index $t$ at image $i$, and $N_I, N_T$ are normalizing factors for $i$ and $t$, respectively; $W_{\text{DLA}}$ and $W_{\text{head}}$ denotes weight parameters for DLA backbone and regression heads, respectively; $\alpha, \beta$ are balancing factors and are set as $\alpha = 0.00001$ and $\beta = 0.0001$. The $l_1$ term is the ordinary Mean Square Error (MSE), and the $l_2$ term can be viewed as a regularization term. The $l_3$ term, however, models the label correlations with a symmetric positive semidefinite matrix $\Omega$ that can be jointly optimized together with parameters of the heads:

$$\Omega = \frac{\left(W_{\text{head}}^T W_{\text{head}}\right)^{1/2}}{\text{trace}\left(\left(W_{\text{head}}^T W_{\text{head}}\right)^{1/2}\right)} \qquad (10)$$

### 2.3 Cardiac Phase Recognition

Cardiac phase recognition can be regarded as a sequence modeling task, as the temporal dynamics is important for determining the cardiac phase. Considering that Recurrent Neural Networks (RNNs) are widely used in sequence modeling tasks, especially the success in natural language processing, it is natural to utilize RNNs for the task of cardiac phase recognition. However, we found RNNs are prone to make un-continuous predictions like "diastole-systole-diastole-systole". More labeled data or dedicated architecture may help fix the issue, but in this paper we apply another strategy for cardiac phase recognition. Firstly we smooth out the cavity area predictions by polynomials,

$$a_1', a_2', ..., a_n' = \text{PolynomialFitting}(a_1, a_2, ..., a_n) \qquad (11)$$

where $a_1, a_2, ..., a_n$ and $a_1', a_2', ..., a_n'$ are the cavity area of continuous $n$ frames for one subject before and after polynomial fitting, respectively. Then we search the maximum and minimum frame with respect to the smoothed cavity area. Finally we derive the cardiac phase based on the maximum and minimum frame, diastole for frames from minimum frame to maximum frame and systole for frames from maximum frame to minimum frame. Compared with RNNs, the proposed method has far less parameters and will never make un-continuous predictions.

## 3    Experiments

We evaluate the proposed model in dataset provided in Left Ventricle Full Quantification Challenge MICCAI2018 (LVQuan18). The dataset consists of 145 subjects, 20 frames per subject. We apply five fold cross validation to better evaluate the performance. The five folds are constructed as shown in Fig. 5: Firstly we calculate the gray scale histogram of each normalized image, and concatenate the histogram and labels as the descriptive feature for the image. The feature for the subject is thus the concatenation of features for all frames of the subject. Next we utilize k-means clustering algorithm to divide the subjects into five categories, and finally evenly split each category to five folds. We believe the construction can better address the possible bias in the evaluation introduced by imbalance distribution among five folds. For model training, we utilize Adam [2] optimizer and set the learning rate as 0.001 for the first 100 epochs, 0.0001 from 100 to 200 and 0.00001 from 200 to 300 epochs, to make the convergence stable. The model is implemented with PyTorch framework in Ubuntu 14.4.

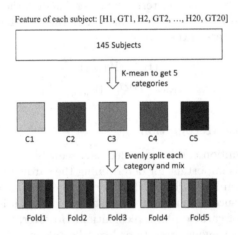

**Fig. 5.** The construction of five folds in cross validation.

The quantifications of two areas, three LV dimensions and six RWTs are measured by Mean Absolute Error (MAE); the cardiac phase recognition task is measured by Error Rate (ER). In Table 1 we presents five fold cross validation performance compared with existing state-of-the-art methods [1,4–7,9–11]. Experiments demonstrate the superiority of the proposed method in performance.

**Table 1.** Five fold cross validation performance compared with existing state-of-the-art methods.

| Method | Max flow | Multi-features + RF | SDL + AKRF | MCDBN + RF | Indices-Net | FullLVNet | DMTRL | ours |
|---|---|---|---|---|---|---|---|---|
| Area(mm$^2$) | | | | | | | | |
| A-cav | 156 | 231 | 198 | 208 | 185 | 181 | 172 | **135** |
| A-myo | 339 | 291 | 286 | 269 | 223 | 199 | 189 | **177** |
| Average | 247 | 261 | 242 | 239 | 204 | 190 | 180 | **156** |
| Dimension(mm) | | | | | | | | |
| dim1 | 2.81 | 3.53 | 2.99 | 2.88 | - | 2.62 | 2.47 | **2.04** |
| dim2 | 2.60 | 3.49 | 2.55 | 2.45 | - | 2.64 | 2.59 | **2.02** |
| dim3 | 2.49 | 3.91 | 3.10 | 2.93 | - | 2.77 | 2.48 | **2.05** |
| Average | 2.65 | 3.64 | 2.88 | 2.75 | - | 2.68 | 2.51 | **2.03** |
| RWT(mm) | | | | | | | | |
| IS | 1.53 | 1.70 | 1.98 | 1.78 | 1.39 | 1.32 | **1.26** | 1.39 |
| I | 3.23 | 1.71 | 1.67 | 1.68 | 1.51 | **1.38** | 1.40 | 1.41 |
| IL | 4.15 | 1.97 | 1.88 | 1.92 | 1.65 | 1.57 | 1.59 | **1.48** |
| AL | 5.08 | 1.82 | 1.87 | 1.66 | 1.53 | 1.60 | 1.57 | **1.46** |
| A | 3.47 | 1.55 | 1.65 | 1.20 | 1.30 | 1.34 | 1.32 | **1.24** |
| AS | 1.76 | 1.68 | 2.04 | 1.63 | 1.28 | 1.26 | **1.25** | 1.31 |
| Average | 3.21 | 1.73 | 1.85 | 1.65 | 1.44 | 1.41 | 1.39 | **1.38** |
| Phase(%) | | | | | | | | |
| phase | - | 22.2 | 22.6 | 16.17 | - | 10.4 | 8.2 | **8.1** |

## 4   Conclusion

We propose a deep-learning based approach for left ventricle full quantification, including 11 indices regression and cardiac phase recognition. We utilize Deep Layer Aggregation as backbone, perform 11 indices regression simultaneously supervised by multitask relationship loss, and then derive the cardiac phase by searching maximum and minimum frame from polynomial-fitted cavity area. Experiments demonstrate the superiority of the proposed method in performance.

## References

1. Ayed, I.B., Chen, H.M., Punithakumar, K., Ross, I., Li, S.: Max-flow segmentation of the left ventricle by recovering subject-specific distributions via a bound of the bhattacharyya measure. Med. Image Anal. **16**(1), 87–100 (2012)
2. Kingma, D., Ba, J.: Adam: a method for stochastic optimization. Computer Science (2014)
3. Ronneberger, O., Fischer, P., Brox, T.: U-Net: convolutional networks for biomedical image segmentation. In: Navab, N., Hornegger, J., Wells, W.M., Frangi, A.F. (eds.) MICCAI 2015. LNCS, vol. 9351, pp. 234–241. Springer, Cham (2015). https://doi.org/10.1007/978-3-319-24574-4_28
4. Xue, W., Islam, A., Bhaduri, M., Li, S.: Direct multitype cardiac indices estimation via joint representation and regression learning. IEEE Trans. Med. Imaging **36**(10), 2057–2067 (2017)

5. Xue, W., Brahm, G., Pandey, S., Leung, S., Li, S.: Full left ventricle quantification via deep multitask relationships learning. Med. Image Anal. **43**, 54–65 (2018)

6. Xue, W., Lum, A., Mercado, A., Landis, M., Warrington, J., Li, S.: Full quantification of left ventricle via deep multitask learning network respecting intra- and inter-task relatedness. In: Descoteaux, M., Maier-Hein, L., Franz, A., Jannin, P., Collins, D.L., Duchesne, S. (eds.) MICCAI 2017. LNCS, vol. 10435, pp. 276–284. Springer, Cham (2017). https://doi.org/10.1007/978-3-319-66179-7_32

7. Xue, W., Nachum, I.B., Pandey, S., Warrington, J., Leung, S., Li, S.: Direct estimation of regional wall thicknesses via residual recurrent neural network. In: Niethammer, M., et al. (eds.) IPMI 2017. LNCS, vol. 10265, pp. 505–516. Springer, Cham (2017). https://doi.org/10.1007/978-3-319-59050-9_40

8. Yu, F., Wang, D., Shelhamer, E., Darrell, T.: Deep layer aggregation. Computer Vision and Pattern Recognition (2018)

9. Zhen, X., Wang, Z., Islam, A., Bhaduri, M., Chan, I., Li, S.: Multi-scale deep networks and regression forests for direct bi-ventricular volume estimation. Med. Image Anal. **30**, 120 (2015)

10. Zhen, X., Islam, A., Bhaduri, M., Chan, I., Li, S.: Direct and simultaneous four-chamber volume estimation by multi-output regression. In: Navab, N., Hornegger, J., Wells, W.M., Frangi, A.F. (eds.) MICCAI 2015. LNCS, vol. 9349, pp. 669–676. Springer, Cham (2015). https://doi.org/10.1007/978-3-319-24553-9_82

11. Zhen, X., Wang, Z., Islam, A., Bhaduri, M., Chan, I., Li, S.: Direct estimation of cardiac Bi-ventricular volumes with regression forests. In: Golland, P., Hata, N., Barillot, C., Hornegger, J., Howe, R. (eds.) MICCAI 2014. LNCS, vol. 8674, pp. 586–593. Springer, Cham (2014). https://doi.org/10.1007/978-3-319-10470-6_73

# Convexity and Connectivity Principles Applied for Left Ventricle Segmentation and Quantification

Elias Grinias and Georgios Tziritas[✉]

Department of Computer Science, University of Crete, Heraklion, Greece
tziritas@csd.uoc.gr

**Abstract.** We propose an unsupervised method for MRI image segmentation, and global and regional shape quantification, based on pixel labeling using image analysis, connectivity constraints and near convex region requirements for the LV cavity and the epicardium. The proposed method is developed in the framework of the MICCAI Left Ventricle Full Quantification Challenge. At first the LV cavity is approximately localized based on the strong intensity contrast in the myocardium region between the two ventricles (left and right). The requirement of a near convex connected component is then applied. The image intensity statistical parameters are extracted for three classes: LV cavity, myocardium and chest space. Even if the whole background is completely inhomogeneous, the application of topological, connectivity and shape constraints permits to extract in two steps the LV cavity and the myocardium. For the later two approaches are proposed: regularization using B-spline smoothing and adaptive region growing with boundary smoothing using Fourier coefficients. On the segmented images are measured the significant clinical global and regional shape LV indices. We consider that we have obtained good results on indices related to the endocardium for both Training and Test datasets. There is place for improvements concerning the myocardium global and regional shape indices.

**Keywords:** Cardiac MRI · LV segmentation · Shape prior · LV quantification

## 1 Introduction

Cardiac Magnetic Resonance Imaging is widely used for heart morphology assessment and cardiac functional analysis. Cardiac imaging has an important role in preventive cardiology, in pathology diagnosis and in the management of operations. The accurate morphological and functional analysis of the left ventricular myocardium in 3D for the whole cardiac cycle is very important for cadiac function assessment.

Automatic left ventricle (LV) quantification in Short Axis MRI is an active research domain and remains a challenging problem in medical image analysis. Even if deep learning methods opened new research possibilities, as in the

© Springer Nature Switzerland AG 2019
M. Pop et al. (Eds.): STACOM 2018, LNCS 11395, pp. 389–401, 2019.
https://doi.org/10.1007/978-3-030-12029-0_42

recent work by Xue *et al.* [1], where very good results have been reported, the domain is still open due to the variability of the data and the difficulties of the problem. Existing work for automatic LV quantification falls into two categories: segmentation-based and direct regression. In the last approach a fully learning-based framework for direct bi-ventricular volume estimation is presented in [2] for global functional analysis. In [3] is proposed a real-time machine-learning approach to automating the detection and localization of regional myocardial abnormalities.

Even if direct regression methods have many advantages, in particular in case of deep learning by end-to-end optimization, we consider that interesting insights could be obtained using image segmentation, having as objective the design and implementation of a fully automatic method. Our work aims to contribute in this domain.

A recent review [4] of the main existing algorithms for the whole heart segmentation shows that the state-of-the-art image analysis methods use prior models, either atlas-based or generic deformable models. More specifically, a review on segmentation in short axis (SAX) cardiac MR images shows also the importance of prior knowledge [5] and emphasizes in different levels of information for different categories of methods. In our work we exploit high level weak prior knowledge.

In a more recent review paper [6] are summarised the most recent advances in cardiac image segmentation methods, which can be employed for the assessment of cardiac structure and function with MRI. In this regard, our approach is image-driven with weak priors and shares in a new perspective fundamental methods of region-growing, clustering and classification.

Recently, in the MICCAI "Automatic Cardiac Diagnosis Challenge" (ACDC 2017) the demonstration is given that deep learning methodologies could lead to highly-accurate and fully-automatic cardiac segmentation [7]. The proposed methods have been evaluated on global clinical indices, such as the LV and RV volumes and the myocardium mass. Regional indices have not been measured in this framework.

## 2  Motivation and Contributions

For cardiac function analysis the segmentation algorithm should extract the LV cavity, maybe the RV cavity too, and the myocardium from their surroundings. All these relevant segments are connected image components in 2-D SAX, in 3-D and during the cardiac cycle. Therefore the connectivity principle provides constraints for the segmentation algorithm, allowing to use different visual grouping methods. The only 'holes' of the myocardium are the cavities, which have not 'holes'. In addition, the endocardium and the epicardium for the LV cavity are near convex curves, if not convex, and smooth. These are strong constraints for resolving many ambiguities in pixel labeling for segmentation.

The requirement for smooth boundaries is taken into consideration in our work using a Fourier series with a small number of coefficients. The coordinates of the boundary (endocardium or epicardium) are transformed in the polar space, with reference to the center of the LV cavity, where the radius $\rho$ is expressed as a function of the direction $\theta$,

$$\rho = b_0 + \sum_{n=1}^{N} b_{2n-1} \cos n\theta + b_{2n} \sin n\theta, \quad -\pi < \theta \leq \pi \quad (1)$$

We have checked the validity of this approximation on the ground truth of the LVQuan18 Training dataset and obtained a median Jaccard similarity coefficient of 0.98 for both the endocardium and the epicardium using the above approximation with 7 coefficients ($N = 3$). If we measure the area using the formula $A = \pi b_0^2$, we have an error in mm$^2$ of 3.45 (resp. 5.34) on the LV cavity (resp. myocardium), which is quite small. The relative error in average is less than 1%. About at this percentage are also the relative errors for the dimensions of the LV cavity.

The LVQuan18 Training dataset contains 20 mid-cavity slices throughout a cardiac cycle for 145 subjects. The Test dataset contains images from 30 subjects. All cardiac images have been preprocessed and approximately aligned, and have the same size 80 × 80 pixels.

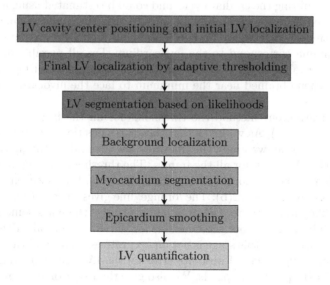

**Fig. 1.** Flow chart of the proposed LV segmentation and quantification algorithm: the objective of the 'orange' modules is to extract the LV cavity, of the "blue" the myocardium, and in the last module the clinical indices are measured. (Color figure online)

The flow chart of the proposed method is given in Fig. 1. We adopt a sequential step-by-step approach guided by the principle of high confidence decision first. At each stage we care to take almost sure decisions in very uncertain conditions. We start with a rough localization of the LV cavity presented in Sect. 3.1. The visual grouping used is based on topological constraints and visual similarity criteria. In Sect. 3.2 we give our LV segmentation method based on data analysis and the connectivity and convexity principles. The discrimination of the myocardium from its surroundings is presented in Sect. 4. The measurement of global and regional shape indices is given in Sect. 5.1 and the results obtained on the LVQuan18 datasets are presented in Sect. 5.2.

## 3   LV Segmentation

### 3.1   Automatic Localization of the Left Ventricle

The first step of our method consists on estimating the position of the LV cavity center. This task takes into account that the images in the LVQuan18 datasets are approximately aligned. In addition, at this step we need only an approximated localization of the center of the cavity. The vertical coordinate of the LV cavity center is near, if not at the same position, with the vertical coordinate of the image center. Therefore it will be not prejudiciable if we fix this coordinate for the whole dataset. The deviation of the horizontal coordinate of the LV cavity center is small during the cardiac cycle, and could be estimated using a smoothed signal obtained from the central image line. At first the local maxima and minima of the smoothed intensity on the central line signal are extracted. Two of the local minima correspond to the myocardium. For all possible myocardium positions the local contrast is measured from the two neighbouring maxima. A threshold is then obtained near the minimum to face the myocardium intensity inhomogeneities.

For illustrating our method the thresholds for all the frames of subject 75 are shown in Fig. 2(a). As we know that the center position is nearly stable during the cardiac cycle, we apply the majority law for finding the more probable position of the LV cavity for all the frames. The threshold-crossing line provides an estimation of the horizontal cavity dimension and of its position. The result ('blue' line) is given in Fig. 2(b). The 'orange' line gives the center position measured from the ground truth. Due to inhomogeneities there are some significant errors, which however are not prejudiciable for the subsequent algorithmic steps. Even if the objective at this stage is not an accurate estimation of the cavity center, the mean (resp. median) absolute error on the LVQuan18 Training dataset was only 1.54 (resp. 0.99) in pixels. We also give the result on one subject of the Test dataset in Fig. 2(c).

The next task is to find a region which is almost surely part of the LV cavity. We assume that given the center and considering a sequence of increasing concentric disks, the sites with intensity value above the median value in the disk belong almost surely to the LV cavity. Knowing also that the cavity is a convex region, for each considered disk the resulting convex hull including the detected sites is likely a subset of the LV cavity. The sequence of growing disks is bounded by the median intensity value on the resulting convex hull. This means

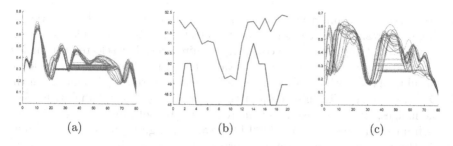

**Fig. 2.** Subject 75 (Training dataset): (a) smoothed intensity of the central line and threshold-crossing line for all the frames, (b) estimated center vertical position (blue) with ground truth for the 20 frames of the cardiac cycle. (c) Same as (a) for the subject 1 of the Test dataset. The intensity value is in the [0, 1] interval. (Color figure online)

that the growing process continues until the median intensity starts to decrease, possibly when the disk starts to include points from the myocardium. Four results are given in Fig. 3 in 'white', all satisfactory, because the important assessment criterion here is the precision. The two first frames are from the Training dataset. For these frames the ground truth for the LV cavity is illustrated in 'green'. On the whole Training dataset the minimum value of the precision was 0.95.

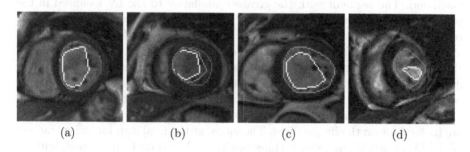

**Fig. 3.** Initial LV localization ('white'), final LV localization ('red') and endocardium ground truth ('green') (a) Image 0001 and (b) Image 1488 (Training dataset) and (c) Image 001 and (d) Image 373 (Test dataset). (Color figure online)

Then the LV cavity localization is refined by adaptive thresholding. Knowing that the strongest intensity contrast between the myocardium and the two ventricles is located in the septal region, and using the initial LV localization, we identify a rectangular piece of the image lying on the myocardium in the septal region and extended at least on the LV cavity too. Using the *k-means* algorithm we find a binary quantization of the intensity values. The threshold separating the two classes is estimated by a sequential technique trying possible thresholds and measuring criteria of validation. Thus we propose a region growing method guided by decreasing the intensity threshold for separating the myocardium from the two cavities. As far as we obtain a LV cavity satisfying the constraints of connectivity for the myocardium and for the shape of the LV cavity, the threshold decreases in order to detect the largest possible area of the LV cavity, satisfying always the convexity requirement. The selected thresholds, in the trial process, partition the interval between the two quantization levels in 10 subintervals with equal occurrence frequency on the whole image. On the LVQuan18 challenge Training dataset the precision is greater than 0.95 for 84% of the images, while the median of the recall is 0.91, both measured on the separate frames. The results of the final LV localization given in Fig. 3 are in 'red'. Again the precision is much more important than the recall.

### 3.2  LV Segmentation

From the localized LV cavity and its neighbouring surroundings, corresponding probably to the myocardium, are estimated, from the 20 frames of the whole cardiac cycle, statistical parameters for the intensity distribution for two classes: RV and LV (blood pools, label '2') and myocardium (muscle, label '1'). For reasons of robustness we assume Gaussian distributions for all the intensity classes.

Then for each image the dominant intensity distribution of the background (chest space, label '0') is estimated. For the three classes, thus defined, are estimated on the whole image the likelihoods, and the *a posteriori* probabilities. Then all the image pixels are labeled according to the principle of maximum likelihood. The segment with the greatest similarity to the LV localized in the previous step is extracted as the final LV cavity. Then the time coherence of the resulting regions is checked throughout the cardiac cycle. The outlying results are rejected and replaced by morphological interpolation in the time axis.

Knowing that the shape of the LV cavity is always almost convex and that, due to the papillary muscles in the segmented region, relevant points could be not retrieved by only the intensity, the final result is obtained by the convex hull of the region. In addition, the endocardium is an almost smooth surface, and so are its sections on the image slices. Therefore, at the final step for the extraction of the LV cavity, we use a smoothing procedure based on Fourier coefficients as in Eq. (1) with $N = 3$. On the Training dataset we obtain for the $F_1$ score a mean value of 0.93, with a median value at 0.95, a minimum at 0.69 and a maximum at 0.99. Four results of segmentation from the Training dataset are given in Fig. 4, the two first are good, and the other two are among the worst we obtained for images in difficult conditions. The estimated *a posteriori* probabilities are also shown, for measuring the difficulties and assessing also the weaknesses of the approach. The 'red' channel is the *a posteriori* probability for the 'blood pool', the 'green' for the 'muscle' and the 'blue' for the 'chest space'. Are also given two results from the Test dataset with the corresponding *a posteriori* probabilities.

## 4  Myocardium Segmentation

The intensity in the myocardium is very often inhomogeneous and furthermore in the background, *i.e.* the region in the exterior of the epicardium, where appear

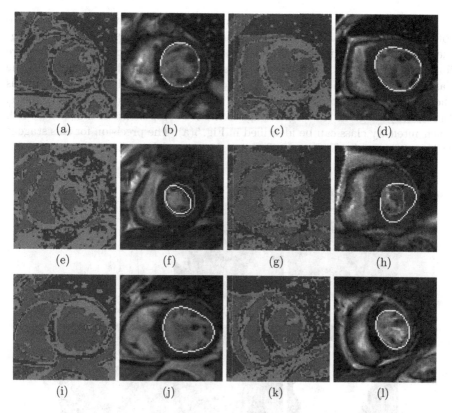

**Fig. 4.** *A posteriori* probabilities: (a) frame 2372, (c) frame 59, (e) frame 107, (g) frame 651 (Training dataset), (i) frame 1, (k) frame 373 (Test dataset). LV localization ('red'), final LV segmentation ('white') and ground truth ('green'): (b) frame 2372, (d) frame 59, (f) frame 107, (h) frame 651 (Training dataset). LV segmentation: (j) frame 1, (l) frame 373 (Test dataset). (Color figure online)

all the intensity classes. On the other hand the epicardium is almost always a convex and smooth curve and its size varies slightly during the cardiac cycle for mid-cavity frames. Even if we have meaningful constraints, their accurate quantitative expression is not obvious and the task of myocardium segmentation is always difficult.

## 4.1  Region of Interest Localization

At first we extract for each image a region belonging almost surely to the background. For obtaining this region we consider points lying at least a certain distance away from the LV cavity. This distance results from a first estimation of the maximum myocardium wall thickness, knowing also that the wall is often thicker in the septal region, where in addition the intensity edges are sharper, facilitating direct measurements of the thickness.

When for a point the *a posteriori* probability of the class corresponding to 'blood' is greater than 0.8, and the point is sufficiently far away from the segmented LV cavity, this point is labeled as being in the 'background'. Therefore for the scope of this work the RV is labeled 'background'. Of course all the 'large' components labeled as 'chest space' maintain their label. An example is shown in Fig. 5, where at top-left is the image of the *a posteriori* probabilities. In Fig. 5(b) are given the large enough extracted components of the background. Their intensity class can be identified in Fig. 5(a). The precision for this stage is at least 0.94 on the LVQuan18 Training dataset, with median value near to 1.

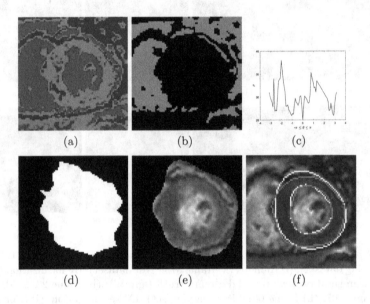

**Fig. 5.** Image 1943: (a) *a posteriori* probabilities, (b) background connected components, (c) distance from the LV cavity center, (d) polygon including the region of interest, (e) region of interest with smooth border, (f) final endo- and epicardial segmentation.

We then determine the largest polygon that could define the interior boundary of the background, in order to obtain one connected background component. For this reason we find the boundaries of all the extracted background components, and we measure the distance of all boundary points from the LV cavity center and the corresponding angle. After quantizing the angle, if more than one points are found at the same direction, the nearest is retained. The remaining points define a polygon which is the initial border of the background. Such a result is shown in Fig. 5(c), where the vertices of the polygon are given by their distance from the LV cavity center versus their angular position in the interval $(-\pi, \pi]$. The resulting polygon is also shown in Fig. 5(d). For angular positions where background components are not found, a linear interpolation is done.

In the plot given in Fig. 5(c) an interpolation is done for the interval $(0.4\pi, 0.7\pi)$. Assuming that the epicardium is smooth, we could also assume that the border defined above should be smooth. Smoothing is done using a small number of Fourier coefficients ($N = 7$) according to Eq. (1).

Knowing that the epicardium size varies slightly, we then obtain a region including the epicardium for the whole cardiac cycle, and compare the resulting regions on the successive frames. In case of important discrepancy a correction is applied. More precisely, we compute the median region of interest $\mathcal{R}_m$, and if the region extracted for a slice has a Jaccard similarity coefficient to $\mathcal{R}_m$ less than 0.8, this region is replaced by the median region of interest.

Finally, on the Training dataset we obtain for the $F_1$ score a median value of 0.92, with a minimum at 0.76 and a maximum at 0.98. This result is yet satisfying as the recall is more important here than the precision. The mean recall rate is 0.98 and the median near to 1. We may consider that we have extracted an almost sure region of interest with satisfactory precision. Considering also satisfactory the segmentation of the LV cavity we have to localize the epicardium. We propose two methods for segmenting the myocardium.

### 4.2   Epicardial Boundary Extraction Using B-Spline Smoothing

The finally segmented myocardium is obtained from regions labeled '1', which are contained to the region of interest. The region obtained is irregular and needs regularization. Our first method is only based on the boundary's shape.

The final epicardial boundary is obtained by fitting a cubic B-spline [9] of weighted minimum square error on boundary points. The computation of the weights for the boundary points is based on the mean and the variance of the wall thickness. The variance estimate is used to progressively factor out the impact of boundary points as their distance from the mean wall thickness increases. Cubic spline minimization takes place in the polar space defined by the centroid of the LV cavity. The detailed description of this method is given in [10]. In Fig. 5(f) a final segmentation result is given, in 'white', while the endocardium from the ground truth is shown in 'green' and the epicardium in 'yellow'.

### 4.3   Myocardium Segmentation Using Adaptive Region Growing

Our second method is mainly based on the intensity distribution using as shape priors the connectivity of the segment and the smoothness of its boundary. An initial segment is extracted in the septal region, because the intensity contrast is stronger there and the discrimination could be unambiguous. The initial region is growing then according to the intensity distribution and the connectivity principle, until to obtain an annular segment. This is a first, but yet inaccurate segmentation of the myocardium. However the accuracy is sufficient for localizing the center of the whole LV from the initial epicardial boundary. As the epicardial curve has nearly a circular shape, fitting a circle to the extracted boundary gives the initial annular segment for a final adaptive region growing step based on the intensity distribution. Finally, the boundary is smoothed using

Fourier coefficients. The mean and median epicardium (whole LV) $F_1$ score is 0.96, while the score on the myocardium is at the median 0.86 and in mean 0.84. The method is illustrated in Fig. 6 for a frame in the Test dataset.

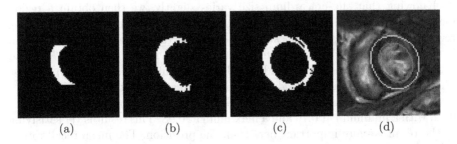

(a)               (b)               (c)               (d)

**Fig. 6.** Image 373 (Test dataset): (a) initial segment in the *Septal* region, (b) first connected component, (c) annular segment resulting from the adaptive region growing, (d) final segmentation result.

## 5     Experimental Results on LVQuan18 Datasets

### 5.1     LV Quantification

The quantification of the global and regional LV morphology is defined by the LVQuan18 challenge. It contains 11 measures: the areas of the LV cavity and the myocardium, three dimensions of the cavity and 6 measurements of the wall thickness. In our work the areas were measured by the number of points in the segmented regions corresponding to the LV cavity and the myocardium.

For measuring the three dimensions (IS-AL, I-A, IL-AS) of the LV cavity we find the boundary of the extracted LV cavity and then obtain a dense interpolation of the boundary in polar coordinates around the centroid of the cavity. Thus we can obtain smooth measurements for any direction. For measuring the six regional wall thicknesses we obtain again firstly a dense interpolation of the epicardium. Empirically we found that the thickness at every direction could be estimated by the difference of the two radii multiplied by 0.9.

For the phase classification, as systolic or diastolic, we use a simple method. We consider that from the frame with maximum area of LV cavity to the frame with minimum area, the phase is systolic.

### 5.2     LV Quantification Performance

Our method is evaluated on the LVQuan18 Challenge datasets. All the segmentation results obtained by our method on the LVQuan18 Test dataset are given in https://www.csd.uoc.gr/~tziritas/demos/LVQuan_demo/demoLVQuan.html.

The results on all the LV indices are given in Table 1 for the Training dataset, and in Table 2 for the Test dataset. For the Training dataset are also given the results of [1] (DMTRL) and [8] (DLAM). As evaluation criteria are used the

**Table 1.** Performance on the LVQuan18 Training dataset.

|  | MAE Spline | MAE RG | MAE DMTRL | MAE DLAM | Corr DMTRL | Corr Spline | Corr RG |
|---|---|---|---|---|---|---|---|
| LV area | 139 | 138 | 172 | 135 | 0.94 | 0.97 | 0.97 |
| Myocardium area | 213 | 205 | 189 | 177 | 0.95 | 0.87 | 0.88 |
| Mean area | 176 | 172 | 180 | 156 | 0.95 | 0.94 | 0.94 |
| Dim1 LV (IS-AL) | 2.39 | 2.39 | 2.47 | 2.04 | 0.96 | 0.95 | 0.95 |
| Dim2 LV (I-A) | 2.33 | 2.30 | 2.59 | 2.02 | 0.89 | 0.96 | 0.97 |
| Dim3 LV (IL-AS) | 1.95 | 1.94 | 2.48 | 2.05 | 0.94 | 0.97 | 0.97 |
| Mean Dim LV | 2.23 | 2.21 | 2.51 | 2.03 | 0.93 | 0.96 | 0.96 |
| RWT1 (IS) | 1.61 | 1.35 | 1.32 | 1.39 | 0.86 | 0.82 | 0.86 |
| RWT2 (I) | 1.82 | 1.69 | 1.38 | 1.41 | 0.75 | 0.68 | 0.71 |
| RWT3 (IL) | 1.84 | 1.64 | 1.57 | 1.48 | 0.69 | 0.68 | 0.71 |
| RWT4 (AL) | 1.86 | 1.94 | 1.60 | 1.46 | 0.66 | 0.68 | 0.70 |
| RWT5 (A) | 1.52 | 1.59 | 1.34 | 1.24 | 0.78 | 0.74 | 0.75 |
| RWT6 (AS) | 1.82 | 1.46 | 1.26 | 1.31 | 0.88 | 0.86 | 0.87 |
| Mean RWT | 1.75 | 1.61 | 1.41 | 1.38 | 0.77 | 0.71 | 0.76 |

**Table 2.** Performance on the LVQuan18 Test dataset.

|  | MAE Spline | Corr RG | MAE Spline | Corr RG |
|---|---|---|---|---|
| LV area | 250 | 0.97 | 250 | 0.97 |
| Myocardium area | 392 | 0.76 | 396 | 0.80 |
| Mean area | 321 | 0.86 | 322 | 0.89 |
| Dim1 LV (IS-AL) | 2.70 | 0.95 | 2.70 | 0.95 |
| Dim2 LV (I-A) | 4.70 | 0.95 | 4.65 | 0.94 |
| Dim3 LV (IL-AS) | 3.21 | 0.96 | 3.21 | 0.96 |
| Mean Dim LV | 3.52 | 0.95 | 3.52 | 0.95 |
| RWT1 (IS) | 2.79 | 0.70 | 2.46 | 0.72 |
| RWT2 (I) | 1.88 | 0.59 | 2.09 | 0.60 |
| RWT3 (IL) | 1.79 | 0.62 | 2.24 | 0.64 |
| RWT4 (AL) | 2.08 | 0.60 | 1.83 | 0.70 |
| RWT5 (A) | 2.26 | 0.59 | 2.33 | 0.60 |
| RWT6 (AS) | 3.75 | 0.72 | 3.22 | 0.72 |
| Mean RWT | 2.41 | 0.64 | 2.36 | 0.66 |

mean absolute error (MAE) and the correlation coefficient (Corr). On the LV cavity the correlation coefficient is almost the same for the Training and the Test datasets, while there is a loss of performance for the myocardium indices. Concerning the error rate on the phase, we have obtained 10.1% on the Training dataset and 11.2% on the Test dataset.

Finally, an important strong point of our algorithm is its computational efficiency. Processing the 20 slices of the whole cardiac cycle in Matlab on a laptop Intel Core i-7 2.6 GHz takes about 2 s.

## 6   Conclusion

We developed a fully automatic cardiac image segmentation method using a pipeline processing guided by weak shape priors. Even if our method is unsupervised, the functional indices obtained are not substantially worse than those of deep learning methods. We consider that we have obtained good results on the LV segmentation, with good measurements for indices related to the LV cavity. We have proposed two approaches for the myocardium segmentation, both with satisfactory results, better for the adaptive region growing method. Our work being still in progress, and knowing that we have good results for the epicardium localization, it could be possible to improve the accuracy of the myocardium related indices, area and wall thickness, using shape priors and time coherence. Finally, we think that there is place for learning algorithms for some important hyperparameters of the various segmentation modules and that some of our intermediate results could be used in deep learning approaches instead of pretraining or data augmentation.

## References

1. Xue, W., Brahm, G., Pandey, S., Leung, S., Li, S.: Full left ventricle quantification via deep multitask relationships learning. Med. Image Anal. **43**, 54–65 (2018)
2. Zhen, X., Wang, Z., Islamd, A., Bhadurie, M., Chane, I., Li, S.: Multi-scale deep networks and regression forests for direct bi-ventricular volume estimation. Med. Image Anal. **30**, 120–129 (2016)
3. Afshin, M., et al.: Regional assessment of cardiac left ventricular myocardial function via MRI statistical features. IEEE Trans. Med. Imag. **33**, 481–494 (2014)
4. Zhuang, X.: Chalenges and methodologies of fully automatic whole heart segmentation: a review. J. Healthc. Eng. **4**, 371–407 (2013)
5. Petitjean, C., Dacher, J.-N.: A review of segmentation methods in short axis cardiac MR images. Med. Image Anal. **15**, 169–184 (2011)
6. Peng, P., Lekadir, K., Gooya, A., Shao, L., Petersen, S.E., Frangi, A.F.: A review of heart chamber segmentation for structural and functional analysis using cardiac magnetic resonance imaging. Magn. Reson. Mater. Phys. Biol. Med. **29**, 155–195 (2016)
7. Bernard, O., et al.: Deep learning techniques for automatic MRI cardiac multi-structures segmentation and diagnosis: is the problem solved? IEEE Trans. Med. Imag. **37**, 2514–2525 (2018)

8. Li, J., Hu, Z.: Left Ventricle Full Quantification using Deep Layer Aggregation based Multitask Relationship Learning, MICCAI Left Ventricle Quantification Challenge (2018)

9. Bartles, R.H., Beatty, J.C., Barsky, B.A.: An Introduction to Splines for use in Computer Graphics and Geometric Modeling. Morgan Kaufmann Publishers, Los Altos (1987)

10. Grinias, E., Tziritas, G.: Fast fully-automatic cardiac segmentation in MRI using MRF model optimization, substructures tracking and B-spline smoothing. In: Pop, M., et al. (eds.) STACOM 2017. LNCS, vol. 10663, pp. 91–100. Springer, Cham (2018). https://doi.org/10.1007/978-3-319-75541-0_10

# Calculation of Anatomical and Functional Metrics Using Deep Learning in Cardiac MRI: Comparison Between Direct and Segmentation-Based Estimation

Hao Xu[1]([✉]), Jurgen E. Schneider[2], and Vicente Grau[1]

[1] Department of Engineering Science, University of Oxford, Oxford, UK
hao.xu@eng.ox.ac.uk
[2] Leeds Institute of Cardiovascular and Metabolic Medicine,
University of Leeds, Leeds, UK

**Abstract.** In this paper we propose a collection of left ventricle (LV) quantification methods using different versions of a common neural network architecture. In particular, we compare the accuracy obtained with direct calculation (regression) of the desired metrics, a segmentation network and a novel combined approach. We also introduce temporal dynamics through the use of a Long Short-Term Memory (LSTM) network. We train and evaluate our methods on MICCAI 2018 Left Ventricle Full Quantification Challenge dataset. We perform 5-fold cross-validation on the training dataset and compare our results with the state-of-the-art methods evaluated on the same dataset. In our experiments, segmentation-based methods outperform all the other variants as well as current state of the art. The introduction of LSTM does produces only minor improvements in accuracy. The novel segmentation/estimation network improves the results on estimation-only but does not reach the accuracy of segmentation-based metric calculation.

**Keywords:** Left ventricle quantification · Deep learning

## 1 Introduction

LV quantification has significant clinical importance for identification and diagnosis of cardiac diseases [2]. Existing methods for LV quantification can be classified into one of the three types: (1) manual quantification; (2) automated segmentation-based (SB) quantification; and (3) direct regression (DR) [13]. Quantification of LV anatomy from manual contouring of myocardial borders (manual delineation) is the gold standard in clinical practice [7], and includes two steps: segmentation and indices calculation. SB methods follow the same two-step strategy, with the development focusing on the segmentation algorithms. DR methods aim to evaluate LV indices directly from images, without intermediate segmentation. A graphical overview of these three approaches is shown in

© Springer Nature Switzerland AG 2019
M. Pop et al. (Eds.): STACOM 2018, LNCS 11395, pp. 402–411, 2019.
https://doi.org/10.1007/978-3-030-12029-0_43

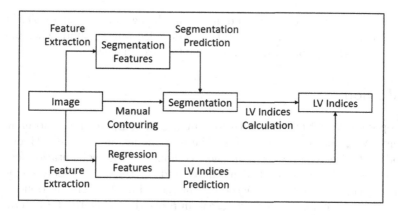

**Fig. 1.** Graphical overview of existing LV quantification methods.

Fig. 1. In this paper, we compare methods (2) and (3) and we propose a new combined approach. An advantage of using segmentation for indices evaluation is that the results can be verified by direct comparison to manual segmentation. DR methods intend to skip the segmentation process, introducing a fundamental limitation that the evaluation can only be made through statistical analysis, providing no explanations or possible correction approaches for failed cases.

The recent development of deep learning methods has improved the performance of both segmentation methods and DR methods. Variations of the fully convolutional neural network (FCN) [2] show promising results for segmentation of LV in cine-MR images [4,5,8,9]. Convolutional neural networks (CNNs) have also been applied to direct quantification methods and shown steady improvements [10–13]. Unfortunately, there has been very limited direct comparison between these two types of methods, and the MICCAI 2018 Left Ventricle Full Quantification Challenge provides a great platform for evaluating different methods on the same benchmark dataset. We propose a neural network with multiple modes corresponding to different types of methods, and we develop and evaluate our methods on this challenge dataset.

## 1.1   Key Contributions

The main contributions of this work include:

To our best knowledge, this is the first paper that directly compares the performance of SB and DR methods for quantifying LV, with state of the art deep learning algorithms applied to both types. The experiments are performed under a common neural network framework for better comparison.

We also propose a novel type of feature for regression: segmentation-feature, which shows great potential by comparing with features learnt through image reconstruction and direct regression tasks.

Finally, we introduce temporal dynamic modeling (TDM) using LSTM units for our SB method and show its limitation in deep learning methods when manual references is used as gold standard for quantitative analysis.

## 2    Method

We propose a neural network with four modes to compare the performance of different LV quantification methods. The four modes are: (1) Reg Mode (DR method); (2) Seg-Reg Mode, regression combined with segmentation-features (DR method); (3) Seg Mode (SB method); and (4) Seg-TDM Mode, segmentation combined with TDM (SB method). The second mode, segmentation-feature regression is a novel method we propose to improve the DR method. In literature, TDM appears to improve on the DR method [11–13]; furthermore, the application of recurrent neural networks (RNN) along the third spatial dimension has shown improvement on LV segmentation results [5].

### 2.1    Multi-mode Network

The network we propose is summarized in Fig. 2. The Feature Extraction Net (FE-Net) consists of two stages (an encoder and a decoder shown in Fig. 2) connected with short paths between layers with the same resolutions, using a FCN with skip connections inspired by the U-Net architecture [6]. The encoder transforms the input image into the Middle Features (M.F.), while the decoder

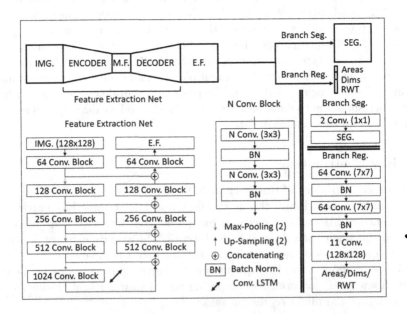

**Fig. 2.** A schematic diagram of our multi-mode network.

transforms these into the End Features (E.F), at the original image resolution. This is connected to two output branches (Branch Seg. and Branch Reg.). For the Seg-TDM Mode, the FE-Net becomes an RFCN, by including LSTM blocks [1] applied to the M.F., in a similar way to [5].

The segmentation output layer has sigmoid activation functions and all the other FCN activations are rectified linear units (ReLU). Two convolutional layers and two batch normalisation (BN) layers form a convolutional block, with the number of kernels for each block specified in Fig. 2. Input feature maps of convolutional layers are padded with zeros so that the output feature maps have the same size. Branch Reg. includes two convolutional layers (64 kernels with size of $7 \times 7$ for each layer) and a convolutional layer connecting to the regression output, which has 11 kernels with size of $128 \times 128$. All the other convolutional layers have kernel size of $3 \times 3$. Max-pooling and up-sampling layers have the stride of 2 in both dimensions. Branch Seg. has only one convolutional layer with 2 kernels of size $1 \times 1$ giving two output layers corresponding to regions enclosed by endocardial and epicardial contours. When LSTM is used we change the middle convolutional block (1024 Conv. Block in Fig. 2) to a convolutional LSTM layer with 1024 kernels of size $3 \times 3$.

The overall loss function is $L_T = \lambda_R \times L_R + \lambda_S \times L_S$, where $L_S$ and $L_R$ are segmentation output and regression output losses respectively. $L_R$ is mean square root error (MSE) and $L_S$ is binary cross-entropy. $\lambda_S$ and $\lambda_R$ are the corresponding loss weights, which are set based on the mode: (1) Reg Mode has $\lambda_S = 0$ and $\lambda_R = 1$; (2) Seg-Reg Mode alternates between $\lambda_S = 1$ and $\lambda_R = 0$ for odd epochs and $\lambda_S = 1$ and $\lambda_R = 1$ for even epochs; (3) Seg Mode has $\lambda_S = 1$ and $\lambda_R = 0$; and (4) Seg-TDM has the same losses as (3). For Seg-Reg Mode, by more frequently updating weights of the FE-Net with segmentation losses, the end-feature is primarily optimized for the segmentation task. This novel approach learns segmentation-features for the regression task.

## 2.2    Calculation of LV Indices from Segmentation

For our segmentation method, we normalize the intensity of the input images in three steps: (1) initial segmentation for each unit (one slice for Seg Mode and one sequence for Seg-TDM Mode); (2) by assuming reasonable segmentation results, we calculate the mean intensity values of the cavity and myocardial sections on the image ($i_{\text{cav}}$ and $i_{\text{myo}}$), and the corresponding intensity distribution including population means ($mean_{\text{cav}}$ and $mean_{\text{myo}}$) and population standard deviations ($std_{\text{cav}}$ and $std_{\text{myo}}$) of the data used to train and validate the neural network; (3) if $i_{\text{cav}}$ is out of the range $mean_{\text{cav}} \pm std_{\text{cav}}$ or $i_{\text{myo}}$ is out of the range $mean_{\text{myo}} \pm std_{\text{myo}}$, the intensity value of the unit is divided by a factor of $(\frac{i_{\text{cav}}}{mean_{\text{cav}}} + \frac{i_{\text{myo}}}{mean_{\text{myo}}})/2$ for normalization.

The segmentation outputs are converted to binary masks (by thresholding at 0.5) and cropped to the size of $80 \times 80 \times 2$. The gold standard LV indices we were provided with the training data had been originally calculated from segmentations; however the methods used for this were not provided and thus we developed our own. Cavity and myocardium binary masks ($bm_{\mathrm{cav}}$ and $bm_{\mathrm{myo}}$) are calculated and for each of them the biggest object is extracted with holes in it filled. The areas are then normalized by $\frac{\text{number of pixel with class 1}}{\text{total number of pixels per image (6400)}}$. The LV dimensions are calculated in three steps: (1) calculating the centroid of the cavity; (2) calculating the distance from the centroid to the endocardial border along integer angles in degrees with zero corresponding to the positive horizontal axis; (3) averaging the values of 60 distances for each region and adding opposing regions to get the three LV dimensions (Dims). The regional wall thickness (RWT) values are calculated in three steps: (1) sampling the middle contour of the myocardium with 360 points; (2) adding the minimum distances from each point to the endocardial and epicardial contours; (3) averaging the values of 60 points for each region. Both distances are normalised by $\frac{\text{distance}}{\text{image size (80)}}$.

## 2.3   Phase Error

Our method uses the cavity segmentation results to estimate phase using a simple procedure. The estimated cavity area sequence is first smoothed using a moving average filter with a span of 5 frames. End-diastolic (ED) and end-systolic (ES) frame are then identified as the frames with maximum and minimum cavity areas respectively. The remaining frames are then assigned phase numbers sequentially.

# 3   Dataset

To develop and evaluate our methods we use the MICCAI 2018 Left Ventricle Full Quantification Challenge dataset, which contains images collected from three hospitals with a wide range of subject age and presence of diverse pathologies. The pixel spacings of the MR images range from 0.68 mm/pixel to 2.08 mm/pixel, with mode of 1.56 mm/pixel. Each subject contains 20 frames throughout a cardiac cycle, and all cardiac images undergo several pre-processing steps before resulting in approximately aligned images with dimensions of 80 × 80 pixels. These images are manually contoured to obtain the epicardial and endocardial borders, which are double-checked by two experienced cardiac radiologists [11, 13]. Due to the resizing, the actual pixel spacings on the 80 × 80 images are in a range of [1.13, 2.29] mm/pixel, with mean and median values of 1.64 and 1.63 mm/pixel respectively.

The gold standard is included in the dataset as two sets of binary masks, representing the region enclosed by manual contours. The LV indices include: areas of the myocardium and LV cavity, three directional LV cavity dimensions, six regional wall thicknesses, and one cardiac phase, with the reference values of these quantifications are also included in the training dataset [11,13].

## 3.1   Data Augmentation

To preserve as much information from the cropped images (80 × 80) as possible, we decided to have a bigger input-image size of the neural networks, and

$128 \times 128$ is a reasonable size providing extra space for data augmentation. We only apply augmentation for training and validation data, by initiating a background with zeros of size $128 \times 128$, and then applying random rotation, $\pm 10\%$ additional resizing and random flipping to the $80 \times 80$ images and placing the resultant image on a random position of the background while keeping the complete ventricle appearing on the expanded image. The binary references are augmented correspondingly, and the LV indices are also rescaled accordingly. The augmentation process is online, i.e. for each epoch the augmented data is different. Comparing with further cropping from the original images, our augmentation method preserves more information from the original images and in theory would help to prevent over-fitting, as the number of trained parameters remains the same. The main draw back of our method is higher memory consumption, but we do not have any issues for training and testing on a single modern GPU.

## 4    Experiments and Results

### 4.1    Implementation

We implement neural networks in Keras with Tensorflow as backend, and the networks are trained with Adadelta [14]. We use one NVIDIA GeForce GTX 1080 Ti GPU for all training and testing experiments. For the record, an estimated average processing time of Seg Mode is less than 25 ms for each input image.

### 4.2    Training Data Experiments

For all the training data experiments we perform five-fold cross validation (29 on each fold), and randomly assign 100 training data and 16 validating data from the remaining dataset. The results are shown in Table 1 compared with the state-of-the-art methods applied on the same datasets [10,12,13]. In our initial experiments, SB methods performed best and thus we trained two additional sets of networks for both of them and ensemble the segmentation results by averaging the predictions, yielding an overall improvement of around 3%.

Since the method used for indices evaluation from segmentation was not available to us as part of the challenge information, we develop our quantification method as detailed in Sect. 2.2 and evaluated it by applying to the manual segmentation references. This introduced a small discrepancy between our calculation of indices from segmentations and those provided in the challenge. This discrepancy is shown in Table 1 as Seg Ref, and is likely to have affected the results of our Seg and Seg-TDM methods negatively, albeit to a small extent.

The Reg Mode achieves comparable results to the state-of-the-art methods, as expected due to their common use of end-to-end training using indices. When combined with segmentation in our novel method, the Seg-Reg Mode produces better results for all indices. Indice-Net uses image reconstruction features from

**Table 1.** Average results of compared methods. Values are shown as mean error $\pm$ standard deviation, followed by the correlation coefficient shown in brackets.

| Dataset | Method | Avg. Areas (mm$^2$) | Avg. Dims (mm) | Avg. RWT (mm) | Phase error (%) |
|---|---|---|---|---|---|
| Training and Validating data | Indice-Net [10] | 204 ± 133 (0.903) | NA | 1.44 ± 0.71 (0.758) | NA |
| | FullLVNet [11] (intra/inter) | 190 ± 128 (0.937) | 2.68 ± 1.64 (0.917) | 1.41 ± 0.72 (0.761) | 10.4 |
| | DMTRL [13] | 180 ± 118 (0.945) | 2.51 ± 1.58 (0.925) | 1.39 ± 0.68 (0.768) | 8.2 |
| | Reg Mode | 158 ± 162 (0.907) | 2.08 ± 1.84 (0.917) | 1.51 ± 1.28 (0.802) | 9.4 |
| | Seg-Reg Mode | 142 ± 125 (0.926) | 1.94 ± 1.59 (0.935) | 1.42 ± 1.17 (0.817) | 8.6 |
| | Seg Mode | 120 ± 105 (0.951) | 1.25 ± 1.06 (0.972) | 1.04 ± 0.85 (0.908) | **7.8** |
| | Seg-TDM Mode | **116 ± 103** (**0.952**) | **1.23 ± 1.11** (**0.972**) | **1.03 ± 0.86** (**0.908**) | 8.2 |
| | Seg Ref | 7 ± 6 (1.000) | 0.19 ± 0.16 (0.999) | 0.17 ± 0.13 (0.998) | 6.3 |
| Testing data | Seg/Seg-TDM Ensemble | 341 (0.953) | 2.67 (0.979) | 2.28 (0.829) | 10.3 |

an auto-encoder, and also has two convolutional layers for regression [10]. Comparing Indice-Net and Seg-Reg Mode, we see a clear advantage in using segmentation features for area evaluation and slight improvement on RWT evaluation. Our best performing DR method (Seg-Reg Mode) shows comparable results for RWT and Phase estimation with the best existing method (DMTRL), and more accurate results in area and dimension indices calculation.

SB methods perform better than DR methods for all indices. Comparing the SB methods, the mean absolute error (MAE) values of Areas, Dims, and RWT favor Seg-TDM Mode, but when we calculate the p-values (0.05, 0.71, $1.1e-6$ respectively) only RWT values show a significant (though small) difference between the two methods. Seg Mode produces a more accurate phase prediction.

## 4.3  Analysis of TDM

From the quantitative results we see very small improvements after introducing TDM, by comparing Seg Mode and Seg-TDM Mode. We then evaluate the results of Seg-TDM mode qualitatively by visualising the iso-surface of $2D+time$ volume of both reference binary segmentation and Seg-TDM Mode predicted binary segmentation iso-surfaces at 0.5 for each slice. We show three pairs of endocardial and epicardial surfaces in Figs. 3 and 4, and the vertical axis is the time dimension ascending from top to bottom.

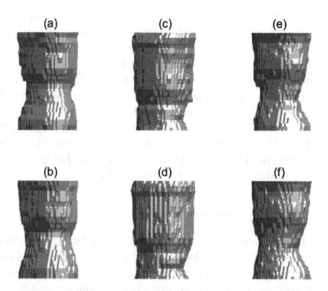

**Fig. 3.** Visualisation of endocardial surfaces. (a) and (b), (c) and (d), (e) and (f) are for the same slice respectively. (a), (c) and (e) are generated from reference binary segmentation volume, and (b), (d) and (f) are generated from predicted binary segmentation of one Seg-TDM Mode results.

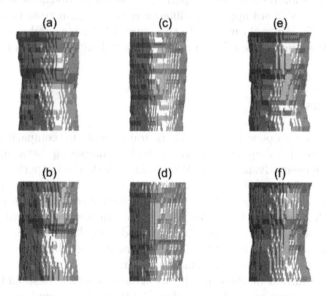

**Fig. 4.** Visualisation of epicardial surfaces. (a) and (b), (c) and (d), (e) and (f) are for the same slice respectively. (a), (c) and (e) are generated from reference binary segmentation volume, and (b), (d) and (f) are generated from predicted binary segmentation of one Seg-TDM Mode results.

We can see that there are sudden changes in the reference surfaces from one frame to the next one, and this is caused by fundamental limitations of manual segmentation, such as only one frame can be segmented at a time. We can see that the Seg-TDM networks learn a more general model and produce temporally smoother results comparing to the reference surfaces. Because the quantitative analysis is comparing with the manual reference, unlike Seg-TDM Mode, Seg Mode segments each frame independently and therefore would likely to produce more similar results to the reference in those cases. By increasing the LSTM-convolution kernel sizes and numbers or the number of recurrent layers, more complected patterns could be learnt by the network, but that will rise the chance of over-fitting or gradient vanishing at the same time.

### 4.4   Testing Data Experiments

We ensembled the segmentations from all SB networks (15 Seg Mode and 15 Seg-TDM Mode) and calculated the LV indices using the method described in Sect. 2.2. We normalized the indices following the procedure described by the organisers. Surprisingly, compared to our cross validation results on training data, the testing data results are 2 to 3 time larger in MAE values, while the correlation coefficients are similar or even slightly better for areas and dimensions. The combination of large correlation and increased error suggests a constant bias between our result and the reference values. We performed a visual inspection of the segmentation results and comparison between distributions of normalized indices, and we could not appreciate differences between our cross validation and the final results. We could not locate the source of the bias and suspect that it might have been introduced due to potential underlying differences between the two datasets.

## 5   Conclusion

In this paper, we propose a multi-mode neural network for comparison between different types of LV quantification methods. Comparing between our four modes, we see clear advantages of SB methods, which also outperform the state of the art at the time of paper submission. The evaluated indices are more accurate, and they can be verified with segmentation results which is a further advantage. Between two DR modes, we see that our novel segmentation-feature improves the accuracy of the results. By introducing TDM, we do not see significant improvement of the SB method performance, and by visualising the LV contours in time we see temporal smoothness introduced by TDM.

For generalization of these results to clinical practice, it should be aware that all of the images are mid-cavity slices, while for full quantification of LV, a complete stack of short-axis slices and a pair of long-axis slices are necessary.

Our Seg-Reg Mode shows improvement on the regression task, but not on the segmentation task. This might be due in part to the fact that training was not fully simultaneous but alternated between epochs. The use of a recursive network did not improve results significantly, and this is partly caused by referring to 2D manual segmentation. In our experiments, we only apply recursive units for one feature level, and further integration may produce improved results.

# References

1. Hochreiter, S., Schmidhuber, J.: Long short-term memory. Neural Comput. **9**(8), 1735–1780 (1997)
2. Karamitsos, T.D., Francis, J.M., Myerson, S., Selvanayagam, J.B., Neubauer, S.: The role of cardiovascular magnetic resonance imaging in heart failure. J. Am. Coll. Cardiol. **54**(15), 1407–1424 (2009)
3. Long, J., Shelhamer, E., Darrell, T.: Fully convolutional networks for semantic segmentation. In: Proceedings of the IEEE CVPR, pp. 3431–3440 (2015)
4. Oktay, O., et al.: Anatomically constrained neural networks (ACNNs): application to cardiac image enhancement and segmentation. IEEE TMI **37**(2), 384–395 (2018)
5. Poudel, R.P.K., Lamata, P., Montana, G.: Recurrent fully convolutional neural networks for multi-slice MRI cardiac segmentation. In: Zuluaga, M.A., Bhatia, K., Kainz, B., Moghari, M.H., Pace, D.F. (eds.) RAMBO/HVSMR 2016. LNCS, vol. 10129, pp. 83–94. Springer, Cham (2017). https://doi.org/10.1007/978-3-319-52280-7_8
6. Ronneberger, O., Fischer, P., Brox, T.: U-Net: convolutional networks for biomedical image segmentation. In: Navab, N., Hornegger, J., Wells, W.M., Frangi, A.F. (eds.) MICCAI 2015, Part III. LNCS, vol. 9351, pp. 234–241. Springer, Cham (2015). https://doi.org/10.1007/978-3-319-24574-4_28
7. Suinesiaputra, A., et al.: Quantification of LV function and mass by cardiovascular magnetic resonance: multi-center variability and consensus contours. JCMR **17**(1), 63 (2015)
8. Tran, P.V.: A fully convolutional neural network for cardiac segmentation in short-axis MRI (2016). arXiv preprint arXiv:1604.00494
9. Vigneault, D.M., Xie, W., Ho, C.Y., Bluemke, D.A., Noble, J.A.: $\Omega$-Net (Omega-Net): fully automatic, multi-view cardiac MR detection, orientation, and segmentation with deep neural networks. Med. image Anal. **48**, 95–106 (2018)
10. Xue, W., Islam, A., Bhaduri, M., Li, S.: Direct multitype cardiac indices estimation via joint representation and regression learning. IEEE TMI **36**(10), 2057–2067 (2017)
11. Xue, W., Lum, A., Mercado, A., Landis, M., Warrington, J., Li, S.: Full quantification of left ventricle via deep multitask learning network respecting intra- and inter-task relatedness. In: Descoteaux, M., Maier-Hein, L., Franz, A., Jannin, P., Collins, D.L., Duchesne, S. (eds.) MICCAI 2017, Part III. LNCS, vol. 10435, pp. 276–284. Springer, Cham (2017). https://doi.org/10.1007/978-3-319-66179-7_32
12. Xue, W., Nachum, I.B., Pandey, S., Warrington, J., Leung, S., Li, S.: Direct estimation of regional wall thicknesses via residual recurrent neural network. In: Niethammer, M., et al. (eds.) IPMI 2017. LNCS, vol. 10265, pp. 505–516. Springer, Cham (2017). https://doi.org/10.1007/978-3-319-59050-9_40
13. Xue, W., Brahm, G., Pandey, S., Leung, S., Li, S.: Full left ventricle quantification via deep multitask relationships learning. MIA **43**, 54–65 (2018)
14. Zeiler, M.D.: ADADELTA: an adaptive learning rate method (2012). arXiv preprint arXiv:1212.5701

# Automated Full Quantification of Left Ventricle with Deep Neural Networks

Lihong Liu, Jin Ma$^{(\boxtimes)}$, Jianzong Wang, and Jing Xiao

Ping An Technology (Shenzhen) Co., Ltd., Shenzhen, China
{liulihong493,majin730,wangjianzong347,xiaojing661}@pingan.com.cn

**Abstract.** Accurate cardiac left ventricle (LV) quantification is among the most clinically important and most frequently demanded tasks for identification and diagnosis of cardiac diseases and is of great interest in the research community of medical image analysis. However, it is still a task of great challenge due to the high variability of cardiac structure across subjects and the complexity of temporal dynamics of cardiac sequences. Full quantification of cardiac LV includes simultaneously quantifying, for every frame in the whole cardiac cycle, multiple types of cardiac indices, such as cavity and myocardium areas, regional wall thicknesses, LV dimension and cardiac phase. Accurate quantification of these indices will support comprehensive global and regional cardiac function assessment. In this paper, we propose a newly designed multitask learning network which combines the segmentation task and the cardiac indices quantification task. It comprises a segmentation network based on the U-net for image representation, and then followed by a simple convolutional neural network for feature extraction of cardiac indices, two parallel recurrent neural network models are then added for temporal dynamic modeling. Then we use multitask learning to capture the existing correlations among different tasks. Experiments of 5-fold validation results show that the proposed framework achieves high accurate prediction, with average mean absolute error of $173\,\text{mm}^2$, $2.44\,\text{mm}$, $1.37\,\text{mm}$ for areas, dimensions, RWT and phase error rate 7.8%.

**Keywords:** Multitask learning · Left ventricle quantification · Segmentation · LSTM

## 1 Introduction

Cardiovascular disease is one of the major causes of death in the world [1]. Accurate cardiac left ventricle (LV) quantification is among the most clinically important and most frequently demanded tasks for identification and diagnosis of cardiac diseases and is of great interest in the research community of medical image analysis [2,3]. In clinical practice, a traditional way for LV quantification is segmenting the LV myocardium from other structures, after which measuring the corresponding indices, which is time consuming and tedious. In the past few

M. Pop et al. (Eds.): STACOM 2018, LNCS 11395, pp. 412–420, 2019.
https://doi.org/10.1007/978-3-030-12029-0_44

years, a number of different methods for automatizing this procedure have been proposed. However, it is still a task of great challenge due to the high variability of cardiac structure across subjects and the complexity of temporal dynamics of cardiac sequences.

The approaches for left ventricle quantification presented in the literatures can be roughly divided into two categories: (1) segmentation-based quantification; (2) direct regression methods. Segmentation methods segment out cardiac myocardium from the complex surroundings and calculate the LV indices from the segmentation mask. Traditional segmentation methods [3–5] for cardiac images require strong prior information and user interaction to obtain reliable results. Current state of the art LV segmentation approaches rely on deep neural networks. [1] proposed an end to end neural network architecture to detect and segment the LV jointly from the entire stack of short-axis images, which produced state of the art results.

Recently, direct methods without segmentation have grown in popularity in cardiac volumes estimation and obtained effective performance. This method can be further divided into two classes: two-phase volume-only method [6–10] and end to end deep neural network-based methods [1,2,11–14]. For two phases method, it is inadequate to achieve accurate estimation for multitype cardiac indices, for the reason that image representation and indices regression are separately handled [13]. The later method employs deep neural networks in an end to end manner. Some methods also take the temporal information and multitask correlation information into account to leverage the results. [12] proposed Residual Recurrent Neural Network to utilize the spatial and temporal dynamics of LV myocardium to achieve accurate frame-wise RWT estimation. [3] proposes a method which learns multitask relationships automatically.

Multitask relationship learning aims at improving the performance of the main task or all tasks by transferring knowledge among these tasks, which is widely researched in other topics. In this paper, we propose a framework that combines the segmentation task and quantification task to better utilize the label information to leverage the performance.

The rest of this paper is organized as follows. In Sect. 2, the detail of the proposed framework is reviewed along with their advantages and disadvantages. In Sect. 3, experiment results evaluating the proposed methods are given. We drew our conclusions in Sect. 4.

## 2    Method

### 2.1    Architecture Overview

The overview of our architecture is shown in Fig. 1. The Left Ventricle Full Quantification Challenge offers good quality of binary mask of the endocardium and epicardium, which are double-checked by two experienced cardiac radiologists. Making good use of the binary mask can enhance the expressiveness of image representation with regard to cardiac indices and benefit the final cardiac indices quantification task. In our proposal, we first design a convolutional deep neural segmentation network based on the U-net for cardiac image representation. Then a shallow convolutional neural network (CNN) is followed after the segmentation network to further extract cardiac index related features. Two parallel recurrent neural network models are then added for temporal dynamic modeling. Then we use multitask learning to capture the existing correlations among different tasks.

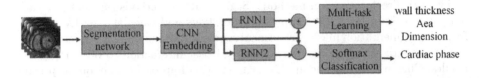

**Fig. 1.** Overview architecture

## 2.2 Segmentation Network

To enhance the expressiveness of image representation with respect to cardiac indices, we design a deep learning segmentation network. To capture larger context and multi-scale features, we extend the DenseNets and dilation block to work as U-net [15].

Figure 2 shows a simple example of our architecture. The diagram is composed of a downsampling path with 2 Transitions Down (TD) and an upsampling path with 2 Transitions Up (TU). The circle represents concatenation and the arrows represent connectivity patterns in the network. The gray horizontal arrows represent skip connections, and the feature maps from the down-sampling path are concatenated with the corresponding feature maps in the upsampling path. Figure 3 shows the details of dilation Block (DL). The DLs with dilation rate of 1, 3, 5 are concatenated as the input of the following convolutional layer, to capture multi-scale feature. The dense block (DB) is constructed as described in [16,17].

In order to recover the input spatial resolution, fully convolutional networks (FCNs) introduce an upsampling path composed of convolution, upsampling operations (transposed convolutions or unpooling operations) and skip connections. However, the process of reducing the image size and then restoring resolution can result in information loss, while dilation convolution [18] can expand the receptive field without losing information. With the same convolution kernel size, different dilation rates can be used to obtain multi-scale features and more contextual information. But the simple superposition of multiple dilation convolution produces gridding effect, and long-ranged information might be not relevant. So this article uses the combination of dilation convolution [19], at the same time satisfies the requirement of the small target and large target segmentation. Small dilation rate focuses on the characteristics of the small object, as vice verse.

The advantages of DB structure enable us to build a deeper and more efficient network with fewer parameters and to achieve feature reuse.

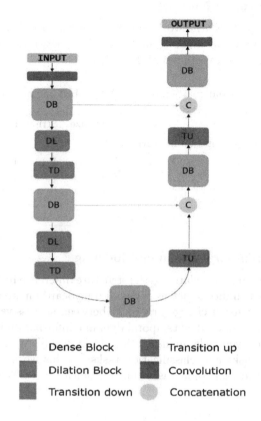

**Fig. 2.** The architecture of segmentation network

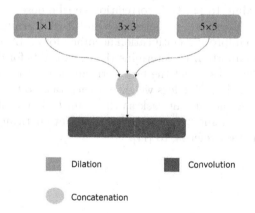

**Fig. 3.** The architecture of dilation block

## 2.3    CNN Embedding Network

We designed a simple CNN network to extract index-specific feature from segmentation feature maps, as is shown in Table 1. The network maps each cardiac image into low dimension vectors with fixed lengths.

Table 1. Configuration of the CNN embedding network.

| Input | Kernel size | Stride | Pad | Output size |
|---|---|---|---|---|
| Feature map from segmentation network | 3 * 3 | 1 | 1 | 80 * 80 * 16 |
| conv1 | 3 * 3 | 1 | 1 | 80 * 80 * 16 |
| conv2 | 3 * 3 | 1 | 1 | 80 * 80 * 1 |
| fc1 | | | | 256 |
| fc2 | | | | 128 |

## 2.4    RNN for Temporal Dynamic Modeling

The cardiac phase is mainly related to the structure difference between successive frames, other cardiac indices, such as cavity and myocardium areas, regional wall thicknesses, LV dimension change gradually between successive frames. Therefore, making good use of cardiac temporal dynamic information is very useful for accurate cardiac left ventricle quantification. In [2,3], LSTM units are deployed for the regression tasks and classification tasks. In this work, we also use two LSTM to model the regression tasks and classification tasks separately as is shown in Fig. 1.

## 2.5    Multitask Learning

Multitask learning aims at utilizing the correlation among different tasks and learning several tasks simultaneously to boost the performance of the main task or all tasks. According to [2], task correlation can be observed for the cardiac qualification tasks. Therefore, we introduce multitask learning to leverage the tasks correlation to improve quantification accuracy for all LV indices. In Multitask learning, it is important to determine the loss weights for different tasks, as it is hard for all the tasks to get the best performance in the same loss weights. Therefore, we assigned specific loss weight for each task to train the model. For each task, we regard the current task as the main task and other task as side tasks. We use the dynamic-weighting scheme to automatically assign the loss weight to each side task refering to [11].

## 3    Experiment

### 3.1    Dataset

The dataset is offered by Left Ventricle Full Quantification Challenge MICCAI 2018. The database includes 2D short axis cine MR images of 145 for training phase and 30 for testing phase subjecting from routine clinic patients in 3 hospitals affiliated with two healthcare centers (London Healthcare Center, ON, Canada, and St. Josephs Health Care). The collected data covers a wide range of diversity in ages, pathologies, image quality and pixel-spacing. The temporal resolution is 20 frames per cardiac cycle. All images are coupled with two set of reference values: values of the cardiac indices and manual segmentation of the myocardium.

### 3.2 Experiment Setup

We trained our models from scratch without using any extra-data or postprocessing module. The model is implemented with pytorch. The models are trained using Adam optimization with default parameters, $\beta 1 = 0.9$ and $\beta 2 = 0.999$. The initial learning rate is set to $2 * 10^{-4}$. We use a server with 4 Tesla M60 GPUs to train the models. We divide the data into five equal parts for 5-fold cross-validation. The training set provided by the competition is divided into parts. The first one contains 80% of the data for training, while the second part contains the rest for validation as well as model selection. For the final submission, we use the whole training set to finetune the best model chosen by the validation phase.

### 3.3 Training Method

We adopt a three-stage training procedure. Firstly, the segmentation network is trained with the dice loss with per-pixel binary label of the endocardium and epicardium. Secondly the weight of segmentation part is fixed. The regression part including the CNN embedding and the first one of the parallel RNNs, is trained with no feedback information from phase part. In the third stage, the regression part is frozen, and the second one of the parallel RNNs is learned with feedback from phase part. The second and the third stages are iterated for several times to get a better result.

### 3.4 Results

Extensive experiments have been conducted to evaluate the effectiveness of our architecture from the following aspects and the results are shown in Table 2. The first column of Table 2 lists all the parameters the competition requires to measure. Results reported by (DMTRL) [3] are shown in the second column of Table 2 and the results of our proposed method are shown in the last column.

The effects of segmentation is explored, namely Test 1. The segmentation enhances the expressiveness of the representation model. To demonstrate this, we trained another network to do the regression and phase task from scratch instead of training the segmentation network with perpixel binary label of the endocardium and epicardium which is proposed in this work. The third column of Table 2 lists the results.

We researched the effects of different loss weight that each task brings, namely Test 2. Compared with our proposed multitask training we trained the specific model for each task and combine the task results as the final results, as is described in Sect. 2.5. Test 2 trained a single model for all the task. The results are shown in the fourth column of Table 2.

For the final results of the cavity and myocardium areas, we use the binary mask generated by the segmentation networks to calculate, thereafter ensembled the results with the proposed method with direct regression. Regarding to the mean of all the parameters of the LV, our proposed method shows better results compared with DMTRL.

**Table 2.** Performance of full LV quantification. Mean average error (MAE) and $\rho$ are shown in each cell.

|  | DMTRL | Test 1 | Test 2 | Proposed method |
|---|---|---|---|---|
| Area $(mm^2)$ |  |  |  |  |
| A1 | 172 (0.943) | 183 (0.939) | 160 (0.957) | 163 (0.954) |
| A2 | 189 (0.947) | 228 (0.817) | 205 (0.84) | 184 (0.944) |
| A_mean | 180 (0.945) | 205 (0.878) | 182 (0.899) | 173 (0.949) |
| Dim (mm) |  |  |  |  |
| Dim1 | 2.47 (0.957) | 2.45 (0.931) | 2.35 (0.939) | 2.35 (0.939) |
| Dim2 | 2.59 (0.894) | 2.53 (0.923) | 2.41 (0.950) | 2.41 (0.950) |
| Dim3 | 2.48 (0.943) | 2.66 (0.938) | 2.56 (0.933) | 2.56 (0.933) |
| Dim_mean | 2.51 (0.925) | 2.54 (0.930) | 2.44 (0.940) | 2.44 (0.940) |
| Dim (mm) |  |  |  |  |
| RWT1 | 1.26 (0.856) | 1.48 (0.757) | 1.26 (0.820) | 1.21 (0.850) |
| RWT2 | 1.40 (0.747) | 1.70 (0.470) | 1.73 (0.475) | 1.51 (0.645) |
| RWT3 | 1.59 (0.693) | 1.97 (0.450) | 1.76 (0.563) | 1.43 (0.739) |
| RWT4 | 1.57 (0.659) | 1.57 (0.550) | 1.67 (0.560) | 1.64 (0.529) |
| RWT5 | 1.32 (0.777) | 1.42 (0.641) | 1.21 (0.754) | 1.14 (0.701) |
| RWT6 | 1.25 (0.877) | 1.55 (0.726) | 1.30 (0.852) | 1.31 (0.858) |
| RTW_mean | 1.39 (0.78) | 1.61 (0.599) | 1.49 (0.671) | 1.37 (0.720) |
| Phase (%) |  |  |  |  |
| Phase | 8.2 | 12.2 | 9.4 | 7.8 |

## 4    Conclusions

In this paper, we proposed an integrated deep multitask relationship leaning framework for full quantification of cardiac LV. The proposed framework combines the advantages of segmentation task and cardiac indices quantification task. It comprises a segmentation network which is based on the U-net to enhance the expressiveness of image representation with respect to cardiac indices and benefits the final cardiac indices quantification task. Two parallel recurrent neural network models are integrated for temporal dynamic modeling. Multitask learning is added to capture the existing correlations among different tasks.

# References

1. Poudel, R.P., Lamata, P., Montana, G.: Recurrent fully convolutional neural networks for multi-slice MRI cardiac segmentation. arXiv preprint arXiv:1608.03974 (2016)
2. Xue, W., Brahm, G., Pandey, S., Leung, S., Li, S.: Full left ventricle quantification via deep multitask relationships learning. Med. Image Anal. **43**, 54–65 (2018)
3. Xue, W., Lum, A., Mercado, A., Landis, M., Warrington, J., Li, S.: Full quantification of left ventricle via deep multitask learning network respecting intra- and inter-task relatedness. In: Descoteaux, M., Maier-Hein, L., Franz, A., Jannin, P., Collins, D.L., Duchesne, S. (eds.) MICCAI 2017. LNCS, vol. 10435, pp. 276–284. Springer, Cham (2017). https://doi.org/10.1007/978-3-319-66179-7_32
4. Ayed, I.B., Chen, H.M., Punithakumar, K., Ross, I., Li, S.: Max-flow segmentation of the left ventricle by recovering subject-specific distributions via a bound of the Bhattacharyya measure. Med. Image Anal. **16**(1), 87–100 (2012)
5. Petitjean, C., Dacher, J.N.: A review of segmentation methods in short axis cardiac MR images. Med. Image Anal. **15**(2), 169–184 (2011)
6. Afshin, M., et al.: Regional assessment of cardiac left ventricular myocardial function via MRI statistical features. IEEE Trans. Med. Imaging **33**(2), 481–494 (2014)
7. Wang, Z., Ben Salah, M., Gu, B., Islam, A., Goela, A., Li, S.: Direct estimation of cardiac biventricular volumes with an adapted Bayesian formulation. IEEE Trans. Biomed. Eng. **61**(4), 1251–1260 (2014)
8. Zhen, X., Wang, Z., Islam, A., Bhaduri, M., Chan, I., Li, S.: Direct estimation of cardiac bi-ventricular volumes with regression forests. In: Golland, P., Hata, N., Barillot, C., Hornegger, J., Howe, R. (eds.) MICCAI 2014. LNCS, vol. 8674, pp. 586–593. Springer, Cham (2014). https://doi.org/10.1007/978-3-319-10470-6_73
9. Zhen, X., Wang, Z., Islam, A., Bhaduri, M., Chan, I., Li, S.: Multiscale deep networks and regression forests for direct biventricular volume estimation. Med. Image Anal. **30**, 120–129 (2016)
10. Zhen, X., Islam, A., Bhaduri, M., Chan, I., Li, S.: Direct and simultaneous four-chamber volume estimation by multi-output regression. In: Navab, N., Hornegger, J., Wells, W.M., Frangi, A.F. (eds.) MICCAI 2015. LNCS, vol. 9349, pp. 669–676. Springer, Cham (2015). https://doi.org/10.1007/978-3-319-24553-9_82
11. Xi, Y., Liu, X.: Multi-task convolutional neural network for Pose-Invariant face recognition. arXiv preprint arXiv:1702.04710
12. Xue, W., Nachum, I.B., Pandey, S., Warrington, J., Leung, S., Li, S.: Direct Estimation of Regional Wall Thicknesses via Residual Recurrent Neural Network. In: Niethammer, M., et al. (eds.) IPMI 2017. LNCS, vol. 10265, pp. 505–516. Springer, Cham (2017c). https://doi.org/10.1007/978-3-319-59050-9_40
13. Xue, W., Islam, A., Bhaduri, M., Li, S.: Direct multitype cardiac indices estimation via joint representation and regression learning. IEEE Trans. Med. Imaging (2017)

14. Xue, W., Lum, A., Mercado, A., Landis, M., Warrington, J., Li, S.: Full quantification of left ventricle via deep multitask learning network respecting intra- and inter-task relatedness. In: Descoteaux, M., Maier-Hein, L., Franz, A., Jannin, P., Collins, D.L., Duchesne, S. (eds.) MICCAI 2017. LNCS, vol. 10435, pp. 276–284. Springer, Cham (2017b). https://doi.org/10.1007/978-3-319-66179-7_32

15. Ronneberger, O., Fischer, P., Brox, T.: U-Net: convolutional networks for biomedical image segmentation. In: Navab, N., Hornegger, J., Wells, W.M., Frangi, A.F. (eds.) MICCAI 2015. LNCS, vol. 9351, pp. 234–241. Springer, Cham (2015). https://doi.org/10.1007/978-3-319-24574-4_28

16. Huang, G., Liu, Z., Weinberger, K.Q.: densely connected convolutional networks. In: Proceedings of the IEEE Conference on Computer Vision and Pattern Recognition (CVPR), pp. 4700–4708 (2016)

17. Jégou, S., Drozdzal, M., Vazquez, D., Romero, A., Bengio, Y.: The one hundred layers Tiramisu: fully convolutional DenseNets for semantic segmentation. arXiv preprint arXiv:1611.09326 (2016)

18. Yu, F., Koltun, V.: Multi-scale context aggregation by dilated convolutions (2015)

19. Wang, P., Chen, P., Yuan, Y., et al.: Understanding convolution for semantic segmentation (2017)

# ESU-P-Net: Cascading Network for Full Quantification of Left Ventricle from Cine MRI

Wenjun Yan[1], Yuanyuan Wang[1(✉)], Shaoxiang Chen[2],
Rob J. van der Geest[3], and Qian Tao[3(✉)]

[1] Department of Electrical Engineering, Fudan University, Shanghai, China
yywang@fudan.edu.cn
[2] Department of Computer Science, Fudan University, Shanghai, China
[3] Department of Radiology, Leiden University Medical Center,
Leiden, the Netherlands
Q.Tao@lumc.nl

**Abstract.** Left ventricle (LV) quantification is of great clinical importance for diagnosing and monitoring cardiac diseases. Full quantification of LV indices includes: (1) two areas of LV cavity and myocardium, (2) six regional wall thicknesses (RWT), (3) three LV dimensions, and (4) phase identification (diastole or systole). However, due to the large variability in the object shape and imaging quality, it is time-consuming and user-dependent to quantify LV parameters manually. In this work, we propose a cascading deep neural network, including an enhanced supervision U-net followed a recurrent neural network (RNN) type of phase-prediction net called P-net, abbreviated as ESU-P-net, for full LV quantification in a fully automated manner. The proposed ESU-P-net framework is dedicated to the full quantification of LV for all four types of indices.

Experiments on MR sequences of 145 subjects provided by MICCAI 2018 STACOM Challenge showed that the proposed network achieved highly accurate LV quantification, with an average mean absolute error (MAE) of 62 $mm^2$, 1.14 mm, 0.96 mm for LV areas, RWT, dimensions, respectively, and an error rate of 8.0% for cardiac phase identification.

**Keywords:** Full quantification · LV segmentation · U-net · RNN

## 1 Introduction

Left ventricle (LV) quantification is of great clinical importance for diagnosing and monitoring cardiac disease [1]. To comprehensively assess cardiac function, evaluation of a range of cardiac indices from short-axis MRI cine images are required, including (1) LV cavity and myocardium areas, (2) regional wall thicknesses, (3) LV dimension, and (4) cardiac phase, as depicted in Fig. 1. Conventionally, assessment of these indices is based upon manual or semi-automated segmentation of the LV, the accuracy of which determines the quality of LV assessment.

© Springer Nature Switzerland AG 2019
M. Pop et al. (Eds.): STACOM 2018, LNCS 11395, pp. 421–428, 2019.
https://doi.org/10.1007/978-3-030-12029-0_45

**Fig. 1.** Illustration of three types of LV indices and one classification to be quantified for cardiac image. (1) Cavity (light blue) and myocardium (deep blue) areas. (2) Directional dimensions of cavity (black arrows). (3-1) Regional wall thicknesses (black arrows). IS: inferoseptal; I: inferior; IL: inferolateral; AL: anterolateral; A: anterior; AS: anterospetal. (3-2) Phase (systole or diastole). (Color figure online)

Manual delineation of LV from the cine MRI, with hundreds of images per scan, is time-consuming and subjective. Moreover, variations in LV morphology across slices, phases, vendors, and patients make it virtually impossible to reach a clinically acceptable balance of accuracy and robustness for conventional atlas-based or shape-based computer methods.

Recently, deep convolutional neural networks (CNN) show remarkable advantages in solving complex data-driven problems. Emerging as a new state-of-art, CNN demonstrates unique capability to work accurately on data of high variability. This work is dedicated to the exploration of an effective CNN structure to segment LV from short-axis cine MRI, in response to the Left Ventricle Full Quantification Challenge (LVQuan18), the STACOM workshop in MICCAI 2018. For this purpose, we developed and validated an enhanced-supervision U-net (ESU-net) followed by a recurrent neural network (RNN) P-net to obtain full quantification of all four types of LV indices, in a fully automated manner.

## 2   Data

The proposed method was developed and validated using data provided by the LVQuan18 Challenge. A training dataset was collected including a dataset of short-axis cine MR images of 145 subjects from 3 hospitals affiliated with two health care centers (London Healthcare Center and St. Josephs HealthCare). Age of the subjects ranges from 16 years to 97 years, with an average of 58.9 years. The in-plane resolution of the MR images ranges from 0.6836 mm to 2.0833 mm, with the mode being 1.5625 mm. Diverse pathologies are present in the dataset including regional wall motion abnormalities, myocardial hypertrophy, mildly enlarged LV, atrial septal defect, LV dysfunction, etc. Each subject contains 20 frames in a whole cardiac cycle at the mid-cavity level.

All cardiac images underwent several preprocessing steps, including landmark labeling, rotation, ROI cropping, and resizing. The resulted images were approximately aligned with a dimension of $80 \times 80$. Then, these cardiac images were manually contoured to obtain the epicardial and endocardial borders, which were double-checked by two experienced cardiac radiologists. The ground truth of LV indices and cardiac phase can be calculated from the endocardial and epicardial borders.

## 3   Method

We propose a fully automatic workflow for segmentation and frame classification in cardiac cine MR images. For the fact that the classification of cardiac phase is closely related to deformation of cardiac, cardiac phase can be predicted over statistics of shape parameters like area and dimensions. The method utilized a U-net to segment the LV cavity and the myocardium in short-axis cine MRI. Then quantitative indices, i.e. cavity and myocardium areas, regional wall thicknesses (RWT), LV dimension, are directly calculated from the segmentation results, while cardiac phase is classified by feeding the calculated indices into a RNN. Figure 2 illustrates the architecture of the proposed CNN.

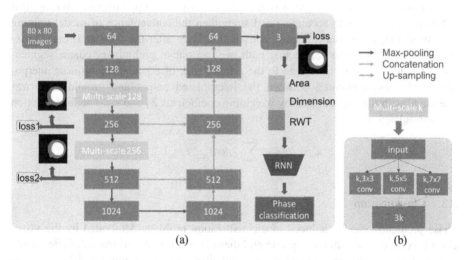

**Fig. 2.** The proposed ESU-P-Net. (a) The architecture of the network. (b) The structure of multi-scale module in the network. 3 k indicates 3 channels.

### 3.1   Segmentation

Since U-net is a popular segmentation architecture for medical images [2], we take advantage of this encoder-decoder fashion to produce delicate pixel-wise prediction. As shown in Fig. 2, the encoder is equivalent to a contracting path that extracts features at both local and global levels. At shallow layers, low-level features focus on structural or gray-scale details of heart; at deep layers, high-level features provide localization information.

(1)   Multi-reception

In view of the fact that the resolution of 80 × 80 image is relatively low, the border of myocardium is ambiguous even for manual delineation. Nevertheless, the boundary is recognized by experts knowing the nature of LV structure. This inspired us to explore more global characteristics of LV. We applied the multi-reception field mechanism into network to enhance the high-level feature learning, inspired by the inception module of GoogLenet with modification of kernel size [3]. After the second pooling layers, the kernel sizes of convolution layers vary from $3 \times 3$ to $7 \times 7$ and different outputs are combined through concatenation to aggregate the multi-scale information, as shown in Fig. 2.

(2) Multi-up-sampling

In the decoder known as expanding path, we used not only the transpose convolution but also the bilinear interpolation as up-sampling operations to obtain high-resolution result.

(3) Loss function for enhance supervision

The loss function we used was the general cross entropy function. We applied enhanced supervision to accelerate and strengthen the convergence of model training [4]. By adding additional transpose convolution layer after the third and fourth max-pooling operation in the contracting path, we can obtain preliminary coarse segmentation results, which were resized to the original size of $80 \times 80$ by bilinear interpolation. The loss computed between the interpolated output and ground truth was combined with the original loss by weighting coefficients $\lambda_1$ and $\lambda_2$. The loss function was formulated as:

$$\mathcal{L} = loss + \sum \lambda_i loss_{\text{sup}}^i, i = 1, 2.$$

### 3.2   Classification

Phase prediction can be reduced to a classification problem. Motivated by the visual clues used in classification of systole and diastole, we propose to learn the inter-frame state transitions. The state transitions are explicitly depicted by multiple LV indices as we obtained from the previous segmentation. Compared to the massive feature maps, the limited number of indices as input largely reduced the complexity of computation and memory. We used a specific RNN architecture called Prediction-net (abbreviated as P-net) to address the phase classification problem [5]. During training, we first train the end-to-end ESU-net to obtain relatively accurate indices and then feed it into P-net as training data.

## 4   Experiments and Results

### 4.1   Implementation Details

The proposed method was evaluated using the LVQuan18 training set in a five-fold cross-validation experiment. The dataset is divided into 5 groups, each containing 29 subjects. For each fold the system was trained using 4 groups and tested in the remaining group. The network is implemented by Google TensorFlow, using the NVIDIA GeForce GTX TITAN X GPU. Two sub-networks have respective configuration:

(1) ESU-net: the learning rate was set to $10^{-4}$ with a decaying factor of 0.5 every 10 epochs. The batch size was 10. Both $\lambda_1$ and $\lambda_2$ were set to 0.5.
(2) P-net: learning rate was set to $10^{-2}$. The batch size was 5. The number of neuron in hidden layer was 40.

### 4.2   Evaluation Criteria

The performance of the proposed network was evaluated in terms of estimation accuracy for full LV quantification, i.e., three types of LV indices and cardiac phase, for all frames in the cardiac cycle. To assess the estimation accuracy for the three LV indices, correlation coefficient $\rho$ and mean absolute error (MAE) between the ground truth $g$ and the prediction $p$ were computed. For cardiac phase identification, error rate (ER) is computed. For $g \in R^M$ and $p \in R^M$, the three parameters were computed as follows:

$$\rho(g,p) = \frac{2\sum_{i=1}^{M}(g_i - m_g)(p_i - m_p)}{\sum_{i=1}^{M}(g_i - m_g)^2 + (p_i - m_p)^2} \tag{1}$$

$$MAE(g,p) = \frac{1}{M}\sum_{i=1}^{M}|g_i - p_i| \tag{2}$$

$$ER(g,p) = \frac{\sum_{i=1}^{M}1(g \neq p)}{M} \tag{3}$$

where g is ground truth, p is predicted result, $m_g = \frac{1}{M}\sum_{i=1}^{M}g_i$, and $m_p = \frac{1}{M}\sum_{i=1}^{M}p_i$. An example of measurement is shown in Fig. 3.

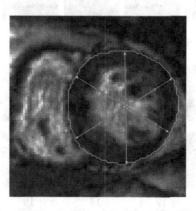

**Fig. 3.** Illustration on LV quantification. The blue bi-directional arrows are three types of dimensions, and the yellow unidirectional arrows are six types of regional wall thicknesses. (Color figure online)

### 4.3   Quantitative Analysis

We validated our proposed network in twofold: (1) LV quantification performance by the three types of LV indices and cardiac phase identification. (2) Comparison with three state-of-art deep learning methods, including one segmentation-based method Max-flow [6], one semi-automated method Indice-net [7], and one end-to-end method DMTRL [8].

Performance of Full LV Quantification

The performance of full LV quantification is reported in Table 1. Compared to the ground truth, the proposed ESU-net achieves average MAE of $62 \pm 54.9$ mm$^2$, $1.14 \pm 0.84$ mm, and $0.96 \pm 0.70$ mm, and an average correlation of 0.961, 0.960,

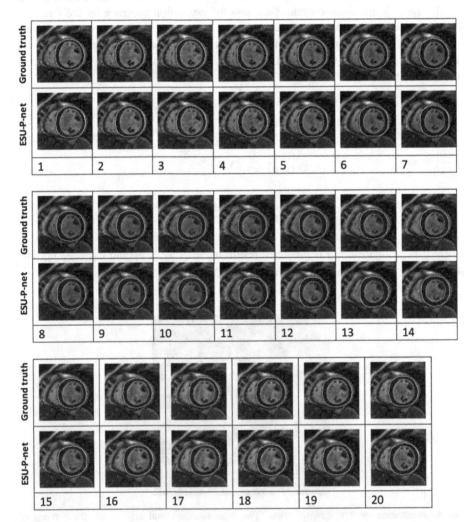

**Fig. 4.** An example of the automated LV segmentation by the ESU-net. All 20 cardiac phases are shown. In each row, above is the ground truth, and below is the automated segmentation.

**Table 1.** Performance comparison between the proposed ESU-P-net and the other three state-of-the-art methods on full LV quantification. In each cell, both MAE (above) and ρ (below) are reported.

| Method | Max-flow | Indice-net | DMTRL | ESU-P-net |
|---|---|---|---|---|
| Area (mm$^2$) | | | | |
| cav | 156 ± 193 | 185 ± 162 | 172 ± 148 | 34.4 ± 29.8 |
|  | 0.958 | 0.953 | 0.943 | 0.970 |
| myo | 339 ± 272 | 223 ± 193 | 189 ± 159 | 89.5 ± 60.2 |
|  | 0.851 | 0.853 | 0.947 | 0.952 |
| Average | 247 ± 201 | 204 ± 133 | 180 ± 118 | **62.0 ± 54.9** |
|  | 0.904 | 0.903 | 0.945 | **0.961** |
| Dimension (mm) | | | | |
| Dim1 | 2.81 ± 2.76 | / | 2.47 ± 1.95 | 0.955 ± 0.819 |
|  | 0.937 | / | 0.957 | 0.960 |
| Dim2 | 2.60 ± 2.62 | / | 2.59 ± 2.07 | 1.33 ± 0.99 |
|  | 0.946 | / | 0.894 | 0.951 |
| Dim3 | 2.49 ± 2.88 | / | 2.48 ± 2.34 | 1.36 ± 0.93 |
|  | 0.945 | / | 0.943 | 0.968 |
| Average | 2.65 ± 2.33 | / | 2.51 ± 1.58 | **1.14 ± 0.84** |
|  | 0.943 | / | 0.925 | **0.960** |
| RWT (mm) | | | | |
| IS | 1.53 ± 1.73 | 1.39 ± 1.13 | 1.26 ± 1.04 | 1.25 ± 0.65 |
|  | 0.796 | 0.824 | 0.856 | 0.947 |
| I | 3.23 ± 2.83 | 1.51 ± 1.21 | 1.40 ± 1.10 | 0.78 ± 0.70 |
|  | 0.720 | 0.701 | 0.747 | 0.829 |
| IL | 4.15 ± 3.17 | 1.65 ± 1.36 | 1.59 ± 1.29 | 0.72 ± 0.57 |
|  | 0.743 | 0.671 | 0.693 | 0.863 |
| AL | 5.08 ± 3.95 | 1.53 ± 1.25 | 1.57 ± 1.34 | 0.87 ± 0.72 |
|  | 0.706 | 0.698 | 0.659 | 0.866 |
| A | 3.47 ± 3.25 | 1.30 ± 1.12 | 1.32 ± 1.10 | 1.045 ± 0.74 |
|  | 0.724 | 0.781 | 0.777 | 0.895 |
| AS | 1.76 ± 1.80 | 1.28 ± 1.00 | 1.25 ± 1.01 | 1.13 ± 0.65 |
|  | 0.785 | 0.871 | 0.877 | 0.952 |
| Average | 3.21 ± 1.98 | 1.44 ± 0.71 | 1.39 ± 0.68 | **0.96 ± 0.70** |
|  | 0.746 | 0.758 | 0.768 | **0.892** |
| Phase | | | | |
| Error(%) | / | / | 8.2 | 8.0 |

and 0.892, for areas, dimensions and RTWs, respectively. As a reference, the average values of these indices in LVQuan18 dataset are 1942 mm$^2$, 40.7 mm, 7.2 mm. The total error rate for cardiac phase identification was 8.0%. Figure 4 shows an example of the segmentation results by the ESU-net, compared to the ground truth segmentation.

428     W. Yan et al.

(1) Performance comparison

The performance of different methods is reported in Table 1.

The proposed ESU-P-net shows to outperform the existing segmentation-based and non-segmentation-based methods. Compared to the Max-flow method, it achieved an average MAE reduction of 75.1%, 57% and 70.1% for areas, dimensions, and RWTs, respectively. Moreover, Max-flow requires user-interaction, i.e. segmentation of the first frame. As for the other semi-automated method indice-net, the proposed method obtained 69.6% and 33.3% performance improvement for areas and RWTs. ESU-P-net also outperformed the end-to-end CNN methods DMTRL. It achieved an average MAE reduction of 65.6%, 54.6% and 30.9% for areas, dimensions, and RWTs, respectively.

## 5 Conclusion

In this work, we proposed an ESU-P-net, an enhanced-supervision U-net followed by a RNN type of P-net, for full quantification of LV. This framework combined the advantages of CNN-based semantic segmentation and RNN-based sequential prediction. Evaluated on the MICCAI 2018 STACOM Challenge database, the proposed method achieved highly accurate results in LV quantification.

**Acknowledgement.** This work was supported by National Key Research and Development Program of China (No. 2018YFC0116303).

## References

1. Karamitsos, T.D., Francis, J.M., Myerson, S., Selvanayagam, J.B., Neubauer, S.: The role of cardiovascular magnetic resonance imaging in heart failure. J. Am. Coll. Cardiol. **54**(15), 1407–1424 (2009)
2. Tran, P.V.: A fully convolutional neural network for cardiac segmentation in short-axis MRI (2016). arXiv:1604.00494
3. Szegedy, C., Vanhoucke, V., Ioffe, S., Shlens J., Wojna, Z.: Rethinking the inception architecture for computer vision (2015). arXiv preprint arXiv:1512.00567
4. Dou, Q., et al.: 3D deeply supervised network for automated segmentation of volumetric medical images. Med. Image Anal. **41**, 40–54 (2017)
5. Zhang, X., Lu, L., Lapata, M.: Tree recurrent neural networks with application to language modeling (2015). arXiv preprint arXiv:1511.00060
6. Ayed, I.B., Chen, H.-M., Punithakumar, K., Ross, I., Li, S.: Max-flow segmentation of the left ventricle by recovering subject-specific distributions via a bound of the Bhattacharyya measure. Med. Image Anal. **16**(1), 87–100 (2012)
7. Xue, W., Islam, A., Bhaduri, M., Li, S.: Direct multitype cardiac indices estimation via joint representation and regression learning. IEEE Trans. Med. Imaging **36**(10), 2057–2067 (2017)
8. Xue, W., Brahm, G., Pandey, S., Leung, S., Li, S.: Full left ventricle quantification via deep multitask relationships learning. Med. Image Anal. **43**, 54–65 (2018)

# Left Ventricle Full Quantification
# via Hierarchical Quantification Network

Guanyu Yang[1,4], Tiancong Hua[1(✉)], Chao Lu[3], Tan Pan[1],
Xiao Yang[3], Liyu Hu[1], Jiasong Wu[1,4], Xiaomei Zhu[2,4],
and Huazhong Shu[1,4]

[1] LIST, Key Laboratory of Computer Network and Information Integration
(Southeast University), Ministry of Education, Nanjing 210096, China
yang.list@seu.edu.cn
[2] Department of Radiology, The First Affiliated Hospital of Nanjing
Medical University, Nanjing, China
[3] National Mobile Communications Research Laboratory,
Southeast University, Nanjing, China
[4] Centre de Recherche en Information Biomédicale Sino-Français (CRIBs),
Southeast University, Nanjing, China

**Abstract.** Automatic quantitative analysis of cardiac left ventricle (LV) function is one of challenging task for heart disease diagnosis. Four different parameters, i.e. regional wall thicknesses (RWT), area of myocardium and LV cavity, LV dimensions in different direction and cardiac phase, are used for evaluating the LV function. In this paper, we implemented a novel multi-task quantification network (HQNet) to simultaneously quantify the four different parameters. The network is mainly constituted by a customized convolutional neural network named Hierarchical convolutional neural network (HCNN) which includes different pyramid-like 3D convolution blocks with different kernel sizes for efficient feature embedding; and two long-short term memory (LSTM) networks for temporal modeling. Respecting inter-task correlations, our proposed network uses multi-task constraints for phase to improve the final estimation of phase. Selu activation function is selected instead of relu, which can bring better performance of model in experiments. Experiments on MR sequences of 145 patients show that HQNet achieves high accurate estimation by means of 7-fold cross validation. The mean absolute error (MAE) of average areas, RWT, dimensions are $197\,mm^2, 1.51\,mm, 2.57\,mm$ respectively. The error rate of phase classification is 9.8%. These results indicate that the approach we proposed has a promising performance to estimate all four parameters.

**Keywords:** Left ventricle quantification ·
Hierarchical convolutional neural network · Multi-task learning ·
Multi-task constraints

M. Pop et al. (Eds.): STACOM 2018, LNCS 11395, pp. 429–438, 2019.
https://doi.org/10.1007/978-3-030-12029-0_46

# 1 Introduction

Quantitative analysis of cardiac left ventricle (LV) images is one of challenging task for heart disease identification and diagnosis [4], However, it is difficult due to the variations of cardiac left ventricle structures. Full quantification [1, 2] refers to the simultaneous quantification of six different regional wall thicknesses (RWT), two areas, three LV dimensions and one phase (see Fig. 1). These indices fully describe the characteristics of cardiac left ventricle and provide reliable data support for the diagnosis and analysis of heart disease. However, the prediction and acquisition of these quantitative indices is still a challenging task. It is necessary to establish a reasonable connection for them because of the inter relatedness of various quantitative data. In this work, we propose a customized deep-multitasking learning network [1, 2, 12, 13], HQNet. This approach can be used to complete full quantification tasks and establish a reasonable connection between tasks.

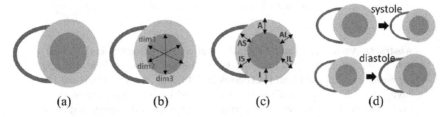

(a)              (b)              (c)              (d)

**Fig. 1.** Demonstration of the cardiac indices. (a) Area of myocardium and LV cavity. (b) Three LV cavity dimension. (c) Regional wall thickness. (d) Cardiac phase [22].

In clinical practice, quantitative analysis is often performed by segmentation of cardiac images [6, 9, 10], and the segmentation results are often obtained by manual labeling or using machine learning algorithms. What do matter are that manual label accuracy depends on the professional level of the maker and a well-performing LV segmentation algorithm is still a problem. The existing LV segmentation algorithms [6, 9, 10] can't obtain accuracy with diagnostic criteria so that they cannot be used for clinical diagnosis [2].

In recent years, methods for performing cardiac analysis directly through quantitative indices have become increasingly popular [3, 5, 8, 11]. However, some of these methods have some deficiencies. (1) Lack of temporal modeling, unable to build connection for adjacent frames. (2) Not end-to-end learning. (3) Not fully quantified, cardiac capacity is insufficient for global/local dynamic functional assessment. In addition, more sophisticated methods have been proposed in [1, 2], these methods make full use of the inter- and intra-task relatedness, and well-modeled of temporal dynamic modeling, which achieve stable and great performance.

In this article, we propose a more advanced method, called HQNet, which consists of two different hierarchies. From the overall, a customized CNN (HCNN) for better feature extraction, and two LSTMs are used for temporal modeling. Through the data augmentation and multi-task constraints for phase, admirable results are achieved.

## 2  Hierarchical Quantification Network

### 2.1  Network Architecture

The network (HQNet) is inspired by FullLVNet [2] and DTMRL [1], and it can be divided into three parts (see Fig. 2). The first part is a customized CNN named hierarchical convolutional neural network (HCNN) which can be deployed for extracting representative features of cardiac images. The second part is LSTM [7, 14, 15, 17], the variation of the RNN for temporal modeling. The third part is the linear layer used to obtain the final output. More details will be described in the below.

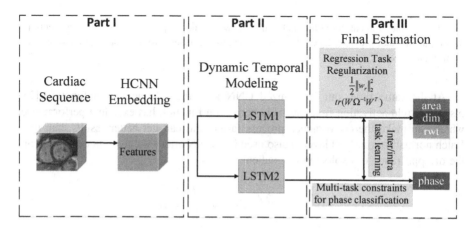

**Fig. 2.** The network architecture of HQNet, two LSTM are used for temporal dynamic modeling, and multi-task learning are used to achieve a high performance.

**Hierarchical Convolutional Neural Network (HCNN).** Different from DTMRL [1], HQNet uses a 3D convolution method to perform convolution on cardiac sequences. This method uses 'time' as the third dimension instead of the third dimension 'depth' in conventional 3D data, since the spatial information on each frame are equivalent with image registration, and the only difference is that the time node corresponding to each frame is successive, so it's reasonable to consider the 'time' dimension as the third dimension and perform convolution operations and this idea has been verified in experiments.

We customized a module which called HCNN block (see Fig. 4), this module convolves with different sized regions, and extracts different shallow and deep features during this process [16]. Different from Inception module [23], each unit in HCNN block shares the same kernel size, in order to keep the same receptive fields inside, which does contribute to better feature extraction. After processing of HCNN block, the shapes of feature maps will become smaller.

Besides, HCNN has three different HCNN blocks which use different convolution kernel sizes in the convolution process (see Fig. 3). By using different convolution kernel sizes, different sizes of receptive fields can be obtained [18], and different types of features will be obtained.

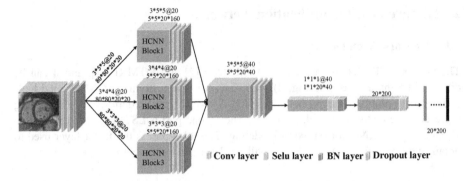

**Fig. 3.** The hierarchical convolutional neural network, the shape on the bottom represents for the output shape of previous layer, the value on the top represents for the convolution kernel size of previous layer.

At the same time, our customized CNN uses the selu activation function [21] instead of relu activation function. Selu activation function has excellent performance, which can make the convergence process faster, and can get better results than relu. Batch normalization (BN) layer is also used in HCNN after selu layer, in order to avoid the disappearance or explosion of gradient.

$$\text{selu}(x) = \lambda \begin{cases} x & \text{if } x > 0 \\ \alpha e^x - \alpha & \text{if } x \leq 0 \end{cases} \tag{1}$$

**Long Short Time Memory for Time Series Modeling.** LSTM is a variation of RNN with special units and has better performance than initial RNN. In this work, LSTM units are employed for temporal dynamic modeling which implement the need of classification task of phase and three regression tasks of the other three quantification indices [17].

**Final Estimation.** In the last part of HQNet, linear layers receive the outputs of LSTM as input, and the output of linear layer are the estimation of LV indices and phase:

$$\begin{cases} \hat{y}_{t1}^{s,f} = w_{t1}h_{LSTM1}^{s,f} + b_{t1}, where \ t1 \in \{area, dim, rwt\} \\ phase_{es} = argmax\left[softmax\left(w_{t2}h_{LSTM2}^{s,f} + b_{t2}\right)\right], phase_{es} = 0, 1 \end{cases} \tag{2}$$

## 2.2 Loss Function

Dataset used in training HQNet model has quantification indices named areas, RWT, and dimensions (dims) for regression tasks, with phase for classification task. There is a correlation between these indices, we try to establish equations for modeling the relationships of these indices, in order to constrain the loss and drive it to a low level.

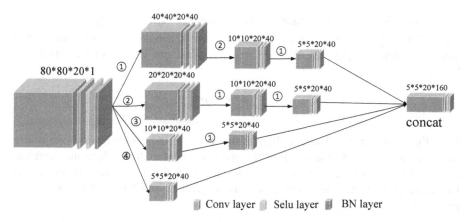

**Fig. 4.** The HCNN block we customized. Four different branches of convolution layer are used to extract the features of the cardiac image. Each branch has same pool size of [3, 3, 3] but different stride, ① represents for stride of [2, 2, 1], with ②, ③, ④ represent for [4, 4, 1], [8, 8, 1], [16, 16, 1] in order. All the convolution kernel size is equivalent and shown in Fig. 3.

**Loss Function for Phase Classification.** The proposed HQNet framework for phase classification has a softmax layer as its last layer, which could change the input into probability. We use cross entropy as major loss to minimize:

$$L_1 = \mathrm{H}(\mathrm{p}, \mathrm{q}) = -\sum_x p(x) \log q(x) \tag{3}$$

Cardiac phase indicates the temporal variety of LV myocardium in a cardiac cycle. Cavity area and LV dimensions increase during diastole and decreased during systole; myocardium area and RWT decrease during diastole and increase during systole. With the prior information we construct new functions called multi-task constraints for phase and it's an inverse version of the functions proposed in FullLVNet [2].

$$R_{inter}^{area} = \frac{1}{2S \times F} \sum_{s,f} \left| phase_{gd} - phase_{es} \right| \left[ \delta \left( phase_{gd} \right) \left( relu \left( -z_{area}^{s,f,1} \right) + relu \left( z_{area}^{s,f,2} \right) \right) \right.$$
$$\left. + \delta \left( phase_{gd} - 1 \right) \left( relu \left( z_{area}^{s,f,1} \right) + relu \left( -z_{area}^{s,f,2} \right) \right) \right] \tag{4}$$

$$R_{inter}^{dim} = \frac{1}{S \times F} \sum_{s,f} \left| phase_{gd} - phase_{es} \right| \left[ \delta \left( phase_{gd} \right) relu \left( -\bar{z}_{dim}^{s,f} \right) \right.$$
$$\left. + \delta \left( phase_{gd} - 1 \right) relu \left( \bar{z}_{dim}^{s,f} \right) \right] \tag{5}$$

$$R_{inter}^{rwt} = \frac{1}{S \times F} \sum_{s,f} \left| phase_{gd} - phase_{es} \right| \left[ \delta \left( phase_{gd} \right) relu \left( \bar{z}_{rwt}^{s,f} \right) \right.$$
$$\left. + \delta \left( phase_{gd} - 1 \right) relu \left( -\bar{z}_{rwt}^{s,f} \right) \right] \tag{6}$$

Where $\delta(\cdot)$ is the dirac delta function, $z_t^{s,t} = y_t^{s,f} - y_t^{s,f-1}$, for $t \in$ {area, dim, rwt}, $z_t^{s,t,i}$ denotes the $i$th input of $z_t$ and $\bar{z}_t$ denotes the average value of $z_t$ across its multiple inputs [2]. With this regularization, our loss for classification task becomes:

$$L_c = \lambda_1 L_1 + \lambda_2 \left( R_{inter}^{area} + R_{inter}^{dim} + R_{inter}^{rwt} \right) \tag{7}$$

**Loss Function for LV Indices Regression.** For LV indices regression, we utilize mean square error (MSE) as loss function and it's a measure that reflects the degree of difference between ground truth and an estimation. In Eq. 8, $s$ represents for cardiac sequences, $f$ represents for frame in cardiac sequences, and $t$ represents for type of parameters.

$$L_2 = MSE = \sum_{s,f} (y_t(s,f) - \hat{y}_t(s,f))^2, t \in \{area, dim, rwt\} \tag{8}$$

In addition, another constraint used in regression tasks is based on Bayesian multitask relationship learning [1], this method is based on Bayesian theorem and contributes a lot in training model:

$$R_3 = tr\left(W\Omega^{-1}W^T\right) \tag{9}$$

Where $W = (w_{rwt}, w_{area}, w_{dim})$, and $\Omega$ represents for column covariance matrix of $W$. Equation 9 is a simplified version which could achieve stable performance in the process and avoid calculation errors. For feasibility, $tr(\Omega) = 1$ is set to constrain the $\Omega$.

L2 norm is proven to be utilized to normalize the loss, which also called ridge regression and Tikhonov norm.

$$R_4 = \frac{1}{2}\|w\|_2^2 \tag{10}$$

So the final loss function for regression tasks is showed in Eq. 11, we choose three different $\lambda$ for each separated part of final loss function, in order to mediate the weights of each term.

$$L_r = \lambda_3 L_2 + \lambda_4 R_3 + \lambda_5 R_4 \tag{11}$$

## 3  Dataset and Experiments

The proposed HQNet is validated with processed SAX MR sequences of 145 subjects from clinical environment provided by LVQuan18 challenge[1], for each subject, the subjects age from 16 years to 97 years, with average of 58.9 years. 20 frames are

---

[1] LVQuan18 challenge, website: https://lvquan18.github.io/.

included for the whole cardiac cycle which is labeled as either Diastole phase or Systole phase, and three LV indices, RWT, dim, area, are included in the initial dataset. Each cardiac image is resized to $80 \times 80$. With pixel spacing and reversing-resizing ratio we can convert estimation into physical measurement. More details could be found in the LVQuan18 challenge website. In our experiments, we implement HQNet by tensorflow with Adam Optimizer and Nvidia GTX 1070 with 8 GB Memory was used in training process. We choose 0.0006 as learning rate, $\lambda_1$ *and* $\lambda_4$ share the same value of 1e−4, with $\lambda_2, \lambda_5$ are 1e−5, and $\lambda_3$ is 0.01, batch size is set to 8. Seven-fold cross validation is employed for performance evaluation and comparison. Data augmentation is conducted by two methods: (1) For phase task, we utilize cyclic shift for each frame in different shift steps. (2) For regression tasks, we select random cropping as our training strategy. We choose ensemble learning strategy [19, 20] and train five different models with different methods, which include DTMRL [1] 2D and 3D version, HQNet naïve version, HQNet and HQNet 2D version. Majority voting was selected for phase classification in our experiments.

## 4  Results and Analysis

HQNet we proposed achieve a stable and high performance with the convincing results which were obtained by using 7-fold cross validation (see Table 1). We use Mean Absolute Error (MAE) and Pearson correlation coefficient (PCC) $\rho$ to assess the performance of the algorithms for estimation of areas, dimensions and regional wall thicknesses. In addition, Error Rate (ER) is used to assess the performance of the algorithms for cardiac phase identification [1, 22].

$$\text{MAE}(y, \hat{y}) = \frac{1}{N} \sum_{i=1}^{N} |y^i - \hat{y}^i| \tag{12}$$

$$\rho(y, \hat{y}) = \frac{2 \sum_{i=1}^{N} (y^i - m_1)(\hat{y}^i - m_2)}{\sum_{i=1}^{N} \left( (y^i - m_1)^2 + (\hat{y}^i - m_2)^2 \right)} \tag{13}$$

$$ER(y, \hat{y}) = \frac{\sum_{i=1}^{N} 1(y^i \neq \hat{y}^i)}{N} \times 100\% \tag{14}$$

Where $y^i, \hat{y}^i$ are ground truth and estimation, and $m_1 = \frac{1}{N} \sum_{i=1}^{N} y^i$, $m_2 = \frac{1}{N} \sum_{i=1}^{N} \hat{y}^i$.

As shown in the Table 1, HQNet shows admirable results with error rate (ER) of 9.8%, average Mean Absolute Error (MAE) of 1.51 mm, 2.57 mm, 197 mm$^2$ and Pearson correlation coefficient (PCC) 0.805, 0.921, 0.935 for RWT, dimension, and areas.

**Table 1.** Performance of HQNet with 7-fold cross validation. Values of first row represent for the order of 7-fold cross validation. Values of first column represent for the type of indices. Mean absolute error (MAE) and Pearson correlation coefficient (PCC) are used for the three regression tasks and prediction error rate is used for phase classification task.

| Index | Results | | | | | | | |
|---|---|---|---|---|---|---|---|---|
| Order | 1 | 2 | 3 | 4 | 5 | 6 | 7 | Mean |
| **Area (mm²)** | | | | | | | | |
| Cavity | 160 ± 112 | 173 ± 113 | 180 ± 130 | 214 ± 145 | 204 ± 184 | 146 ± 95 | 145 ± 103 | 175 ± 126 |
| Myocardium | 206 ± 142 | 237 ± 161 | 213 ± 152 | 150 ± 138 | 318 ± 246 | 239 ± 189 | 171 ± 151 | 219 ± 168 |
| Mean MAE | 183 ± 127 | 205 ± 137 | 197 ± 141 | 182 ± 142 | 261 ± 215 | 193 ± 142 | 158 ± 127 | **197 ± 147** |
| Mean PCC | 0.933 | 0.936 | 0.929 | 0.942 | 0.932 | 0.939 | 0.936 | **0.935** |
| **Dimension (mm)** | | | | | | | | |
| dim1 | 2.09 ± 1.51 | 2.84 ± 2.28 | 2.33 ± 1.88 | 2.51 ± 2.02 | 2.93 ± 2.34 | 2.11 ± 1.83 | 2.19 ± 1.69 | 2.43 ± 1.94 |
| dim2 | 2.41 ± 1.81 | 2.47 ± 1.85 | 2.62 ± 2.21 | 2.94 ± 2.46 | 2.79 ± 2.23 | 2.30 ± 1.76 | 2.09 ± 1.58 | 2.52 ± 1.99 |
| dim3 | 2.74 ± 2.23 | 2.88 ± 2.15 | 2.45 ± 1.78 | 3.69 ± 2.62 | 3.3 ± 2.64 | 2.13 ± 1.58 | 2.13 ± 1.65 | 2.76 ± 2.09 |
| Mean MAE | 2.41 ± 1.81 | 2.73 ± 1.96 | 2.47 ± 1.64 | 3.05 ± 2.37 | 3.01 ± 2.40 | 2.18 ± 1.72 | 2.14 ± 1.64 | **2.57 ± 2.00** |
| Mean PCC | 0.938 | 0.921 | 0.926 | 0.928 | 0.901 | 0.913 | 0.921 | **0.921** |
| **RWT (mm)** | | | | | | | | |
| IS | 1.47 ± 1.01 | 1.71 ± 1.14 | 2.08 ± 1.72 | 1.48 ± 1.09 | 1.53 ± 1.18 | 1.35 ± 0.88 | 1.22 ± 0.74 | 1.55 ± 1.11 |
| I | 1.55 ± 1.19 | 1.26 ± 0.92 | 1.77 ± 1.44 | 1.09 ± 0.68 | 1.72 ± 1.57 | 1.25 ± 0.83 | 1.53 ± 1.22 | 1.45 ± 1.12 |
| IL | 1.46 ± 1.01 | 1.36 ± 0.98 | 1.66 ± 1.21 | 1.46 ± 1.08 | 2.26 ± 1.95 | 1.62 ± 1.45 | 1.29 ± 0.88 | 1.59 ± 1.22 |
| AL | 1.59 ± 1.02 | 1.43 ± 1.08 | 1.67 ± 1.43 | 1.39 ± 1.03 | 1.97 ± 1.57 | 1.59 ± 1.23 | 1.30 ± 1.01 | 1.56 ± 1.20 |
| A | 1.33 ± 0.96 | 1.68 ± 1.21 | 1.27 ± 0.74 | 1.37 ± 0.98 | 1.74 ± 1.33 | 1.25 ± 0.87 | 1.16 ± 0.87 | 1.40 ± 0.99 |
| AS | 1.33 ± 1.14 | 1.61 ± 1.26 | 1.65 ± 1.13 | 1.68 ± 1.32 | 1.68 ± 1.37 | 1.67 ± 1.05 | 1.03 ± 0.65 | 1.52 ± 1.13 |
| Mean MAE | 1.46 ± 1.06 | 1.51 ± 1.10 | 1.68 ± 1.28 | 1.41 ± 1.03 | 1.82 ± 1.50 | 1.46 ± 1.05 | 1.26 ± 1.17 | **1.51 ± 1.13** |
| Mean PCC | 0.801 | 0.813 | 0.795 | 0.821 | 0.781 | 0.801 | 0.823 | **0.805** |
| Phase (%) | 6.9 | 10.48 | 12.38 | 7.62 | 10.00 | 10.75 | 10.5 | **9.8** |

## 5 Conclusion

We propose a new approach for multitask learning network called HQNet for LV full quantification, which includes three regression tasks and one classification task. By customizing a CNN called HCNN, stable and accurate feature embeddings is achieved. After efficient dynamic temporal modeling by LSTMs and linear calculation, admirable results are obtained. With L2 normalization, Bayesian multitask relationship learning and multi-task constraints for phase, intra- and inter-task correlations are taken into consideration. HQNet is enforced to deliver admirable performance for all the tasks considered.

**Acknowledgement.** This research was supported by National Natural Science Foundation under grants (31571001, 61828101), the National Key Research and Development Program of China (2017YFC0107903) and the Science Foundation for The Excellent Youth Scholars of Southeast University.

## References

1. Xue, W., Brahm, G., Pandey, S., Leung, S., Li, S.: Full left ventricle quantification via deep multitask relationships learning. Med. Image Anal. **43**, 54–65 (2018). https://doi.org/10.1016/j.media.2017.09.005
2. Xue, W., Lum, A., Mercado, A., Landis, M., Warrington, J., Li, S.: Full quantification of left ventricle via deep multitask learning network respecting intra- and inter-task relatedness. In: Descoteaux, M., Maier-Hein, L., Franz, A., Jannin, P., Collins, D.L., Duchesne, S. (eds.) MICCAI 2017. LNCS, vol. 10435, pp. 276–284. Springer, Cham (2017). https://doi.org/10.1007/978-3-319-66179-7_32
3. Xue, W., Nachum, I.B., Pandey, S., Warrington, J., Leung, S., Li, S.: Direct estimation of regional wall thicknesses via residual recurrent neural network. In: Niethammer, M., et al. (eds.) IPMI 2017. LNCS, vol. 10265, pp. 505–516. Springer, Cham (2017). https://doi.org/10.1007/978-3-319-59050-9_40
4. Karamitsos, T.D., Francis, J.M., Myerson, S., Selvanayagam, J.B., Neubauer, S.: The role of cardiovascular magnetic resonance imaging in heart failure. J. Am. Coll. Cardiol. **54**(15), 1407–1424 (2009)
5. Attili, A.K., Schuster, A., Nagel, E., Reiber, J.H., van der Geest, R.J.: Quantification in cardiac MRI: advances in image acquisition and processing. Int. J. Cardiovasc. Imaging **26**(1), 27–40 (2010)
6. Ayed, I.B., Chen, H.M., Punithakumar, K., Ross, I., Li, S.: Max-flow segmentation of the left ventricle by recovering subject-specific distributions via a bound of the Bhattacharyya measure. Med. Image Anal. **16**(1), 87–100 (2012)
7. Graves, A.: Supervised sequence labelling. In: Graves, A. (ed.) Supervised Sequence Labelling with Recurrent Neural Networks. SCI, vol. 385, pp. 5–13. Springer, Heidelberg (2012). https://doi.org/10.1007/978-3-642-24797-2_2
8. Kong, B., Zhan, Y., Shin, M., Denny, T., Zhang, S.: Recognizing end-diastole and end-systole frames via deep temporal regression network. In: Ourselin, S., Joskowicz, L., Sabuncu, M.R., Unal, G., Wells, W. (eds.) MICCAI 2016. LNCS, vol. 9902, pp. 264–272. Springer, Cham (2016). https://doi.org/10.1007/978-3-319-46726-9_31

9. Petitjean, C., Dacher, J.N.: A review of segmentation methods in short axis cardiac MR images. Med. Image Anal. **15**(2), 169–184 (2011)
10. Poudel, R.P., Lamata, P., Montana, G.: Recurrent fully convolutional neural networks for multi-slice MRI cardiac segmentation. arXiv:1608.03974 (2016)
11. Wang, Z., Ben Salah, M., Gu, B., Islam, A., Goela, A., Li, S.: Direct estimation of cardiac biventricular volumes with an adapted bayesian formulation. IEEE TBE **61**(4), 1251–1260 (2014)
12. Zhang, Y., Yeung, D.Y.: A convex formulation for learning task relationships in multi-task learning. In: UAI, pp. 733–742 (2010)
13. Zhang, Z., Luo, P., Loy, C.C., Tang, X.: Facial landmark detection by deep multi-task learning. In: Fleet, D., Pajdla, T., Schiele, B., Tuytelaars, T. (eds.) ECCV 2014. LNCS, vol. 8694, pp. 94–108. Springer, Cham (2014). https://doi.org/10.1007/978-3-319-10599-4_7
14. Shi, X., Chen, Z., Wang, H., Yeung, D.-Y.: Convolutional LSTM network: a machine learning approach for precipitation nowcasting. In: Neural Information Processing Systems (NIPS) (2015)
15. Gers, F.A., Schmidhuber, J., Cummins, F.A.: Learning to forget: continual prediction with LSTM. Neural Comput. **12**(10), 2451–2471 (2000)
16. Zhao, H., Shi, J., Qi, X., Wang, X., Jia, J.: Pyramid scene parsing network. arXiv:1612.01105 (2016)
17. Hochreiter, S., Schmidhuber, J.: Long short-term memory. Neural Comput. **9**(8), 1735–1780 (1997)
18. Szegedy, C., et al.: Going deeper with convolutions. In: CVPR (2015)
19. Liu, Y., Yao, X.: Ensemble learning via negative correlation. Neural Netw. **12**, 1399–1404 (1999)
20. Rosen, B.: Ensemble learning using decorrelated neural networks. Connect. Sci. **8**(3/4), 373–383 (1996)
21. Klambauer, G., Unterthiner, T., Mayr, A., Hochreiter, S.: Self-normalizing neural networks. arXiv preprint arXiv:1706.02515 (2017)
22. Objective, LVQuan18 challenge. https://lvquan18.github.io/2018/03/12/objective.html. Accessed 26 June 2018
23. Szegedy, C., Vanhoucke, V., Ioffe, S., Shlens, J., Wojna, Z.: Rethinking the inception architecture for computer vision. In: CVPR (2016). https://arxiv.org/abs/1512.00567

# Automatic Left Ventricle Quantification in Cardiac MRI via Hierarchical Refinement of High-Level Features by a Salient Perceptual Grouping Model

Angélica Atehortúa[1,2]([✉]), Mireille Garreau[1], David Romo-Bucheli[2], and Eduardo Romero[2]

[1] Univ Rennes, Inserm, LTSI - UMR 1099, 35000 Rennes, France
[2] Universidad Nacional de Colombia, Bogotá, Colombia
amatehortual@unal.edu.co

**Abstract.** An accurate segmentation of the left ventricle (LV) from cardiac magnetic resonance imaging (MRI) provides reliable cardiac indexes such as the ventricular volume, the ejection fraction or regional wall thicknesses (RWT). This paper introduces an automated method to compute such indexes in 2D MRI slices from a semantic segmentation obtained in two steps. A first coarse segmentation is obtained by applying an encoder-decoder neural network architecture that assigns a probability value to each pixel. Afterwards, this segmentation is refined by a spatio-temporal saliency analysis. The method was evaluated in MR sequences of 175 subjects divided in two groups: training (145 subjects) and test (30 subjects). For the training data set, using a K-cross validation setup, the method achieves an average Pearson correlation coefficient of 0.98, 0.92, 0.95 and 0.75 with the set of indexes LV cavity, myocardium areas, cavity dimensions and region wall thicknesses, respectively, while classification of the cardiac phase yielded a rate of 10.01%. For the same set of indexes, evaluated in the test dataset, an average Pearson correlation coefficient of 0.98, 0.87, 0.97 and 0.66 was obtained. Additionally, the cardiac phase classification error rate was 9%. The method provides a reliable LV segmentation and quantification of cardiac indexes.

**Keywords:** Cardiac MRI · Intensity profiles · Saliency · LV cardiac indicators · Deep learning

## 1 Introduction

Treatment, early diagnosis and follow-up of cardiac diseases requires the assessment of the cardiac function by computing cardiac indexes such as ventricular volume, ejection fraction, left ventricular mass, wall thickness and the analysis of motion abnormalities in the cardiac wall [8]. The mentioned indexes are often obtained via manual delineation of the epicardium and endocardium of the left

© Springer Nature Switzerland AG 2019
M. Pop et al. (Eds.): STACOM 2018, LNCS 11395, pp. 439–449, 2019.
https://doi.org/10.1007/978-3-030-12029-0_47

ventricle (LV) in short axis magnetic resonance images (MRI) for two phases of the cardiac cycle: end of the diastole and end of the systole. However, manual delineation is a tedious and challenging task that requires expertise to deal with the complexity of cardiac structures. In addition, these delineations present a high inter- and intra-observer variability. In this context, computational tools might help reduce such variability and obtain more accurate cardiac indicators.

Several works have attempted to quantify cardiac LV. After Xue et al. [14] these methods can be divided in three groups: (a) manual quantification; (b) segmentation-based quantification; and (c) direct regression methods. The first group (a) is related to the clinical practice, case in which endocardium and epicardium are delineated manually. However, this task is time-consuming and prone to errors, above all if performed along the whole cycle. The second group of methods in (b) segment the endocardium and epicardium contours using several approaches, such as image-driven techniques, deformable methods and recently deep neural networks. [9,11]. Nevertheless, these methods require large dataset, user interaction and strong assumptions of the LV to obtain accurate segmentations. The third group, (c), cardiac indexes are obtained without segmentation using several machine learning techniques [14,15].

A main contribution of this work is a fully and accurate automatic method that segments the left ventricle (LV) along the whole cardiac cycle and computes LV indexes using a spatio-temporal saliency analysis that refines an inferred left ventricle from a deep learning model in 2D cardiac MRI (high-level features). Unlike recent segmentation-based quantification strategies, based only on structural information such as deep neural networks or surface deformations, the proposed method produces accurate segmentation results without either user interaction or a large training data sets.

The paper continues as follows. A detailed description of the proposed method is presented in Sect. 2. Experimental results are reported in Sect. 3. Finally, some conclusions about this work are highlighted in Sect. 4.

## 2   Methods

The proposed strategy segments the whole LV, i.e., myocardium and cavity by refining hierarchically coarsely segmented high-level features extracted from a deep learning model. This refinement is defined as a problem of perceptually grouping salient points of the left ventricle constrained by the inferred shape of this learned model. Once a LV segmentation is obtained along the whole sequence, cardiac indexes are automatically computed using a radial analysis. The following subsections completely describe this approach.

### 2.1   High-Level Features Extraction

The analysis is performed with the short axis view. High-level features (LV myocardium and cavity segmentation) for the whole cardiac cycle are extracted via pixel-wise semantic segmentation by using an encode-decoder neural network architecture (SegNet) [3]. SegNet architecture was selected since this provides a precise re-localization of features by its symmetrical architecture and the elegant use of a correlated pooling/unpooling combination. In addition to SegNet, preliminary experiments were carried out with U-Net [6] and FCN [13]. Results reported the best accuracy and shape smoothness for SegNet model. The network was initialized with weights of a pre-trained VGG-16 on ImageNet [12]. An image resolution of 80 × 80 pixels was used. A batch size of 4, 20 epochs and Adadelta optimizer [16] were the parameters set in the network. Three classes were used: LV myocardium, LV cavity and background. The network was implemented by Tensorflow [1] in keras framework [5][1].

## 2.2   Refinement of the High-Level Features

The noise introduced by the high heart variability and data acquisition may lead to wrong segmentations in small annotated training sets using only deep learning. In this sense, the obtained high-level features need to be regularized to produce reliable LV myocardium and cavity delineation, as shown in Fig. 1. For doing so, this refinement is divided in four steps: first myocardium and cavity refinement; end of the diastole (ED) localization; ED refinement by a saliency analysis; and temporal refinement.

Fig. 1. MRI image and its corresponding high-level features. Purple color highlights the LV myocardium wall and brown color the LV cavity. The left panel illustrates a good segmentation, the mid-panel shows how the myocardium has disappeared while segmentation is too noisy in the third panel. (Color figure online)

**First Myocardium and Cavity Refinement.** From the high-level features, LV myocardium and cavity are slightly refined to eliminate segmentation noise by selecting the shape with the highest number of connected pixels for each of two classes. This process is performed along the whole cardiac cycle.

**End of the Diastole (ED) Localization.** The end of the diastole provides the best visualization of the cardiac structures. Therefore, the ED is set if the cardiac phases are unknown. The refined left ventricular cavity area is computed for each frame of the cardiac sequence since this is the easiest region to segment. Those frames with minimum and maximum values correspond to the ones at the end of the systole (ES) and end of the diastole (ED) respectively.

---

[1] By modifying the implementation available in https://github.com/divamgupta/image-segmentation-keras.

**LV Refinement by a Saliency Analysis at the ED.** Once the ED is identified, it is necessary to verify how closed the myocardium shape is in this phase. For doing so, rays are traced from the endocardium centroid in the binary image of the LV myocardium. Consequently, the myocardium is closed if there are non null pixels in all these radial direction. The refinement is described in two steps: (i) Closure of the myocardium contour; and (ii) salient perceptual grouping model.

(i) Closure of the myocardium contour. The LV myocardium is skeletonized to obtain a contour. Afterwards, this contour is mapped to a radial space (radial distance $= r$, angle $= \theta$) using the LV cavity (endocardium) centroid as origin. The LV cavity contour is also mapped to this same space, as observed in Fig. 2d. Then, the missing myocardium values are set to the average radial distance of the four closest radii with endocardium and myocardium values. In fact the approach takes two closer radii with bigger angles and two with smaller ones. Finally, the skeletonized image is reverted to obtain the inferred LV myocardium now closed.

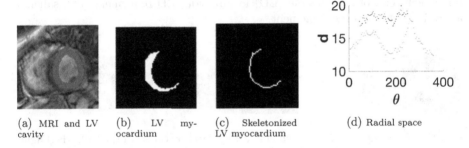

(a) MRI and LV cavity

(b) LV myocardium

(c) Skeletonized LV myocardium

(d) Radial space

**Fig. 2.** Figure (a) presents the LV endocardium contour (red line) overlapped to the MR image. Figure (b) shows the binary image of the LV myocardium wall. Figure (c) illustrates the skeletonized LV myocardium image. Figure (d) shows the skeletonized LV myocardium (blue line) superimposed to the endocardium contour (red line) in the radial space (radial distance $= r$, angle $= \theta$). (Color figure online)

(ii) Salient perceptual grouping model. A radial saliency analysis is performed only for the ED frame. For doing so, first the magnetic resonance image is filtered to enhance edges and remove noise [7]. Afterwards, rays $r$ are traced from the endocardium centroid $c$ and radial intensity profiles correspond to the intensity values of the pixels along the ray, as described in [2] (Figs. 3a and b illustrate an intensity profile). This radial profile provides an intensity pattern of the myocardium and the salient points $p$ correspond to sharp changes in the profile curve (This saliency is characteristic of the myocardium boundary). In this paper, these salient points are detected as inflection points (sharp changes) of the intensity profile. The selection of the salient points that belong to the boundary LV myocardium wall is an ill-posed problem, herein regularized by (a) proximity to the a-priori myocardium wall $\hat{p}$; (b) smoothness of the intensity values in the myocardium segment; (c) continuity of the neighbor intensity profiles in the radial direction; and d) symmetry between the inner-outer layers of LV myocardium contours. Herein, it is taken a neighborhood $\Delta\theta$ of $10°$. Every radius within this range of possible angles contains a series of potential boundary points. A family of angular curves is constructed by connecting these radial points in the angular direction as illustrated in Fig. 3c. These candidate myocardium points $p$ are then pair-wise (endocardium $j$, epicardium $i$)[2] combined at each $k \times \Delta\theta$ angle. The problem of finding a suitable combination of radial points (a good approximation to the ventricle) is solved by finding the smoother angular curve. The problem is read as

$$\underset{p_\theta}{\text{argmin}} \left[ \|S^{\Delta\theta}(p_i - p_j)\|_{NCC} + \|p_i - p_j\|_{Euclidean} + \right.$$

$$\left. \|(p_i, p_j) - (\hat{p}_i, \hat{p}_j)\|_{Euclidean} \right], \forall r \in \Delta\theta \tag{1}$$

where $NCC$ is the normalized cross correlation, used as the similarity measure of myocardium segments $S$ composed of the candidate points $p \in \Delta\theta$. $p_\theta$ are the final selected salient points.

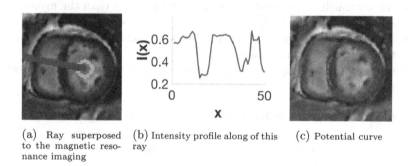

(a) Ray superposed to the magnetic resonance imaging    (b) Intensity profile along of this ray    (c) Potential curve

**Fig. 3.** The left panel show a single radius trace from the LV centroid, the mid panel shows its intensity profile and the right panel displays the a potential curve, in a segment, of the family of angular curves for the endocardium (red) and epicardium (green). (Color figure online)

**Temporal Refinement.** The complex heart temporal dynamic close to the end of the systole complicates any extraction of accurate high-level features. Therefore, a temporal refinement is applied for each radius $r$, in the radial space $(r, \theta)$, using the whole cardiac sequence. For doing so, a spatio-temporal plane description of the radial profile and the myocardium boundary are constructed.

---

[2] A pair of endocardium-myocardium couple in the radial direction.

The analysis described in Subsect. 2.2-ii is then modified by adding them and obtaining a final LV shape. This is further explained hereafter

(i) **The Spatio-temporal plane and the temporal myocardium curves.**
    A spatio-temporal plane is constructed for every radius by tracking their intensity profile along the time, i.e., every radius at the ED is superimposed to the whole sequence as illustrated in Fig. 4a and their radial intensities are stored for the salient analysis. Likewise, a pair of temporal curves are also obtained for every radius at the ED by tracking only the inner-outer myocardium points of the first refined segmentation[3] along the cardiac cycle, as shown in Fig. 4b.

(ii) **Saliency-based refinement.** The obtained spatio-temporal plane shows the intensities in the $y$ axis and the time in the $x$ axis. The temporal inner-outer curves are within this plane and serve as a restriction of the search of the actual cardiac boundaries. In fact, for every time ($x$ axis), those points (candidate to salient points) in the radial direction ($y$ axis) with an important intensity change are selected and only those ones close to the inner-outer interval ($p$) are analyzed. The refinement phase consists then in selecting the best pair of radial points that matches a set of constraints, namely:

- their intensity have to be similar among them and similar to the intensity of the inner-outer curves ($\hat{p}$)
- the myocardium segment width can not be higher than the myocardium segment at the ED ($\hat{p}^{ED}$)
- the obtained epicardium $i$ can not be smaller than the endocardium $j$ in the radial ($y$ axis) direction
- the curve formed by any combination of them has to be as smooth as possible.

Formally the temporal refinement problem reads as

$$\underset{p_t}{\text{argmin}} \left[ \| \langle p_i, p_j \rangle^t \|_{NCC} + \| p_i - p_j \|_{Euclidean} + \right.$$
$$\left. \| (p_i, p_j) - (\hat{p}_i, x\hat{p}_j) \|_{NCC} \right] \forall t \in T$$

Subject to

$$\| (p_i - p_j) \|_{Euclidean} \leq \| (\hat{p}_i^{ED} - \hat{p}_j^{ED}) \|_{Euclidean},$$
$$\text{and } p_j \geq \hat{p}_i \tag{2}$$

being NCC the normalized cross correlation of the temporal series of the points $p_i$ and $p_j$ (first term of the functional), $\hat{p}_i, \hat{p}_j$ the corresponds to the inner-outer points of the temporal myocardium curves and the superindexes $ED$ and $t$ stand for points at the end of the diastole and the temporal series, respectively.

---

[3] The one obtained by the deep learning model, see Sect. 2.2.

## 2.3   Quantification of LV Indexes

**LV Myocardium and Cavity Areas.** The LV myocardium and cavity areas are computed in the final LV segmentation.

**Cardiac Phases.** There are two cardiac phases: systolic and diastolic. In the systolic phase, the LV contracts and ejects blood into the aorta. On the other hand, the LV relaxes and receives blood in the diastolic phase. In the cardiac function analysis, the set of temporal frames is labeled according to the computed area of the LV cavity along the cardiac cycle. The maximum area value corresponds to the end of the diastole (ED) and the minimum represents the end of the systole (ES). Consequently, the systolic phase corresponds to the set of frames between the ED and the ES. The remaining frames are part of the diastolic phase.

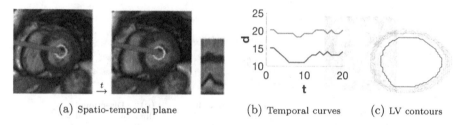

(a) Spatio-temporal plane          (b) Temporal curves      (c) LV contours

**Fig. 4.** Figure (a) shows a radius superimposed to two different views of the cardiac cycle. The spatio-temporal plane (mid panel) is constructed from the profiles of this radius along the time. Figure (b) shows outer-points of the LV myocardium (green line) and cavity (red line) for each time (x-axis). Figure (c) presents all epicardium contours from the cardiac cycle superimposed to the endocardium contour at the ED (red line). (Color figure online)

**Regional Wall Thickness.** Regional wall thickness (RWT) is the LV myocardial width of a LV segment. In the mid-short axis slice, the LV is divided into six regional segments, as proposed by the American Heart Association (AHA) [4], namely: anterior (A), antero-septal (AS), infero-septal (IS), inferior (I), infero-lateral (IL) and antero-lateral (AL), and illustrated in Fig. 5b. The regional myocardial width is then obtained from from six radii crossing the inner-outer layer points, as follows

(i) The LV centroid and the anterior wall between the left and right (RV) ventricles (points A and B in Fig. 5a) are required. Points A and B are set by constructing a new labeled image with three classes: LV, RV and background, being the LV class the LV myocardium and the RV class the shape (left side of the LV) with the highest number of connected pixels after thresholding the MR image. This thresholded image is obtained by applying the Otsu [10] algorithm to a low resolution version of MR image which is then mapped to the original image. Afterward, in the labeled image, a set of radii is traced every degree between $180°$ and $270°$. The point $A$ is assigned to the coordinates of the LV epicardium boundary of the first radius with not RV class. The $B$ point is similarly detected, but the radii are now traced between $180°$ and $90°$.

Afterwards, three lines crossing the centroid define the different regions: one horizontal line, one line crossing point A and a third one, crossing point B point. Therefore, each AHA segment corresponds to myocardium region between two adjacent lines, as illustrated in Fig. 5.

(ii) The myocardial width is then estimated by applying a radial analysis to the binary image. For each AHA segment, six radii define the six AHA regions, as shown by red arrows in Fig. 5b. Afterwards, the distance between inner-outer points of the LV myocardium corresponds to the myocardial thickness of the segment.

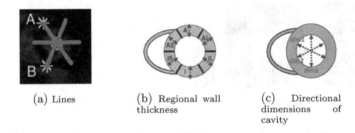

(a) Lines

(b) Regional wall thickness

(c) Directional dimensions of cavity

**Fig. 5.** Figure (a) lines (magenta color) traced using the antero-septal, infero-septal and the LV centroid points (green color). Figures (b) and (c) are taken from [15]. (Color figure online)

**Directional Dimensions of the LV Cavity.** These measurements correspond to three diameters inside the LV cavity, formed by the radios used to measure the myocardial width.

## 3　Results

Experiments were carried out using the data set provided by the Left Ventricle Full Quantification Challenge MICCAI 2018. The dataset is composed of processed short axis MR sequences of 175 subjects. Each sequence comprises a set of 20 MR images acquired along the whole cardiac cycle.

For the training set (145 subjects), several N-fold cross validation setups evaluated the algorithm performance, for instance for $N = 3$ the number of subjects was 49, 48 and 48; for $N = 5$ the number was 29, 29, 29, 29 and 29; and for $N = 7$ the number was 21, 21, 21, 21, 21, 20 and 20. Results are shown in Tables 1 and 2. The average classification error rate of the cardiac phases was

**Table 1.** Method performance with the training data set. Pearson correlation coefficient is shown. Mean and standard deviation (Std) of several N-fold cross validation setups (NFCV).

| NFCV | Area (A1) | Directional dimensions | | |
|---|---|---|---|---|
| | LV cavity | Dim 1 | Dim 2 | Dim 3 |
| | Mean (Std) | Mean (Std) | Mean (Std) | Mean |
| N = 3 | 0.98 (0.01) | 0.96 (0.02) | 0.96 (0.02) | 0.95 (0.02) |
| N = 5 | 0.99 (0.01) | 0.95 (0.02) | 0.96 (0.02) | 0.95 (0.03) |
| N = 7 | 0.98 (0.01) | 0.95 (0.03) | 0.95 (0.07) | 0.95 (0.02) |

9.552%, 10.035% and 10.451% for experiments with $N = 3$, $N = 5$ and $N = 7$ folds, respectively. For all these folds, the observed difference was less than 1%.

Results with test dataset (30 subjects) are shown in Table 3. The Pearson correlation coefficient evaluated the method, particularly the next indexes were estimated: area of myocardium (column A2) and regional wall thickness (columns RWT). These measurements are much more difficult to estimate than indexes computed over the total LV cavity (columns A1 and Dim) since the myocardium in regions such as the antero-lateral or infero-lateral show highly noisy edges. Regarding classification of cardiac phases for this test dataset, an average error rate of 9.0% was obtained. This last result shows the proposed strategy detects the diastolic and systolic phases during the cardiac cycle with a very small error.

**Table 2.** Quantification of the LV in the training data set using the Pearson correlation coefficient. Mean and standard deviation (Std) of several N-fold cross validation setups (NFCV).

| NFCV | Area (A2) | Regional wall thickness | | | | | |
|---|---|---|---|---|---|---|---|
| | LV myocardium | A (RWT1) | AS (RWT2) | IS (RWT3) | I (RWT4) | IL (RWT5) | AL (RWT6) |
| | Mean (Std) | Mean (Std) | Mean (Std) | Mean (Std) | Mean (Std) | Mean (Std) | Mean (Std) |
| N = 3 | 0.91 (0.04) | 0.74 (0.11) | 0.70 (0.11) | 0.73 (0.139) | 0.69 (0.15) | 0.73 (0.11) | 0.70 (0.13) |
| N = 5 | 0.93 (0.03) | 0.81 (0.06) | 0.78 (0.10) | 0.78 (0.074) | 0.80 (0.10) | 0.73 (0.11) | 0.78 (0.07) |
| N = 7 | 0.92 (0.03) | 0.76 (0.24) | 0.89 (0.09) | 0.89 (0.09) | 0.74 (0.07) | 0.66 (0.10) | 0.77 (0.05) |

**Table 3.** Quantification of the entire LV in the test data set. This table shows next indexes: Area for LV cavity (A1) and LV myocardium wall (A2), directional dimensions (Dim), regional wall thickness (RWT). Mean absolute error (MAE) and Pearson correlation coefficient (PCC) were used as similarity measures.

| | Area | | | Dim | | | | RWT | | | | | | |
|---|---|---|---|---|---|---|---|---|---|---|---|---|---|---|
| | A1 | A2 | Mean | Dim1 | Dim2 | Dim3 | Mean | RWT1 | RWT2 | RWT3 | RWT4 | RWT5 | RWT6 | Mean |
| MAE | 118.89 | 818.69 | 468.79 | 3.20 | 2.51 | 3.14 | 2.95 | 7.34 | 4.34 | 4.58 | 5.18 | 4.03 | 3.15 | 4.77 |
| PCC | 0.98 | 0.87 | 0.93 | 0.96 | 0.97 | 0.97 | 0.97 | 0.71 | 0.66 | 0.59 | 0.55 | 0.66 | 0.80 | 0.66 |

# 4   Conclusions

The proposed method provides fast delineation of the LV myocardium wall by applying a deep learning model, which is regularized using a saliency analysis of the myocardium. This refinement provides a more accurate cardiac delineation that obtains reliable cardiac indexes. This methodology might be potentially used as a support tool for quantification of the cardiac function and analysis of the cardiac wall motion abnormalities.

**Acknowledgments.** This work was supported by Région Bretagne and Colciencias-Colombia, Grant No. 647 (2015 call for National PhD studies) in the framework of the Investissement d'Avenir Program through Labex CAMI (ANR-11-LABX-0004).

# References

1. Abadi, M., et al.: TensorFlow: a system for large-scale machine learning. In: OSDI vol. 16, pp. 265–283 (2016)
2. Atehortúa, A., Zuluaga, M.A., García, J.D., Romero, E.: Automatic segmentation of right ventricle in cardiac cine MR images using a saliency analysis. Med. Phys. **43**(12), 6270–6281 (2016)
3. Badrinarayanan, V., Kendall, A., Cipolla, R.: SegNet: a deep convolutional encoder-decoder architecture for image segmentation. IEEE Trans. Pattern Anal. Mach. Intell. **39**(12), 2481–2495 (2017)
4. Cerqueira, M.D., et al.: Standardized myocardial segmentation and nomenclature for tomographic imaging of the heart: a statement for healthcare professionals from the cardiac imaging committee of the council on clinical cardiology of the american heart association. Circulation **105**(4), 539–542 (2002)
5. Chollet, F., et al.: Keras (2015). https://github.com/fchollet/keras
6. Çiçek, Ö., Abdulkadir, A., Lienkamp, S.S., Brox, T., Ronneberger, O.: 3D U-Net: learning dense volumetric segmentation from sparse annotation. In: Ourselin, S., Joskowicz, L., Sabuncu, M.R., Unal, G., Wells, W. (eds.) MICCAI 2016. LNCS, vol. 9901, pp. 424–432. Springer, Cham (2016). https://doi.org/10.1007/978-3-319-46723-8_49
7. Gastal, E.S., Oliveira, M.M.: Adaptive manifolds for real-time high-dimensional filtering. ACM Trans. Graph. (TOG) **31**(4), 33 (2012)
8. Karamitsos, T.D., Francis, J.M., Myerson, S., Selvanayagam, J.B., Neubauer, S.: The role of cardiovascular magnetic resonance imaging in heart failure. J. Am. Coll. Cardiol. **54**(15), 1407–1424 (2009)
9. Ngo, T.A., Lu, Z., Carneiro, G.: Combining deep learning and level set for the automated segmentation of the left ventricle of the heart from cardiac cine magnetic resonance. Med. Image Anal. **35**, 159–171 (2017)
10. Otsu, N.: A threshold selection method from gray-level histograms. IEEE Trans. Syst. Man Cybern. **9**(1), 62–66 (1979)
11. Peng, P., Lekadir, K., Gooya, A., Shao, L., Petersen, S.E., Frangi, A.F.: A review of heart chamber segmentation for structural and functional analysis using cardiac magnetic resonance imaging. Magn. Res. Mater. Phys. Biol. Med. **29**(2), 155–195 (2016)
12. Simonyan, K., Zisserman, A.: Very deep convolutional networks for large-scale image recognition. arXiv preprint arXiv:1409.1556 (2014)

13. Tran, P.V.: A fully convolutional neural network for cardiac segmentation in short-axis MRI. arXiv preprint arXiv:1604.00494 (2016)
14. Xue, W., Brahm, G., Pandey, S., Leung, S., Li, S.: Full left ventricle quantification via deep multitask relationships learning. Med. Image Anal. **43**, 54–65 (2018)
15. Xue, W., Lum, A., Mercado, A., Landis, M., Warrington, J., Li, S.: Full quantification of left ventricle via deep multitask learning network respecting intra- and inter-task relatedness. In: Descoteaux, M., Maier-Hein, L., Franz, A., Jannin, P., Collins, D.L., Duchesne, S. (eds.) MICCAI 2017. LNCS, vol. 10435, pp. 276–284. Springer, Cham (2017). https://doi.org/10.1007/978-3-319-66179-7_32
16. Zeiler, M.D.: ADADELTA: an adaptive learning rate method. arXiv preprint arXiv:1212.5701 (2012)

# Cardiac MRI Left Ventricle Segmentation and Quantification: A Framework Combining U-Net and Continuous Max-Flow

Fumin Guo[1,2]([✉]), Matthew Ng[1,2], and Graham Wright[1,2]

[1] Sunnybrook Research Institute, University of Toronto, Toronto, Canada
fumin.guo@sri.utoronto.ca
[2] Department of Medical Biophysics, University of Toronto, Toronto, Canada

**Abstract.** Cardiac magnetic resonance imaging (MRI) is routinely used for cardiovascular disease diagnosis and therapy guidance. Left ventricle (LV) segmentation is typically required as a first step to quantify cardiac indices. In this work, we developed an automatic approach for LV segmentation and indices quantification of cardiac MRI. We employed a U-net convolutional neural network to generate LV segmentation probability maps. The initial probability maps were used to provide the labeling cost measurements of a continuous min-cut segmentation model and the final segmentation was regularized using image edge information. The continuous min-cut segmentation model was solved globally and exactly through convex relaxation and dual optimization on a GPU. We applied our approach to a clinical dataset of 45 subjects and achieved a mean DSC of $89.4 \pm 5.0\%$ and average symmetric surface distance of $0.81 \pm 0.31$ mm for LV myocardium segmentation. For LV indices quantification, we observed a mean absolute error of $114.8$ mm$^2$ for LV cavity, $168.6$ mm$^2$ for LV myocardium, $\sim 1.8$ mm for LV cavity dimensions, and $1.2 \sim 1.6$ mm for LV myocardium wall thickness measurements. These results suggest that our framework provide the potential for LV function quantification using cardiac MRI.

**Keywords:** Cardiac MRI segmentation · LV function quantification · Deep learning · Continuous max-flow

## 1 Introduction

Cardiac magnetic resonance imaging (MRI) is the method of choice for diagnosis, characterization and treatment guidance of patients with cardiovascular disease [6]. Cardiac MRI permits high resolution visualization of cardiac anatomy, non-invasive evaluation of cardiac function, and guidance of cardiac interventions as well as assessment of treatment efficacy. Clinical use of cardiac MRI typically requires segmentation of cardiac structures and here we focus on the left ventricle

© Springer Nature Switzerland AG 2019
M. Pop et al. (Eds.): STACOM 2018, LNCS 11395, pp. 450–458, 2019.
https://doi.org/10.1007/978-3-030-12029-0_48

(LV) that is widely used for quantification of cardiac indices. Unfortunately, cardiac MRI LV segmentation is challenging because of weak image edges, image signal intensity inhomogeneity, large shape variations, partial volume and motion artefacts [13].

Various algorithms have been developed for cardiac MRI LV segmentation and efforts have been continuing [9]. Briefly, these algorithms can be categorized as intensity, statistical model and atlas-based approaches, *a.k.a.* traditional segmentation algorithms. These algorithms mainly employed low-level hand-crafted image features, e.g., image signal intensities, and were optimized with limited training dataset. Although easy to implement, the performance of these traditional LV segmentation algorithms is limited likely due to the use of shallow image features and feature classification strategies [11]. Recently, deep learning using convolutional neural network (CNN) has witnessed tremendous progress in various areas including medical image analysis [11]. The achieved unprecedented success might be attributed to the unique ability of discovering high-level image features with greater representation power through self-learning [5]. Originally developed for high-level sparse classification and object detection, CNN has also been adopted for medical image segmentation and yielded tremendous promise. Although promising, one of the main problems associated with CNN for dense prediction tasks such as image segmentation is the abstraction and suboptimal use of spatial information [2]. Accordingly, recent investigations have attempted to combine the sparse classification power of CNN and the dense prediction power of conventional energy optimization-based image segmentation algorithms [1]. For example, a recent study [8] employed a deep belief network for LV endocardium and epicardium detection followed by a level-set-based refinement step.

In this work, we employed a basic U-net CNN for rough detection of LV myocardium followed by a continuous max-flow refinement step. The resulting CNN prediction probability maps were used to initialize the pixel labeling cost measurements of a continuous min-cut segmentation model and the final segmentation was spatially regularized using image edge information. The challenging min-cut segmentation model was solved globally and exactly through convex relaxation. In addition, we introduced a linear and convex continuous max-flow module that was dual to the convex relaxed continuous min-cut formulation, for which an efficient continuous max-flow algorithm with high computational performance was developed and implemented on a GPU to boost the computational performance of our framework.

## 2   Methods

Given a cardiac cine MR image $I(x)$, $x \in \Omega$, we aimed to segment $I(x)$ into the LV myocardium ($R_{myo}$) and the complementary background ($R_{bk} = \Omega \backslash R_{myo}$). Here we employed a U-net CNN for rough LV myocardium detection and the derived probability maps were refined using a continuous max-flow segmentation model with a spatial regularization prior.

## 2.1  A U-Net Network for LV Myocardium Segmentation Prediction

We employed the 2D U-net convolutional neural network with a symmetric contracting and expanding path [10] for LV myocardium prediction. The contracting
path consisted of five layers each comprising two blocks of $3 \times 3$ convolution filters
and a rectified linear unit, followed by $2 \times 2$ max-pooling (stride $= 2$, doubles
feature maps). The expanding path consisted of five layers each comprising $2 \times 2$
up-sampling (stride $= 2$, halves feature channels), concatenation of the feature
maps from the corresponding contracting path, followed by repeated $3 \times 3$ convolution and a rectified linear unit. The concatenation of feature maps from the
contracting and expanding paths enables combination of abstract context features and corresponding local image information. The final feature maps at the
end of the expanding path was processed by a $1 \times 1$ convolution filter followed
by a softmax operation, resulting in pixel-wise probability maps $\rho_l(x) \in [0,1]$
for each class $l \in L = \{myo, bk\}$.

## 2.2  Continuous Max-Flow Refinement

**U-Net-Guided Continuous Min-Cut Segmentation and Convex Relax-
ation:** The U-net prediction maps $\rho_l(x)$, $l \in L$, represent the probability and
inherently the cost of assigning a label $l$ to each pixel $x \in \Omega$. Let $u(x) \in \{0,1\}$
be the indicator function of $R_{myo}$ such that $u(x) = 1$ for $x \in R_{myo}$ and $u(x) = 0$
for $x \in R_{bk}$. To this end, we formulated the myocardium segmentation problem
as a spatially continuous min-cut energy model [4] as follows:

$$\min_{u(x)\in\{0,1\}} \langle 1 - u(x), 1 - \rho_{myo}(x)\rangle + \langle u(x), \rho_{myo}(x)\rangle + \int_\Omega g(x)\,|\nabla u(x)|\,dx, \quad (1)$$

where $g(x) \geq 0$ denotes the weight function associated with image edges and the
weighted total variation function measures the surface area of $R_{LV}$.

Direct optimization of the *continuous min-cut* model (1) is challenging due
to the non-convex indicator function $u(x)$. In this work, we proposed to solve the
challenging combinatorial optimization problem Eq. (1) by optimizing its convex
relaxed formulation:

$$\min_{u(x)\in[0,1]} \langle 1 - u(x), 1 - \rho_{myo}(x)\rangle + \langle u(x), \rho_{myo}(x)\rangle + \int_\Omega g(x)\,|\nabla u(x)|\,dx, \quad (2)$$

where the binary labeling function $u(x) \in \{0,1\}$ in Eq. (1) was relaxed to the
convex set $u(x) \in [0,1]$. The convexity of Eq. (2) resulted in a simpler convex
optimization problem, for which the global optimum solution exists.

**U-Net-Guided Continuous Max-Flow Segmentation Approach:** Now
we introduced a continuous max-flow model that is dual to the convex relaxed
continuous min-cut model (2). The continuous max-flow configuration is the

same as in [4] and omitted here. That said, instead of dealing with the non-linear and non-smooth combinatorial optimization problem Eq. (2), we proposed to optimize the linear and simpler convex flow-maximization problem as follows:

$$\max_{ps,pt,q} \int_\Omega ps \, dx, \tag{3}$$

subject to:

– flow capacity constraints:
  $ps(x) \leq 1 - \rho_{myo}(x), pt(x) \leq \rho_{myo}(x)$ and $|q(x)| \leq g(x), \quad \forall x \in \Omega$;
– flow conservation constraints:
  $(\text{div } q - ps + pt)(x) = 0, \quad \forall x \in \Omega$.

Previous studies [3,4] have shown that the continuous max-flow model (3) is mathematically equivalent to the convex relaxed continuous min-cut model (2), e.g., (2) $\Longleftrightarrow$ (3). By solving the max-flow problem, the individual flows are maximized while the minimum cut $u(x)$ is achieved in the meantime. The globally optimal binary segmentation $u^*(x)$ can be obtained by thresholding $u(x)$ with any scale $\gamma \in [0, 1)$. Clearly, optimizing the max-flow model (3) demonstrated great advantages: it avoids tackling the challenging combinatorial optimization problem (2) and leads to a high performance continuous max-flow algorithm (see [3,4] for the details of the algorithm) that was parallelized on a GPU for speed-up.

## 3  Experiments and Results

### 3.1  Cardiac MRI Dataset

We evaluated our algorithm on the Left Ventricle Full Quantification Challenge MICCAI 2018 (LVQuan18) dataset of 145 subjects from three local hospitals [7]. The dataset consists of diverse cardiovascular pathologies including regional wall motion abnormalities, myocardial hypertrophy, mildly enlarged LV, atrial septal defect and LV dysfunction. For each subject, 20 frames of short axis cine MR images were acquired throughout a cardiac cycle at mid-cavity (isotropic inplane voxel size of 0.6836~2.0833 mm). The original 2D images were preprocessed following LV-RV intersection points identification, image rotation, region-of-interest cropping and resizing, resulting in final images with matrix size of $80 \times 80$ [12] that were used for our experiments. The preprocessed images were manually segmented to obtain the epicardium and endocardium contours and the manual segmentation outcomes were approved by two experienced radiologists as the reference standard.

### 3.2  Algorithm Implementation

The initial number of feature channels of the 5-layer U-net network was set to 32. We trained the network with 100 subjects (20 2D frames each) for 300 epochs each comprising 100 updates with a batch size of 2 subjects. We also performed a series of data augmentation including random translation ($-10{\sim}10$ pixels), rotation ($-10{\sim}10°$), and scaling ($0.9{\sim}1.1$ times) to match the variability of clinical MR scans and to improve the robustness of the learning process. We employed a pixel-wise cross-entropy as the loss function and the ADAM solver to optimize the network parameters with a learning rate of $1 \times 10^{-4}$. The derived labeling probability maps were provided to the continuous max-flow segmentation module as the data term and the segmentation regularization term was designed based on image edge information, i.e., $g(x) = \lambda_1 + \lambda_2 \cdot exp(-\lambda_3 |\nabla I(x)|)$ in (1), where $\lambda_{1,2,3} = \{0.1, 1, 30\}$ were optimized on a separate dataset. The optimized segmentation framework was applied to the remaining 45 subjects and the final segmentation outputs were used to calculate the indices including LV cavity and myocardium area, dimensions and regional wall thickness [12].

### 3.3    Evaluation Methods

**LV Segmentation Performance:** Algorithm segmentation results were compared with manual segmentation using Dice similarity coefficient ($DSC$) to measure region overlap and average symmetric surface ($ASSD$) to measure to contour agreement [4]. In addition, we reported the runtime of the U-net training, U-net testing, continuous max-flow refinement components as well as the whole pipeline to evaluate the computational efficiency of our approach.

**LV Indices Quantification:** The derived LV segmentation results were used to quantify the LV indices (Fig. 2) in each frame as follows:

- **A_cav, A_myo** (mm$^2$): Area of left ventricle cavity and myocardium, respectively;
- **Dim1, Dim2, Dim3** (mm): Dimensions of LV cavity in: inferoseptal-anterolateral (IS-AL), inferior-anterior (I-A), inferolateral-anteroseptal (IL-AS) directions, respectively;
- **IS, I, IL, AL, A**, and **AS** (mm): Left ventricle myocardium wall thickness in IS, I, IL, AL, A and AS directions, respectively.

In particular, **A_cav** and **A_myo** were calculated by counting the number of pixels enclosed by endo and epicardium, respectively. **Dim1, Dim2** and **Dim3** were determined by casting a line from the centroid of LV cavity in IS-AL, I-A and IL-AS directions and measuring the distance between the intersections of the casted lines and the LV endocardium contour. Similarly, **IS, I, IL, AL, A**, and **AS** were determined as the distance between the intersections of the casted lines and the endo and epicardium contours in the respective directions (see Fig. 2 for illustration of these indices). The derived algorithm quantification results ($m_a$) were compared with reference measurements ($m_r$) using mean absolute error ($MAE = \frac{1}{N} \sum_{i=1}^{N} |m_a^i - \hat{m}_r^i|$ [12]) and Pearson correlation coefficient ($r$) with GraphPad Prism version 7.00 (GraphPad Software, Inc., San Diego, CA). Results were considered significant when the probability of making a two-tailed type I error was less than 5% (p $< 0.05$).

## 3.4   Results

Figure 1 shows representative segmentation results of 20 cardiac MR images across one cardiac cycle. For 45 subjects, we obtained a mean $DSC$ of $89.4 \pm 5.0\%$ and $MAD$ of $0.81 \pm 0.31$ mm for LV myocardium.

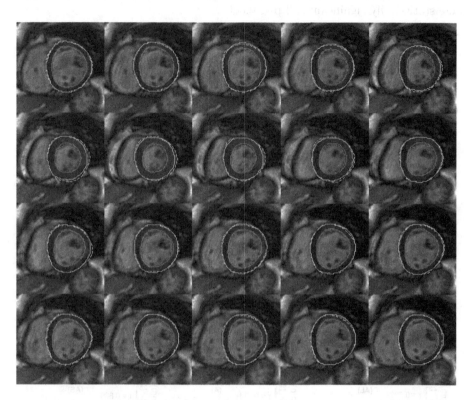

**Fig. 1.** Representation LV myocardium segmentation results across a cardiac cycle of a single subject. Algorithm segmentation results are shown in solid contours and manual delineations are shown in dashed contours. **First** row: phase 1–5, **Second** row: phase 6–10, **Third** row: phase 11–15, **Fourth** row: phase 16–20.

Table 1 summarizes the LV indices quantification results. For example, we obtained a MAE of 114.8 mm$^2$ for **A_cav** and 168.6 mm$^2$ for **A_myo** by comparing algorithm results with reference measurements. Similarly, we achieved MAE

**Table 1.** LV indices quantification results for n = 45 subjects.

|        | A_cav  | A_myo  | Dim1   | Dim2   | Dim3   | IS     | I      | IL     | AL     | A      | AS     |
|--------|--------|--------|--------|--------|--------|--------|--------|--------|--------|--------|--------|
| MAE    | 114.8  | 168.6  | 1.72   | 1.87   | 1.72   | 1.55   | 1.43   | 1.47   | 1.38   | 1.17   | 1.36   |
| Corr.  | 0.984* | 0.949* | 0.983* | 0.972* | 0.976* | 0.958* | 0.840* | 0.809* | 0.792* | 0.871* | 0.889* |

Corr.: Pearson correlation coefficient ($r$); *: p-value < 0.001.

of 1.72, 1.87 and 1.72 mm for **Dim1, Dim2** and **Dim3**, respectively. For regional myocardium wall thickness measurements, the resulting MAE were 1.55, 1.43, 1.47, 1.38, 1.17 and 1.36 mm for **IS, I, IL, AL, A**, and **AS**, respectively. As shown in Fig. 2, algorithm LV indices quantification results were strongly correlated (all Pearson $r > 0.8$) with reference measurements and all the correlations were statistically significant (all $p < 0.0001$).

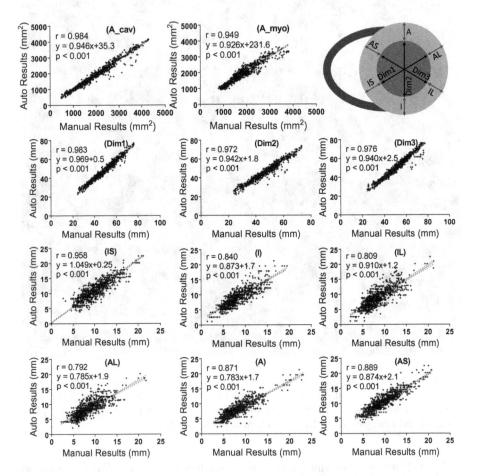

**Fig. 2.** Relationship of algorithm (Auto. Results) and reference (Manual Results) LV indices quantification results for **A_cav, A_myo, Dim1, Dim2, Dim3, IS, I, IL, AL, A**, and **AS**, as illustrated in upper right. Solid red lines indicate the mean and dotted red lines represent the upper and lower limits. (Color figure online)

Finally, our approach required ~95 min for U-net training with 100 subjects (2000 frames, 80 × 80 matrix), ~1 s for U-net prediction, ~1 s for continuous max-flow refinement and ~0.5 s for LV indices quantification for each subject.

## 4  Discussion and Conclusions

In this work, we developed a fully automatic approach for LV indices quantification in cardiac MRI. Our approach employed a U-net convolutional neural network for rough LV myocardium detection and a continuous max-flow module for segmentation refinement followed by LV cavity/regional wall thickness measurements. For a diverse group of 45 test patients, we demonstrated: (1) high agreement between algorithm and expert myocardium manual segmentation, (2) low error and excellent correlation between algorithm and reference LV indices measurements, suggesting the potential for clinical applications of our approach.

The developed approach required LV myocardium segmentation to generate LV indices and we note that it is also possible to directly quantify these indices without performing segmentation [12]. However, we think that the generated myocardium segmentation may provide added values for other application including image-guided cardiac interventions. Compared with the previous study [12], we achieved low error in LV myocardium and cavity area and dimension measurements while the error of regional wall thickness measurements was comparable, as shown in Table 1. However, we must acknowledge that we evaluated our approach on 45 patients while the previous work performed 5-fold cross validation and this will be our future work. In addition, our approach required myocardium segmentation and the segmentation error may affect LV indices quantification although high segmentation accuracy was achieved. Due to the sparse classification nature of CNN, the U-net-derived initial segmentation may be "noisy" and this may affect the final quantification. The continuous max-flow module provides a way to refine the initial segmentation by removing small "islands" (or filling small "holes") and encouraging better edge alignment. Another aspect is that the regional wall thickness measurements requires two enclosed contours representing the endo and epicardium. This may be achieved by using a multi-region segmentation approach and enforcing that the LV cavity is inside the myocardium. Future work will involve multi-fold cross validation, evaluation of the influence of myocardium segmentation on LV indices quantification, and improving the segmentation performance by incorporating prior information.

## References

1. Avendi, M., Kheradvar, A., Jafarkhani, H.: A combined deep-learning and deformable-model approach to fully automatic segmentation of the left ventricle in cardiac MRI. Med. Image Anal. **30**, 108–119 (2016)
2. Chen, L.C., Papandreou, G., Kokkinos, I., Murphy, K., Yuille, A.L.: DeepLab: semantic image segmentation with deep convolutional nets, atrous convolution, and fully connected CRFs. IEEE Trans. Pattern Anal. Mach. Intell. **40**(4), 834–848 (2018)
3. Guo, F., Svenningsen, S., Eddy, R., Capaldi, D., Sheikh, K., Fenster, A., Parraga, G.: Anatomical pulmonary magnetic resonance imaging segmentation for regional structure-function measurements of asthma. Med. Phys. **43**(6 Part 1), 2911–2926 (2016)

4. Guo, F., et al.: Globally optimal co-segmentation of three-dimensional pulmonary 1H and hyperpolarized 3He MRI with spatial consistence prior. Med. Image Anal. **23**(1), 43–55 (2015)
5. LeCun, Y., Bengio, Y., Hinton, G.: Deep learning. Nature **521**(7553), 436 (2015)
6. Lima, J.A., Desai, M.Y.: Cardiovascular magnetic resonance imaging: current and emerging applications. J. Am. Coll. Cardiol. **44**(6), 1164–1171 (2004)
7. MICCAI: LVQuan18 dataset (2018). https://lvquan18.github.io/
8. Ngo, T.A., Lu, Z., Carneiro, G.: Combining deep learning and level set for the automated segmentation of the left ventricle of the heart from cardiac cine magnetic resonance. Med. Image Anal. **35**, 159–171 (2017)
9. Petitjean, C., Dacher, J.N.: A review of segmentation methods in short axis cardiac MR images. Med. Image Anal. **15**(2), 169–184 (2011)
10. Ronneberger, O., Fischer, P., Brox, T.: U-Net: convolutional networks for biomedical image segmentation. In: Navab, N., Hornegger, J., Wells, W.M., Frangi, A.F. (eds.) MICCAI 2015. LNCS, vol. 9351, pp. 234–241. Springer, Cham (2015). https://doi.org/10.1007/978-3-319-24574-4_28
11. Shen, D., Wu, G., Suk, H.I.: Deep learning in medical image analysis. Ann. Rev. Biomed. Eng. **19**, 221–248 (2017)
12. Xue, W., Brahm, G., Pandey, S., Leung, S., Li, S.: Full left ventricle quantification via deep multitask relationships learning. Med. Image Anal. **43**, 54–65 (2018)
13. Zhuang, X.: Challenges and methodologies of fully automatic whole heart segmentation: a review. J. Healthc. Eng. **4**(3), 371–407 (2013)

# Multi-estimator Full Left Ventricle Quantification Through Ensemble Learning

Jiasha Liu[1], Xiang Li[2(✉)], Hui Ren[2,4], and Quanzheng Li[1,2,3]

[1] Peking University, Beijing 100080, China
[2] MGH/BWH Center for Clinical Data Science, Boston, MA 02115, USA
xli60@mgh.harvard.edu
[3] Laboratory for Biomedical Image Analysis,
Beijing Institute of Big Data Research, Beijing 100871, China
[4] Heart Center, Peking University People's Hospital, Beijing 100044, China

**Abstract.** Cardiovascular disease accounts for 1 in every 4 deaths in United States. Accurate estimation of structural and functional cardiac parameters is crucial for both diagnosis and disease management. In this work, we develop an ensemble learning framework for more accurate and robust left ventricle (LV) quantification. The framework combines two 1st-level modules: direct estimation module and a segmentation module. The direct estimation module utilizes Convolutional Neural Network (CNN) to achieve end-to-end quantification. The CNN is trained by taking 2D cardiac images as input and cardiac parameters as output. The segmentation module utilizes a U-Net architecture for obtaining pixel-wise prediction of the epicardium and endocardium of LV from the background. The binary U-Net output is then analyzed by a separate CNN for estimating the cardiac parameters. We then employ linear regression between the 1st-level predictor and ground truth to learn a 2nd-level predictor that ensembles the results from 1st-level modules for the final estimation. Preliminary results by testing the proposed framework on the LVQuan18 dataset show superior performance of the ensemble learning model over the two base modules.

**Keywords:** Semantic segmentation · Direct estimation · Ensemble learning

## 1 Introduction

Cardiac MRI is a widely-adopted, accurate non-ionizing method for the assessment of cardiac disease, especially in evaluating cardiac function and morphology. Quantitative measures from short axes (SAX) cine are recommended in reporting guideline to describe left ventricular (LV) size and function, such as volume, wall thickness and regional wall motion of left ventricle [1]. Manually quantitative measurements are time-consuming, less reproducible and increasing the burden of radiologists. Automatic quantification of the cardiac parameters has become an increasingly important problem, especially in recent years due to the success in applying computer vision techniques on

---

J. Liu and X. Li—Joint first authors.

© Springer Nature Switzerland AG 2019
M. Pop et al. (Eds.): STACOM 2018, LNCS 11395, pp. 459–465, 2019.
https://doi.org/10.1007/978-3-030-12029-0_49

various image processing tasks. In this work, we present an ensemble learning framework which combines the results from two well-established quantification methods: direct estimation-based method and segmentation-based method, and ensembles them through a linear regression model, to further improve the prediction power. Preliminary results on LVQuan18 dataset using 3-folds cross validation scheme shows that the proposed ensemble learning framework can achieve accurate and robust left ventricle quantification for both structural and functional cardiac parameters.

**Related Works**

Previous literatures on cardiac parameters estimation mainly belong to two categories: the segmentation-based methods [2–4] which perform LV segmentation first, then measure the parameters based on segmentation results. This category of methods is advantages as the results (usually as segmentation masks) are easier to interpret which can provide more insight into the analytics procedure. The latter category of direct estimation-based methods [5–8] perform estimation of the cardiac parameters directly from the image and/or image features, without relying on explicit segmentation procedure. It is advantages as the training error can be directly back-propagated to the feature selection and regression process, thus normally resulting in higher accuracy. Also, direct estimation methods do not require manual annotation of cardiac regions, thus are more feasible to be used in real practice.

## 2    Ensemble Learning Framework

### 2.1    Overview and Data Description

In this work, we utilize two 1st-level ("base") modules for LV quantification: a direct estimation module which is based on an end-to-end CNN structure to map the input 2D cardiac image to its quantitative measurements; and a semantic segmentation module which is based on a U-Net structure to map the same input image to a binary mask indicating its epicardium and endocardium, which are then fed into a separate CNN to obtain the measurements. In order to effectively and robustly combine them, we then train a 2nd-level ensemble classifier based on linear regression to map the results from these two modules to the ground truth measurements. The proposed framework is evaluated on cardiac MR dataset provided by LVQuan18. The dataset consists processed SAX MR sequences of 145 subject, where for each subject there are 20 frames to capture the whole cardiac cycle. Ground truth measurements as well as annotations for epicardium and endocardium are provided for every single frame.

The goal of full left ventricle quantification is estimating three types of measurement indices, including; (1) $A1$, $A2$: cavity area and myocardium area; (2) $D1$-$D3$: dimensions of cavity of three directions (AS-IL, IS-AL, and I-AL); and (3) $RWT1 \sim RWT6$: regional wall thickness, staring from the anterior-septal segment in counter clockwise direction. In addition, the cardiac phase as a binary value (1 for systolic and 0 for diastolic) will also be inferred from the quantifications. Evaluation (training and testing) of the proposed system is performed using 3-folds cross-validation scheme.

## 2.2   Direct Estimation Module

Structure of CNN in the direct estimation module consists of four convolution layers (conv1 ~ 4) and two fully-connected (fc1 ~ 2) layers, which is visualized in Fig. 1. The direct estimation module essentially performs non-linear regression between the input image (high-dimensional, pixel-wise representation of a 2-D object) and its characterizations (low-dimensional, semantic information). The effective mapping between high and low-dimensional data is achieved through convolutional filters, which perform dimension reduction and feature extraction on images based on the premise of shift-invariance, local connectivity and compositionality.

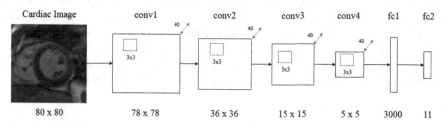

**Fig. 1.** Diagram of the direct estimation network. Each convolution layer is followed by a rectified linear unit layer (ReLU), a batch normalization layer and a max-pooling layer.

In addition, we also analyzed the outputs of each convolutional kernel in conv1-conv4 from a given image and obtained their visualizations (i.e. "feature maps") in Fig. 2. It can be observed that convolutional kernels are capable of effectively representing structure information and local variability of the input images.

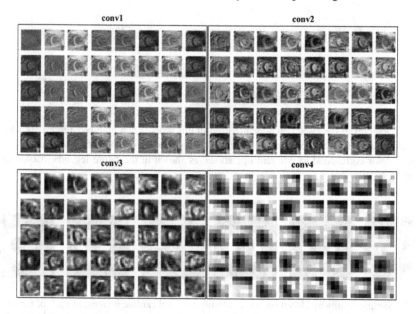

**Fig. 2.** Visualization of conv1-conv4 layers output (i.e. feature maps) of the direct estimation CNN on a sample image.

## 2.3   Segmentation Module

In segmentation module, we apply U-Net for cardiac image segmentation task, as shown in Fig. 3. U-Net has been successfully applied in many visual tasks, especially in biomedical image processing. It is based on encoder-decoder architecture, where the spatial dimension is reduced by pooling layers in encoder, and the object details and spatial dimension are gradually recover in decoder [9]. In contrast to direct estimation module, here the input cardiac image is explicitly mapped to the manual annotation masks through U-Net. After the segmentation results (i.e. binary mask of the cardiac image) are obtained, we use a CNN with three convolution layers and two fully-connected layers to map the binary mask to the quantification measurements. Sample segmentation results are shown in Fig. 4, illustrating that U-Net has very robust performance on images with noises and varying cardiac structures: DICE coefficient of the segmentation reaches higher than 0.9 in the test data. In both direct estimation module and segmentation module, the networks are implemented by Keras, using Adam optimizer. Batch size and epochs are determined by cross-validation results combined with grid search.

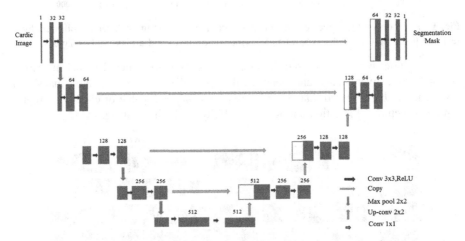

**Fig. 3.** Diagram of the U-Net architecture used in this work. Each blue box represents a multi-channel feature map. Number on top of the blue box shows number of channels. Arrows in different color correspond to different operations, as shown in the figure legends. (Color figure online)

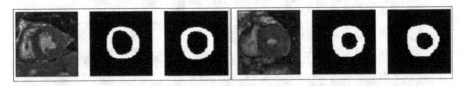

**Fig. 4.** Sample results by U-Net on two randomly-selected cardiac images. In each of the image, Left column: input cardiac images; Middle column: ground truth annotations; Right column: segmentation results.

### 2.4    2nd-Level Ensemble Learning from Base Modules

Prediction results from the two 1st-level base modules, represented as two $11 \times 1$ vectors for each subject, are then used for training a 2nd-level ensemble learning framework. The ensemble predictor is based on linear regression between the 1st-level predictor results and ground truth measurements in the training dataset. Learned linear ensemble model thus consists of weights between the two base predictors for each quantification, which will be used to predict the final estimation in the testing dataset. Final output of the ensemble framework is the quantification measurements as one $11 \times 1$ vector for each subject. Cardiac phase is inferred separately from other measurements, which is based on a thresholding for the predicted area parameters and regularization rules (at most two change points for the systolic/diastolic transition through the whole cycle).

## 3    Experiments and Results

In this work, we use 3-folds cross-validation scheme for performance evaluation and comparison, where the 145-subjects LVQuan18 dataset is divided into three groups with (49, 48, 48) subjects. Cross-validation performances of the 1st-level direct estimation module and the segmentation module, as well as the 2nd-level ensemble learning framework which serves as the final output of the proposed system, are summarized in the tables below. In addition, average accuracy of the threshold-based phase estimator based on the 2nd-level prediction results is 88%.

In Table 1, value in each cell is the MAE (mean average error) of the prediction, plus/minus standard deviation. Best prediction accuracy (i.e. lowest MAE) is highlighted in bold. It should be noted that direct estimation module and segmentation module have different levels of prediction power for different quantifications. Specifically, segmentation module performs better for area and dimension estimation, which are more directly related to the LV segmentation; direct estimation module performs better for RWT (i.e. LV regional wall thicknesses) estimation which is a more complicated and abstract measurement. While the ensemble learning framework outperforms the two base modules in all fields, indicating that it successfully combines the results from two methods with different underlying mechanisms and utilizes results from weaker predictors to further improve the quantification accuracy.

In addition to the cross-validation experiments, we also obtain the average MAEs of the prediction in different time frames. Visualizations for the temporal changes of the prediction performance, using area parameters estimation as example, are shown in Fig. 5. From the temporal analysis it can be observed that our ensemble learning framework can obtain smoother estimation for the quantifications, reducing impacts from frames with large prediction errors, which indicates that by ensemble predictions from multiple sources, we can improve the model robustness on more difficult/complex instances.

**Table 1.** Performance comparison between the 1st-level modules and 2nd-level ensembles in the cross-validation experiment.

|  | Direct estimation | Segmentation | Ensemble |
|---|---|---|---|
| *Area* (mm$^2$) | | | |
| *Cavity* | 112.61 ± 89.29 | 95.22 ± 80.46 | **88.75 ± 72.72** |
| *Myocardium* | 172.97 ± 145.39 | 163.11 ± 119.31 | **159.50 ± 125.14** |
| Average | 142.79 ± 117.34 | 129.17 ± 99.88 | **124.13 ± 98.93** |
| *Dimension* (mm) | | | |
| *dims1* | 2.81 ± 2.22 | 2.37 ± 1.91 | **2.22 ± 1.80** |
| *dims2* | 2.62 ± 2.12 | 2.52 ± 2.03 | **2.28 ± 1.87** |
| *dims3* | 2.82 ± 2.29 | 2.38 ± 1.93 | **2.32 ± 1.81** |
| Average | 2.75 ± 2.21 | 2.56 ± 0.96 | **2.27 ± 1.83** |
| *RWT* (mm) | | | |
| *IS* | 1.68 ± 1.41 | 1.82 ± 1.33 | **1.64 ± 1.32** |
| *I* | 1.65 ± 1.31 | 1.62 ± 1.22 | **1.57 ± 1.22** |
| *IL* | 1.90 ± 1.61 | 1.90 ± 1.50 | **1.83 ± 1.49** |
| *AL* | 1.76 ± 1.50 | 1.79 ± 1.38 | **1.69 ± 1.39** |
| *A* | 1.42 ± 1.29 | 1.54 ± 1.20 | **1.41 ± 1.22** |
| *AS* | 1.65 ± 1.32 | 1.60 ± 1.26 | **1.56 ± 1.25** |
| Average | 1.68 ± 1.41 | 1.71 ± 1.31 | **1.62 ± 1.32** |

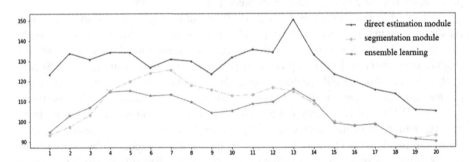

**Fig. 5.** Average frame-wise estimation errors of the area quantification estimated by the two 1st-level module (direct estimation in blue and segmentation in orange), as well as the ensemble learning framework (green). X-axis shows the indices of temporal frames, from 1st frame to the 20th frame. Y-axis shows the MSE of predictions, measured in mm$^2$. (Color figure online)

## 4 Conclusions

In this work, we propose a multi-estimator, ensemble learning-based framework for full LV quantification. Preliminary results on the LVQuan18 dataset shows that the proposed framework can outperform the commonly applied quantification methods by integrating predictions from multiple models in a learning-based approach. Currently only two 1st-level base modules are included in the framework, in our next step of investigation we are aiming to employ more quantification methods, especially the unsupervised learning-based image processing algorithms, into our framework to further improve its robustness in different image settings.

# References

1. Hundley, W.G., et al.: Society for Cardiovascular Magnetic Resonance guidelines for reporting cardiovascular magnetic resonance examinations. J. Cardiovasc. Magn. Reson. **11**, 5 (2009)
2. Peng, P., Lekadir, K., Gooya, A., Shao, L., Petersen, S.E., Frangi, A.F.: A review of heart chamber segmentation for structural and functional analysis using cardiac magnetic resonance imaging. Magn. Reson. Mater. Phys. Biol. Med. **29**, 155–195 (2016)
3. Ben Ayed, I., Chen, H.-M., Punithakumar, K., Ross, I., Li, S.: Max-flow segmentation of the left ventricle by recovering subject-specific distributions via a bound of the Bhattacharyya measure. Med. Image Anal. **16**, 87–100 (2012)
4. Petitjean, C., Dacher, J.-N.: A review of segmentation methods in short axis cardiac MR images. Med. Image Anal. **15**, 169–184 (2011)
5. Afshin, M., et al.: Regional assessment of cardiac left ventricular myocardial function via MRI statistical features. IEEE Trans. Med. Imaging **33**, 481–494 (2014)
6. Wang, Z., Salah, M.B., Gu, B., Islam, A., Goela, A., Li, S.: Direct estimation of cardiac biventricular volumes with an adapted Bayesian formulation. IEEE Trans. Biomed. Eng. **61**, 1251–1260 (2014)
7. Xue, W., Lum, A., Mercado, A., Landis, M., Warrington, J., Li, S.: Full quantification of left ventricle via deep multitask learning network respecting intra- and inter-task relatedness. In: Descoteaux, M., Maier-Hein, L., Franz, A., Jannin, P., Collins, D., Duchesne, S. (eds.) MICCAI 2017. LNCS, vol. 10435, pp. 276–284. Springer, Cham (2017). https://doi.org/10.1007/978-3-319-66179-7_32
8. Xue, W., Islam, A., Bhaduri, M., Li, S.: Direct multitype cardiac indices estimation via joint representation and regression learning. IEEE Trans. Med. Imaging **36**, 2057–2067 (2017)
9. Ronneberger, O., Fischer, P., Brox, T.: U-Net: convolutional networks for biomedical image segmentation. In: Navab, N., Hornegger, J., Wells, William M., Frangi, Alejandro F. (eds.) MICCAI 2015. LNCS, vol. 9351, pp. 234–241. Springer, Cham (2015). https://doi.org/10.1007/978-3-319-24574-4_28

# Left Ventricle Quantification Through Spatio-Temporal CNNs

Alejandro Debus and Enzo Ferrante[✉]

Research Institute for Signals, Systems and Computational Intelligence,
sinc(i), FICH-UNL/CONICET, Santa Fe, Argentina
eferrante@sinc.unl.edu.ar

**Abstract.** Cardiovascular diseases are among the leading causes of death globally. Cardiac left ventricle (LV) quantification is known to be one of the most important tasks for the identification and diagnosis of such pathologies. In this paper, we propose a deep learning method that incorporates 3D spatio-temporal convolutions to perform direct left ventricle quantification from cardiac MR sequences. Instead of analysing slices independently, we process stacks of temporally adjacent slices by means of 3D convolutional kernels which fuse the spatio-temporal information, incorporating the temporal dynamics of the heart to the learned model. We show that incorporating such information by means of spatio-temporal convolutions into standard LV quantification architectures improves the accuracy of the predictions when compared with single-slice models, achieving competitive results for all cardiac indices and significantly breaking the state of the art [10] for cardiac phase estimation.

**Keywords:** Left ventricle quantification ·
Spatio temporal convolutional neural network

## 1 Introduction

In 2015, around 17.7 million people died worldwide due to heart diseases. Left ventricle (LV) quantification is a key factor for the identification and diagnosis of such pathologies [2]. However, the estimation of cardiac indices remains a very complex task due to its intricated temporal dynamics and the inter-subject variability of the cardiac structures. Indices such as cavity and myocardium area, regional wall thickness, cavity dimensions, among others, provide useful information to diagnose various types of cardiac pathologies. Cardiovascular magnetic resonance (CMR) is one of the preferred modalities for LV related studies since it is non invasive, presents high spatio-temporal resolution, has a good signal-to-noise ratio and allows to clearly identify the tissues and muscles of interest [6].

The classical approach to LV quantification consists in estimating such indices by means of automatic segmentation [3–7,9]. Segmentation is usually performed following supervised learning approaches, which require expert manual annotations contouring the edges of the myocardium for training. Once the segmentation is performed, the indices are computed from the resulting mask.

© Springer Nature Switzerland AG 2019
M. Pop et al. (Eds.): STACOM 2018, LNCS 11395, pp. 466–475, 2019.
https://doi.org/10.1007/978-3-030-12029-0_50

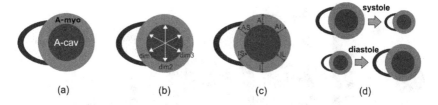

**Fig. 1.** Illustration of indices of the left cardiac ventricle (based on Fig. 1 from [10]). (a) Cavity area (brown) and myocardial area (orange). (b) Directional dimensions of cavity (white arrows). (c) Regional wall thicknesses. A: anterior; AS: anteroseptal; IS: inferoseptal; I: inferior; IL: inferolateral; AL: anterolateral. (d) Cardiac phase (systole or diastole) (Color figure online)

Therefore, the accuracy of the predicted indices is conditioned on the quality of the segmentation. In this work, we follow an alternative strategy that directly estimates the indices of interest from the input image sequence. Inspired by the work of [10–12], our model is based on a convolutional neural network directly operating on images and regressing the target indices. Different from previous approaches like [10] where the temporal dynamics of cardiac sequences is incorporated using recurrent neural networks (RNNs), we propose a simple but effective strategy based on the use of spatio-temporal convolutions [8]. In the context of video analysis, spatio-temporal convolutions are standard 3D convolutions that operate on spatio-temporal video volumes [7]. Here we employ them to process subsets of temporally contiguous CMR slices, leveraging temporal information towards improving prediction accuracy (Fig. 1).

We investigate the use of spatio-temporal convolutions for estimating cardiac phase, directional dimensions of the cavity, regional wall thicknesses and area of cavity and myocardium under the hypothesis that such indices may be better explained when taking into account the temporal dynamics of the heart. We benchmark the proposed architecture using the LVQuan Challenge 2018[1] dataset, which provides CMR sequences with annotations for the aforementioned indices, and provide empirical evidence that incorporating the temporal dynamics of the heart through 3D spatio-temporal convolutions improves prediction accuracy when compared with single-slice models.

## 2   Materials and Methods

### 2.1   Architecture

An overview of the proposed CNN architecture is presented in Fig. 2. The network takes sequences of $\kappa$ slices and outputs the corresponding indices *only* for the central slice. In such way, we incorporate information from the surrounding slices, easing the prediction task. In what follows, we describe in detail the main components of the proposed architecture.

---

[1] LVQuan Challenge website: https://lvquan18.github.io/.

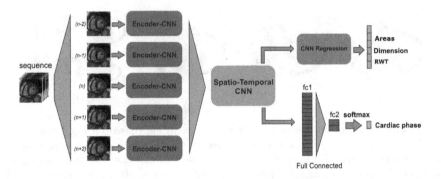

**Fig. 2.** Overview of proposed architecture.

**Encoder-CNN.** We use a first CNN (referred as encoder-CNN in Figs. 2 and 3) to extract informative features from individual slices. Inspired by [11], we designed the per-slice encoding phase using a two-layers CNN where the convolutional and pooling kernels are of size $5 \times 5$, instead of the frequently used $3 \times 3$, to introduce more shift invariance (see Fig. 3 for more details). We use ReLU activation function and batch normalization to alleviate the training process.

**Spatio-Temporal CNN.** After the encoding phase, the 40 filters generated for every individual encoder-CNN are used to construct a spatio-temporal volume with 40 channels per temporal slice. This volume is then processed using 3D convolutions that operate on the temporal and spatial dimensions (see Fig. 3), producing compound feature maps that incorporate information from both of them. This module is composed of two 3D convolutional layers with kernels of size $3 \times 5 \times 5$ and $2 \times 5 \times 5$ when considering $\kappa = 5$ slices. When considering $\kappa = 1, 3, 7$ slices, the proposed architecture is modified by using padding in the temporal dimension ($\kappa = 1, 3$) and adding an extra convolution ($\kappa = 7$) so that the shape of the output tensor matches $1 \times 6 \times 6$, the size required by the CNN Regression and Fully Connected modules. ReLU activations and batch normalization are also used in this module.

**Final Parallel Branches.** After fusing the spatio-temporal features, two parallel branches are derived: (i) the first branch corresponds to a shallow CNN coupled after the spatio-temporal module, acting as a regressor of the directional dimensions, wall thickness and areas; (ii) in the second branch, a third convolutional layer is coupled to the spatio-temporal module, followed by a fully connected multi layer perceptron (MLP) with 640 neurons in the hidden layer and 2 output neurons encoding the probability for the cardiac phase (systole or diastole).

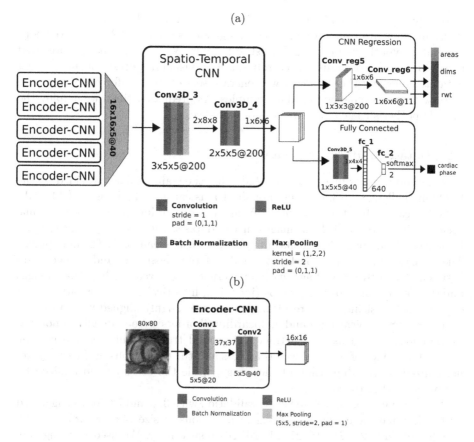

**Fig. 3.** (a) Detailed overview of the spatio-temporal CNN based on 3D convolutions. (b) Zoomed version of the individual encoder-CNNs: for a single input slice of size $80 \times 80$ it outputs 40 filters of size $16 \times 16$ which are then fed to the spatio-temporal CNN.

**Training Procedure and Loss Function.** We train the proposed network by minimizing a loss function over sets of $\kappa$ slices where annotations are provided only for the central slice. Given a set of $\kappa$ slices $\boldsymbol{x}^i = \{x_0, ..., x_\kappa - 1\}$, ground-truth annotations for the central slice $\boldsymbol{y}^i = \{y_{dim}, y_{areas}, y_{rwt}, y_{phase}\}$ and corresponding predictions from the proposed neural network $\phi_{phase}$ and $\phi_{dim}, \phi_{areas}, \phi_{rwt}$ the loss function is defined as:

$$
\begin{aligned}
\mathcal{L}(\boldsymbol{x}^i, \boldsymbol{y}^i) = {} & \mathcal{L}_{mse}(\phi_{areas}, y_{areas}) + \mathcal{L}_{mse}(\phi_{dim}, y_{dim}) \\
& + \mathcal{L}_{mse}(\phi_{rwt}, y_{rwt}) + \mathcal{L}_{ce}(\phi_{phase}, y_{phase}) + \lambda \mathcal{L}_{reg},
\end{aligned}
\tag{1}
$$

where $\mathcal{L}_{mse}$ is the mean squared error between predictions and ground truth, $\mathcal{L}_{ce}$ is the cross-entropy loss, $\mathcal{L}_{reg}$ is the regularizer (L2 norm of the network weights) and $\lambda$ is a weighting factor. We minimize this loss using stochastic gradient descent with momentum, with mini-batches of size $s = 20$.

**Circular Hypothesis.** Since we require sets of temporally contiguous slices as input for our spatio-temporal architecture, given a sequence of $N$ slices, we adopt a circular hypothesis meaning that slice number $N - 1$ is temporally followed by slice 0. This hypothesis was corroborated by visual inspection of the training dataset. Following this strategy, we generate sets of $\kappa$ slices for every sequence and use them as independent data samples. At prediction time, we employ the same hypothesis to generate the sets of test slices.

## 2.2    Dataset and Experimental Setting

Our method is experimentally validated using the training data provided by the LVQuan challenge 2018, composed of short axis cardiac MR images of 145 subjects. For each subject, it contains 20 frames corresponding to a complete cardiac cycle (giving a total of 2900 images in the dataset with pixel spacing ranging from 0.6836 mm/pixel to 2.0833 mm/pixel, with a mean of 1.5625 mm/pixel). The images have been collected from 3 different hospitals and subjects are between 16 and 97 years of age, with an average of 58.9 years. All cardiac images undergo several preprocessing steps (including historical tagging, rotation, ROI clipping, and resizing). The resulting images are roughly aligned with a dimension of $80 \times 80$. Epicardium and endocardium borders were manually annotated by radiologists, and used to extract the ground truth LV indices and cardiac phase. The values of regional wall thickness and the dimensions of the cavity are normalized by the dimension of the image, while the areas are normalized by the pixel number (6400).

In our experiments, we used cross validation with 3, 5 and 7 folds as suggested by the LVQuan organizers, resulting in partitions of size (49, 48, 48), (29, 29, 29, 29, 29) and (21, 21, 21, 21, 21, 20, 20) respectively. We used learning rate = 1e–4, momentum = 0.5 and $\lambda$ = 0.005 (these parameters were obtained by grid-search).

The model was implemented in Python[2], using PyTorch and trained in GPU.

**Evaluation Criteria.** Pearson correlation coefficient (PCC) and Mean Absolute Error (MAE) were used to assess the performance of the algorithms for estimation of areas, dimensions and regional wall thicknesses. Error Rate (ER) was used to assess the performance for cardiac phase classification.

$$PCC_{ind} = \frac{\sum_{i=1}^{N}(\phi_{ind}^{(i)} - \bar{\phi}_{ind})(y_{ind}^{(i)} - \bar{y}_{ind})}{\sqrt{\sum_{i=1}^{N}(\phi_{ind}^{(i)} - \bar{\phi}_{ind})^2 \sum_{i=1}^{N}(y_{ind}^{(i)} - \bar{y}_{ind})^2}}, \tag{2}$$

$$MAE_{ind} = \frac{1}{N}\sum_{i=1}^{N}|\phi_{ind}^{(i)} - y_{ind}^{(i)}|, \tag{3}$$

---

[2] The source code for the proposed architecture is publicly available at https://github.com/alejandrodebus/SpatioTemporalCNN_lvquan.

where $ind \in (A_1, A_2, D_1...D_3, RWT_1...RWT_6)$, $y_{ind}$ is the ground-truth value and $\phi_{ind}$ is the estimated value. $\bar{y}_{ind}$ and $\bar{\phi}_{ind}$ are their mean values, respectively.

$$ER_{phase} = \frac{\sum_{i=1}^{N} \mathbf{1}(\phi_{phase}^{(i)} \neq y_{phase}^{(i)})}{N} 100\% \qquad (4)$$

where $\mathbf{1}()$ is the indication function, $\phi_{phase}$ and $y_{phase}$ are the estimated and ground truth value of the cardiac phase, respectively.

**Table 1.** Sensitivity analysis for the parameter $\kappa$ (number of neighbouring slices) when using the spatio-temporal model based on 3D convolutions with 5-folds cross validation, compared with the state of the art DMTRL proposed in [10]. Note that incorporating the temporal dynamics by considering multiple slices ($\kappa = 3, 5, 7$) makes a significant different with respect the single slice case ($\kappa = 1$). However, considering $\kappa = 5$ and $\kappa = 7$ slices present a similar performance. Therefore, we consider $\kappa = 5$ as enough temporal context for the remaining experiments. When comparing with [10] we observe similar results for most indices, except for the phase, where the proposed model breaks the state of the art significantly (from 8.2% to 3.2%).

| | Model | | | | |
|---|---|---|---|---|---|
| | $\kappa = 1$ | $\kappa = 3$ | $\kappa = 5$ | $\kappa = 7$ | DMTRL [10] |
| Areas (mm$^2$) | | | | | |
| a-cav | $239 \pm 198$ | $194 \pm 188$ | $181 \pm 130$ | $180 \pm 145$ | $\mathbf{172 \pm 148}$ |
| | $0.861 \pm 0.053$ | $0.922 \pm 0.035$ | $0.940 \pm 0.014$ | $0.923 \pm 0.016$ | $\mathbf{0.943}$ |
| a-myo | $301 \pm 243$ | $223 \pm 179$ | $199 \pm 138$ | $207 \pm 141$ | $\mathbf{189 \pm 159}$ |
| | $0.852 \pm 0.047$ | $0.892 \pm 0.029$ | $0.923 \pm 0.016$ | $0.931 \pm 0.017$ | $\mathbf{0.947}$ |
| average | $270 \pm 154$ | $208 \pm 141$ | $190 \pm 122$ | $193 \pm 115$ | $\mathbf{180 \pm 118}$ |
| | $0.857 \pm 0.049$ | $0.907 \pm 0.033$ | $0.932 \pm 0.015$ | $0.927 \pm 0.018$ | $\mathbf{0.945}$ |
| Dimensions (mm) | | | | | |
| dim1 | $3.05 \pm 2.84$ | $2.63 \pm 2.01$ | $\mathbf{2.27 \pm 1.79}$ | $2.31 \pm 1.81$ | $2.47 \pm 1.95$ |
| | $0.861 \pm 0.031$ | $0.925 \pm 0.018$ | $\mathbf{0.961 \pm 0.012}$ | $0.952 \pm 0.015$ | $0.957$ |
| dim2 | $3.23 \pm 3.02$ | $2.80 \pm 1.89$ | $\mathbf{2.38 \pm 1.90}$ | $2.41 \pm 2.03$ | $2.59 \pm 2.07$ |
| | $0.879 \pm 0.061$ | $0.932 \pm 0.023$ | $0.957 \pm 0.012$ | $\mathbf{0.961 \pm 0.013}$ | $0.894$ |
| dim3 | $3.27 \pm 3.12$ | $2.56 \pm 1.75$ | $\mathbf{2.22 \pm 1.78}$ | $2.23 \pm 1.67$ | $2.48 \pm 2.34$ |
| | $0.912 \pm 0.047$ | $0.939 \pm 0.021$ | $\mathbf{0.963 \pm 0.011}$ | $0.959 \pm 0.010$ | $0.943$ |
| average | $3.18 \pm 2.54$ | $2.66 \pm 1.75$ | $\mathbf{2.29 \pm 1.59}$ | $2.31 \pm 1.62$ | $2.51 \pm 1.58$ |
| | $0.884 \pm 0.044$ | $0.932 \pm 0.022$ | $\mathbf{0.960 \pm 0.011}$ | $0.957 \pm 0.012$ | $0.925$ |
| Regional wall thickness (mm) | | | | | |
| wt1 (IS) | $2.02 \pm 1.32$ | $1.89 \pm 1.04$ | $\mathbf{1.23 \pm 1.14}$ | $1.24 \pm 1.17$ | $1.26 \pm 1.04$ |
| | $0.625 \pm 0.063$ | $0.793 \pm 0.056$ | $0.854 \pm 0.014$ | $0.846 \pm 0.011$ | $\mathbf{0.856}$ |
| wt2 (I) | $2.67 \pm 1.69$ | $2.45 \pm 1.48$ | $1.44 \pm 1.22$ | $1.43 \pm 1.87$ | $\mathbf{1.40 \pm 1.10}$ |
| | $0.618 \pm 0.055$ | $0.751 \pm 0.037$ | $0.797 \pm 0.011$ | $\mathbf{0.801 \pm 0.014}$ | $0.747$ |

**Table 1.** (*continued*)

| | Model | | | | |
|---|---|---|---|---|---|
| | $\kappa = 1$ | $\kappa = 3$ | $\kappa = 5$ | $\kappa = 7$ | DMTRL [10] |
| wt3 (IL) | 2.95 ± 2.01 | 1.74 ± 1.56 | **1.57 ± 1.41** | 1.60 ± 1.59 | 1.59 ± 1.29 |
| | 0.595 ± 0.049 | 0.735 ± 0.033 | **0.765 ± 0.013** | 0.740 ± 0.010 | 0.693 |
| wt4 (AL) | 2.77 ± 1.65 | 1.66 ± 1.17 | 1.48 ± 1.13 | **1.46 ± 1.45** | 1.57 ± 1.34 |
| | 0.603 ± 0.052 | 0.763 ± 0.024 | **0.785 ± 0.022** | 0.782 ± 0.018 | 0.659 |
| wt5 (A) | 3.06 ± 2.12 | 1.49 ± 1.35 | 1.35 ± 1.19 | 1.39 ± 1.21 | **1.32 ± 1.10** |
| | 0.642 ± 0.061 | 0.808 ± 0.029 | 0.842 ± 0.019 | **0.851 ± 0.015** | 0.777 |
| wt6 (AS) | 2.25 ± 1.72 | 1.65 ± 1.11 | 1.46 ± 1.32 | 1.49 ± 1.37 | **1.25 ± 1.01** |
| | 0.651 ± 0.047 | 0.825 ± 0.032 | 0.870 ± 0.015 | 0.866 ± 0.013 | **0.877** |
| average | 2.62 ± 2.10 | 1.81 ± 1.05 | 1.42 ± 0.65 | 1.43 ± 0.71 | **1.39 ± 0.68** |
| | 0.622 ± 0.054 | 0.779 ± 0.033 | **0.819 ± 0.015** | 0.814 ± 0.014 | 0.768 |
| Phase (ER %) | | | | | |
| phase | 28.45 ± 5.50 | 14.67 ± 3.65 | **3.85 ± 2.82** | 3.91 ± 2.76 | 8.2 |

## 3  Results and Discussion

The effectiveness of the proposed method was validated under the experimental setting discussed in Sect. 2.2. We measured the influence of the parameter $\kappa$ (number of contiguous slices fed to the network) for $\kappa = 1$ (single slice), 3, 5 and 7 for the proposed spatio-temporal model based on 3D convolutions, and compare with the state of the art method recently proposed in [10]. Results are presented in Table 1 for a 5-fold cross validation setting (the same experimental setting and dataset was used in [10]). Note that using sets of $\kappa = 5$ slices significantly outperforms the configurations $\kappa = 1, 3$ for all the indices, highlighting the importance of the temporal dynamics. However, considering $\kappa = 5$ and $\kappa = 7$ slices achieves a similar performance. Therefore, we consider $\kappa = 5$ as enough temporal context for the remaining experiments.

In quantitative terms, we reduce the error rate from 28.45.06% to 3.85% for cardiac phase estimation and the MAE from 270 to 190 mm$^2$, 3.18 to 2.29 mm and 2.62 to 1.42 mm in average for the areas, directional dimensions of the cavity and regional wall thickness when comparing the performance for $\kappa = 1$ and $\kappa = 5$ slices respectively. Moreover, considering the baseline [10] we observe similar results for most indices, expect for the phase, where our model improves over the state of the art by a significant margin (reducing the error rate from 8.2% to 3.2%)

Finally, Table 2 presents these results for 3 different cross-validation configurations (3, 5 and 7 folds) as required by the LVQuan challenge organizers, together with the results for phase, directional dimensions, regional wall thicknesses and area of cavity and myocardium obtained with the best performing spatio-temporal model ($\kappa = 5$). Note that performance is consistent across folds.

**Table 2.** Results obtained for the LVQuan challenge dataset using the proposed spatio-temporal model (areas of LV cavity and myocardium (mm$^2$), directional dimensions (mm), wall thicknesses (mm) and cardiac phase) for $N$-folds cross validation with $N = 3, 5$ and 7.

| N-fold cross validation as required by LVQuan Challenge | | | | | | |
|---|---|---|---|---|---|---|
| | MAE | | | PCC | | |
| | N = 3 | N = 5 | N = 7 | N = 3 | N = 5 | N = 7 |
| Areas (mm$^2$) | | | | | | |
| a-cav | $185 \pm 125$ | $181 \pm 130$ | $183 \pm 115$ | 0.932 | 0.940 | 0.939 |
| a-myo | $204 \pm 143$ | $199 \pm 138$ | $198 \pm 145$ | 0.915 | 0.923 | 0.930 |
| average | $194 \pm 131$ | $190 \pm 122$ | $190 \pm 110$ | 0.924 | 0.932 | 0.935 |
| Dimensions (mm) | | | | | | |
| dim1 | $2.71 \pm 2.11$ | $2.27 \pm 1.79$ | $2.26 \pm 1.82$ | 0.938 | 0.961 | 0.959 |
| dim2 | $2.65 \pm 2.09$ | $2.38 \pm 1.90$ | $2.32 \pm 2.01$ | 0.926 | 0.957 | 0.954 |
| dim3 | $2.51 \pm 2.20$ | $2.22 \pm 1.78$ | $2.24 \pm 1.91$ | 0.933 | 0.963 | 0.958 |
| average | $2.62 \pm 1.87$ | $2.29 \pm 1.59$ | $2.27 \pm 1.52$ | 0.932 | 0.960 | 0.957 |
| Regional wall thickness (mm) | | | | | | |
| wt1 (IS) | $1.31 \pm 1.16$ | $1.23 \pm 1.14$ | $1.25 \pm 1.15$ | 0.831 | 0.854 | 0.857 |
| wt2 (I) | $1.58 \pm 1.10$ | $1.44 \pm 1.22$ | $1.43 \pm 1.41$ | 0.768 | 0.797 | 0.802 |
| wt3 (IL) | $1.62 \pm 1.22$ | $1.57 \pm 1.41$ | $1.56 \pm 1.56$ | 0.743 | 0.765 | 0.755 |
| wt4 (AL) | $1.60 \pm 1.08$ | $1.48 \pm 1.13$ | $1.50 \pm 1.11$ | 0.776 | 0.785 | 0.797 |
| wt5 (A) | $1.43 \pm 1.12$ | $1.35 \pm 1.19$ | $1.33 \pm 1.24$ | 0.829 | 0.842 | 0.861 |
| wt6 (AS) | $1.52 \pm 1.29$ | $1.46 \pm 1.32$ | $1.46 \pm 1.09$ | 0.857 | 0.870 | 0.873 |
| average | $1.51 \pm 0.98$ | $1.42 \pm 0.65$ | $1.42 \pm 0.61$ | 0.801 | 0.819 | 0.824 |
| Phase (ER %) | | | | | | |
| | N = 3 | | N = 5 | | N = 7 | |
| phase | $5.10 \pm 3.72$ | | $3.85 \pm 2.82$ | | $3.81 \pm 3.05$ | |

## 4    Conclusions

In this work, we proposed a new CNN architecture for LV quantification that incorporates the dynamics of the heart by means of spatio-temporal convolutions. Differently from other methods that rely on more complex mechanisms (like recurrent neural networks [10]) we employ simple 3D convolutions to fuse information coming from temporally contiguous CMR slices. We generated training samples following a circular hypothesis, meaning that first and last slices of the sequences are considered as temporally contiguous. Validation was performed using CRM sequences provided by the LVQuan challenge organizers. Results show that incorporating temporal information through spatio-temporal convolutions significantly boosts prediction performance for all the indices. Moreover, when compared with the RNN based model presented in [10], we observe a significant reduction in error rate for phase estimation (from 8.2% to 3.85%) while keeping equivalent results for the other indices. More importantly, our method achieves state of the art results employing simple 3D convolutions instead of the more complex parallel RNN and Bayesian based multitask relationship learning module proposed in [10].

In this work we incorporated the spatio-temporal dynamics by means of 3D convolutions. However, if we consider the slices as multiple channels of a standard 2D architecture, conventional 2D convolutions could also be used, reducing the complexity of the model. Moreover, temporal information encoded by inter-slice deformation fields (obtained trough deep learning based image registration methods [1]) could also be considered to improve model performance. In the future, we plan to explore the performance of these models when compared with the proposed architecture.

**Acknowledgements.** The present work used computational resources of the Pirayu Cluster, acquired with funds from the Santa Fe Science, Technology and Innovation Agency (ASACTEI), Government of the Province of Santa Fe, through Project AC-00010-18, Resolution No. 117/14. This equipment is part of the National System of High Performance Computing of the Ministry of Science, Technology and Productive Innovation of the Republic of Argentina. We also thank NVidia for the donation of a GPU used for this project. Enzo Ferrante is a beneficiary of an AXA Research Fund grant.

# References

1. Ferrante, E., Oktay, O., Glocker, B., Milone, D.H.: On the adaptability of unsupervised CNN-based deformable image registration to unseen image domains. In: Shi, Y., Suk, H.-I., Liu, M. (eds.) MLMI 2018. LNCS, vol. 11046, pp. 294–302. Springer, Cham (2018). https://doi.org/10.1007/978-3-030-00919-9_34

2. Karamitsos, T.D., Francis, J.M., Myerson, S., Selvanayagam, J.B., Neubauer, S.: The role of cardiovascular magnetic resonance imaging in heart failure. J. Am. Coll. Cardiol. **54**(15), 1407–1424 (2009)

3. Peng, P., Lekadir, K., Gooya, A., Shao, L., Petersen, S.E., Frangi, A.F.: A review of heart chamber segmentation for structural and functional analysis using cardiac magnetic resonance imaging. Magn. Reson. Mater. Phys. Biol. Med. **29**(2), 155–195 (2016)

4. Petitjean, C., Dacher, J.N.: A review of segmentation methods in short axis cardiac MR images. Med. Image Anal. **15**(2), 169–184 (2011)

5. Poudel, R.P.K., Lamata, P., Montana, G.: Recurrent fully convolutional neural networks for multi-slice MRI cardiac segmentation. In: Zuluaga, M.A., Bhatia, K., Kainz, B., Moghari, M.H., Pace, D.F. (eds.) RAMBO/HVSMR-2016. LNCS, vol. 10129, pp. 83–94. Springer, Cham (2017). https://doi.org/10.1007/978-3-319-52280-7_8

6. Suinesiaputra, A., et al.: Quantification of LV function and mass by cardiovascular magnetic resonance: multi-center variability and consensus contours. J. Cardiovasc. Magn. Reson. **17**(1), 63 (2015)

7. Tan, L.K., Liew, Y.M., Lim, E., McLaughlin, R.A.: Convolutional neural network regression for short-axis left ventricle segmentation in cardiac cine MR sequences. Med. Image Anal. **39**, 78–86 (2017)

8. Tran, D., Bourdev, L., Fergus, R., Torresani, L., Paluri, M.: Learning spatiotemporal features with 3D convolutional networks. In: IEEE International Conference on Computer Vision (ICCV), pp. 4489–4497. IEEE (2015)

9. Tran, P.V.: A fully convolutional neural network for cardiac segmentation in short-axis MRI. arXiv preprint arXiv:1604.00494 (2016)

10. Xue, W., Brahm, G., Pandey, S., Leung, S., Li, S.: Full left ventricle quantification via deep multitask relationships learning. Med. Image Anal. **43**, 54–65 (2018)

11. Xue, W., Lum, A., Mercado, A., Landis, M., Warrington, J., Li, S.: Full quantification of left ventricle via deep multitask learning network respecting intra- and inter-task relatedness. In: Descoteaux, M., Maier-Hein, L., Franz, A., Jannin, P., Collins, D.L., Duchesne, S. (eds.) MICCAI 2017. LNCS, vol. 10435, pp. 276–284. Springer, Cham (2017). https://doi.org/10.1007/978-3-319-66179-7_32

12. Xue, W., Nachum, I.B., Pandey, S., Warrington, J., Leung, S., Li, S.: Direct estimation of regional wall thicknesses via residual recurrent neural network. In: Niethammer, M., et al. (eds.) IPMI 2017. LNCS, vol. 10265, pp. 505–516. Springer, Cham (2017). https://doi.org/10.1007/978-3-319-59050-9_40

# Full Quantification of Left Ventricle Using Deep Multitask Network with Combination of 2D and 3D Convolution on 2D + t Cine MRI

Yeonggul Jang[1(✉)], Sekeun Kim[2], Hackjoon Shim[2],
and Hyuk-Jae Chang[2,3]

[1] Brain Korea 21 PLUS Project for Medical Science,
Yonsei University, Seoul, South Korea
jygl722@gmail.com
[2] Integrative Cardiovascular Imaging Research Center,
Yonsei University College of Medicine, Seoul, South Korea
[3] Division of Cardiology, Severance Cardiovascular Hospital,
Yonsei University College of Medicine, Seoul, South Korea

**Abstract.** Accurate quantification of left ventricle (LV) from cardiac image are valuable to evaluate ventricular function information such as stroke volume and ejection fraction. In this paper, we proposed a novel FCN architecture, which is trained in end-to-end manner, for full quantification of cardiac LV on 2D + t cine MR images. Considering 3D information as features for temporal modeling can improve performance of the model for temporal-related task. The proposed FCN with the alternate 3D-2D convolutional module addresses each sequence with assistance from adjacent sequences and shows the comparable results compared with the state-of-the-art method.

**Keywords:** LV quantification · Convolutional neural network · Cardiac MRI

## 1 Introduction

With the advancement of the non-invasive imaging techniques such as the echocardiogram, computed tomography (CT), and magnetic resonance imaging (MRI), the imaging of the beating heart can be acquired. Cardiac image quantification plays an important role for the identification the diagnosis of cardiac diseases and image-guided interventions [1, 2]. Accurate quantification of left ventricle (LV) from cardiac image are particularly valuable to evaluate ventricular function information such as stroke volume and ejection fraction.

Various methods for the quantification of the ventricles are developed with and without segmentation in cardiac MRI [3–6]. Recently, the advancement of the deep learning shows high performances in object detection, recognition, as well as segmentation. There are many attempts to use deep learning technique, especially convolutional neural networks (CNN), in cardiac image segmentation problems. 2D CNN was applied in cardiac images for LV segmentation with auto-encoder [7]. Recurrent neural network (RNN) is applied for LV segmentation in multi-slice MR images [8].

© Springer Nature Switzerland AG 2019
M. Pop et al. (Eds.): STACOM 2018, LNCS 11395, pp. 476–483, 2019.
https://doi.org/10.1007/978-3-030-12029-0_51

Multi-scale convolutional deep belief network is proposed to estimate bi-ventricular volume in cardiac MRI [9].

In this paper, we introduce a novel multi-task network for full quantification of cardiac LV. The proposed network is tailored for cine MRI images of which each consists of twenty 2D MRI sequences throughout a cardiac cycle. The proposed network employs 3D convolution to consider both spatial and temporal information simultaneously. Addressing each sequence with assistance from adjacent sequences can help capturing features for the temporal-related task. The cine MRI images have quite excellent temporal resolution and therefore, it's reasonable to use temporal information through 3D convolution on 2D + t cine MRI. In predicting LV indices and cardiac phase, the proposed network uses 2D and 3D convolution alternatively to effectively extract features for the myocardium and LV cavity.

This paper is organized as follows. In Sect. 2, we present the proposed architecture in detail. We show the experimental results from five-fold cross-validation using given 145 training datasets in Sect. 3. Discussions and conclusions are given in Sect. 4.

## 2 Method

### 2.1 Dataset and Preprocessing

The training dataset with processed short-axis MR sequences of 145 subjects from 3 hospitals affiliated with two health care centers (London Healthcare Center and St. Josephs HealthCare) is provided by Left Ventricle Full Quantification Challenge in MICCAI challenge 2018. For each subject, 20 frames are included for the whole cardiac cycle and all MR image frames are aligned with dimension of 80 × 80. All ground truth values for LV indices and one cardiac phase, which should be predicted, are provided for every single frame. The LV indices include two areas (cavity and myocardium), six regional wall thickness, and three LV dimensions. The cardiac phase is binary variable for cardiac phase, 1 means systolic phase and 0 means diastolic.

The 16-bit MRI datasets provided by challenge have a wide range of voxel intensity that could result from different scanner types or acquisition protocols. This variety could affect the performance of the inference model. We clip the image pixel intensities between the 1st and 99th percentile intensities and normalize the voxel intensity of each image by subtracting its mean then dividing it by its standard deviation.

### 2.2 Architecture

We propose a fully convolutional network (FCN) architecture, which is trained in end-to-end manner, to model full quantification of cardiac LV. This network employs both 2D and 3D convolution to consider not only spatial but also temporal information. The provided datasets in this challenge have short-term temporal dynamic between adjacent images in the sequence over one cardiac cycle. For this reason, we consider 3D information as features for temporal modeling on the 2D + t MR images. The 3D convolution addresses each image with assistance from adjacent images and can represent more robust structural features as well as temporal information.

Figure 1 illustrates our proposed FCN architecture with combination of 2D and 3D convolutional filters. Our architecture consists of three main paths: one shared contraction path and two task-specific inference paths.

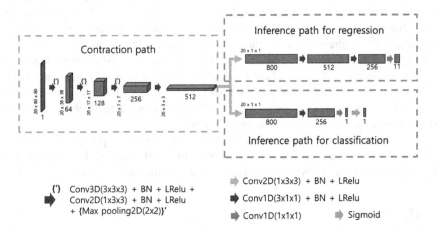

**Fig. 1.** The proposed FCN architecture

Contraction path has 4 cascade steps. Each step in this path has 2 convolutional layers, which uses 3D and 2D filters of size $3 \times 3 \times 3$ and $1 \times 3 \times 3$ respectively in order, and max-pooling layer of size $2 \times 2$. This path reduces the x and y size of input with sequence kept during convolution and max-pooling and allows network to capture task-related contextual features for each image on 2D + t cine MRI. Contraction path is shared by two task-specific inference paths.

Two task-specific inference paths for regression and classification has the similar layers each other. Inference path removes spatial dimensions for the input of size $20 \times 3 \times 3$, which is from the last layer of contraction path, with $1 \times 3 \times 3$ 3D convolutional filters. The subsequent layers are composed of 3D convolution of size $3 \times 1 \times 1$, which is equivalent to 1D convolution of size 3, with the different numbers of filters for compressing information while considering temporal modeling. In short, this layer learns task-specific representation from LV-related spatial features on each image in the sequence that are encoded by contraction path. The numbers of the final layer filters of inference paths for regression and classification are eleven (LV indices) and one (cardiac phase), respectively.

We pad input with the reflection operation at the first and last edges in the sequence before each 3D convolution to keep the number of images in the sequence when performing 3D convolutional operation. Batch normalization layers are applied after each convolutional layer before Leaky-ReLU activation. Dropout with probability 0.5 is applied to the last step of contraction path only twice.

As shown in Eq. 1, the loss functions for regression and classification are averages of L2 distance and cross-entropy with sigmoid over all images on the 2D + t MR sequence, respectively. Considering that two tasks have the different target output size, total loss function is the weighted sum of two loss functions with weight factor $\lambda$. In this paper, $\lambda$ is set to 7.0 empirically. Finally, the proposed network is trained by minimizing it.

$$\left\{ \begin{array}{l} L_{indice} = \frac{1}{S}\sum_i \sum_s \parallel \hat{y}_{i,s} - y_{i,s} \parallel_2^2 \\ L_{phase} = -\frac{1}{S}\sum_s y_s \log(\hat{y}_s) + (1 - y_s)\log(1 - \hat{y}_s) \end{array} \right\} \quad (1)$$

$$L_{total} = L_{indice} + \lambda\, L_{phase} \quad (2)$$

$$, \ where \ i \in (area,\ dim,\ rwt) \ and \ s \leq S \quad (3)$$

where $i \in (area,\ dim,\ rwt)$, $y_s \in (0,1)$, and F and $y_s$ denote the number of images in the sequence and the two cardiac phases *Diastolic* as 0 and *Systolic* as 1, respectively.

## 3  Experimental Results

### 3.1  Implemented Details

We train our FCN in end-to-end manner with five-fold cross validation to evaluate its performance. The varieties for the scale and rotation on the provided datasets, which were already pre-processed, are small considerably unlike the intensity and then, we blur datasets randomly to add the intensity variety as data augmentation during training.

The proposed FCN was trained on NVIDIA TITAN Xp, with 12 GB of RAM with batch size 15 for 1,000 epochs, which took 2 h for training. The FCN was implemented by Tensorflow r1.4 and trained using RMSprop Optimizer with following hyper parameters: learning rate $= 10^{-4}$, decay $= 0.9$, momentum $= 0.0$, and epsilon $= 10^{-10}$.

### 3.2  Results and Quantitative Analysis with the State-of-the-Art Methods

The performance for the LV indices and the cardiac phase is evaluated for all images in the sequence. Mean absolute error (MAE) and Pearson correlation coefficient (PCC) between the estimated values and the ground truth values are computed for LV indices [Eqs. 4 and 5]. Error rate (ER) is computed for cardiac phase [Eq. 6].

$$MAE(y,\hat{y}) = \frac{1}{N}\sum_{i=1}^{N} |y_i - \hat{y}_i| \quad (4)$$

$$\rho(y,\hat{y}) = \frac{\frac{1}{N}\sum_{i=1}^{N}(y_i - m)(y_i - \hat{m})}{\sum_{i=1}^{N}\left\{(y_i - m)^2 + (\hat{y}_i - \hat{m})^2\right\}} \quad (5)$$

$$ER(y,\hat{y}) = \frac{1}{N}\sum_{i=1}^{N} 1(y_i \neq \hat{y}_i) * 100 \quad (6)$$

where $m = \frac{1}{N}\sum_{i=1}^{N} y_i$, $\hat{m} = \frac{1}{N}\sum_{i=1}^{N} \hat{y}_i$, and $1(y_i \neq \hat{y}_i)$ returns zero if $y_i$ and $\hat{y}_i$ has the same value and one, otherwise.

The MAE and PCC values of LV indices and ER of cardiac phase for five fold cross validation and its average are listed in Table 1. The proposed FCN shows the stable results for three LV indices and one cardiac phase on all cross validation (CV) set except for CV set #2.

**Table 1.** The five fold cross-validation (CV) results of the proposed FCN on the 145 subjects. Values correspond to the mean absolute error (MAE), its standard deviation, and correlation coefficient (PCC) for three LV indices and error rate (ER) for cardiac phase with ground truth.

| | CV#1 | CV#2 | CV#3 | CV#4 | CV#5 | Average |
|---|---|---|---|---|---|---|
| **Area (mm$^2$)** | | | | | | |
| LV cavity | 148 ± 118 | 232 ± 204 | 164 ± 124 | 152 ± 130 | 153 ± 125 | **170 ± 147** |
| | 0.960 | 0.960 | 0.938 | 0.966 | 0.958 | **0.956** |
| Myo cavity | 196 ± 182 | 213 ± 159 | 184 ± 176 | 236 ± 186 | 205 ± 161 | **207 ± 174** |
| | 0.890 | 0.932 | 0.832 | 0.854 | 0.864 | **0.874** |
| **Dimension (mm)** | | | | | | |
| Dim1 | 2.13 ± 1.54 | 3.39 ± 2.88 | 2.45 ± 1.81 | 2.44 ± 1.93 | 2.32 ± 1.40 | **2.55 ± 2.08** |
| | 0.958 | 0.938 | 0.909 | 0.939 | 0.940 | **0.937** |
| Dim2 | 2.22 ± 1.84 | 2.94 ± 2.53 | 2.24 ± 1.55 | 1.64 ± 1.37 | 1.78 ± 1.66 | **2.16 ± 1.85** |
| | 0.953 | 0.962 | 0.937 | 0.973 | 0.966 | **0.958** |
| Dim3 | 2.37 ± 1.82 | 2.97 ± 2.57 | 2.65 ± 2.02 | 2.63 ± 2.15 | 2.17 ± 1.07 | **2.56 ± 2.08** |
| | 0.950 | 0.957 | 0.905 | 0.940 | 0.950 | **0.941** |
| **RWT (mm)** | | | | | | |
| IS | 1.42 ± 1.14 | 1.39 ± 1.03 | 1.35 ± 1.29 | 1.22 ± 1.00 | 1.26 ± 1.07 | **1.33 ± 1.11** |
| | 0.783 | 0.897 | 0.822 | 0.861 | 0.852 | **0.843** |
| I | 1.48 ± 1.17 | 1.42 ± 1.15 | 1.12 ± 0.88 | 1.27 ± 0.99 | 1.57 ± 1.30 | **1.37 ± 1.12** |
| | 0.747 | 0.830 | 0.775 | 0.772 | 0.646 | **0.754** |
| IL | 1.47 ± 1.18 | 1.46 ± 1.28 | 1.82 ± 1.54 | 1.71 ± 1.40 | 1.63 ± 1.19 | **1.62 ± 1.33** |
| | 0.705 | 0.824 | 0.458 | 0.656 | 0.668 | **0.668** |
| AL | 1.50 ± 1.29 | 1.60 ± 1.32 | 1.58 ± 1.27 | 1.77 ± 1.69 | 1.47 ± 1.18 | **1.58 ± 1.36** |
| | 0.684 | 0.766 | 0.546 | 0.611 | 0.614 | **0.644** |
| A | 1.35 ± 1.11 | 1.40 ± 1.28 | 1.35 ± 1.08 | 1.50 ± 1.46 | 1.05 ± 0.87 | **1.33 ± 1.19** |
| | 0.793 | 0.855 | 0.710 | 0.755 | 0.760 | **0.775** |
| AS | 1.43 ± 1.17 | 1.30 ± 0.89 | 1.52 ± 1.26 | 1.16 ± 1.03 | 1.11 ± 0.82 | **1.30 ± 1.06** |
| | 0.806 | 0.912 | 0.839 | 0.902 | 0.892 | **0.870** |
| **Phase (%)** | | | | | | |
| Phase | 6.21 | 12.41 | 8.62 | 7.24 | 9.31 | **8.76** |

To validate the effectiveness of the proposed alternate 3D-2D convolution module, the proposed FCN is compared with the networks in which either 2D or 3D convolution module is only used instead of the alternate module. As listed in Table 2, the proposed FCN with alternate 3D-2D convolution module shows most stable performance overall.

Also, we compared the results estimated by the proposed FCN with the state of the art methods [10, 11] that trained with the same cross validation for fair comparison. Based on these results, it is shown that the proposed FCN architecture outperforms the

**Table 2.** Comparison between the proposed FCN and the state of the art methods for full LV quantification results.

| | The proposed FCN | | | Full LVNet [10] | DMTRL [11] |
|---|---|---|---|---|---|
| | Only 2D | Only 3D | 2D and 3D | | |
| **Area (mm$^2$)** | | | | | |
| LV cavity | 170 ± 142 0.960 | 174 ± 145 0.958 | 170 ± 147 0.956 | 181 ± 155 0.940 | 172 ± 148 0.943 |
| Myo cavity | 214 ± 170 0.866 | 216 ± 184 0.859 | 207 ± 174 0.874 | 199 ± 174 0.935 | 189 ± 159 0.947 |
| Average | 192 ± 158 0.931 | 195 ± 167 0.928 | 188 ± 162 0.931 | 190 ± 128 0.937 | **180 ± 118** **0.945** |
| **Dimension (mm)** | | | | | |
| Dim1 | 2.52 ± 2.05 0.940 | 2.57 ± 2.09 0.938 | 2.55 ± 2.08 0.937 | 2.62 ± 2.09 0.952 | 2.47 ± 1.95 0.957 |
| Dim2 | 2.07 ± 1.71 0.964 | 2.28 ± 1.92 0.954 | 2.16 ± 1.85 0.958 | 2.64 ± 2.12 0.881 | 2.59 ± 2.07 0.894 |
| Dim3 | 2.64 ± 1.96 0.946 | 2.63 ± 2.16 0.941 | 2.56 ± 2.08 0.941 | 2.77 ± 2.22 0.935 | 2.48 ± 2.34 0.943 |
| Average | 2.41 ± 1.93 0.951 | 2.49 ± 2.07 0.946 | **2.42 ± 2.01** **0.948** | 2.68 ± 1.64 0.917 | 2.51 ± 1.58 0.925 |
| **RWT (mm)** | | | | | |
| IS | 1.36 ± 1.09 0.841 | 1.40 ± 1.21 0.813 | 1.33 ± 1.11 0.843 | 1.32 ± 1.09 0.840 | 1.26 ± 1.04 0.856 |
| I | 1.48 ± 1.13 0.720 | 1.43 ± 1.14 0.732 | 1.37 ± 1.12 0.754 | 1.38 ± 1.10 0.751 | 1.40 ± 1.10 0.747 |
| IL | 1.76 ± 1.33 0.663 | 1.67 ± 1.35 0.658 | 1.62 ± 1.33 0.668 | 1.57 ± 1.35 0.691 | 1.59 ± 1.29 0.693 |
| AL | 1.67 ± 1.30 0.650 | 1.63 ± 1.38 0.627 | 1.58 ± 1.36 0.644 | 1.60 ± 1.36 0.651 | 1.57 ± 1.34 0.659 |
| A | 1.32 ± 1.10 0.778 | 1.34 ± 1.18 0.765 | 1.33 ± 1.19 0.775 | 1.34 ± 1.11 0.768 | 1.32 ± 1.10 0.777 |
| AS | 1.26 ± 1.03 0.876 | 1.32 ± 1.11 0.860 | 1.30 ± 1.06 0.870 | 1.26 ± 1.10 0.864 | 1.25 ± 1.01 0.877 |
| Average | 1.47 ± 1.18 0.797 | 1.46 ± 1.24 0.789 | 1.42 ± 1.21 **0.801** | 1.41 ± 0.72 0.761 | **1.39 ± 0.68** 0.768 |
| **Phase (%)** | | | | | |
| Phase | 11.41 | 8.72 | 8.76 | 10.4 | **8.2** |

method [10] with intra-inter relatedness in Table 2. Compared to the advanced version [11] of that method, the proposed FCN shows the superior results for dimension and the comparable results for the rest.

The CPU and GPU run times to analyze one 2D + t cine MR sequence using the proposed FCN are approximately 0.61 s and 0.013 s, respectively.

## 4 Discussion and Conclusion

In this paper, we proposed a novel FCN architecture, which is trained in end-to-end manner, for full quantification of cardiac LV on 2D + t cine MR sequence. The datasets provided by challenge have short-term temporal dynamic between adjacent images in the sequence over one cardiac cycle and therefore, we consider 3D information as features for both spatial and temporal modeling. The proposed FCN architecture employs the alternate 3D-2D convolutional module in the shared contraction path to extract features for inference. This module addresses each image with assistance from adjacent images in the sequence. The last two different paths make task-specific inferences from the extracted features with temporal modeling using 3D convolution of size $3 \times 1 \times 1$ equivalent to 1D convolution of size 3. As a result, the proposed FCN shows comparable performance with the state of the art method [11].

The proposed FCN could represent more robust structural features and temporal information by taking spatial information from adjacent images using alternate 3D-2D convolution modules and 3D convolution of various sizes. While in the state of the art methods [10, 11], the combination of CNN and RNN modules for spatial and temporal modeling was used with task-related regularizations and a two-step strategy for training. In that CNN module has the better capability for parallel computing compared with RNN, the proposed FCN only with convolutional module might have better performance in the aspect of run-time.

We do not apply any task-related regularization. Nonetheless, the proposed FCN shows the comparable results in comparison with the state-of-the-art methods [10, 11]. In the future, we will apply further task-related regularization terms on the outputs produced by the proposed FCN for additional performance improvement.

**Acknowledgements.** This work was supported by Institute for Information & communications Technology Promotion (IITP) grant funded by the Korea government (MSIT) (No. 2018-0-00861, Intelligent SW Technology Development for Medical Data Analysis).

## References

1. Kang, D., et al.: Heart chambers and whole heart segmentation techniques. J. Electron. Imaging. **21**(1), pp. 010901-1–010901-16 (2012)
2. Karamitsos, T.D., et al.: The role of cardiovascular magnetic resonance imaging in heart failure. J. Am. Coll. Cardiol. **54**(15), 1407–1424 (2009)
3. Petitjean, C., et al.: A review of segmentation methods in short axis cardiac MR images. Med. Image Anal. **15**(2), 169–184 (2011)
4. Bai, W., et al.: A probabilistic patch-based label fusion model for multi-atlas segmentation with registration refinement: application to cardiac MR images. IEEE Trans. Med. Imaging **32**(7), 1302–1315 (2013)
5. Bai, W., et al.: Multi-atlas segmentation with augmented features for cardiac MR images. Med. Image Anal. **19**(1), 98–109 (2015)
6. Afshin, M., et al.: Regional assessment of cardiac left ventricular myocardial function via MRI statistical features. IEEE TMI **33**(2), 481–494 (2014)

7. Avendi, M., et al.: A combined deep-learning and deformable-model approach to fully automatic segmentation of the left ventricle in cardiac MRI. Med. Image Anal. **30**, 108–119 (2016)
8. Poudel, R.P.K., Lamata, P., Montana, G.: Recurrent fully convolutional neural networks for multi-slice MRI cardiac segmentation. In: Zuluaga, M.A., Bhatia, K., Kainz, B., Moghari, M.H., Pace, D.F. (eds.) RAMBO/HVSMR-2016. LNCS, vol. 10129, pp. 83–94. Springer, Cham (2017). https://doi.org/10.1007/978-3-319-52280-7_8
9. Zhen, X., et al.: Multi-scale deep networks and regression forests for direct bi-ventricular volume estimation. Med. Image Anal. **30**, 120–129 (2016)
10. Xue, W., Lum, A., Mercado, A., Landis, M., Warrington, J., Li, S.: Full quantification of left ventricle via deep multitask learning network respecting intra- and inter-task relatedness. In: Descoteaux, M., Maier-Hein, L., Franz, A., Jannin, P., Collins, D.L., Duchesne, S. (eds.) MICCAI 2017. LNCS, vol. 10435, pp. 276–284. Springer, Cham (2017). https://doi.org/10.1007/978-3-319-66179-7_32
11. Xue, W., et al.: Full left ventricle quantification via deep multitask relationships learning. Med. Image Anal. **43**, 54–65 (2018)

# Author Index

Abdi, Vahid  265
Aguado, Ainhoa  58, 200
Albà, Xènia  114
Al-Kadi, Omar S.  77
Aly, Abdullah H.  142
Aly, Ahmed H.  142
An, Yunqiang  122
Andalò, Alice  329
Aslanidi, Oleg  103
Atehortúa, Angélica  439
Attar, Rahman  114
Axel, Leon  181

Baek, Stephen  161
Bai, Wenjia  292
Banerjee, Abhirup  12
Bhalodia, Riddhish  357
Bian, Cheng  237
Borra, Davide  329
Butakoff, Constantine  302

Camara, Oscar  32, 58, 200, 302
Carlinet, Edwin  339
Cates, Joshua  357
Chang, Hyuk-Jae  476
Chen, Chen  292
Chen, Shaoxiang  421
Choudhury, Robin P.  12
Clough, James  371
Cochet, Hubert  221
Corsi, Cristiana  329

Dai, Yangyang  311
Dangi, Shusil  21
De Backer, Ole  32
De Craene, Mathieu  94
de Potter, Tom  58, 200
de Vente, Coen  348
Debus, Alejandro  466
Delingette, Hervé  221
Despinasse, Antoine  221
Du, Yubo  131
Duan, Yongjie  85
Duncan, James S.  77

Elhabian, Shireen  357
Elrakhawy, Mahmoud  142
Engelhardt, Sandy  265
Esposito, Lorena  329

Fabbri, Claudio  329
Fadil, Hakim  40
Feng, Jianjiang  85, 122, 131
Feng, Zishun  122
Fernandez, Alvaro  58, 200
Fernandez, Justin  282
Ferrante, Enzo  466
Firmin, David  152
Fok, Wilson  282
Frangi, Alejandro F.  114
Freixa, Xavier  58, 200

Gao, Yang  122
Garreau, Mireille  439
Genua, Ibai  58, 200
Géraud, Thierry  339
Gooya, Ali  114
Gorman III, Joseph H.  142
Gorman Jr., Robert C.  142
Gorman, Robert C.  142
Grau, Vicente  12, 171, 402
Grinias, Elias  389
Guo, Fumin  49, 450

Hao, Aimin  211
Haridasan, Shyamalakshmi  265
Haroun, Kirlos  142
Heng, Pheng-Ann  237, 246, 255
Hu, Liyu  429
Hu, Zhiqiang  211, 381
Hua, Tiancong  429

Jaïs, Pierre  221
Jamart, Kevin  282
Jang, Yeonggul  476
Jia, Shuman  221
Jin, Cheng  131
Juhl, Kristine A.  32

Karim, Rashed   348
Keegan, Jenny   152
Kerfoot, Eric   371
Kharbanda, Rajesh K.   12
Khoudli, Younes   339
Kim, Sekeun   476
King, Andrew P.   94, 371
Kofoed, Klaus F.   32

Lacotte, Jérôme   339
Lamata, Pablo   273
Langet, Hélène   94
Lee, Deukhee   161
Lee, Jack   371
Li, Caizi   255
Li, Jiahui   381
Li, Lei   152, 302
Li, Mengyuan   49
Li, Quanzheng   459
Li, Xiang   459
Li, Ziyan   122
Liao, Xiangyun   255
Linte, Cristian A.   21
Liu, Bo   181
Liu, Jiang   131
Liu, Jiasha   459
Liu, Lihong   412
Liu, Yashu   311
Liu, Yu-An   237
Lu, Allen   77
Lu, Bin   122
Lu, Chao   429
Lu, Jiwen   85, 131

Ma, Jianqiang   237
Ma, Jin   412
MacGillivray, Tom   191
Macnaught, Gillian   191
Maier, Andreas   319
Marchesseau, Stephanie   40
Marrouche, Nassir   357
Masci, Alessandro   329
Metaxas, Dimitris   181
Mill, Jordi   58, 200
Mohiaddin, Raad   152
Monaci, Sofia   103

Montana, Giovanni   273
Moore, John   67

Neubauer, Stefan   114
Newby, David   191
Nezafat, Reza   237, 246
Ng, Matthew   49, 450
Ni, Dong   246
Niederer, Steven   348
Nordsletten, David   103
Nuñez-Garcia, Marta   58, 200, 302

Oksuz, Ilkay   371
Olivares, Andy L.   32, 58, 200

Pan, Tan   429
Paulsen, Rasmus R.   32
Pennec, Xavier   221
Pereañez, Marco   114
Peters, Terry   67
Petersen, Steffen E.   114
Piechnik, Stefan K.   114
Piro, Paolo   94
Pluim, Josien   348
Pop, Mihaela   49
Pouch, Alison M.   142
Preetha, Chandrakanth Jayachandran   265
Prieto-Riascos, Luis   142
Puybareau, Élodie   339
Puyol-Antón, Esther   94

Qiao, Mengyun   230

Ravikumar, Nishant   319
Razeghi, Orod   348
Ren, Hui   459
Rhode, Kawal   348
Romero, Eduardo   439
Romo-Bucheli, David   439
Rueckert, Daniel   292
Ruijsink, Bram   94

Saito, Yoshiaki   142
Samari, Babak   3
Sanroma, Gerard   302
Savioli, Nicoló   273
Schnabel, Julia A.   94, 371

Schneider, Jurgen E. 402
Sermesant, Maxime 221
Shim, Hackjoon 476
Shu, Huazhong 429
Si, Weixin 255
Siddiqi, Kaleem 3
Sigvardsen, Per E. 32
Silva, Etelvino 58, 200
Sinusas, Albert J. 77
Slipsager, Jakob M. 32
Sodergren, Tim 357
Sun, Yinzi 255
Syed, Tabish A. 3

Tao, Qian 230, 421
Tong, Qianqian 255
Totman, John J. 40
Tziritas, Georgios 389

van der Geest, Rob J. 230, 421
Vannelli, Claire 67
Vesal, Sulaiman 319
Veta, Mitko 348
Villard, Benjamin 171

Wang, Chengjia 191
Wang, Jianzong 412
Wang, Kuanquan 311
Wang, Na 246
Wang, Qiong 255
Wang, Xu 246
Wang, Yi 246
Wang, Yuanyuan 230, 421
Wang, Zihao 221
Whitaker, Ross 357

Wong, Tom 152
Wright, Graham 49, 450
Wu, Fuping 152
Wu, Jiasong 429

Xia, Qing 211
Xia, Wenyao 67
Xiao, Jing 412
Xu, Hao 402
Xu, Lingchao 152, 302
Xu, Yongchao 339

Yan, Cong 311
Yan, Wenjun 421
Yang, Dong 181
Yang, Guang 152, 191
Yang, Guanyu 429
Yang, Xiao 429
Yang, Xin 237, 246
Yaniv, Ziv 21
Yao, Yuxin 211
Yoon, Siyeop 161
Yushkevich, Natalie 142
Yushkevich, Paul A. 142

Zacur, Ernesto 171
Zhang, Le 114
Zhao, Jichao 282
Zhao, Zhou 339
Zheng, Shen 237
Zheng, Yefeng 237
Zhou, Jie 85, 122, 131
Zhu, Xiaomei 429
Zhuang, Xiahai 152, 302

Printed in the United States
by Bookmasters

Printed in the United States
By Bookmasters